MICROWAVE DEVICES, CIRCUITS AND SUBSYSTEMS FOR COMMUNICATIONS ENGINEERING

MICROWAVE DEVICES, CIRCUITS AND SUBSYSTEMS FOR COMMUNICATIONS ENGINEERING

Edited by

I. A. Glover, S. R. Pennock and P. R. Shepherd

All of
Department of Electronic and Electrical Engineering
University of Bath, UK

John Wiley & Sons, Ltd

Telephone (+44) 1243 779777

Email (for orders and customer service enquiries): cs-books@wiley.co.uk
Visit our Home Page on www.wiley.com

Other Wiley Editorial Offices

John Wiley & Sons Inc., 111 River Street, Hoboken, NJ 07030, USA

Jossey-Bass, 989 Market Street, San Francisco, CA 94103-1741, USA

Wiley-VCH Verlag GmbH, Boschstr. 12, D-69469 Weinheim, Germany

John Wiley & Sons Australia Ltd, 33 Park Road, Milton, Queensland 4064, Australia

John Wiley & Sons (Asia) Pte Ltd, 2 Clementi Loop #02-01, Jin Xing Distripark, Singapore 129809

John Wiley & Sons Canada Ltd, 22 Worcester Road, Etobicoke, Ontario, Canada M9W 1L1

Wiley also publishes its books in a variety of electronic formats. Some content that appears in print may not be available in electronic books.

British Library Cataloguing in Publication Data

A catalogue record for this book is available from the British Library

ISBN 0-471-89964-X (HB)

Typeset in 10/12pt Times by Graphicraft Limited, Hong Kong, China.
Printed and bound in Great Britain by Antony Rowe Ltd, Chippenham, Wiltshire.
This book is printed on acid-free paper responsibly manufactured from sustainable forestry in which at least two trees are planted for each one used for paper production.

Contents

List of Contributors

E. Artal	ETSIIT – DICOM, Av. Los Castros, 39005, Santander, Cantabria, Spain.
L. de la Fuente	ETSIIT – DICOM, Av. Los Castros, 39005, Santander, Cantabria, Spain.
D. Dernikas	Formerly University of Bradford, U.K. Currently Aircom International, Grosvenor House, 65–71 London Road, Redhill, Surrey, RH1 1LQ, U.K.
T. C. Edwards	Engalco, 3, Georgian Mews, Bridlington, East Yorkshire, YO15 3TG, U.K.
T. Fernandez	ETSIIT – DICOM, Av. Los Castros, 39005, Santander, Cantabria, Spain.
I. A. Glover	Department of Electronic and Electrical Engineering, University of Bath, Claverton Down, Bath, BA2 7AY, U.K.
N. J. McEwan	Filtronic PLC, The Waterfront, Salts Mill Road, Saltaire, Shipley, West Yorkshire, BD18 3TT, U.K.
A. Mediavilla	ETSIIT – DICOM, Av. Los Castros, 39005, Santander, Cantabria, Spain.
J. P. Pascual	ETSIIT – DICOM, Av. Los Castros, 39005, Santander, Cantabria, Spain.
S. R. Pennock	Department of Electronic and Electrical Engineering, University of Bath, Claverton Down, Bath, BA2 7AY, U.K.
J. Portilla	ETSIIT – DICOM, Av. Los Castros, 39005, Santander, Cantabria, Spain.
P. R. Shepherd	Department of Electronic and Electrical Engineering, University of Bath, Claverton Down, Bath, BA2 7AY, U.K.
A. Suarez	ETSIIT – DICOM, Av. Los Castros, 39005, Santander, Cantabria, Spain.
A. Tazon	ETSIIT – DICOM, Av. Los Castros, 39005, Santander, Cantabria, Spain.

Preface

This text originated from a Master's degree in RF Communications Engineering offered since the mid-1980s at the University of Bradford in the UK. The (one-year) degree, which has now graduated several hundred students, was divided into essentially three parts:

Part 1 – RF devices and subsystems
Part 2 – RF communications systems
Part 3 – Dissertation project.

Part 1 was delivered principally in Semester 1 (October to mid-February), Part 2 in Semester 2 (mid-February to June) and Part 3 during the undergraduate summer vacation (July to September). Parts 1 and 2 comprised the taught component of the degree consisting of lectures, tutorials, laboratory work and design exercises. Part 3 comprised an individual and substantial project drawing on skills acquired in Parts 1 and 2 for its successful completion.

In the mid-1990s it was decided that a distance-learning version of the degree should be offered which would allow practising scientists and technologists to retrain as RF and microwave communications engineers. (At that time there was a European shortage of such engineers and the perception was that a significant market existed for the conversion of numerate graduates from other disciplines, e.g. physics and maths, and the retraining of existing engineers from other specialisations, e.g. digital electronics and software design.) In order to broaden the market yet further, it was intended that the University of Bradford would collaborate with other European universities running similar degree programmes so that the text could be expanded for use in all. The final list of collaborating institutions was:

University of Bradford, UK
University of Cantabria, Spain
University of Bologna, Italy
Telecommunications Systems Institute/Technical University of Crete, Greece

Microwave Devices, Circuits and Subsystems for Communications Engineering is a result of this collaboration and contains the material delivered in Part 1 of the Bradford degree plus additional material required to match courses delivered at the other institutions.

In addition to benefiting students studying the relevant degrees in the collaborating institutions, it is hoped that the book will prove useful to both the wider student population and to the practising engineer looking for a refresher or conversion text.

A companion website containing a sample chapter, solutions to selected problems and figures in electronic form (for the use of instructors adopting the book as a course text) is available at ftp://ftp.wiley.co.uk/pub/books/glover.

1

Overview

I. A. Glover, S. R. Pennock and P. R. Shepherd

1.1 Introduction

RF and microwave engineering has innumerable applications, from radar (e.g. for air traffic control and meteorology) through electro-heat applications (e.g. in paper manufacture and domestic microwave ovens), to radiometric remote sensing of the environment, continuous process measurements and non-destructive testing. The focus of the courses for which this text was written, however, is microwave communications and so, while much of the material that follows is entirely generic, the selection and presentation of material are conditioned by this application.

Figure 1.1 shows a block diagram of a typical microwave communications transceiver. The transmitter comprises an information source, a baseband signal processing unit, a modulator, some intermediate frequency (IF) filtering and amplification, a stage of up-conversion to the required radio frequency (RF) followed by further filtering, high power amplification (HPA) and an antenna. The baseband signal processing typically includes one, more, or all of the following: an antialising filter, an analogue-to-digital converter (ADC), a source coder, an encryption unit, an error controller, a multiplexer and a pulse shaper. The antialisaing filter and ADC are only required if the information source is analogue such as a speech signal, for example. The modulator impresses the (processed) baseband information onto the IF carrier. (An IF is used because modulation, filtering and amplification are technologically more difficult, and therefore more expensive, at the microwave RF.)

The receiver comprises an antenna, a low noise amplifier (LNA), microwave filtering, a down-converter, IF filtering and amplification, a demodulator/detector and a baseband processing unit. The demodulator may be coherent or incoherent. The signal processing will incorporate demultiplexing, error detection/correction, deciphering, source decoding, digital-to-analogue conversion (DAC), where appropriate, and audio/video amplification and filtering, again where appropriate. If detection is coherent, phase locked loops (PLLs) or their equivalent will feature in the detector design. Other control circuits, e.g. automatic gain control (AGC), may also be present in the receiver.

The various subsystems of Figure 1.1 (and the devices comprising them whether discrete or in microwave integrated circuit form) are typically connected together with transmission

Microwave Devices, Circuits and Subsystems for Communications Engineering Edited by I. A. Glover, S. R. Pennock and P. R. Shepherd

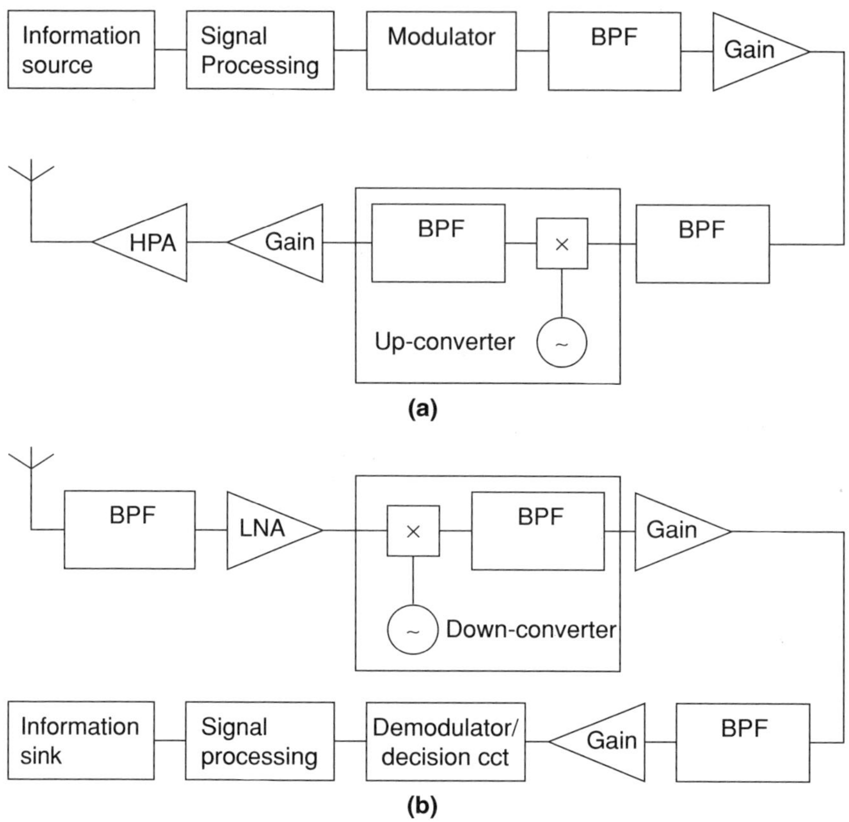

Figure 1.1 Typical microwave communications transmitter (a) and receiver (b)

lines implemented using a variety of possible technologies (e.g. coaxial cable, microstrip, co-planar waveguide).

This text is principally concerned with the operating principles and design of the RF/ microwave subsystems of Figure 1.1, i.e. the amplifiers, filters, mixers, local oscillators and connecting transmission lines. It starts, however, by reviewing the solid-state devices (diodes, transistors, etc.) incorporated in most of these subsystems since, assuming good design, it is the fundamental physics of these devices that typically limits performance.

Sections 1.2–1.7 represent a brief overview of the material in each of the following chapters.

1.2 RF Devices

Chapter 2 begins with a review of semiconductors, their fundamental properties and the features that distinguish them from conductors and insulators. The role of electrons and holes as charge carriers in intrinsic (pure) semiconductors is described and the related concepts of carrier mobility, drift velocity and drift current are presented. Carrier concentration gradients, the diffusion current that results from them and the definition of the diffusion coefficient are also examined and the doping of semiconductors with impurities to increase the concentration of electrons or holes is described. A discussion of the semiconductor energy-band model, which underlies an understanding of semiconductor behaviour, is presented and the important concept of the Fermi energy level is defined. This introductory but fundamental review of semiconductor properties finishes with the definition of mean carrier lifetime and an outline derivation of the carrier continuity equation, which plays a central role in device physics.

Each of the next six major sections deals with a particular type of semiconductor diode. In order of treatment these are (i) simple P-N junctions; (ii) Schottky diodes; (iii) PIN diodes; (iv) step-recovery diodes; (v) Gunn diodes; and (vi) IMPATT diodes. (The use of the term diode in the context of Gunn devices is questionable but almost universal and so we choose here to follow convention.) The treatment of the first three diode types follows the same pattern. The device is first described in thermal equilibrium (i.e. with no externally applied voltage), then under conditions of reverse bias (the P-material being made negative with respect to the N-material), and finally under conditions of forward bias (the P-material being made positive with respect to the N-material). Following discussion of the device's physics under these different conditions, an equivalent circuit model is presented that, to an acceptable engineering approximation, emulates the device's terminal behaviour. It is a device's equivalent circuit model that is used in the design of circuits and subsystems. There is a strong modern trend towards computer-aided design in which case the equivalent circuit models (although of perhaps greater sophistication and accuracy than those presented here) are incorporated in the circuit analysis software. The discussion of each device ends with some comments about its manufacture and a description of some typical applications.

The treatment of the following diode types is less uniform. Step-recovery diodes, being a variation on the basic PIN diode, are described only briefly. The Gunn diode is discussed in some detail since its operating principles are quite different from those of the previous devices. Its important negative resistance property, resulting in its principal application in oscillators and amplifiers, is explained and the relative advantages of its different operating modes are reviewed. Finally, IMPATT diodes are described, that, like Gunn devices, exhibit negative resistance and are used in high power (high frequency) amplifiers and oscillators, their applications being somewhat restricted, however, by their relatively poor noise characteristics. The doping profiles and operating principles of the IMPATT diode are described and the important device equations are presented. The discussion of IMPATT diodes concludes with their equivalent circuit.

Probably the most important solid-state device of all in modern-day electronic engineering is the transistor and it is this device, in several of its high frequency variations, that is addressed next. The treatment of transistors starts with some introductory and general remarks about transistor modelling, in particular, pointing out the difference between small and large signal models. After these introductory remarks three transistor types are addressed in turn, all suitable for RF/microwave applications (to a greater or lesser extent). These are (i) the gallium arsnide metal semiconductor field effect transistor (GaAs MESFET); (ii) the high electron mobility transistor (HEMT); and (iii) the heterojunction bipolar transistor (HBT). In each case the treatment is essentially the same: a short description followed by presentations of the current-voltage characteristic, capacitance-voltage characteristic, the small signal equivalent circuit and the large signal equivalent circuit.

1.3 Signal Transmission and Network Methods

Chapter 3 starts with a survey of practical transmission line structures including those without conductors (dielectric waveguides), those with a single conductor (e.g. conventional waveguide), and those with two conductors (e.g. microstrip). With one exception, all the two-conductor transmission line structures are identified as supporting a quasi-TEM (transverse electromagnetic) mode of propagation – important because this type of propagation can be modelled using classical distributed-circuit transmission line theory. A thorough treatment of this theory is given, starting with the fundamental differential equations containing voltage, current and distributed inductance (L), conductance (G), resistance (R) and capacitance (C), and deriving the resulting line's attenuation constant, phase constant and characteristic impedance. Physical interpretations of the solution of the transmission line equations are given in terms of forward and backward travelling waves and the concepts of loss, dispersion, group velocity and phase velocity are introduced. The frequency-dependent behaviour of a transmission line due to the frequency dependence of its L, G, R and C (due in part to the skin effect) is examined and the special properties of a lossless line (with $R = G = 0$) are derived.

Following the distributed-circuit description of transmission lines, the more rigorous field theory approach to their analysis is outlined. A short revision of fundamental electromagnetic theory is given before this theory is applied to the simplest (TEM) types of transmission line with a uniform dielectric and perfect conductors. The relationship between the time-varying field on the TEM line and the static field solution to Maxwell's equations is discussed and the validity of the solutions derived from this relationship is confirmed. The special characteristics of the TEM propagation mode are examined in some detail. The discussion of basic transmission line theory ends with a physical interpretation of the field solutions and a visualisation of the field distribution in a coaxial line.

Most traditional transmission lines (wire pair, coaxial cable, waveguide) are purchased as standard components and cut to length. Microstrip, and similarly fabricated line technologies, however, are typically more integrated with the active and passive components that they connect and require designing for each particular circuit application. A detailed description of microstrip is therefore given along with the design equations required to obtain the physical dimensions that achieve the desired electrical characteristics, given constraints such as substrate permittivity and thickness that are fixed once a (commercial) substrate has been selected. The limitations of microstrip including dispersion and loss are discussed and methods of evaluating them are presented. The problem of discontinuities is addressed and models for the foreshortened open end (an approximate open circuit termination), vias (an approximate short circuit termination), mitred bends (for reducing reflections at microstrip corners) and T-junctions are described.

In addition to a simple transmission line technology for interconnecting active and passive devices, microstrip can be used as a passive device technology in its own right. Microstrip implementation of low-pass filters is described and illustrated with a specific example. The general theory of coupled microstrip lines, useful for generalised filter and coupler design, is presented and the concepts of odd- and even-modes explained. Equations and design curves for obtaining the physical microstrip dimensions to realise a particular electrical design objective are presented. The directivity of parallel microstrip couplers is discussed and simplified expressions for its calculation are presented. Methods of improving coupler performance by capacitor compensation are described. The discussion of practical microstrip design methods concludes with a brief survey of other microstrip coupler configurations including Lange couplers, branch-line couplers and hybrid rings.

Network methods represent a fundamental way of describing the effect of a device or subsystem inserted between a source and load (which may be the Thévenin/Norton equivalent circuits of a complicated existing system). From a systems engineering perspective, the network parameters of the device or subsystem describe its properties completely – knowledge of the detailed composition of the device/subsystem (i.e. the circuit configuration or values of its component resistors, capacitors, inductors, diodes, transistors, transformers, etc.) being unnecessary. The network parameters may be expressed in a number of different ways, e.g. impedance (z), admittance (y), hybrid (h), transmission line (ABCD) and scattering (s) parameters, but all forms give identical (and complete) information and all forms can be readily transformed into any of the others. Despite being equivalent, there are certain practical advantages and disadvantages associated with each particular parameter set and at RF and microwave frequencies these weigh heavily in favour of using s-parameters. A brief review of all commonly used parameter sets is therefore followed by a more detailed definition and interpretation of s-parameters for both one- and two-port (two- and four-terminal) networks.

The reflection and transmission coefficients at the impedance discontinuities of a device's input and output are described explicitly by the device's s-parameters. One of the central problems in RF and microwave design is impedance matching the input and output of a device or subsystem with respect to its source and load impedances. (This problem may be addressed in the context of a variety of objectives such as minimum reflection, maximum gain or minimum noise figure.) The chapter therefore continues with an account of the most widely used aids to impedance matching, namely the Smith chart and its derivatives (admittance and immitance charts). These aids not only accelerate routine (manual) design calculations but also present a geometrical interpretation of relative impedance that can lead to analytical insights and creative design approaches. Both lumped and distributed element techniques are described including the classic transmission line cases of single and double stub matching. The treatment of matching ends with a discussion of broadband matching, its relationship to quality factor (Q-) circles that can be plotted on the Smith chart, and microstrip line transformers.

Chapter 3 closes with a description of network analysers – arguably the most important single instrument at the disposal of the microwave design engineer. The operating principles of this instrument, which can measure the frequency dependent s-parameters of a device, circuit or subsystem, are described and the sources of measurement error are examined. The critical requirement for good calibration of the instrument is explained and the normal calibration procedures, including the technologies used to make measurements on naked (unpackaged) devices, are presented.

1.4 Amplifiers

Virtually all systems need amplifiers to increase the amplitude and power of a signal. Many people are first introduced to amplifiers by means of low frequency transistor and operational amplifier circuits. At microwave frequencies amplifier design often revolves around terms such as available power, unilateral transducer gain, constant gain and constant noise figure circles, and biasing the transistor through a circuit board track that simply changes its width in order to provide a high isolation connection. This chapter aims to explain these terms and why they are used in the design of microwave amplifiers.

Chapter 4 starts by carefully considering how we define all the power and gain quantities. Microwave frequency amplifiers are often designed using the s-parameters supplied by the device manufacturer, so following the basic definitions of gain, the chapter derives expressions for gain working in terms of s-parameters. These expressions give rise to graphical representations in terms of circles, and the idea of gain circles and their use is discussed.

If we are to realise an amplifier, we want to avoid it becoming an oscillator. Likewise, if we are to make an oscillator, we do not want the circuit to be an amplifier. The stability of a circuit needs to be assessed and proper stability needs to be a design criterion. We look at some basic ideas of stability, and again the resulting conditions have a graphical interpretation as stability circles.

Amplifiers have many different requirements. They might need to be low noise amplifiers in a sensitive receiver, or high power amplifiers in a transmitter. Some applications require narrowband operation while some require broadband operation. This leads to different implementations of microwave amplifier circuits, and some of these are discussed in this chapter.

At microwave frequencies the capacitance, inductance and resistance of the packages holding the devices can have very significant effects, and these features need to be considered when implementing amplifiers in practice. Also alternative circuit layout techniques can be used in place of discrete inductors or capacitors, and some of these are discussed. In dealing with this we see that even relatively simple amplifier circuits are described by a large number of variables. The current method of handling this amount of data and achieving optimum designs is to use a CAD package, and an outline of the use of these is also given.

1.5 Mixers

Mixers are often a key component in a communication or radar system. We generally have our basic message to send, for example, a voice or video signal. This has a particular frequency content that typically extends from very low frequency, maybe zero, up to an upper limit, and we often refer to this as the baseband signal. Many radio stations, TV stations, and mobile phones can be used simultaneously, and they do this by broadcasting their signal on an individually allocated broadcast frequency. It is the mixer circuit that provides the frequency translation from baseband up to the broadcast frequency in the transmitter, and from the broadcast frequency back down to the original baseband in the receiver, to form a superheterodyne system.

A mixer is a non-linear circuit, and must be implemented using a nonlinear component. Chapter 5 first outlines the operation of the commonly used nonlinear components, the diode and the transistor. After that the analysis of these circuits are developed, and the terms used to characterise a mixer are also described. This is done for the so-called linear analysis for small signals, and also for the large signal harmonic balance analysis.

The currently popular transistor mixers are then described, particularly the signal and dual FET implementations that are common in Monolithic Microwave Integrated Circuits (MMICs). The designs of many mixers are discussed and typical performance characteristics are presented. The nonlinear nature of the circuit tends to produce unwanted frequencies at the output of the mixer. These unwanted terms can be 'balanced' out, and the chapter also discusses the operation of single and double balanced mixer configurations.

1.6 Filters

Chapter 6 provides the background and tools for designing filter circuits at microwave frequencies. The chapter begins with a review of two-port circuits and definitions of gain, attenuation and return loss, which are required in the later sections. The various filter characteristics are then described including low-pass, high-pass, band-pass and band-stop responses along with the order number of the filter and how this affects the roll-off of the gain response from the band edges.

The various types of filter response are then described: Butterworth (maximally flat within the passband), Chebyshev (equal ripple response in the pass band), Bessel (maximally flat in phase response) and Elliptic (equal ripple in pass band and stop band amplitude response).

The chapter then addresses the topic of filter realisation and introduces the concept of the low pass prototype filter circuit, which provides the basis for the filter design concepts in the remainder of the chapter. This has a normalised characteristic such that the 3 dB bandwidth of the filter is at a frequency of 1 radian/s and the load impedance is 1 ohm. The four types of filter response mentioned above are then considered in detail, with the mathematical description of the responses given for each.

The chapter then continues with the detail of low pass filter design for any particular value of the order number, N. The equivalent circuit descriptions of the filters are given for both odd and even values of N and also for T- and Π-ladders of capacitors and inductors. Analyses are provided for each of the four filter response types and tables of component values for the normalised response of each is provided.

As these tables are only applicable to the normalised case (bandwidth = 1 radian/s and the filter having a load impedance of 1 ohm), the next stage in the description of the filter realisation is to provide techniques for scaling the component values to apply for any particular load impedance and also for any particular value of low-pass bandwidth. The mathematical relationships between these scalings and the effects on the component values of the filter ladder are derived.

So far, the chapter has only considered the low pass type of filter, so the remainder of the chapter considers the various transformations required to convert the low pass response into equivalent high pass, band pass and band stop responses for each of the filter characteristic types. This therefore provides the reader with all the tools and techniques to design a microwave filter of low pass, high pass, band pass or band stop response with any of the four characteristic responses and of various order number.

1.7 Oscillators and Frequency Synthesisers

Chapter 7 describes the fundamentals of microwave oscillator design, including simple active component realisations using diodes and transistors. The chapter commences with an introduction to solid-state oscillator circuits considered as a device with a load. The fundamental approach is to consider the oscillation condition to be defined so that the sum of the device and load impedances sum to zero. Since the real part of the load impedance must be positive, this implies that the real part of the active device's effective impedance must be negative. This negative resistance is achieved in practice by using a negative resistance diode or a transistor which has a passive feedback network. The active device will have a nonlinear behaviour and its impedance depends on the amplitude of the signal. The balancing condition for the zero impedance condition therefore defines both the frequency and amplitude of oscillation.

The chapter continues with a description of diode realisations of negative resistance oscillators including those based on IMPATT and Gunn devices. This section includes example designs using typical diode characteristics, optimum power conditions and oscillation stability considerations.

Transistor oscillators are then considered. The fundamental design approach is to consider the circuit to be a transistor amplifier with positive feedback, allowing the growth of any starting oscillating signal. This starting signal is most likely to be from ever-present noise in electronic circuits. The feedback circuit is resonant at the desired oscillation frequency, so only noise signals within the bandwidth of the resonant circuit will be amplified and fed back, the other frequencies being filtered out. The possible forms of the resonant feedback circuit are discussed, these include lumped L-C circuits, transmission line equivalents, cavity resonators and dielectric resonators.

The standard topology of transistor feedback oscillators such as the Colpitts, Clapp and Hartley configurations are described and analysed from a mathematical point of view. This section concludes with a discussion of some of the Computer Aided Design (CAD) tools available for the design and analysis of solid-state microwave oscillator circuits.

The next section of the chapter deals with the inclusion of voltage-controlled tuning of the oscillator so that the frequency of oscillation can be varied by the use of a controlling DC voltage. The main implementations for voltage-controlled oscillators (VCOs) use varactor diodes and Yttrium Iron Garnets (YIGs). Varactors are diodes whose junction capacitance can be varied over a significant range of values by the applied bias voltage. When used as part of the frequency selective feedback circuit, variation of this diode capacitance will lead to a variation in the oscillation frequency. YIGs are high-Q resonators in which the ferromagnetic resonance depends (among other factors) on the magnetic field across the device. This in turn can be controlled by an applied voltage. As well as having a high Q value (and therefore a highly stable frequency), YIGs are also capable of a very wide range of voltage control, leading to very broadband voltage control. Examples of practical VCO design complete this section.

The next section considers the characterisation and testing of oscillators. The various parameters used to specify the performance of a particular oscillator include: the oscillator frequency, characterised using a frequency counter or spectrum analyser; the output power, characterised using a power meter; stability and noise (in both amplitude and phase), which can again be analysed using a spectrum analyser or more sophisticated phase noise measurement equipment.

The final section of the chapter deals with phase-locked oscillators, which use phase-locked loops (PLLs) to stabilise the frequency of microwave oscillators. A fundamental description of PLLs is given, along with a consideration of their stability performance. These circuits are then incorporated into the microwave oscillators using a frequency multiplier and harmonic mixers so that the microwave frequency is locked on to a lower, crystal stabilised, frequency so that the characteristics of the highly stable low frequency source are translated on to the microwave frequency.

2

RF Devices: Characteristics and Modelling

A. Suarez and T. Fernandez

2.1 Introduction

Semiconductor transistors and diodes both exhibit a non-linear current and/or voltage input–output characteristic. Such non-linearity can make the behaviour of these devices difficult to model and simulate. It does enable, however, the implementation of useful functions such as frequency multiplication, frequency translation, switching, variable attenuators and power limiting. Transistors and some types of diodes may also be active, i.e. capable of delivering energy to the system, allowing them to be used in amplifier and oscillator designs. Passive non-linear responses are used for applications such as frequency mixing, switching or power limiting.

The aims of this chapter are: (1) to give a good understanding of the operating principles of the devices presented and to convey factual knowledge of their characteristics and limitations so as to ensure their appropriate use in circuit design; (2) to present accurate equivalent circuit models and introduce some efficient modelling techniques necessary for the analysis and simulation of the circuits in which the devices are employed; and (3) to present the most common applications of each device, illustrating the way in which their particular characteristics are exploited.

The chapter starts with a revision of semiconductor physics, including the general properties of semiconductor materials and band theory that is usually used to explain the origin of these properties. This is followed by a detailed description of the two most important semiconductor devices, diodes and transistors, in their various RF/microwave incarnations. In keeping with the practical circuit and subsystem design ethos of the courses on which this text is based, significant emphasis is placed on the devices' equivalent circuit models that are necessary for both traditional and modern computer-aided design.

Microwave Devices, Circuits and Subsystems for Communications Engineering Edited by I. A. Glover, S. R. Pennock and P. R. Shepherd

2.2 Semiconductor Properties

Solids may be divided into three principal categories: metals, insulators and semiconductors [1, 2]. Metals consist of positive ions, surrounded by a cloud of electrons. Free electrons, which are shared by all the atoms, are able to move under the influence of an electric field at 0 K. Metals have only one type of charge carrier, the electron, conduction being due to electron movement only. The concentration of electrons in an electrically neutral metal is always approximately the same whatever the material or temperature, with values between 10^{22} and 10^{23} cm^{-3}.

Insulators are crystalline structures in which electrons are bound closely together in covalent bonds. Electrons in insulators do not move under the influence of an electric field at room temperature, T_0 (= 300 K).

Semiconductors are crystalline structures composed of valence-IV atoms, linked by covalent bonds. They behave as insulators at 0 K and as good conductors at room temperature. The absence of one electron leaves a hole in the covalent bond. When applying an electric field, the filling of this vacancy by another electron, leaving, in turn, another hole, gives rise to an apparent hole movement. A positive charge value may be associated with every hole. Semiconductors have thus two types of charge carriers: electrons and the holes.

2.2.1 Intrinsic Semiconductors

Intrinsic semiconductors may have any of several different forms: single elements with valence IV, such as silicon (Si), germanium (Ge), and carbon (C), and compounds with average valence IV. Among the latter, binary III–V and II–VI associations are the most usual. For example, Gallium Arsenide (GaAs), a III–V compound, is commonly used for microwave devices due to its good conduction properties. In the semiconductor crystal every atom is surrounded by four other atoms and linked to them by four covalent bonds (Figure 2.1).

In compound semiconductors these bonds are formed between the positive and negative ions with four peripheral electrons and are called hetero-polar valence bonds. For instance, in the case of GaAs, four covalent bonds are formed between the negative ions Ga^- and the positive ions As^+.

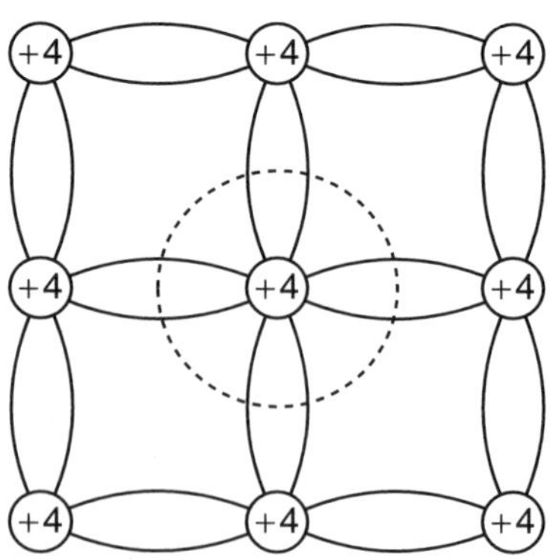

Figure 2.1 Covalent bonding in semiconductor crystal

The charge movement in semiconductors may be due to thermal agitation or to the application of an electric field. In the former case the movement is random, while in the latter case carriers move systematically in a direction, and sense, that depends on the applied electric field, **E**. Electrons are accelerated in the opposite sense to that of the electric field until they reach a final drift velocity, $\mathbf{v}_n$, given by:

$$\mathbf{v}_n = -\mu_n \mathbf{E} \tag{2.1}$$

where μ_n is a proportionality constant, known as the electron mobility. Mobility is defined as a positive quantity so Equation (2.1) is in agreement with the electron movement in opposite sense to the electric field. The holes are accelerated in the same sense as the applied field until they reach a final drift velocity, $\mathbf{v}_p$, given by:

$$\mathbf{v}_p = \mu_p \mathbf{E} \tag{2.2}$$

where μ_p is hole mobility. Mobility therefore depends on the type of particle, electron or hole, and on the material. For electrons, its value is 1500 $cm^2V^{-1}s^{-1}$ in Si and 8500 $cm^2V^{-1}s^{-1}$ in GaAs. The higher mobility value makes the latter well suited for microwave applications.

For relatively low electric field strengths, the mobility μ has a constant value, resulting in a linear relationship between the velocity and the applied electric field. At higher field strengths, however, mobility depends on electric field, i.e. $\mu = \mu(E)$, which gives rise to a nonlinear relationship. For still higher values of electric field, saturation of the drift velocity is observed.

Electron and hole movement due to the application of an electric field, i.e. drift current, may be easily related to the applied field. The current density, $\bar{J}_n$, due to the electron movement, is given by:

$$\mathbf{J}_n = \frac{dI_n}{d\mathbf{S}} = \frac{dQ_n}{dt} \cdot \frac{1}{d\mathbf{S}} \tag{2.3}$$

where I_n is total electron current through the differential surface element dS and Q_n is the total electron charge passing the surface in differential time element dt. For a total electron number N, the total charge will be $Q_n = -Ne$, where e is the (magnitude of the) electron charge, i.e. 1.602×10^{-19} C. The differential charge element dQ_n is given by $dQ_n = -dNe$. Since the electron density per unit volume, n, is given by:

$$n = \frac{dN}{dV} \tag{2.4}$$

the element dN may be written as $dN = ndV$ and substituting into Equation (2.3) gives:

$$\mathbf{J}_n = -ne\frac{dV}{dt}\frac{1}{d\mathbf{S}} = -ne\mathbf{v}_n \tag{2.5}$$

Combining Equations (2.1) and (2.5):

$$\mathbf{J}_n = \mu_n en\mathbf{E} \tag{2.6}$$

By following the same steps a similar result can be obtained for the hole current, i.e.:

$$\mathbf{J}_p = \mu_p ep\mathbf{E} \tag{2.7}$$

where p is the hole concentration (i.e. number per unit volume). The total drift current is, therefore:

$$\mathbf{J}_d = (\mu_n n + \mu_p p)e\mathbf{E} \tag{2.8}$$

Ohm's law can be expressed as:

$$\mathbf{J} = \sigma\mathbf{E} \tag{2.9}$$

where σ is conductivity (in S/m). By comparing Equations (2.8) and (2.9) it is possible to derive an expression relating a material's conductivity with its mobility values, i.e.:

$$\sigma = \mu_n ne + \mu_p pe \tag{2.10}$$

It is clear from Equation (2.10) that increasing the carrier concentration enhances the conductivity properties of the semiconductor. As the temperature rises, more covalent bonds break and for every broken bond a pair of carriers, an electron and a hole, are generated. The conductivity of the intrinsic material, therefore, increases with the temperature.

A second type of current, quite different in origin from the drift current discussed above, may also exist in semiconductor devices. This diffusion current is due to any non-homogeneity in the carrier concentration within the semiconductor material. In order to balance concentrations, carriers move from the higher concentration regions to regions of lower concentration, the movement being governed by the concentration gradient. For a one-dimension device (i.e. a device in which variations only occur in one, the x, dimension), electron diffusion current density, $\mathbf{J}_{Dn}$, is given by [1, 2]:

$$\mathbf{J}_{Dn} = eD_n \frac{\mathbf{d}n}{\mathbf{dx}} \tag{2.11}$$

where D_n is the diffusion coefficient for electrons. D_n is related to electron mobility and temperature through the expression [1, 2]:

$$D_n = \frac{kT}{e}\mu_n \tag{2.12}$$

where k is Boltzmann's constant (1.381×10^{-23} J/K) and T is absolute temperature.

For holes, the diffusion current density, $\mathbf{J}_{D_p}$, is given by:

$$\mathbf{J}_{D_p} = -eD_p \frac{\mathbf{d}p}{\mathbf{dx}} \tag{2.13}$$

where D_p is the diffusion coefficient for electrons given by:

$$D_p = \frac{kT}{e}\mu_p \tag{2.14}$$

In addition to the drift and diffusion currents, a third current component should be considered in the semiconductor when a time variable electric field is applied. This is displacement current. Displacement current density, $\mathbf{J}_d$, given by [1, 2]:

$$\mathbf{J}_d = \varepsilon_{sc} \frac{d\mathbf{E}}{dt} \tag{2.15}$$

where ε_{sc} is the semiconductor dielectric constant.

Self-assessment Problems

2.1 What are the mechanisms that respectively give rise to drift current and diffusion current in a semiconductor material?

2.2 What is the relationship between semiconductor conductivity and carrier mobility?

2.3 Why is GaAs preferred to silicon for microwave circuit design?

2.2.2 Doped Semiconductors

As shown by Equation (2.10), semiconductor conductivity improves with carrier concentration. This may be increased by adding impurities to (i.e. doping) the semiconductor. There are two different types of doping: N-type, when valence-V atoms are added, and P-type, when valence-III atoms are added.

2.2.2.1 N-type doping

A semiconductor material is N-doped, or N-type, when a certain concentration, N_D, of valence-V atoms, such as phosphorus (P), arsenic (As) or antimony (Sb), is introduced into the crystal (Figure 2.2).

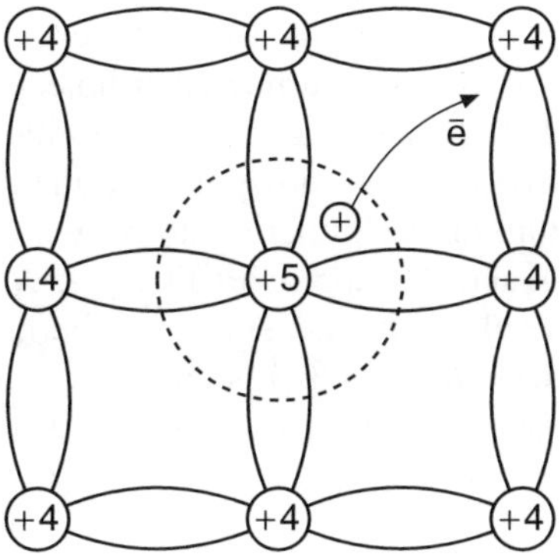

Figure 2.2 N-doped semiconductor crystal

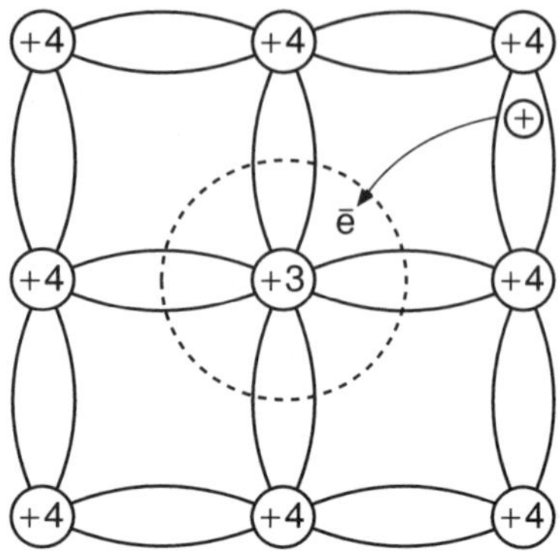

Figure 2.3 P-doped semiconductor crystal

These valence-V impurity atoms are known as donors. Due to the resulting lack of symmetry, one of the five valence electrons in each impurity atom is linked only weakly to the crystal lattice. These electrons therefore need only a small amount of energy to become conductive. At room temperature, in fact, there is sufficient thermal energy to ionise virtually all the impurity atoms. It is therefore possible to assume $n \cong N_D$ at room temperatures, since the impurity electrons are far more numerous than those created by thermal electron-hole pair generation in the intrinsic material. In an N-type material, electrons are thus majority carriers, while holes are minority carriers. As temperature increases, carrier creation by thermal generation of (intrinsic material) electron-hole pairs becomes progressively more important and at very high temperatures (500 to 550 K for Si) the material loses its doped characteristics, behaving essentially as an intrinsic semiconductor.

2.2.2.2 P-type doping

A semiconductor material is P-doped when a certain concentration, N_A, of valence-III atoms, such as boron (B) or gallium (Ga), is introduced into the crystal (Figure 2.3).

These valence-III impurity atoms are known as acceptors. Due to the resulting lack of symmetry, the impurity atoms easily accept electrons from the surrounding covalent bonds, thus becoming ionised. For each ionised impurity, a hole is left in the crystal lattice. The filling of this hole by an electron leaves in turn another hole, equivalent to a positive charge displacement. The hole therefore behaves as a positive charge carrier. Measurements of the ease with which holes can be made to move leads to their being ascribed an equivalent mass value. At room temperature all the impurity atoms, with concentration N_P, are usually ionised and it is possible to make the approximation $p \cong N_P$. In this case, the holes are the majority carriers, while the electrons, generated by thermal effects within the intrinsic material, are the minority carriers. At very high temperature, due to the strong electron-hole creation process, the material basically behaves as an intrinsic semiconductor.

2.2.3 Band Model for Semiconductors

It is well known that the electrons in an atom can neither have arbitrary energy values (but instead can occupy only certain, discrete, allowed energy levels) nor can they share a single

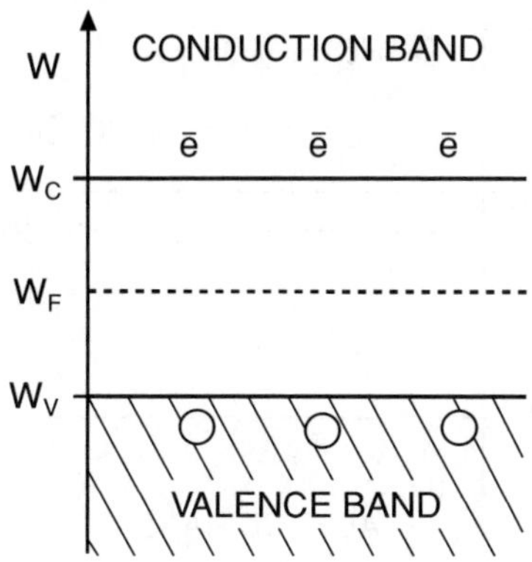

Figure 2.4 Semiconductor energy band diagram

energy value. If two atoms are placed in close proximity, the number of allowed energy levels is doubled in order to host twice the number of electrons without more than one electron occupying a single level. If a large number of electrons are placed close together, as is the case within a solid, the discrete levels become packed so tightly together as to effectively give rise to a continuous allowed energy band. Forbidden energy bands then separate permitted energy bands.

The atoms of an intrinsic semiconductor are, at 0 K, joined by covalent bonds established between their four valence electrons and an equal number of electrons from each of the four neighbouring atoms in the crystal. These electrons may be thought to occupy the discrete levels of an energy band called the valence band. As the temperature rises, some of these bonds may break. The corresponding electrons will be more energetic, being located at higher energy levels. These higher levels, containing electrons capable of moving under the effect of an electric field, form what is known as the conduction band (Figure 2.4).

An electron making the transition to the conduction band leaves a hole in the valence band. Between the valence and conduction bands there is a forbidden energy gap. This energy gap represents the energy needed to break a covalent bond. In good insulators, the valence band is complete at room temperature and the conduction band is empty, with a wide energy gap between the two. In metals, there is no forbidden band, the valence and conduction bands overlap with the electrons, which are the only carriers, being located at the lower levels of the non-saturated (i.e. only partly filled) conduction band.

The Fermi energy level is defined as the energy level at which the probability of finding an electron is 0.5. For an intrinsic material, the Fermi level is always located in the middle of the forbidden energy gap, whatever the temperature [1, 2]. (For 0 K, the probability of finding an electron below the Fermi level is 1.0 and the probability of finding an electron above the Fermi level is zero.) The Fermi level is always defined in the absence of an external force, such as a bias generator. When an external force is present, the so-called pseudo-Fermi level must be considered.

In the case of an N-doped material, one of the electrons in each impurity atom is far more energetic than the other four, which form close valence bonds with the four surrounding atoms. The more energetic electrons will, by definition, be located at higher energy levels in the band diagram. These higher levels form what is called the donor band, located immediately beneath the conduction band (Figure 2.5).

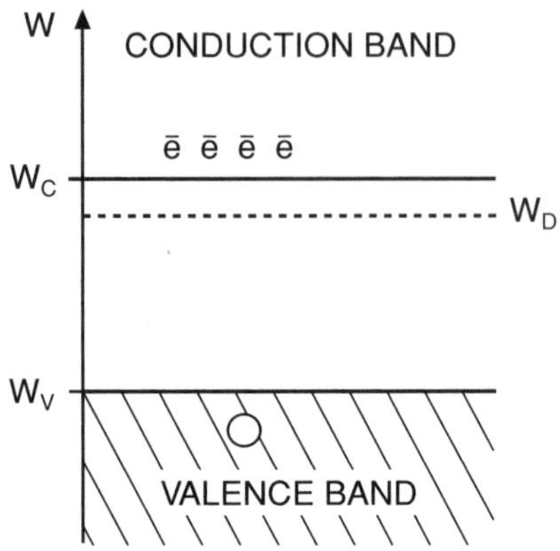

Figure 2.5 Energy band diagram of *N*-doped semiconductor showing donor band below conduction band

The donor band will be complete at 0 K. However, since these electrons only need a small energy supply to become conductive, the donor band will be generally empty at room temperature, all its electrons having moved to the conduction band. It must be noted that these carriers do not leave holes, in contrast to the case for thermal generation that results in electron-hole pairs. In N-doped semiconductors, the Fermi level is located immediately below the conduction band [1, 2] and as temperature increases, it progressively descends towards the middle of the forbidden band. This is consistent with the increasing number of electron-hole pairs as the temperature rises, progressively making the material behave as an intrinsic one. (For an intrinsic material, the Fermi level is always located in the middle of the forbidden energy gap, whatever the temperature.)

In the case of a P-doped material, an acceptor band is located immediately above the valence band (Figure 2.6).

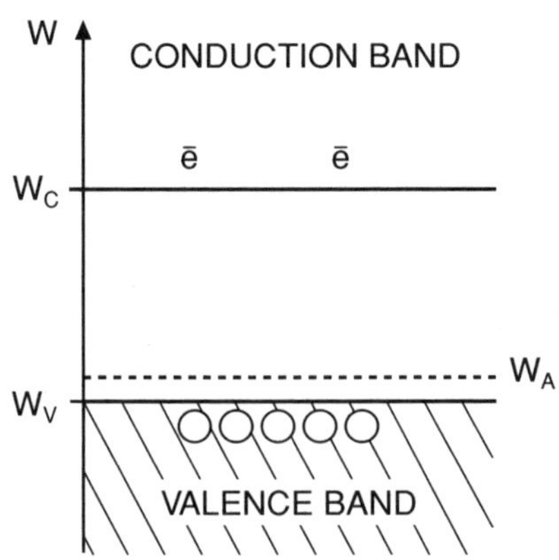

Figure 2.6 Energy band diagram of P-doped semiconductor showing acceptor band above valence band

The acceptor band is empty at 0 K. As the temperature rises, however, electrons in the valence bonds will move to the acceptor band, leaving holes in the valence band, ready for conduction. For a P-doped material, at low temperature, the Fermi level is located immediately above the valence band [1, 2], moving to the middle of the forbidden band as the temperature rises.

Self-assessment Problems

2.4 Why does doping increase the conductivity of semiconductor material?

2.5 Why is the valence band of a semiconductor complete at 0 K?

2.6 Why is the donor band located immediately below the conductor band?

2.2.4 Carrier Continuity Equation

Charge within the semiconductor must be conserved and it follows that its variation with time must equal the current flow, plus the charge creation and recombination loss per unit time. Charge creation may be due to thermal effects, to the illumination of the material, or to the application of a very high electric field. Charge recombination occurs when a free electron, moving through the crystal lattice, fills a hole, both of them disappearing as free charges. The average time between carrier creation and recombination is called the *mean carrier lifetime*, τ. Perhaps surprisingly, the mean carrier lifetime is different for holes and electrons.

In the case of holes, for a semiconductor surface S and differential length dx, the charge loss per unit time (i.e. the current lost, dI_{rp}) due to recombination will be [1, 2, 3]:

$$dI_{rp} = -\frac{p}{\tau_p} eSdx \tag{2.16}$$

where τ_p is mean hole lifetime. If, at the same time, g_p holes are generated per unit volume per unit time, the corresponding charge increase per unit time, i.e. the current gained, will be:

$$dI_{gp} = g_p eSdx \tag{2.17}$$

The differential element for the total hole current, due to drift and diffusion effects is then given by:

$$\begin{aligned} dI_p &= \frac{\partial \mathbf{J}_p}{\partial x} Sdx \\ &= \left(-eD_p \frac{\partial^2 p}{\partial x^2} + e\mu_p \frac{\partial (pE)}{\partial x} \right) Sdx \end{aligned} \tag{2.18}$$

and the total charge variation with time, due to holes, is:

$$\frac{\partial p}{\partial t} eSdx = dI_{rp} + dI_{gp} - dI_p$$

$$= -\frac{p}{\tau_p} eSdx + g_p eSdx + \left(eD_p \frac{\partial^2 p}{\partial x^2} - e\mu_p \frac{\partial (pE)}{\partial x} \right) Sdx \tag{2.19}$$

In Equation (2.19) it is possible to eliminate the term *eSdx*, obtaining the following expression:

$$\frac{\partial p}{\partial t} = -\frac{p}{\tau_p} + g_p + D_p \frac{\partial^2 p}{\partial x^2} - \mu_p \frac{\partial (pE)}{\partial x} \tag{2.20}$$

which is known as the hole continuity equation. A similar equation may be deduced for the electrons, i.e.:

$$\frac{\partial n}{\partial t} = -\frac{n}{\tau_n} + g_n + D_n \frac{\partial^2 n}{\partial x^2} + \mu_n \frac{\partial (nE)}{\partial x} \tag{2.21}$$

Equations (2.20) and (2.21) govern much of semiconductor device physics.

2.3 P-N junction

A P-N junction is obtained when two pieces of semiconductor, one P-type and one N-type, with similar impurity concentrations, are connected. Due to the difference in electron and hole concentration on either side of the junction, electrons from atoms close to the junction in the N region will diffuse to the P side in an attempt to balance the concentrations. Similarly, holes from atoms close to the junction in the P region will diffuse to the N side. The currents due to both these processes have the same sense since they are due to charges of opposite sign moving in opposite directions.

2.3.1 Thermal Equilibrium

The electrons moving from N to P material will leave fixed positive ions, while the holes moving from P to N material will leave fixed negative ions (Figure 2.7).

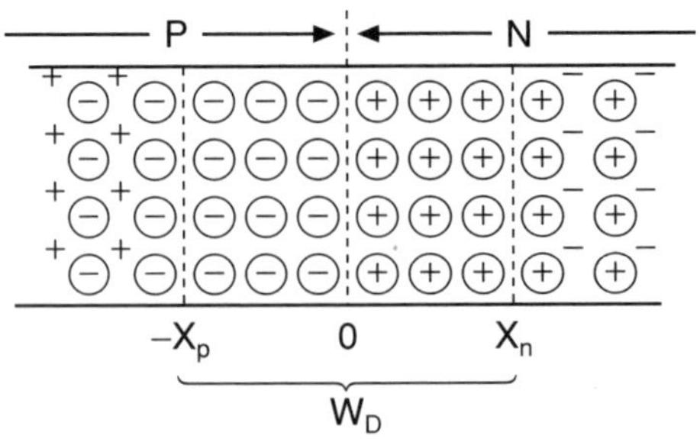

Figure 2.7 Schematic representation of P-N junction

A depletion zone, void of mobile charge is thus created close to, and on both sides of, the junction. This zone is also known as the space-charge region. The fixed charges give rise to an electrostatic potential, opposing the diffusion process, which drastically reduces the current magnitude. At the same time the few minority carriers (electrons in the P region and holes in the N region) created thermally, will be attracted by the fixed charges. Both P and N minority currents add, the total minority current having opposite sense to the current due to majority carriers. Equilibrium is attained when the total current reaches zero. The final value of the potential barrier, also called built-in potential, is denoted by ϕ_0. Its magnitude depends on the semiconductor material and on the doping level, with typical values between 0.5 and 0.9 V.

For a constant acceptor concentration N_a in the P region and a donor concentration N_d in the N region, the fixed charge density along device is shown in Figure 2.8(a).

In Figure 2.8 the junction plane has been taken as the x co-ordinate origin. The limits of the depletion region are defined as $-x_p$ on the P side and x_n on the N side. From the principle

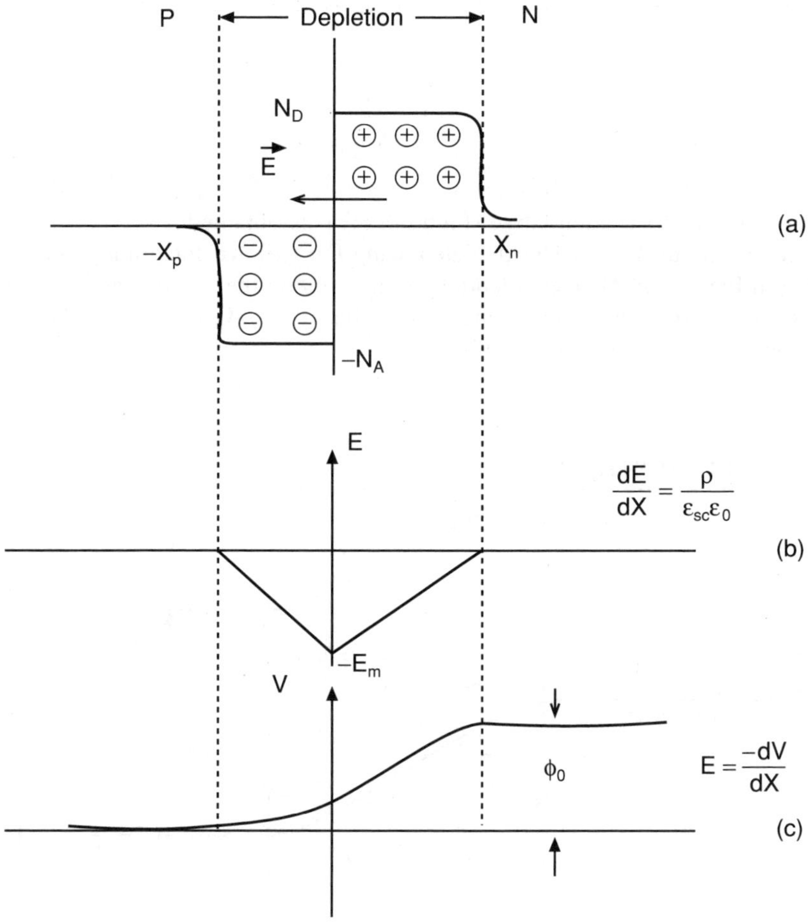

Figure 2.8 Charge (a), field (b) and potential (c) variations across a P-N junction

of electrical neutrality, the absolute value of the total charge at both sides of the junction must be equal, i.e.:

$$N_a e x_p S = N_d e x_n S \tag{2.22}$$

Therefore:

$$N_a x_p = N_d x_n \tag{2.23}$$

According to Equation (2.23), the extension of the space-charge zone is inversely proportional to the doping concentration. Its extension will therefore be smaller at the side of the junction with the highest doping concentration. In the derivation of Equation (2.23) an abrupt junction has been considered, with a constant impurity concentration on both sides of the junction. Other doping profiles are possible with a doping concentration (including the carrier sign) $N(x)$. A linear profile, for example, will have a constant slope value and will satisfy $N(0) = 0$.

The electric field E along the depletion zone can be determined from Poisson's equation, i.e.:

$$\frac{dE}{dx} = \frac{\rho}{\varepsilon_{sc}} \tag{2.24}$$

where ρ is charge density (in C/m^3) and ε_{sc} is semiconductor permitivity. By integrating Equation (2.24) a field variation curve (Figure 2.8(b)) is obtained.

Finally, the potential, V, may be calculated using $E = -dV/dx$. Integrating both sides leads to the curve in Figure 2.8(c). The potential variation along the space-charge region leads to a bending of the energy bands, due to the relationship $W = -eV$, where W is the energy. The resulting energy diagram is shown in Figure 2.9.

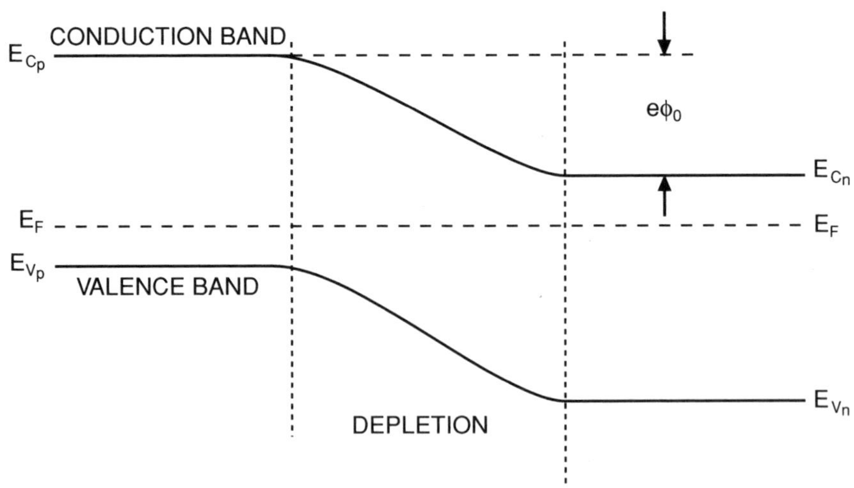

Figure 2.9 Energy band diagram for a P-N junction

Since no external bias is applied, the Fermi level must be constant along the device (which represents another explanation for the band bending). According to Figure 2.9, electrons from the N region and holes from the P region now have great difficulty in traversing the junction plane, and only a few will surmount the resulting energy barrier. The minority electrons in the P region and the minority holes in the N region, however, traverse the junction without opposition. As previously stated, the value of the total current, once equilibrium has been reached, is zero.

2.3.2 Reverse Bias

A P-N junction is reverse-biased when an external voltage V is applied with the polarity as shown in Figure 2.10.

This increases the potential barrier between the P and N regions to the value $\phi_0 + V$ (Figure 2.11), which reduces majority carrier diffusion to almost zero.

The increase in height of the potential barrier does not affect the minority carrier current, since these carriers continue to traverse the junction without opposition. A small reverse current I_s is then obtained, which constitutes the major contribution (under reverse bias

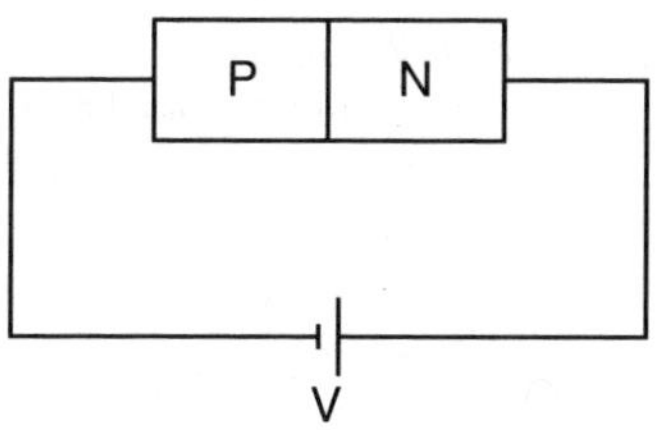

Figure 2.10 P-N junction in reverse bias

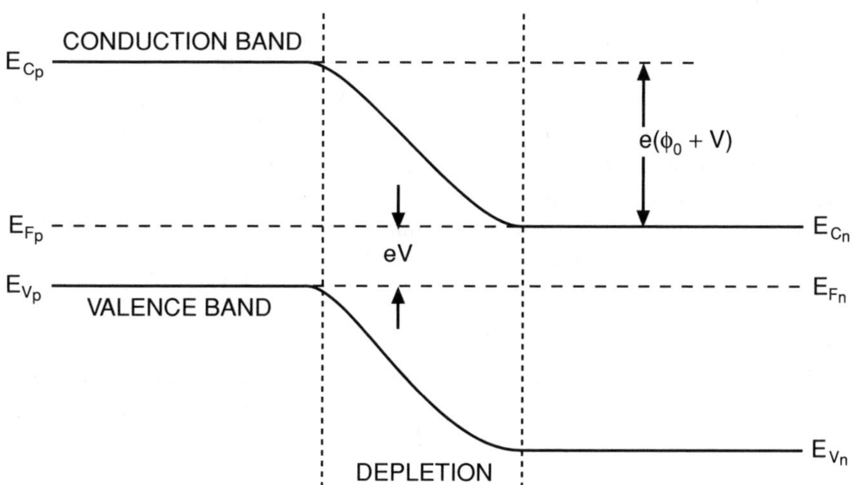

Figure 2.11 Increased potential barrier of P-N junction in reverse bias

conditions) to the total device current. I_s is called the reverse saturation current, its size depending on the temperature and the semiconductor material.

Due to the device biasing, no absolute, Fermi level can be considered, and pseudo-Fermi levels are introduced instead, one on each side of the junction. In terms of potential, the difference between these pseudo-Fermi levels, $V_{Fp} - V_{Fn}$, equals the applied voltage V, with $V > 0$.

As a result of the reverse bias, the width of the depletion region increases compared to that obtained for thermal equilibrium. This is due to electrons moving towards the positive terminal of the biasing source and holes moving towards the negative terminal. The width of the depletion region, given by $w_d = x_n + x_p$, therefore depends on the applied voltage, i.e. $w_d = w_d(V)$.

The depletion region, composed of static ionized atoms, is non-conducting and can be considered to be a dielectric material. This region is bounded by the semiconductor neutral zones, which, due to the doping, are very conducting. This structure therefore exhibits capacitance, the value of which is known as the junction capacitance, C_j. Junction capacitance can be calculated from:

$$C_j(V) = \varepsilon_{sc} \frac{S}{w_d(V)} \tag{2.25}$$

where S is the effective junction area and the dependence on applied voltage, V, is explicitly shown. From Poisson's equation, it is possible to obtain the following expression for the width of the depletion zone, w_d, as a function of the applied voltage V [1, 2]:

$$C_j(V) = \frac{C_{j0}}{\left(1 - \dfrac{V}{\phi_0}\right)^{\gamma}} \tag{2.26}$$

where C_{j0} is the capacitance for $V = 0$, $\gamma = 0.5$ for an abrupt junction and $\gamma = 0.33$ for a graded (i.e. gradual) junction.

As reverse bias is increased, a voltage $V = V_b$ is eventually attained, for which the electric field in the depletion zone reaches a critical value, E_c. At this (high) value of electric field, electrons traversing the depletion region are accelerated to a sufficiently high velocity to knock other electrons out of their atomic orbits during collisions. The newly generated carriers will in turn be accelerated and will release other electrons during similar collisions. This process is known as avalanche breakdown. The breakdown voltage, V_b, imposes an upper limit to the reverse voltage that can normally be applied to a P-N junction.

The breakdown voltage, V_b, can clearly be found from [1, 2]:

$$V_b = E_c w_d \tag{2.27}$$

The breakdown voltage is thus directly proportional to the width of the depletion zone, w_d.

The energy threshold required by an electron or a hole in order to create a carrier pair is higher than the width of the forbidden energy band. The new carriers therefore have a non-zero kinetic energy after their creation. The ionisation coefficient, α, gives the number of pairs that are created by a single accelerated carrier per unit length. This coefficient depends on the applied field and has a lower value for a larger width of the forbidden energy band.

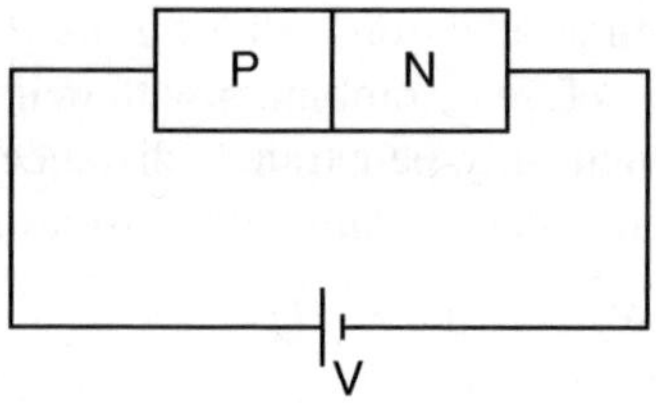

Figure 2.12 P-N junction in forward bias

2.3.3 Forward Bias

For forward bias, the external generator is connected as shown in Figure 2.12.

This reduces the potential difference along the junction (Figure 2.13) to a net value of $\phi_0 - V$.

This reduction of the potential barrier favours the majority carrier current that will grow exponentially as V is increased. Actually, as $V \to \phi_0$, the majority carrier current will be limited only by the device resistance and the external circuit. The device resistance is due to the limited conductivity value of the neutral zones ($\rho = 1/\sigma$). The minority carrier current, I_s, depends only on device temperature and is not affected by the biasing. The current variation as a function of the applied voltage is modelled using [1, 2]:

$$I(V) = I_s(e^{\alpha V} - 1) \tag{2.28}$$

where the parameter α is given by [1, 2]:

$$\alpha = \frac{e}{kT} \tag{2.29}$$

and where k is Boltzmann's constant expressed as 8.6×10^{-5} eV/K.

Since an external bias is applied, pseudo-Fermi levels must be considered and their potential difference will be now $V_{Fp} - V_{Fn} = -V$, with $V > 0$ (Figure 2.13).

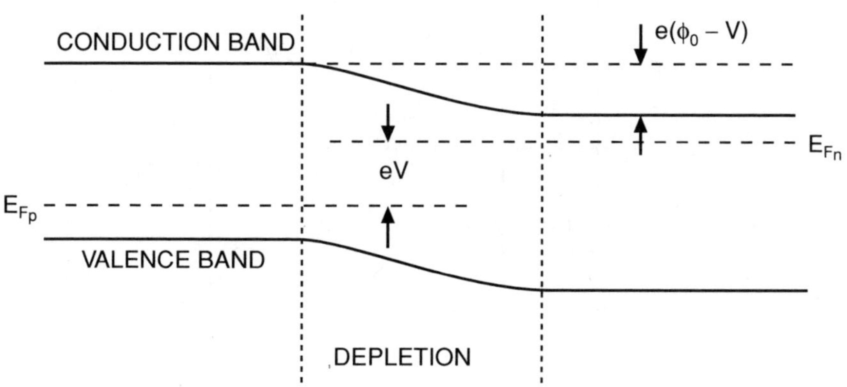

Figure 2.13 Decreased potential barrier of P-N junction in forward bias

Due to the limited carrier lifetime, electrons entering the P region only penetrate a certain distance (the diffusion length, L_n) before recombining with majority holes. The same applies for holes entering the N region, which only penetrate a distance L_p. Both diffusion lengths are related to the respective diffusion constants and carrier lifetimes through the formulae [1, 2]:

$$L_n = \sqrt{D_n \tau_n}$$
$$L_p = \sqrt{D_p \tau_p} \tag{2.30}$$

Under forward bias conditions the depletion region is narrow, with a width, w_d, much smaller than the diffusion lengths, L_n and L_p. This is responsible for a charge accumulation on both sides of the junction, represented by electrons on the P side, and holes on the N side. This charge accumulation, which varies with applied voltage, gives rise to a non-linear capacitance (recall the relationship $C = dQ/dV$). The value of this capacitance is therefore given by:

$$C_d = \frac{dQ}{dV} = \tau \frac{dI}{dV} = \tau \alpha I_S e^{\alpha V} \tag{2.31}$$

τ being the mean carrier lifetime. C_d is known as the diffusion capacitance. It is typically observed in bipolar devices (having both electron and hole conduction), as a direct consequence of conduction through minority carriers.

2.3.4 Diode Model

The electric model for the P-N diode consists of a current source (with the non-linear I–V characteristic given by Equation (2.28)) in parallel with two capacitors; one for the junction capacitance, C_j, and the other for the diffusion capacitance, C_d (Figure 2.14).

The model is completed with a series resistor, R_s, accounting for the loss in the neutral regions. The diffusion capacitance often prevents P-N diodes from being used in microwave applications due to the extremely low impedance value associated with it in this frequency range.

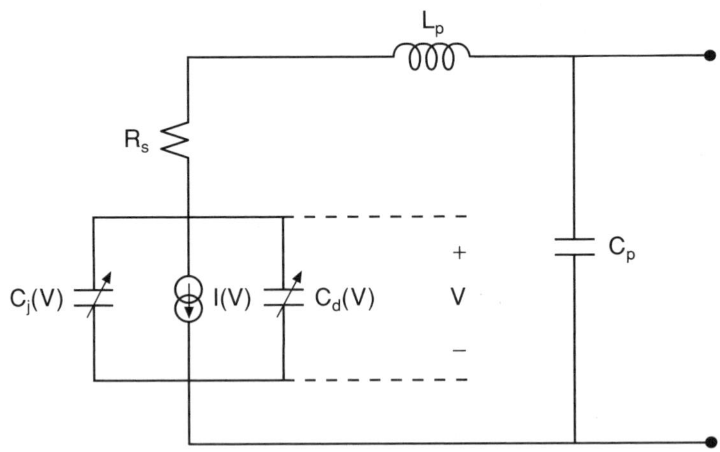

Figure 2.14 Diode equivalent circuit model

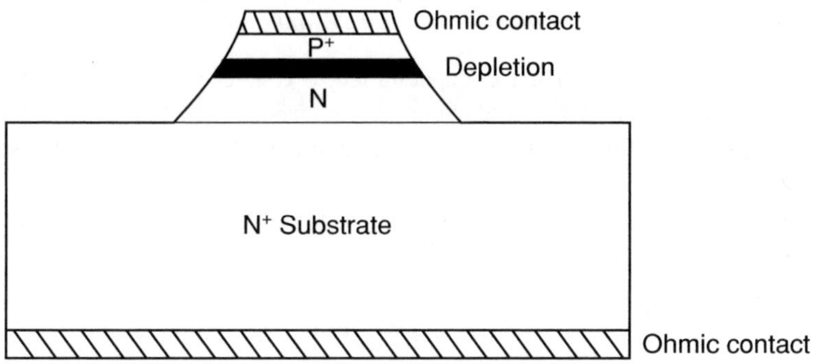

Figure 2.15 Schematic diagram of P-N junction structure

In order to connect a P-N diode in a conventional circuit, metal contacts at the device terminals are necessary. These are fabricated, at each end of the device, by depositing a metallic layer over a highly doped semiconductor section. The intense doping is usually represented by a superscript +, i.e. P^+ on the P side of the junction and N^+ on the N side. The reasons why high doping is necessary are given later.

The final diode may be produced in chip or packaged forms. When the device is packaged, two parasitic elements, due to the packaging, must be added to the model. One element is a capacitance, C_p, which arises due to the package insulator, and the other element is an inductance, L_p, which arises due to the bonding wires.

2.3.5 Manufacturing

P-N diodes are manufactured in *mesa¬* [1, 2]. In this technology the active layers of the device are realised by locally attacking the semiconductor substrate, but leaving some isolated (i.e. non-attacked) regions. The attack may be chemical or mechanical. In P-N junctions, an N layer is deposited by epitaxial growth over a silicon substrate N^+ (Figure 2.15).

Acceptor impurities are then diffused over the surface, obtaining a P^+N junction. Metal contacts are then deposited at the two device ends, over the P^+ layer and below the N^+ layer. As will be seen later, the high doping level is necessary in order to avoid the formation of Schottky junctions.

Self-assessment Problems

2.7 What is the reason for the formation of the potential barrier in the unbiased P-N junction?

2.8 Why does the P-N junction mainly behave as a variable capacitance for reverse bias voltage? What is the phenomenon that gives rise to this variable capacitance?

2.9 Why is the diffusion capacitance due to the bipolar quality of the P-N junction diode?

2.3.6 Applications of P-N Diodes at Microwave Frequencies

At microwave frequencies the diffusion capacitance, C_d, severely limits P-N diode applications which depend on its non-linear current characteristic. Most microwave applications depend instead on the non-linear behaviour of the junction capacitance, C_j. These are known as varactor applications.

The junction capacitance has voltage dependence given by:

$$C_j(V) = \frac{C_{j0}}{\left(1 - \dfrac{V}{\phi_0}\right)^{\gamma}} \tag{2.32}$$

with $\gamma = 0.5$ for an abrupt junction and $\gamma = 0.33$ for a gradual junction (i.e. a junction with linear spatial variation of the impurity concentration).

The capacitance C_{j0} depends on the semiconductor material, on the doping concentrations, N_a and N_d, and on the P-N diode junction area S. Its value is usually between 1 pF and 20 pF. Varactor diodes are often used as variable capacitors to modify the resonant frequency of a tuned circuit. In order to achieve the largest variation range, a high ratio between the maximum and minimum capacitance values must be guaranteed. According to Equation (2.32), the highest capacitance value of the varactor diode (for reverse-bias) is C_{j0}. Minimum values are obtained for high reverse bias, but care must be taken to avoid avalanche breakdown. The quality factor, Q, of a varactor diode, which is a measure of the capacitive reactance compared to series resistance (or stored energy to energy loss per cycle), is given by [4]:

$$Q(V) = \frac{1}{2\pi f R_s C_j(V)} \tag{2.33}$$

where f is frequency. As is clear from Equation (2.33) quality factor has a sensitive frequency dependence. A silicon device, for example, with a bias voltage $V = -4$ volts, has a Q factor of 1800 at 50 MHz and 36 at 2.5 GHz. Q factors are much higher in GaAs devices due to the higher mobility value, which implies a lower R_s. The frequency, f_c, at which the Q takes the value unity is often taken as a device figure of merit [4].

The following example illustrates how the electrical model of a varactor diode can be obtained.

Example 2.1

A packaged varactor diode, with a negligible loss resistance, can be modelled by the circuit shown in Figure 2.16.

The total diode capacitance is measured as a function of reverse bias at 1 MHz, resulting in the curve shown in Figure 2.17.

When the diode is biased at $V = -20$ volt, the input impedance exhibits a series resonance at $f_r = 10$ GHz. The P-N junction is of an abrupt type, with $\phi_0 = 0.6$ V. Determine the value of the model elements.

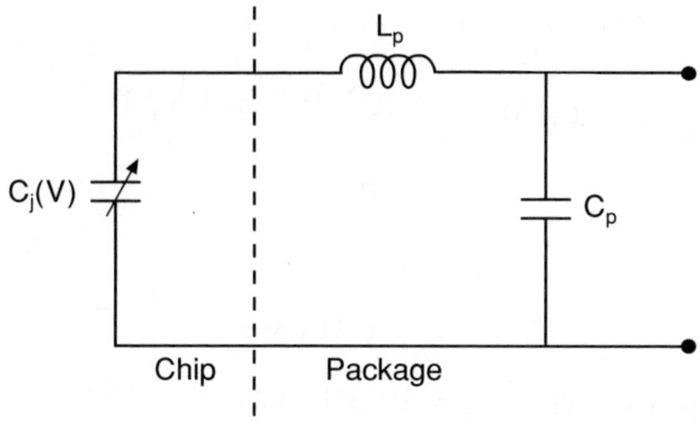

Figure 2.16 Equivalent circuit model of a varactor diode

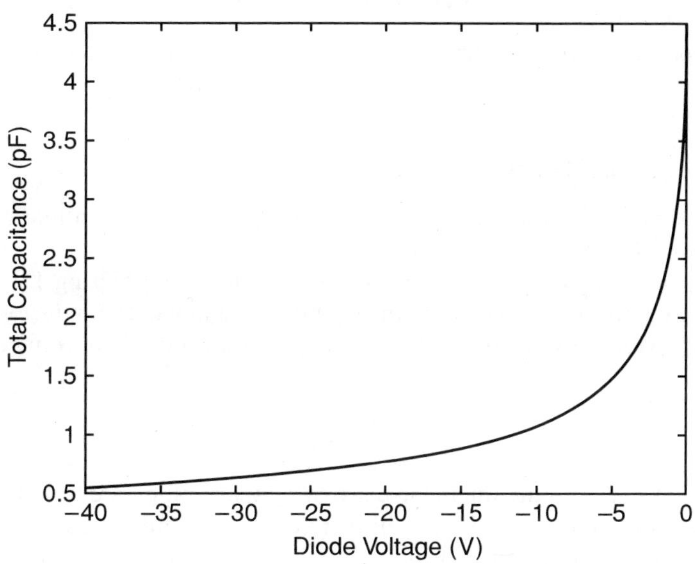

Figure 2.17 Diode capacitance as a function of reverse bias at 1 MHz

The capacitance has been intentionally measured at low frequency, in order to ensure that the inductance is negligible. Two values are taken from the curve:

$$C_T(20) = C_j(20) + C_p = 0.981 \text{ pF}$$

$$C_T(30) = C_j(30) + C_p = 0.843 \text{ pF}$$

Then:

$$C_j(20) - C_j(30) = 0.138 \text{ pF}$$

Using Equation (2.32):

$$C_{j0} = 4.5 \text{ pF}$$

For $V = -20$ volt, the intrinsic diode capacitance will be:

$$C_j(20) = \frac{4.5}{\left(1 + \dfrac{20}{0.6}\right)^{1/2}} = 0.767 \text{ pF}$$

Therefore:

$$C_p = 0.213 \text{ pF}$$

Since the series resonance occurs at $f_r = 10$ GHz, then:

$$L_p = \frac{1}{C_T(20)(2\pi f_r)^2} = 0.258 \text{ nH}$$

Common applications of varactor diodes include amplitude modulators, phase-shifters and frequency multipliers. Some of these are analysed below.

2.3.6.1 Amplitude modulators

Consider the schematic diagram in Figure 2.18 showing a transmission line with a varactor diode and an inductor connected in parallel across it.

The capacitor C_b is a DC-block, used for isolating the diode biasing [4]. Let f_0 be the frequency of the microwave signal at the transmission line input. The inductor is selected in order to resonate with the varactor capacitance at f_0, for a certain bias voltage V_1, i.e.:

$$2\pi f_0 = \frac{1}{LC_j(V_i)} \tag{2.34}$$

The impedance presented to the microwave signal by the parallel circuit will be infinite for $V = V_1$. For this bias voltage, the signal will therefore be transmitted unperturbed along the transmission line. If the diode voltage is now changed to a small value, V_2 (around zero volts), the diode capacitance will increase, presenting a low value of impedance and thus reduce the transmitted amplitude of the microwave signal.

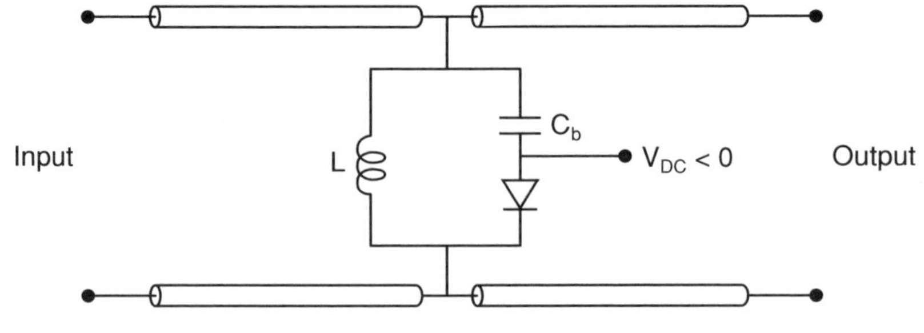

Figure 2.18 Equivalent circuit of varactor diode connected in parallel across a transmission line

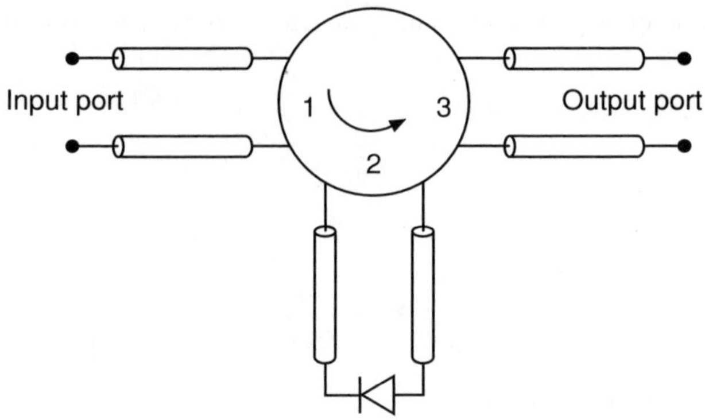

Figure 2.19 Implementation of phase shifter using a circulator and a varactor diode

The circuit shown in Figure 2.18 can be used as an amplitude modulator by biasing the diode at the average value $V_{DC} = (V_1 + V_2)/2$ and superimposing the modulating signal $v_m(t)$.

2.3.6.2 Phase shifters

Varactor diodes may be used as phase shifters by taking advantage of their variable reactance. A possible topology, based on a circulator, is shown in Figure 2.19.

For an ideal circulator, all the input power from port 1 will be transferred to port 2. If the diode resistance loss is neglected, complete reflection will occur at the purely reactive diode termination, the value of reactance will affect the phase of the reflected signal, however. By varying the varactor bias the phase shift can therefore be modified. Finally, the reflected signal is transferred to port 3.

Example 2.2

Determine the phase shift provided by the circuit of Figure 2.19 for input frequency 10 GHz and two bias voltages of the varactor diode: $V_1 = -0.5$ V and $V_2 = -6$ V. Obtain a general expression for the phase shift θ versus the varactor bias. Assume an ideal circulator and varactor parameters $C_{j0} = 0.25$ pF, $\gamma = 0.5$ and $\phi_0 = 0.8$ V.

The varactor capacitance variation, as a function of the applied voltage, is given by:

$$C_j(V) = \frac{C_{j0}}{\left(1 - \dfrac{V}{\phi_0}\right)^{1/2}} = \frac{0.25}{\left(1 - \dfrac{V}{0.8}\right)^{1/2}} \text{ pF}$$

Therefore:

$$C(-0.5) = 0.1961 \text{ pF}$$

and:

$$C(-6) = 0.0857 \text{ pF}$$

Considering the varactor impedance as purely reactive, the reflection coefficient at the diode terminals will be:

$$\Gamma = \frac{jX_{in} - Z_0}{jX_{in} + Z_0}$$

where:

$$jX_{in} = \frac{-j}{C\omega}$$

Replacing the two capacitance values:

$$X_{in}(-0.5) = -81.2\ \Omega$$

$$X_{in}(-6) = -185.7\ \Omega$$

Substituting into the expression for the reflection coefficient:

$$\Gamma(-0.5) = e^{-j1.10}$$

$$\Gamma(-6) = e^{-j0.53}$$

In order to obtain a general equation for the phase shift, the following expressions are considered:

$$-Z_0 + jX_{in} = Ae^{j\alpha}$$

$$Z_0 + jX_{in} = Ae^{-j(\alpha-\pi)}$$

The phase of the reflection coefficient is then given by:

$$\theta = 2\alpha - \pi = 2\ \text{arctg}\ \frac{1}{C_j(V)\omega_0 Z_0}$$

2.3.6.3 Frequency multipliers

The non-linear response of varactor capacitance may be used for harmonic generation. Consider the circuit shown in Figure 2.20.

Assume that the voltage at the diode terminals is a small amplitude sinusoid superimposed on the bias voltage V_{DC}, i.e.:

$$v(t) = V_{DC} + V_1 \cos(\omega t) \tag{2.35}$$

The non-linear capacitance will then vary as:

$$C_j(t) = \frac{C_{j0}}{\left[1 - \dfrac{(V_{DC} + V_1 \cos(\omega t))}{\phi_0}\right]^{\gamma}} \tag{2.36}$$

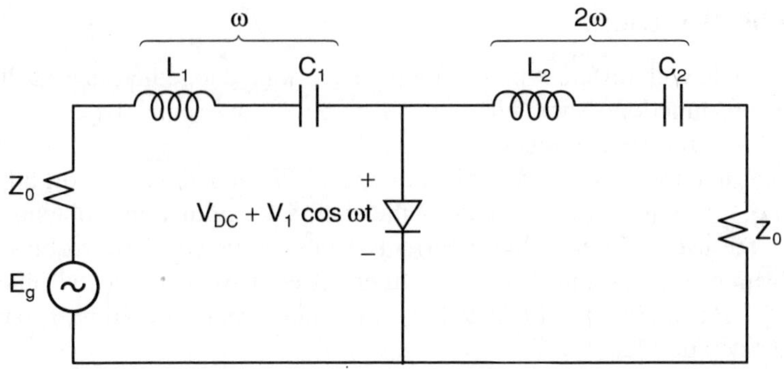

Figure 2.20 Circuit for harmonic generation using varactor diode

Assuming small RF variations about the bias point, the preceding expression may be developed in a first-order Taylor series about V_{DC} giving:

$$C_j(t) = C_j(V_{DC})[1 + a\cos(\omega t)] \tag{2.37}$$

where a is given by:

$$a = \frac{\gamma V_1}{\phi_0 - V_{DC}} \tag{2.38}$$

The current through the diode will be:

$$I_j(V) = C_j(V)\frac{dv}{dt} \tag{2.39}$$

Differentiating the sinusoidal voltage:

$$I(t) = C_j(V_{DC})[1 + a\cos(\omega t)]V_1\omega[-\sin(\omega t)] \tag{2.40}$$

Using trigonometric identities:

$$I(t) = -C_j(V_{DC})V_1\omega\sin(\omega t) - \frac{1}{2}[C_j(V_{DC})aV_1\omega]\sin(2\omega t) \tag{2.41}$$

The ratio between the second harmonic component and the fundamental one is given by:

$$\frac{I_2}{I_1} = \frac{a}{2} = \frac{1}{2}\frac{\gamma V_1}{\phi_0 - V_{DC}} \tag{2.42}$$

Finally, the series resonant circuit (with resonant frequency 2ω) selects the second harmonic component, resulting in the desired frequency doubling action.

2.4 The Schottky Diode

When two materials with different band structures are joined together, the resulting junction is called a hetero-junction. Such junctions may be of two main types: semiconductor-semiconductor or metal-semiconductor.

The Schottky junction is a metal-semiconductor hetero-junction. Its working principle is similar to that of a P-N junction, but without the charge accumulation problems of the latter, which enables the use of its non-linear current characteristic for forward bias applications such a rectification, mixing and detection. Other differences with the PN junction are the lower value of built-in potential ($\phi_0 \cong 0.4$ V), the higher reverse current, I_s, and the lower avalanche breakdown voltage, V_b.

Ohmic contacts represent another form of metal-semiconductor hetero-junction. The current-voltage characteristic resulting from a metal-semiconductor junction may, in fact, be Schottky (rectifying) or ohmic, depending on the semiconductor doping level. The former is obtained for a doping level lower than 10^{17} cm^{-3}. The latter is obtained for a doping level higher than 10^{18} cm^{-3}. For this high doping level, the current-voltage characteristic degenerates to (approximately) a straight line resulting in the ohmic contact [2].

The band structure of a metal is very different from that of a semiconductor. The free electrons (in a very large numbers) occupy the lower levels of a conduction band that is not saturated, i.e. that contains more allowed levels than those occupied. Considering an N-type semiconductor, before metal and semiconductor come into contact, the free energy level, E_0, is the same for both. The respective Fermi levels, however, are different (Figure 2.21).

The Fermi level is lower in the metal, since it has a lower occupation density of energy levels. Three cases are analysed below for the Schottky junction. These cases are thermal equilibrium, reverse bias and forward bias.

2.4.1 Thermal Equilibrium

Although the number of electrons is much higher in the metal than in the N-type semiconductor, the occupation density of energy levels is higher for the semiconductor, since there are less available levels. When the metal and the semiconductor come into contact, the electrons from the semiconductor tend to diffuse into the metal but not the converse. Electrons from the semiconductor atoms that are close to the junction plane will cross it, leaving positively ionised donors (Figure 2.22(a)).

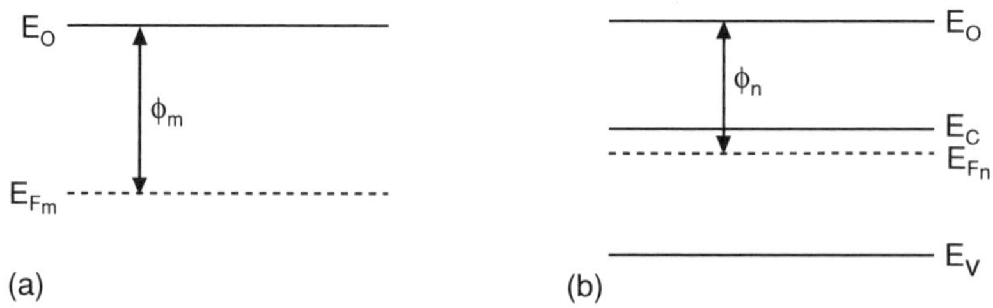

Figure 2.21 Comparison of Fermi levels in (a) metal and (b) N-type semiconductor

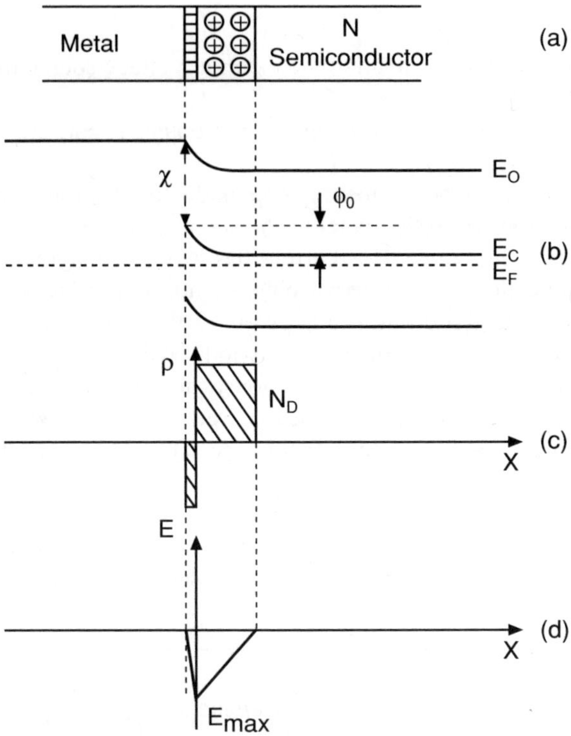

Figure 2.22 Metal semiconductor junction (a) and the resulting energy band diagram (b), charge density (c) and electric field (d)

This static charge is compensated for with electrons from the metal, which gives rise to a static negative charge over the junction plane, not contributing to conduction. A space charge zone therefore appears on both sides of the junction, giving rise to a built-in potential, ϕ_0, opposing electron movement from the semiconductor to the metal. This electron current (noticeably reduced by the built-in potential) is compensated for by the small, thermally generated, electron current flowing from the metal to the semiconductor. Since the free electron density is so high in the metal, the space charge zone has a very short extension in the metal. As expected, in thermal equilibrium, the global current is zero.

From the point of view of energy bands, once thermal equilibrium is attained, the Fermi level must be of equal height across the entire structure. This leads to band bending (Figure 2.22(b)). The same result is obtained when calculating the potential, V, along the device. (As in the case of the P-N junction, V can be found by calculating the electric field using Poisson's equation and integrating.) Figure 2.22(c) and Figure 2.22(d) show the charge density and electric field that exist in the junction region.

The energy diagram of Figure 2.22(b) makes clear the origin of the difficulty that semiconductor electrons experience in traversing the junction plane, as they must surmount the potential barrier. The few electrons moving from the metal to the semiconductor simply descend the potential barrier, attracted by the fixed positive ions. Due to the reduced space charge zone on the metal side of the junction, band bending will be negligible in this region.

2.4.2 Reverse Bias

For reverse biasing of a Schottky junction the external voltage source must be connected as shown in Figure 2.23.

This gives rise to an increase in the height of the potential barrier, which now takes the value $\phi_0 + V$ (Figure 2.24).

This reduces to almost zero the number of semiconductor electrons moving into the metal. The movement of electrons from the metal to the semiconductor is not affected, giving rise to the reverse saturation current I_s. This current has a bigger magnitude than in P-N junctions. Due to the application of an external bias, pseudo-Fermi levels must be considered. For reverse bias, their potential difference is given by $V_{Fm} - V_{Fn} = V$, with $V > 0$.

The space charge zone, containing the positively ionised donors, is equivalent to a dielectric material, bounded by two very conductive regions: the metal and the neutral zone of the N-type semiconductor. The width of the space charge zone varies with the applied reverse voltage, giving rise to a variable junction capacitance. This capacitance has a value given by Equation (2.32) with $\gamma = 0$.

When the reverse voltage is too high, the electrons from the metal traversing the depleted zone will be accelerated enough to knock other electrons out of their atomic orbits. When this process cascades, the current grows very rapidly with voltage. This effect is, of course, identical

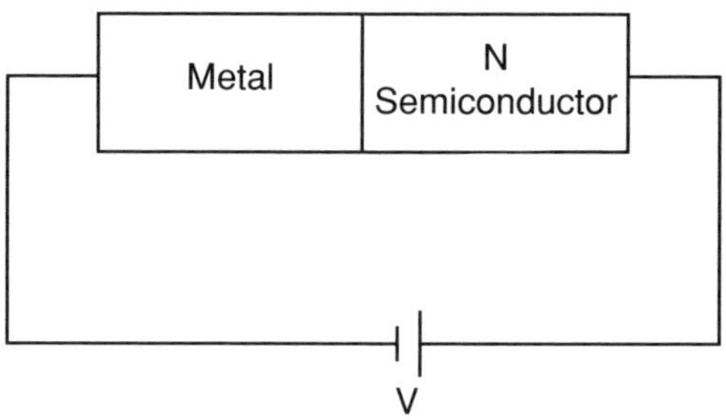

Figure 2.23 Reverse-biased Schottky junction

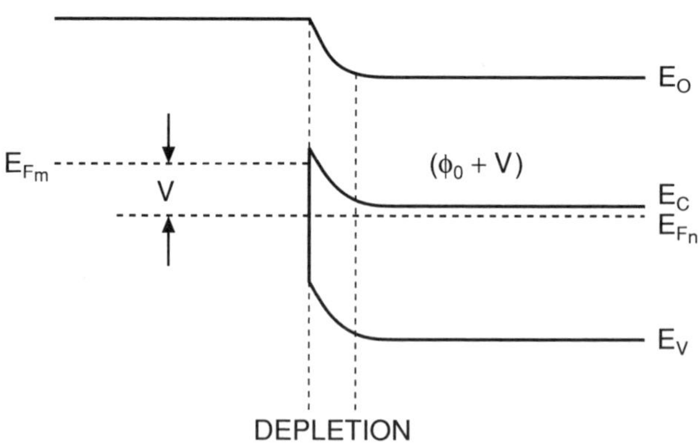

Figure 2.24 Energy band diagram for a Schottky junction

to that observed in reverse-biased P-N junctions as discussed in Section 2.3.2. A value of breakdown voltage (V_b) and critical field strength (E_c), related via Equation (2.27), can therefore be defined. (In this context w_d in Equation (2.27) is the width of the space charge zone.)

The breakdown voltage will be smaller as the doping increases, which is explained by the existence of a larger number of free electrons available for the cascaded process. On the other hand, as can be seen from Equation (2.27), V_b will be higher for a wider space charge zone.

Schottky diodes have breakdown voltages that are lower than those observed in P-N junctions and, as a consequence, smaller capacitance variation ranges may be expected in Schottky varactors. When a Schottky diode is manufactured for varactor applications, the doping is usually reduced, thus increasing W_d. This has the drawback, however, of increasing the device loss resistor, R_s, since the semiconductor conductivity (related to the doping level through the formula $\sigma = \mu en$) decreases. Doping reduction therefore decreases diode quality factor (see Equation (2.33)). For non-linear current applications device resistance must be reduced and diode doping is therefore increased (at the expense of decreased V_b).

2.4.3 Forward Bias

When connecting an external voltage source as shown in Figure 2.25, the Schottky diode will be forward biased.

This reduces the height of the potential barrier to $\phi_0 - V$ (Figure 2.26).

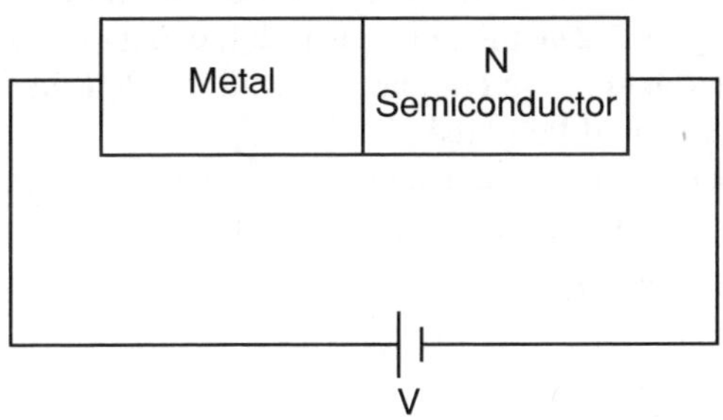

Figure 2.25 Forward-biased Schottky junction

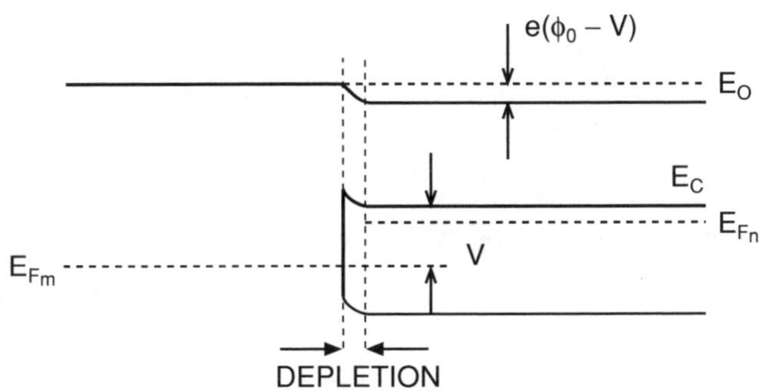

Figure 2.26 Energy band diagram for forward-biased Schottky junction

The difference between the pseudo-Fermi levels is now $V_{Fm} - V_{Fn} = -V$, with $V > 0$. As $V \rightarrow \phi_0$, the device current is limited only by the loss resistance and the external circuit. Like the forward biased P-N junction, the current-voltage characteristic can be modelled by [1, 2]:

$$I(V) = I_s(e^{\alpha V} - 1) \tag{2.43}$$

The parameter α, here, is given by [2]:

$$\alpha = \frac{e}{nkT} \tag{2.44}$$

where n is an empirical factor, used to fit the model to the characteristic measured in practice.

It is important to notice that, since the Schottky diode is a majority carrier device, no charge accumulation takes place under forward bias conditions, there being no diffusion capacitance. The Schottky diode is therefore better suited than the P-N junction to applications depending on the current-voltage characteristic.

2.4.4 Electric Model

The electric model for the Schottky diode comprises a non-linear current source $I(V)$, in parallel with a variable junction capacitance C_j, this parallel combination connected in series with a loss resistor R_s (Figure 2.27), the latter, as usual, being due to the limited conductivity of the semiconductor neutral zone. In comparison with the P-N diode model, the absence of the diffusion capacitance should be noted.

For chip Schottky diodes, a series inductance L_p, accounting for the bonding wire, must be considered. In the case of packaged devices, the parallel capacitance C_p, due to the package insulator, must also be included.

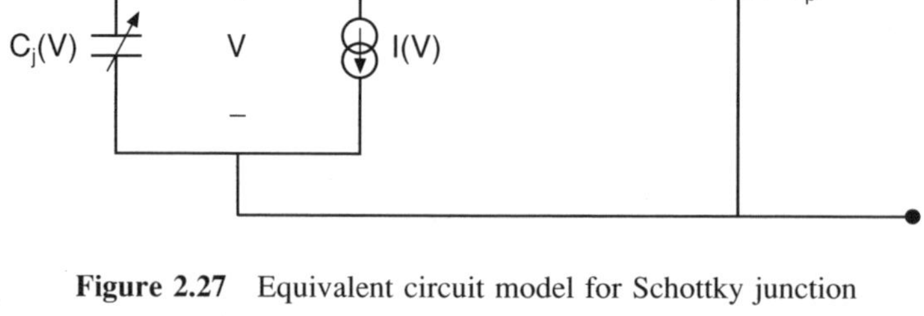

Figure 2.27 Equivalent circuit model for Schottky junction

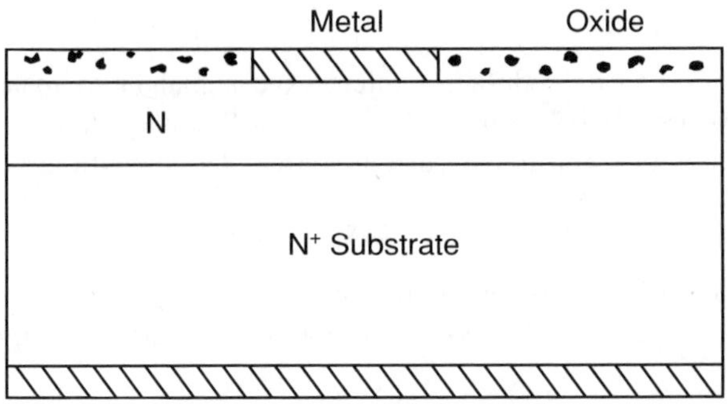

Figure 2.28 Schottky junction structure

2.4.5 Manufacturing

Schottky diodes are manufactured in Si or GaAs. The bulk semiconductor is a highly doped N^+ substrate, with a width of several microns. Over this substrate, a thin N layer, of the desired width, is obtained either through epitaxial growth or by ionic implantation. This layer is protected by means of an oxide, as shown in Figure 2.28.

The metallic strip is realised by vacuum evaporation, through special holes in the oxide. Suitable metals are aluminium and gold. The metallic deposit acts as one of the device terminals. The second terminal is realised using an ohmic contact at the semiconductor base.

Self-assessment Problems

2.10 Before a metal and an N-type semiconductor are joined to form a Schottky junction, why is the Fermi level of the metal located at a lower energy level than that of the N-type semiconductor?

2.11 What is the phenomenon that gives rise to the negative static-charge accumulation at the metal side of the Schottky junction?

2.12 Why do Schottky diodes not present a diffusion capacitance when forward biased?

2.4.6 Applications

The application of Schottky diodes as varactors is similar to that of P-N junctions although, due to the smaller breakdown voltages of the former, smaller capacitance variation ranges are generally obtained.

Other applications of the Schottky diode stem from the non-linear characteristic of its equivalent current source. Among them, applications as detectors and frequency mixers are the most usual.

2.4.6.1 Detectors

A detector can be used to demodulate a microwave signal or to measure its power and microwave detectors are often based on the use of a Schottky diode. Such detectors take advantage of the Schottky's non-linear current source characteristic given by:

$$i(v) = I_s(e^{\alpha v} - 1) \tag{2.45}$$

where v (= $v(t)$) is the (signal) voltage across the diode terminals.

Considering a bias voltage $V = V_0$ and a non-modulated microwave signal, the voltage v is given by:

$$v = V_0 + V \cos \omega t \tag{2.46}$$

where V and ω are, respectively, the microwave signal amplitude and frequency. Provided V is small, it is possible to develop $i(v)$ as a Taylor series around the bias point, i.e.:

$$i(v) = i(V_0) + \left.\frac{di}{dv}\right|_{v0} \Delta v + \frac{1}{2}\left.\frac{d^2 i}{dv^2}\right|_{v0} \Delta v^2 + \ldots \tag{2.47}$$

where $\Delta v = V \cos \omega t$. The derivatives are:

$$\begin{aligned} \left.\frac{di}{dv}\right|_{v0} &= \alpha I_s e^{\alpha V_0} \equiv d_1 \\ \left.\frac{d^2 i}{dv^2}\right|_{v0} &= \alpha^2 I_s e^{\alpha V_0} \equiv d_2 \end{aligned} \tag{2.48}$$

Therefore, $i(v)$ can be expressed as:

$$i(v) = I_0 + d_1 \Delta v + (d_2/2)\Delta v^2 + \ldots \tag{2.49}$$

where $I_0 = i(V_0)$. Replacing the Δv value:

$$i(v) = I_0 + d_1 V \cos(\omega t) + \frac{d_2}{2} V^2 \cos^2(\omega t) + \ldots \tag{2.50}$$

and using the trigonometric identity:

$$\cos^2 \omega t = \frac{1}{2}(1 + \cos 2\omega t) \tag{2.51}$$

gives:

$$i(v) = I_0 + \frac{d_2}{4} V^2 + d_1 V_1 \cos \omega t + \frac{d_2}{4} V^2 \cos 2\omega t + \ldots \tag{2.52}$$

After low pass filtering, a (quasi) DC output may be obtained, which (apart from the offset I_0) is proportional to the square of the microwave signal amplitude V.

The preceding analysis is for the case of a quadratic detector. For larger microwave signal amplitudes DC contributions from other terms in the Taylor series would result in a detector

(DC) output that is proportional to the microwave signal amplitude [3, 4, 5], rather than its power. For still larger signal amplitudes, a saturated behaviour is obtained. On the other hand, for very small microwave signal amplitudes, the detected output would deviate negligibly from I_0 resulting in a noisy operating range if detection is possible at all [3, 4, 5].

2.4.6.2 Mixers

The mixing applications of the Schottky diode also arise from its non-linear current-voltage characteristic. Using a Taylor expansion, this characteristic can be expressed in the form of a power series, i.e.:

$$\Delta i = a_1 \Delta v^1 + a_2 \Delta v^2 + a_3 \Delta v^3 + \ldots \tag{2.53}$$

where Δi and Δv are respectively the RF increments of current and voltage about the DC bias point (V_0 and I_0).

In a mixer, the input voltage is composed of two tones: one corresponding to the input signal, modelled here by $V_{in} \sin \omega_{in}$, and the other corresponding to the local oscillator with amplitude V_{LO} and frequency ω_{LO} radian/s, i.e.:

$$\Delta v = V_{in} \sin(\omega_{in} t) + V_{LO} \sin(\omega_{LO} t) \tag{2.54}$$

Substituting this expression for Δv into Equation (2.53) gives:

$$\Delta i = \frac{a_2}{2}(V_{in}^2 + V_{LO}^2) + a_1 V_{in} \sin(\omega_{in} t) + a_1 V_{LO} \sin(\omega_{LO} t) - \frac{a_2}{2} V_{in}^2 \cos(2\omega_{in} t) +$$
$$+ a_2 V_{in} V_{LO} \cos(\omega_{in} - \omega_{LO})t - V_2 V_{in} V_{LO} \cos(\omega_{in} - \omega_{LO})t - \frac{a_2}{2} V_{LO}^2 \cos(2\omega_{LO}) + \ldots \tag{2.55}$$

where terms involving both ω_{in} and ω_{LO} are called intermodulation products. Provided that ω_{in} and ω_{LO} are sufficiently close in frequency, the intermediate frequency, $\omega_{IF} = |\omega_{in} - \omega_{LO}|$, is small compared to either the input signal frequencies or the local oscillator frequency and, therefore, can be easily selected by filtering. A possible schematic diagram of a mixer circuit is shown in Figure 2.29.

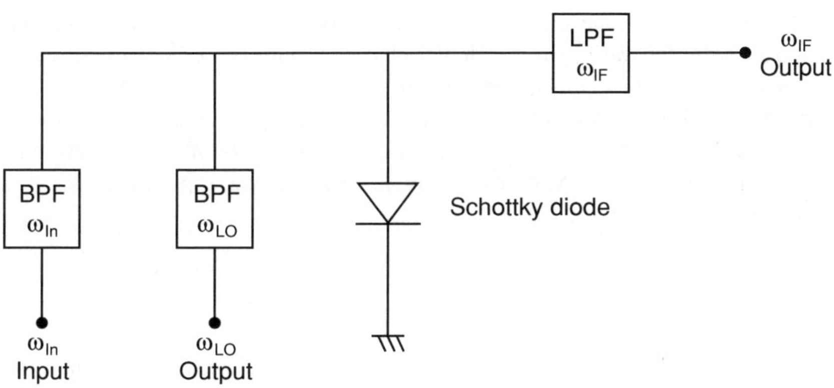

Figure 2.29 Schottky diode used in a mixer implementation

2.5 PIN Diodes

A PIN diode is obtained by connecting a highly doped P^+ layer of semiconductor, a long intrinsic (*I*) layer and a highly doped N^+ layer. (Although, ideally, the I layer is intrinsic, in practice the presence of impurities is unavoidable, making it slightly P or N doped.)

The presence of a long intrinsic section increases the breakdown voltage of the device thus allowing high reverse voltages. This is advantageous when handling high input power. The intrinsic section is also responsible for an almost constant value of reverse bias capacitance, which is also comparatively smaller than that for P-N junctions. On the other hand, the intrinsic semiconductor exhibits a variable resistance as a function of forward bias. Many applications of the PIN diodes stem directly from this property, 'PINs' being often used as variable resistances or variable attenuators. For a relatively high value of forward bias, the resistance of the intrinsic section is noticeably reduced. Switching applications, between reverse biased (no conduction or high resistance) state and forward biased (good conduction or low resistance) state, are also possible.

2.5.1 Thermal Equilibrium

In thermal equilibrium the P^+ and N^+ regions will, respectively, diffuse holes and electrons into the intrinsic region, leaving ionised impurities. In each of these two regions the space charge zone will have a reduced extension, since both are highly doped. Since the intrinsic component is not doped, its corresponding density of net, static, charge is equal to zero. The charge density profile, therefore, develops as shown in Figure 2.30(a).

The electric field variation along the device, obtained by integrating Poisson's equation, is shown in Figure 2.30(b). As can be seen, the electric field, E, takes a constant value along the intrinsic part of the device, due to its zero static charge density.

The potential, V, is obtained by integrating the electric field. A linear variation is obtained in the I region (Figure 2.30(c)), the potential is parabolic between $-x_p$ and 0 and between w_I and $w_I + x_n$, and elsewhere the potential is constant. The energy band diagram is shown in Figure 2.30(d). The resulting potential barrier is higher than in P-N junctions.

2.5.2 Reverse Bias

When the PIN diode is reverse biased, there will be only a small saturation current due to thermal electron-hole generation in the depleted zones, with the electrons moving towards the N^+ region and the holes towards the P^+ region.

In agreement with the principle of electric neutrality, the space charge zone is very narrow in the P^+ and N^+ regions due to the high doping density, while the intrinsic section is entirely depleted. Thus, for any reverse bias it is possible to make the approximation $w_d \cong w_I$, the junction capacitance being given by:

$$C_j = \frac{\varepsilon_{sc} S}{w_I} \tag{2.56}$$

Unlike the case of a P-N junction, the above capacitance is almost independent of the bias voltage and, due to the large width of the intrinsic region w_I, has a small value. This gives

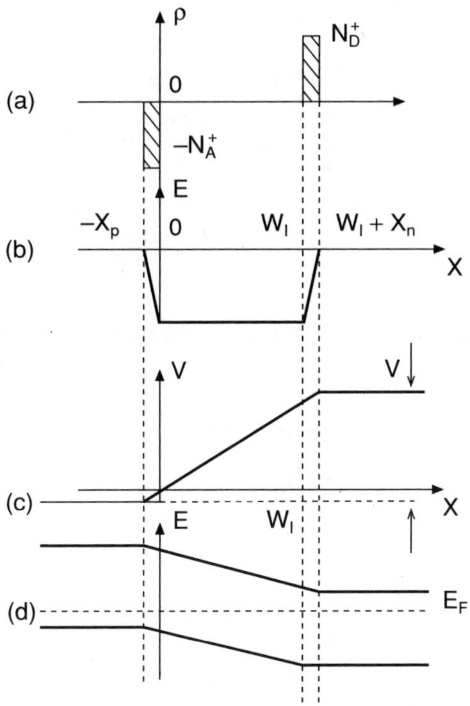

Figure 2.30 PIN diode charge density distribution (a), electric field distribution (b), potential distribution (c), and energy band diagram (d)

rise to a very high reverse impedance, useful for switching applications. The avalanche breakdown voltage, V_b, is given by:

$$V_b = E_c w_I \tag{2.57}$$

where E_c is the critical field strength. Due to the large w_I, the breakdown voltage is high, so the PIN diode can utilise high reverse bias voltages. The critical field strength for silicon is $E_c = 2 \times 10^5$ V. Therefore, for a diode with $w_I = 10$ μm the breakdown voltage is 200 V while for a diode with $w_I = 50$ μm the breakdown voltage is 1 kV.

2.5.3 Forward Bias

In forward bias carriers diffuse into the intrinsic zone from both P^+ and N^- sides, resulting in a forward current due to recombination of these 'injected' carriers. For a low bias voltage the recombination takes place in the intrinsic region, while for a high bias voltage, it takes place in P^+ and N^+ regions.

Ignoring the intrinsic region doping, its carrier concentration will be mainly due to electron and hole injection through the junctions. Assuming an equal concentration of injected electrons and holes, $p = n$, the total drift current will be given by:

$$\mathbf{J} = (\mu_n + \mu_p)en\mathbf{E} \tag{2.58}$$

the conductivity being:

$$\sigma = (\mu_n + \mu_p)en \tag{2.59}$$

Alternatively, the current flow, I_f, due to electron-hole recombination in the I region can be calculated from:

$$I_f = \frac{Q}{\tau} \tag{2.60}$$

where Q is the total injected charge and τ the average carrier life-time. The former is given by:

$$Q = enSw_I \tag{2.61}$$

(Note that the electron and hole, moving in opposite directions before recombining, count as a single charge in the recombination current.)

The resistance of the intrinsic zone can be calculated from its conductivity value. Assuming $\mu_p \cong \mu_n \cong \mu$, the resistivity, ρ, will be:

$$\rho = \frac{1}{2\mu en} \tag{2.62}$$

and the resistance, R_I, of the intrinsic zone is therefore given by:

$$R_I = \frac{\rho w_I}{S} = \frac{w_I}{2\mu enS} \tag{2.63}$$

From Equation (2.61):

$$en = \frac{Q}{Sw_I} \tag{2.64}$$

And substituting Equation (2.64) into Equation (2.63):

$$R_I = \frac{Sw_I}{2\mu Q}\frac{w_I}{S} = \frac{w_I^2}{2\mu Q} \tag{2.65}$$

Expressing the charge as a function of the forward current, $Q = I_f\tau$, a relationship is obtained between the intrinsic resistance and this current, i.e.:

$$R_I = \frac{w_I^2}{2\mu\tau I_f} \tag{2.66}$$

The intrinsic resistance, R_I, is therefore inversely proportional to the forward bias current, I_f, and it is this relationship which enables the diode to be used as a variable resistor.

Since the intrinsic zone is not doped, its conductivity is notably smaller than that of the P^+ and N^+ regions resulting in a capacitance, C_I, [2]:

$$C_I = \frac{\varepsilon_{sc} S}{w_I - w_{dI}} \tag{2.67}$$

where w_{dI} is the extension of the depleted zone in the I region. Note that this capacitance is exclusively due to the intrinsic material, so it vanishes if the I region shrinks to zero width. (For reverse bias $w_{dI} = w_I$ and the capacitance C_I tends to infinity – its effect therefore disappearing in the context of overall device performance. This is more clearly understood with the aid of the circuit electric model given in Section 2.5.4 below.)

In forward bias, the I region stores a charge Q composed of charge carriers. If the diode state is abruptly switched to reverse bias, a non-zero time, τ_s, is required to purge the I region of these charges. This is the diode switching time which will be higher for longer lifetime, τ, and larger intrinsic region width, w_I. (When switching from reverse to forward bias, the time needed for carrier injection is notably smaller.)

2.5.4 Equivalent Circuit

The equivalent circuit for a PIN diode (Figure 2.31) is, principally, composed of two parallel circuits in series combination, one circuit accounting for the two P-N junctions and the other for intrinsic section.

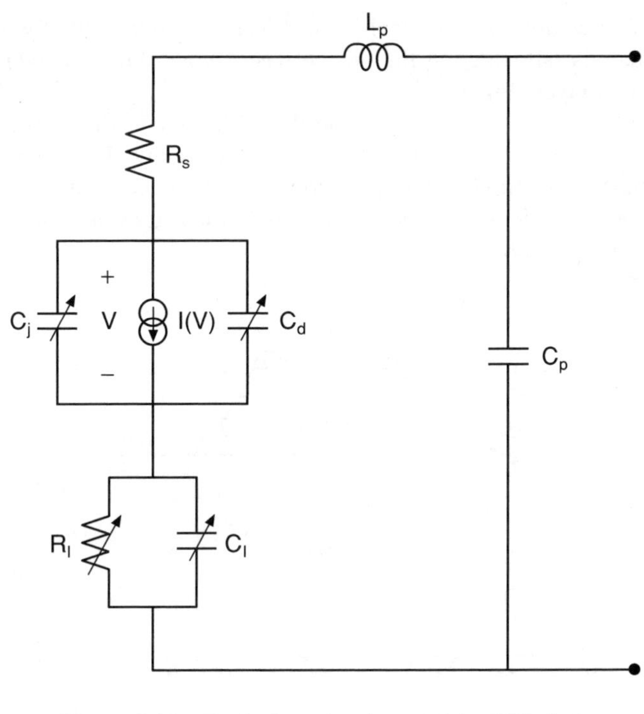

Figure 2.31 Equivalent circuit model for PIN diode

The former contains a non-linear current source in parallel with both junction and diffusion capacitances. (Note that both junctions are accounted for by these capacitors.) The latter contains a variable resistance, R_i, in parallel with the intrinsic capacitance C_i. A constant loss resistance, R_s, is also usually present, accounting for the metallic contact resistance and the resistance of the neutral zones P and N (which have a limited conductivity value).

In addition to the above equivalent circuit elements, packaging parasitics may be included. The equivalent circuit for the packaging comprises a parallel capacitor, C_p, due to the capacitance between the device contacts, and a series inductor, L_p, due to the bonding wires.

In reverse bias, the part of the equivalent circuit representing the junctions is reduced to the junction capacitance, C_j, alone since the diffusion capacitance, C_d, vanishes for negative bias and the current source supplies negligible current. The equivalent circuit representing the intrinsic section also disappears since, for reverse bias, the whole intrinsic section becomes ionised, behaving as a depleted region. As stated in Section 2.5.3 above, the capacitance C_I tends to infinity and short-circuits the resistance R_I.

In forward bias, the diffusion capacitance short-circuits the non-linear current source. The diode equivalent circuit is then reduced to the intrinsic component of the model, i.e. the variable resistor R_I and the parallel capacitor C_I. The value of R_I decreases rapidly with increasing forward bias, so for relatively high bias voltages, the equivalent circuit reduces to a small resistor. This is well suited to switching applications, the high-impedance state being provided by reverse bias.

2.5.5 Manufacturing

PIN diodes are manufactured in mesa technology. The substrate is highly doped silicon N^+. One of the terminals is fabricated at the device base, as an ohmic contact between the N^+ substrate and a metal layer (Figure 2.32).

On the upper surface of the substrate a non-doped layer constitutes the intrinsic region. Due to manufacturing imperfection, this will be either slightly P or N doped in practice. Acceptor impurities are diffused, or implanted, over the surface of the intrinsic layer to realise the P^+ region. The second device terminal is fabricated as an ohmic contact between the P^+ layer and a metal deposit.

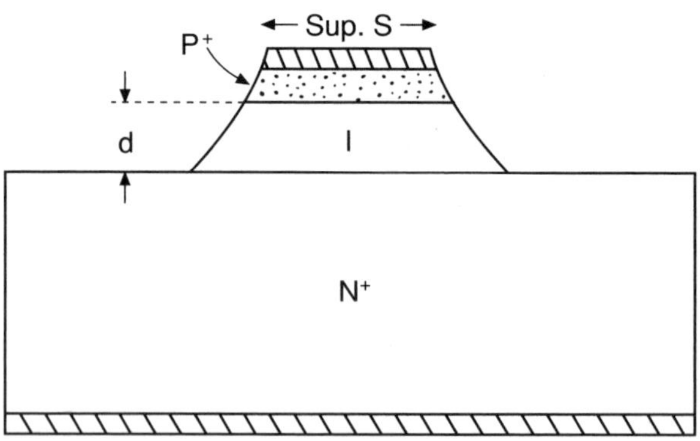

Figure 2.32 PIN diode structure

Self-assessment Problems

2.13 Why does the intrinsic part of resistance in a PIN diode quickly decrease with forward bias?

2.14 Why does the reverse capacitance of a PIN diode remain approximately constant with variations in reverse bias voltage?

2.5.6 Applications

The special applications of PIN diodes are made possible by the long intrinsic region embedded between the P^+ and N^+ zones. This increases the breakdown voltage enabling PIN devices to deal with high input powers. Small reverse bias capacitance and low forward bias resistance make PIN devices useful for switching applications. The variable resistance property in forward bias also enables PIN devices to be used as variable attenuators.

2.5.6.1 Switching

In switching applications, advantage is taken of the PIN diode's low, and almost constant, reverse bias capacitance, C_j, providing very high input impedance. A low-impedance state is obtained by forward-biasing the diode, since the resistance R_I decreases rapidly with the applied voltage. The high breakdown voltage in these devices allows the switching of high power signals. An important PIN diode characteristic that must usually be considered in switching applications is the switching time, τ_s. Switching time is determined by carrier lifetime, shorter switching times being obtained for shorter carrier lifetimes.

In the design of switching circuits, different topologies are possible. Two of these topologies are analysed below.

Topology 1

A simple switch [4] may be realised by connecting a PIN diode in parallel over a transmission line (Figure 2.33).

In the low-impedance state, corresponding to forward bias, there will be high reflection losses and only a small proportion of the input power will reach the load. In the high-impedance state, corresponding to reverse bias, the diode will give rise to only a small insertion loss.

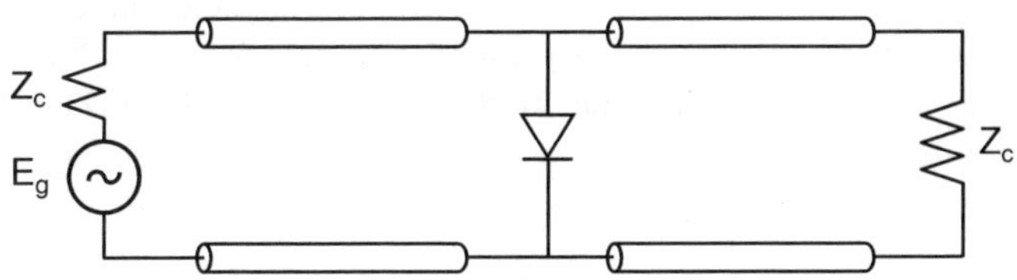

Figure 2.33 Switch realised using a PIN diode and transmission line

For a quantitative analysis of this switch, the attenuation corresponding to both the forward and reverse states must be calculated. Let Y be the diode admittance and Z_0 the characteristic impedance of the transmission line. In order to derive a simple expression for the switch attenuation, the connection of the general admittance, Y, across the transmission line is going to be considered. The input generator has a voltage E_g and internal impedance Z_0 and the load impedance is matched to the characteristic impedance of the line, i.e. $Z_L = Z_0$. The power delivered to the load is given by:

$$P_L = \frac{1}{Z_0} \frac{Eg^2}{2|2 + y|^2} \tag{2.68}$$

where y is the normalised admittance, $y = YZ_0$. If the admittance Y were not connected, then all of the generator's available power would be delivered to the load, i.e.:

$$P_{dg} = \frac{Eg^2}{8Z_0} \tag{2.69}$$

The attenuation, A, due to the presence of admittance, Y, is therefore given by the ratio of Equations (2.69) and (2.68), i.e.:

$$A = \frac{P_{dg}}{P_L} = \frac{|2 + y|^2}{4} \tag{2.70}$$

The forward-biased normalised PIN admittance, y_F, is dominated by the intrinsic resistance, R_I, plus the loss resistance, R_S. Thus:

$$y_F = \frac{Z_0}{R_I + R_S} \tag{2.71}$$

The reverse-biased, normalised, PIN admittance, y_R, is given by the junction capacitance alone. Therefore:

$$y_R = jZ_0C_j\omega \tag{2.72}$$

The attenuation for forward- and reverse-biased PIN states is then found by substituting the corresponding normalised admittance into Equation (2.70).

Topology 2

In Figure 2.34, the input RF signal may be transferred to channel A or channel B, depending on the state of the corresponding PIN diodes [4].

The two transmission lines (of length $\lambda/4$) behave as impedance inverters. The inductors and capacitors in the schematic behave, respectively, as DC-feeds and DC-blocks.

Consider diode D_A in forward bias and diode D_B in reverse bias. Due to the $\lambda/4$ line, channel A will exhibit a very high impedance to the input signal. Since D_B is reverse-biased, its impedance will be high and, connected in parallel across the line, it will have little effect on transmission. The entire input signal will therefore be delivered to channel B. Conversely, when D_A is reverse biased and D_B is forward-biased, the entire input signal is delivered to channel A.

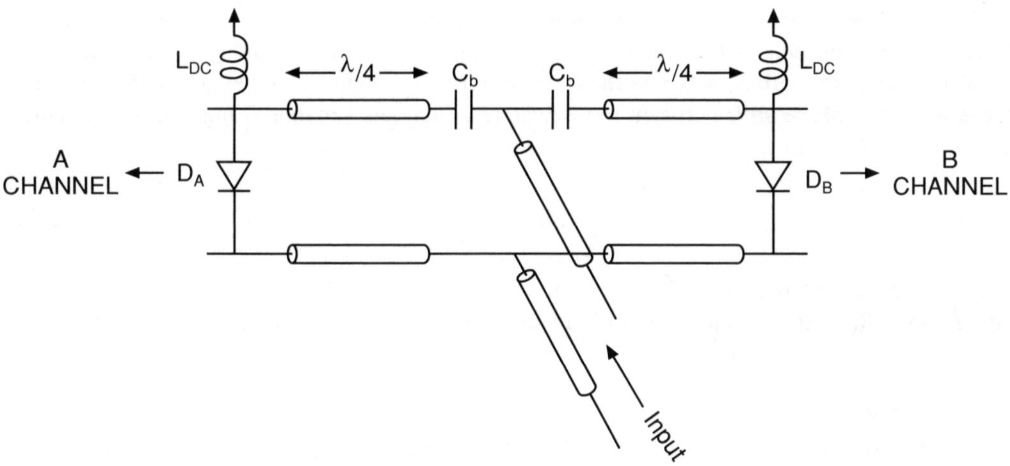

Figure 2.34 Double throw switch implemented by transmission lines and PIN diodes

2.5.6.2 Phase shifting

The phase shifting of high power signals may be obtained with PIN diode circuits. Several circuit topologies are possible, two of which are analysed below.

Topology 1

The topology of Figure 2.35 is based on the transfer characteristics of a transmission line, loaded at both ends by equal susceptance, *jB*.

The phase shift suffered by a microwave signal traversing the complete two-port can be found by calculating the transmission matrix of the two-port and transforming it into a scattering matrix – see Chapter 3 and [3]. The phase of the S_{21} element of the scattering matrix directly provides the phase shift suffered by the signal. For the particular case of a quarter wave line ($l = \lambda/4$), this phase shift is given by:

$$\varphi = \tan^{-1}\left(\frac{2 - B^2Z_0^2}{2BZ_0}\right) \tag{2.73}$$

where Z_0 is the characteristic impedance of the transmission line. It is clear from Equation (2.73) that modifying the susceptance, B, varies the phase shift.

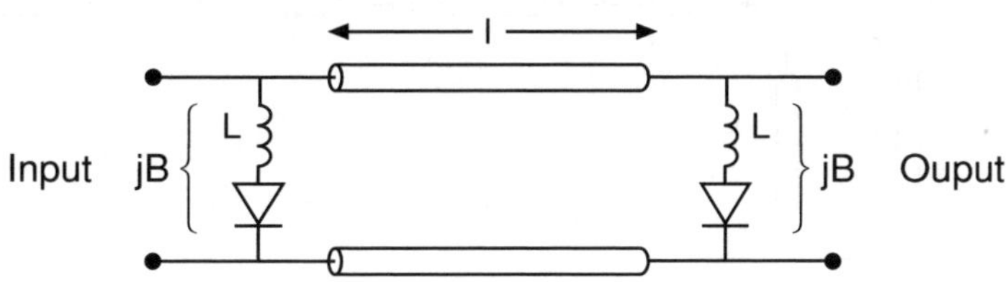

Figure 2.35 High power phase shifter implemented using transmission line and PIN diodes

When a PIN diode with negligible parasitics is connected at each end of the transmission line, two different phase-shift values can be obtained. These respectively correspond to the forward-bias state, with $B_F = 0$, and the reverse-bias state, with $B_R = C_j\omega$. In order to increase the differential phase shift between the two states, each susceptance B may be implemented using a PIN diode in series with an inductor L [4]. The inductance is chosen to satisfy:

$$L\omega = \frac{1}{2C_j\omega} \tag{2.74}$$

With the diode in reverse bias the total reactance is $-1/(2C_j\omega)$ and the corresponding susceptance $B_R = 2C_j\omega$. With the diode in forward bias the susceptance is $B_F = -2C_j\omega$.

Example 2.3

A PIN diode is connected at each end of a 50 Ω quarter-wave transmission line to realise a microwave phase shifter. Each diode has a reverse bias capacitance (C_j) of 0.175 pF and the required operational frequency of the phase shifter is 5 GHz. Find the differential phase shift with, and without, an appropriate series inductor connected to the diodes.

When the diodes are reverse-biased, the susceptance value is $B_R = 5.5\ 10^{-3}$ S and the corresponding phase shift $\varphi_R = 74°$. When the diodes are forward-biased, the susceptance value is $B_F = 0$ and the corresponding phase shift $\varphi_F = 90°$. The differential phase shift is $\Delta\varphi = 16°$. When including a series inductor, calculated according to Equation (2.73), the phase shift in the forward bias state is $\varphi_F = 57°$. In the reverse bias state, the phase shift is $\varphi_R = -57°$. The differential phase-shift is $\Delta\varphi = 114°$.

The principal problem with the above topology is input and output mismatching. The parallel susceptance varies the input and output impedances of the network and this may dramatically increase the reflection losses. In order to avoid this drawback, a different topology (described below), based on the use of a circulator, can be employed.

Topology 2

In the topology of Figure 2.36, assuming an ideal circulator, power incident on port 1 is entirely transferred to port 2. The power reflected by the load of port 2 is transferred to port 3 and thence to the circuit-matched load.

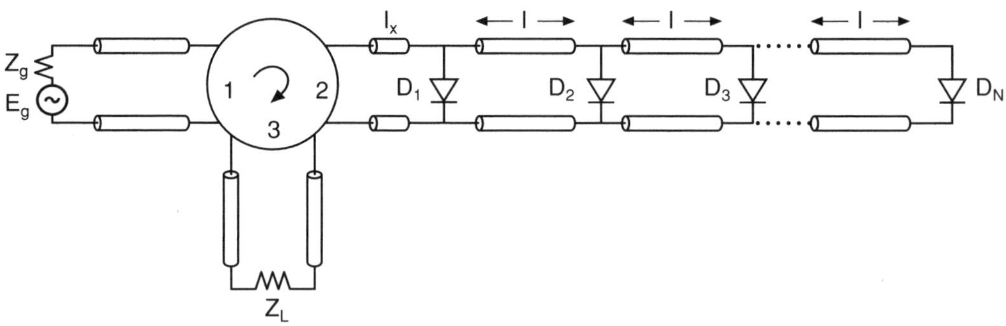

Figure 2.36 High power phase shifter implemented using transmission line, circulator and PIN diodes

A number of PIN diodes are connected in parallel across the transmission line connected to port 2. The first diode is located at a small distance l_x from the circulator [3, 4]. The distance between any other two diodes in the array is constant and equal to l. When the first diode, D_1, is forward-biased, the input-output phase shift is given by:

$$\phi_1 = \pi - 2\beta l_x \tag{2.75}$$

where β is the phase constant (in rad/m) of the line. When all the diodes up to D_n are reverse-biased and D_{n+1} is forward-biased, the phase shift is given by:

$$\phi_{n+1} = \pi - 2\beta l_x - 2\beta nl \tag{2.76}$$

This topology generally provides a larger phase-variation range than topology 1. Its main advantage is the input and output impedance matching provided by the circulator.

Example 2.4

Design a phase-shifter from 0° to 180° in steps of 45°, using PIN diodes. The operating frequency is 4 GHz and the available substrate has an effective dielectric constant $\varepsilon_{eff} = 1.86$.

Since five discrete phase shifts are required, five diodes are needed and the maximum value of n in Equation (2.76) is 4. Ignoring l_x, when the first four diodes are reverse biased (off) and diode D_5 is the first to be forward biased (on), the phase shift is given by:

$$\phi_5 = \pi - 8\beta l$$

The phase shift ϕ_5 is made equal to 0°, so the length of the transmission line sections must have the value:

$$l = \frac{\pi}{8\beta}$$

The phase shift $\phi_1 = \pi$ will be given by the state D_2 to D_5 off and D_1 on. The rest of phase shift values are given by:

$$\phi_{n+1} = \pi - n\frac{\pi}{4}$$

where n is the index of the last diode in off-state.

In order to complete the design, the physical length of the transmission line must be calculated. At $f = 4$ GHz the wavelength value is:

$$\lambda = \frac{c}{f\sqrt{\varepsilon_{eff}}} = 0.055 \text{ m}$$

and:

$$\beta = \frac{2\pi}{\lambda} = 114.25 \text{ m}^{-1}$$

Therefore:

$$l = 3.44 \text{ mm}$$

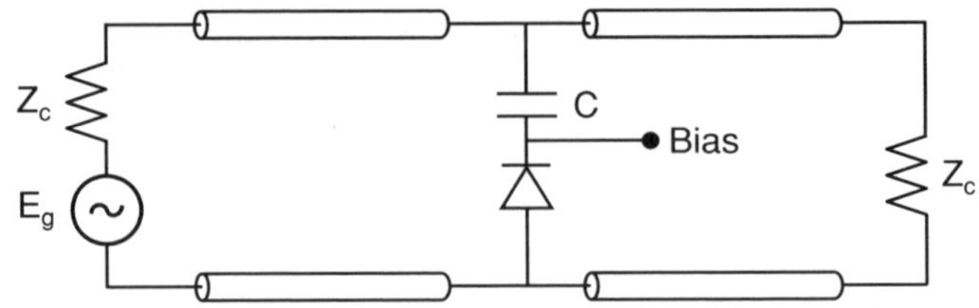

Figure 2.37 PIN-based attenuator

2.5.6.3 Variable attenuation

The variation of the intrinsic resistance, R_I, of the PIN diode as a function of bias current, I_f, allows its application as the active device in a variable attenuator. The resistance/current relationship is:

$$R_I = \frac{\alpha}{I_f} \tag{2.77}$$

where the parameter α is given by:

$$\alpha = \frac{w_I^2}{2\tau\mu} \tag{2.78}$$

and w_I, τ and μ are intrinsic region width, average carrier lifetime and average carrier mobility respectively. A simple implementation of a PIN-based attenuator circuit is shown in Figure 2.37.

The PIN diode with a series (bias blocking) capacitor is connected across a transmission line. When the bias current, I_f, is modified, the corresponding normalised admittance y varies according to:

$$y(I_f) = \frac{Z_0}{\dfrac{\alpha}{I_f} + R_S} \tag{2.79}$$

that results in an attenuation (Equation (2.70)) of:

$$A(I_f) = \frac{|2 + y(I_f)|^2}{4} \tag{2.80}$$

2.5.6.4 Power limiting

For power limiting applications, the PIN diode is connected in parallel with the receiver or device (e.g. detector, mixer, amplifier) to be protected as shown in Figure 2.38.

The inductor, L, enables the DC current circulation. The diode eliminates or attenuates all the signals with amplitude higher than a certain threshold [3, 4, 5].

For small signals, the diode operates around 0 V, exhibiting high impedance, due to the small junction capacitance characteristic of PIN diodes. In this case the signal transmitted to the receiver is almost unperturbed. For large signals (i.e. when the input amplitude is high),

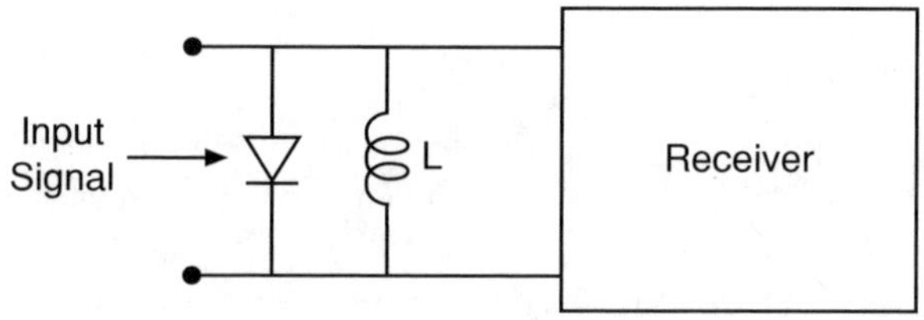

Figure 2.38 PIN-based power limiting for receiver protection

the diode impedance greatly decreases at the signal (positive going) peaks, leading to deep forward bias. For high operating frequencies, the injected carriers are only partially evacuated during the negative going (reverse bias) peaks. Since the diode is still conducting due to the residual carriers, its impedance remains low for the entire cycle of the input waveform. In this way the signal delivered to the receiver is greatly attenuated.

Higher attenuation values are obtained for smaller intrinsic region widths, due to the larger junction capacitance. When handling high input powers, however, the intrinsic region must be sufficiently wide to avoid avalanche breakdown.

2.6 Step-Recovery Diodes

Step-recovery diodes (also called snap-off diodes) are based on a PIN configuration. They are commonly employed in the design of frequency multipliers of high order. Their operating principle is as follows.

When a PIN diode is forward-biased, the two junctions P^+I and IN^+ inject carriers into the I region. The recombination of these carriers after a time τ gives rise to the forward current I_f. If the diode is now reverse-biased through the application of a voltage step, the injected carriers that are present in the I region will have to be removed. There will be a large reverse diffusion current limited only by the external circuit. The carrier concentration, however, does not decrease uniformly along the I region. There will be a faster decrease close to the junctions. After a time, t_a, the concentration level at the junctions becomes zero. However, there will still be stored charge in the I region that is removed, as in a capacitor, with a certain time constant t_b. It is possible to artificially increase t_a, using a special doping process. The aim is to increase t_a reducing the number of carriers still stored in the I region after this time. The smaller number of stored charges gives rise to a much smaller time t_b. The decrease in t_b compared to t_a provides a pulsed response. When using a sinusoidal input voltage, this pulsed current can be used for harmonic generation. It is thus possible to obtain frequency multiplication of high order.

In the design of high-order frequency multipliers, the efficiency of step-recovery diodes is much higher than that of varactor diodes. There is, however, a limit to the output frequency of the multiplier circuit. The total switching time, $t_a + t_b$, must be much smaller than the reciprocal of the output frequency for proper multiplier operation. If this is not the case, very high conversion loss is obtained. A common criterion for the estimation of losses is based on the calculation of the threshold frequency $\omega_{th} = 1.6/(t_a + t_b)$. For output frequencies below ω_{th} the conversion loss increases at approximately 3 dB/octave. Above this frequency it increases at approximately 9 dB/octave.

Self-assessment Problems

2.15 What is the phenomenon giving rise to the high reverse current that is initially observed when switching a PIN diode from forward to reverse bias?

2.16 What limits the output frequency of step-recovery diodes used for frequency multiplication?

2.7 Gunn Diodes

Gunn diodes are manufactured using III–V compound semiconductors, such as GaAs or PIn (Phosphorus-Indium). They exhibit a negative resistance, related to the multi-valley nature of their conduction bands. The oscillating properties of these materials were discovered (by Gunn), in 1963, during his studies of GaAs noise characteristics. (Gunn was working on a biased GaAs sample when he detected RF oscillations in the microwave range.)

Gunn diodes are usually fabricated in GaAs. In the conduction band of this material, a main valley, 1.45 eV above the valence band, is surrounded by six satellite valleys with 0.36 eV higher energy (Figure 2.39).

Electrons residing in the main valley have lower energy and higher mobility compared with electrons in the satellite valleys.

Gunn diodes are manufactured in three layers, with an N-doped layer embedded between two, thinner, N^+-doped layers (Figure 2.40).

The two external layers facilitate the ohmic contacts that provide the device terminals. Consider a Gunn diode connected to an external bias source. As the bias voltage is increased from a low value, the electrons acquire a higher energy and their velocities increase. When the field strength reaches a threshold, E_{th}, electron collisions become sufficiently frequent (due to their enhanced velocities) that their predilection to drift in the electric field becomes

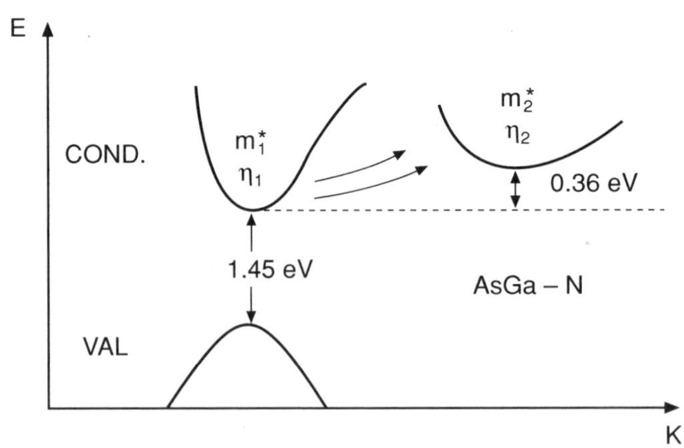

Figure 2.39 Energy band diagram for GaAS

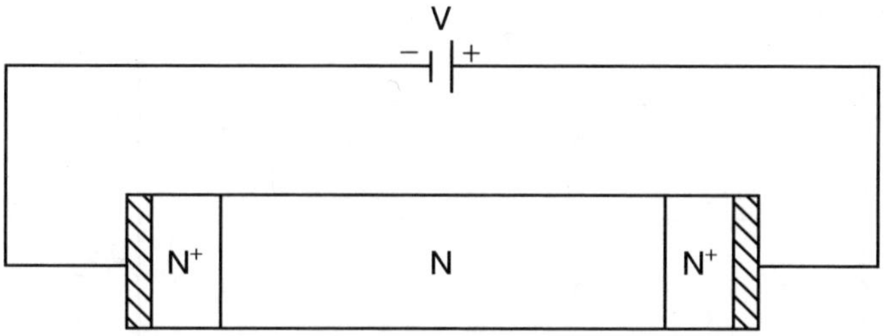

Figure 2.40 Three-layer structure of Gunn diode

significantly impaired; in short, their mobility is reduced. This corresponds to electrons in the lower energy valley being transferred to the higher energy valleys.

The relative concentrations of electrons in the lower and higher energy valleys depend on the electric field strength. These concentrations are denoted by $n_1(E)$ for the lower energy valley and $n_2(E)$ for (all) the surrounding valleys of higher energy. Three different ranges of electric field strength may be considered [1, 2, 3]:

(a) For $0 \leq E \leq E_{th}$ all electrons occupy the lower energy valley, so the total electron concentration $n = n_1$. The electron velocity, in this regime, increases linearly with increasing electric field, i.e. $\mathbf{v} = -\mu_1\mathbf{E}$. This corresponds to the electrons in the lower energy valley acquiring more kinetic energy. The current density is, therefore, given by:

$$\mathbf{J} = \mu_1 e n_1 \mathbf{E} \tag{2.81}$$

(b) For $E_{th} \leq E \leq E_2$ some of the electrons in the lower energy valley acquire sufficient energy to reach the higher valleys. The total number of electrons is then distributed between the low and high energy valleys, i.e. $n(E) = n_1(E) + n_2(E)$. The higher energy valley has a lower mobility, μ_2, due to the collision effect described above and the current density is, therefore, given by:

$$\mathbf{J} = (\mu_1 n_1 + \mu_2 n_2)\mathbf{E} \tag{2.82}$$

The ratio $n_2(E)/n(E)$ increases with increasing electric field. The current density, however, decreases since the reduction in average electron mobility (due to the increasing proportion of slower electrons) outweighs the effect of the increased electric force on each electron.

(c) For $E \geq E_2$ all electrons occupy the higher energy valleys, so $n = n_2$. Since there is now no transference of electrons from lower to higher energy valleys, electron velocity will again increase with applied field, i.e. $\mathbf{v} = -\mu_2\mathbf{E}$. The current density is now given by:

$$\mathbf{J} = \mu_2 e n_2 \mathbf{E} \tag{2.83}$$

Figure 2.41 shows mean electron velocity plotted against applied electric field strength [1].

The behaviour of velocity in this figure arises due to the non-linear variation of mobility as a function of the applied field, i.e. $\mathbf{v} = -\mu(E)\mathbf{E}$. This results in a negative differential

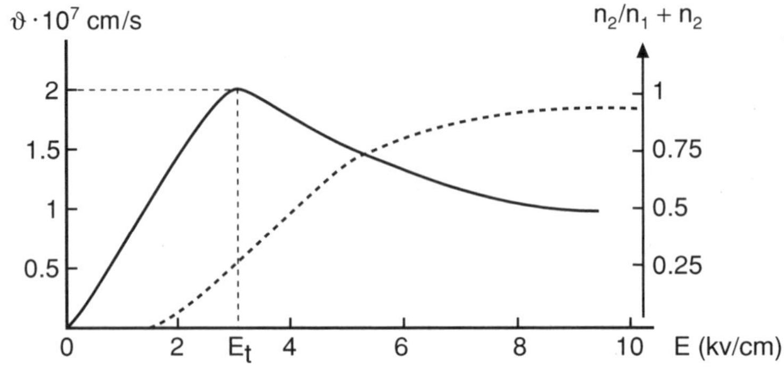

Figure 2.41 Mean electron velocity versus applied electric field

mobility in a certain electric field range and thus in negative resistance. Note that current depends on the carrier velocity, while voltage drop is related to the electric field. A plot of the ratio $n_2(E)/n(E)$ is included in Figure 2.41 showing the close relationship it has with velocity variation. Negative-resistance is only observed if the period is shorter than the time required to transfer electrons from the low-energy to the high-energy valley. The electric field threshold needed to obtain negative differential mobility therefore varies with operating frequency.

2.7.1 Self-Oscillations

Provided the electric field remains below the threshold value E_{th}, the electric potential decreases smoothly along the device (Figure 2.42(a)).

For electric fields above the threshold, however, some electrons will be slowed down due to their transference to the higher energy valleys. The slow electrons will bunch with the

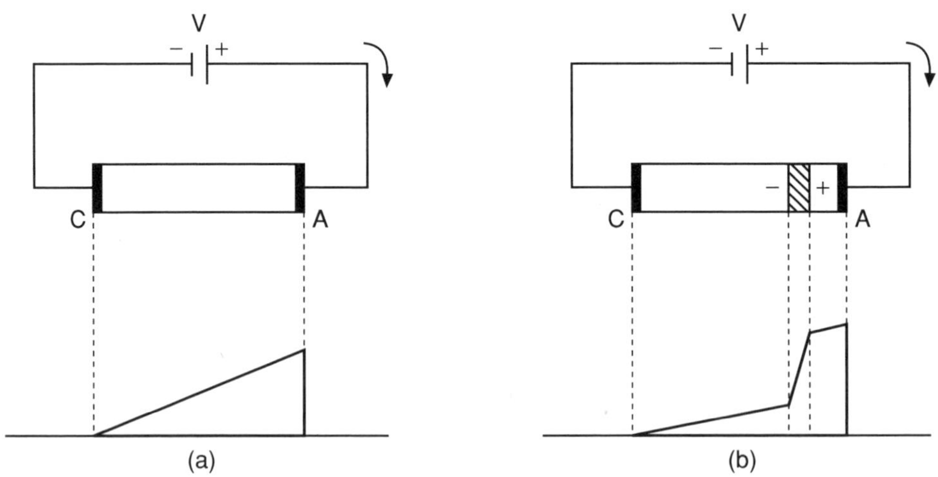

Figure 2.42 Electric potential versus position along Gunn diode: (a) electric field below threshold, (b) electric field above threshold

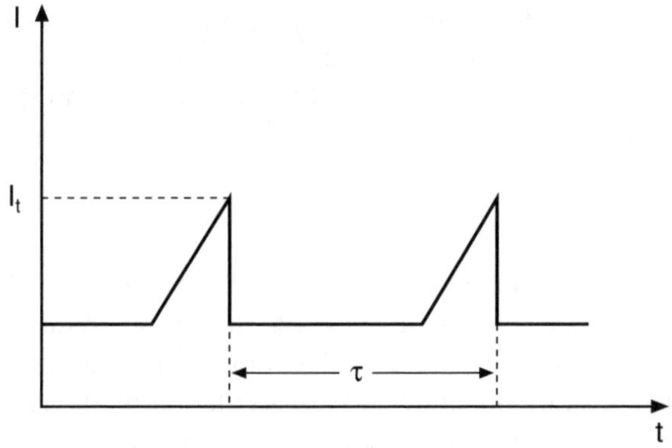

Figure 2.43 Gunn current versus time

following faster electrons giving rise to a negative space charge accumulation [4]. (The precise location where the space charge initially forms is determined by random fluctuations in electron velocity or by a permanent non-uniformity in the doping.) Every delayed electron gives rise to a negative charge deficiency in front of, and an excess negative charge behind, its location had it not been delayed. This effect is cumulative and the negative space charge accumulation is compensated for by a positive space charge of equal magnitude. The result is a space charge dipole that moves from cathode to anode at the electron-saturated velocity (Figure 2.42(b)). The potential now drops unevenly along the diode, the gradient being greatest across the dipole (Figure 2.42(b)) and the electric field along the rest of the device necessarily remaining below the threshold value. The reduced value of electric field away from the space charge dipole prevents more than one dipole forming at any one time.

When the slowly moving dipole reaches the anode, a current peak is detected (Figure 2.43).

The old dipole is now extinguished and a new one will immediately appear, since the electric field in the device is again above the threshold value. This will reduce the detected current. In this way, by simply biasing the semiconductor, it is possible to obtain a non-sinusoidal oscillation of current at the frequency $f_t = 1/\tau$ where τ is transit time across the device of the space charge dipole. (The transit time of the space charge dipole is, of course, equal to the transit time of individual electrons.)

It is sometimes important to appreciate that the load circuit connected to a Gunn diode may reduce the electric field to values below the threshold value. If this is the case, then the space charge dipole will disappear during one part of the high frequency cycle. As will be shown in the following, the load circuit connected to the Gunn device has great influence on the device operating mode.

2.7.2 Operating Modes

Gunn devices can operate in different modes [1], depending on factors such as doping concentration, doping uniformity, length of the active region, type of load circuit and bias range. The formation of a strong space charge instability (i.e. space charge dipole) requires enough charge, and sufficient device length, to allow the necessary charge displacement within the electron transit time.

The boundary between the different modes of operation is denoted by the product $n_0 l$, where n_0 is boundary carrier concentration for a given device length, l [1]. For GaAs and InP devices, the boundary condition between stable field distribution and space charge instability is given by $n_0 l = 10^{12}\ \text{cm}^{-2}$.

2.7.2.1 Accumulation layer mode

A Gunn device with sub-critical nl product ($nl < 10^{12}\ \text{cm}^{-2}$) is now considered. An accumulation layer starts at the cathode, propagates towards the anode and disappears at the anode. The field at the cathode then rises, enabling the formation of another accumulation layer, the process thereafter repeating indefinitely. This oscillation mode has a low efficiency, but advantage can be taken from the negative resistance exhibited by the device in a frequency band near the reciprocal of the electron transit time and its harmonics. In this mode the device can therefore be operated as a stable amplifier. The frequency of the peak gain of such an amplifier increases with applied electric field strength. Measurements show that negative resistance can be obtained over a frequency band greater than one octave.

2.7.2.2 Transit-time dipole layer mode

Transit-time dipole layer operation is obtained for supercritical values of the product nl. Dipole layers form and propagate to the anode. The cyclic formation of these dipoles and subsequent disappearance at the anode constitute the self-oscillations. The efficiency, however, is only a few percentage. To be able to use dipole transit for self-oscillation the quality factor of the load circuit must be low to avoid a reduction in the electric field during the negative part of the cycle. It is this that depresses efficiency and keeps interest in this operating mode low.

The expected frequency of the self-oscillation may be found from $v_s = l/\tau$, where v_s the space charge (saturated) velocity and l the device length. A filter will be necessary in order to eliminate the higher order harmonic components if sinusoidal oscillation is required. Typically the DC bias voltage is three times the threshold voltage, $V_{th} = E_{th} l$.

2.7.2.3 Quenched dipole layer mode

In devices with a supercritical nl product, it is possible to obtain oscillation frequencies higher than the reciprocal of the transit-time frequency by quenching the dipole layer using a resonant circuit. The oscillating voltage from the resonant circuit adds to the bias voltage reducing the total voltage during its negative half cycles. The width of the dipole layer is reduced during what is, effectively, these periods of reduced bias. Dipole layer quenching occurs when the bias voltage is reduced below the threshold value E_{th}. When the effective bias voltage increases above this threshold (during the positive half cycles of the oscillating voltage), a new dipole layer forms, the process then repeating indefinitely. The oscillations therefore occur at the frequency of the resonant circuit rather than the reciprocal of the transit time frequency.

If the quality factor of the load circuit is high, the amplitude at the diode terminals may be as high as the bias voltage. During sign opposition the electric field will decrease to a very low value and the dipoles will disappear during part of the cycle. If dipoles disappear before reaching the anode, the oscillation frequency will be higher than the transit frequency f_t. If the electric field is low, the dipole may traverse the whole device length l, with the appearance of a new dipole being delayed. This would give rise to an oscillation frequency smaller than f_t. The efficiency in this oscillation mode is much higher than the one obtained in the transit-time dipole-layer mode.

2.7.2.4 Limited-space-charge accumulation (LSA) mode

In limited-space-charge mode, the dynamic operating point is located in the regions of positive resistance for most of the oscillation period. The LSA mode assumes large dynamic swings around the operating point. A high frequency, large amplitude, voltage reduces the electric field during the cycle to a value slightly lower than E_c that suppresses the formation of dipoles. The frequency must be high enough to prevent the dipole formation (i.e. there must not be enough time for dipole formation). The electric field across the device rises above, and falls back below, the threshold so quickly that the space charge distribution does not have sufficient time to form. The mean resistance over one cycle, however, must be negative and to fulfil this condition, the electric field must not fall to values much lower than E_c, which requires a limited space charge. The diode behaves like a negative-real-part impedance without producing pulses. Oscillation frequency is determined by the external circuit.

LSA mode operation yields much higher efficiency than that obtained from dipole transit. Furthermore, there is no transit time limitation on the maximum oscillation frequency.

2.7.3 Equivalent Circuit

The most commonly used equivalent circuit for a Gunn diode consists of non-linear current source with an N-shaped current-voltage characteristic, a capacitance and a loss resistance. Series and parallel circuit configurations are possible, the choice depending on bias point, bandwidth and operating frequency. A series configuration is shown in Figure 2.44.

For small signals, and provided that the diode is biased in the negative slope region, the non-linear current source may be replaced by a negative resistance.

In order to implement ohmic contacts, the semiconductor regions close to the metallic terminals are highly doped and thus more conductive than the central region. The capacitor C accounts for the capacitance between these contacts. The loss resistance r arises due to the limited conductivity of the substrate.

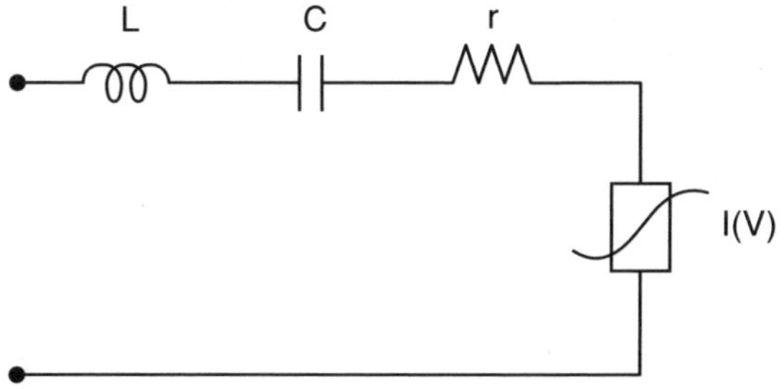

Figure 2.44 Series equivalent circuit model for a Gunn diode incorporating non-linear current source

Self-assessment Problems

2.17 What kind of materials can exhibit the Gunn effect?

2.18 What is the phenomenon that gives rise to the dipole layer formation in the Gunn diode? Why is only one dipole layer observed along the entire device length?

2.19 How does load circuit quality factor influence the operating mode of the Gunn diode?

2.7.4 Applications

Gunn devices are used, principally, in the design of negative resistance amplifiers and oscillators.

2.7.4.1 Negative resistance amplifiers

Consider a Gunn diode, biased with a certain DC voltage. The series equivalent circuit (Figure 2.45) includes a negative resistance, $-R_n < 0$, in series with a capacitor, C_d, and a loss resistance, r.

The negative resistance $-R_n$ is defined by:

$$-R_n = \left[\left. \frac{dI}{dV} \right|_{V_0} \right]^{-1} \tag{2.84}$$

where $I(V)$ is the non-linear current source characteristic and V_0 the device bias voltage.

The net negative resistance is $-R_d = -R_n + r$, with $R_d > 0$ implying a voltage reflection coefficient, Γ, at the device terminals with an amplitude larger than one (i.e. $|\Gamma| > 1$).

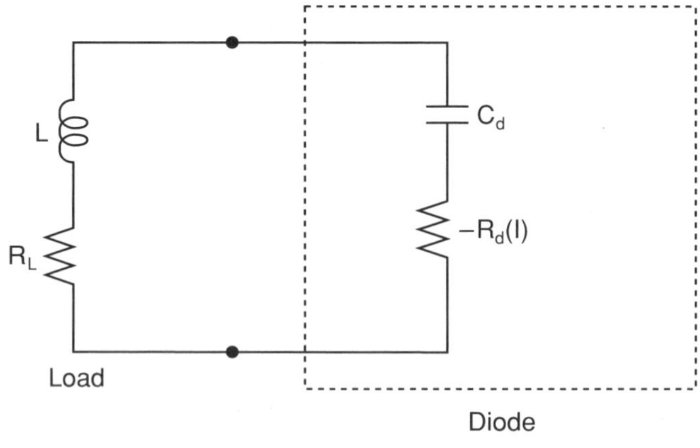

Figure 2.45 Series equivalent circuit of Gunn diode with DC bias and negative resistance

(This implies a reflected power that is higher than the incident power, and thus genuine amplification.) Depending on the value of the passive load, different gains and bandwidths can be obtained. A particularly simple amplifier implementation is realized when incident and reflected powers are separated using a circulator.

2.7.4.2 Oscillators

Gunn diodes are often employed as the active device in microwave oscillators. In the dipole layer mode, device length determines oscillation frequency. Around this frequency, the diode exhibits negative resistance and, when using the accumulation layer mode, the precise oscillation frequency may be fixed by the resonant frequency of an external (passive) circuit. The main attractions of Gunn-based oscillators are their capability to operate over a large frequency band, their low noise performance and their high output power. Gunn oscillations have been obtained up to frequencies of about 150 GHz and output powers of 15 mW at 100 GHz can be realised. Their main disadvantage is low DC-RF efficiency. Another disadvantage is their temperature sensitivity, an effect inherent in their working principle.

A small signal model for the Gunn diode consists of a capacitance C_d and a total negative resistance $-R_d$ (including losses). A simple oscillating configuration is shown in Figure 2.47, in which the diode load is composed of a resistor plus an inductor in series configuration. For oscillation to start, the circuit total resistance (including linear and non-linear contributions) must be negative at the resonance frequency, ω_0, given by:

$$\omega_0 = \frac{1}{\sqrt{LC_T}} \tag{2.85}$$

where:

$$C_T = \frac{CC_d}{C + C_d} \tag{2.86}$$

2.8 IMPATT Diodes

IMPATT (Impact Avalanche and Transit Time) diodes are very powerful microwave sources, providing the highest (solid state device) output power in the millimetre-wave frequency range. They exhibit a dynamic negative resistance based on transit time effects and they are often used in the design of oscillators and amplifiers when high output power is required. They are manufactured in Si, GaAs and InP.

IMPATT diode provides negative resistance using the phase shift between current through the device and the applied voltage. (Negative resistance is obtained when this phase shift is greater than 90°.) The device consists of a reverse-biased P-N junction (operating in avalanche breakdown) and a drift zone. Carriers are injected into the drift zone from the junction in avalanche breakdown where they drifted at saturated velocity. The constant value of the saturated velocity provides a linear relationship between the device length and the current delay. It is, therefore, possible to determine a length that will result in negative resistance within a certain frequency band.

Like Gunn diodes, IMPATT diodes are generally used in the design of negative resistance amplifiers and oscillators. They provide higher output power than Gunn devices, and may be operated at frequencies up to about 350 GHz when manufactured in silicon. They are noisier than the Gunn devices, however, a characteristic inherent in their operating principle. Due to their noisy operation they are seldom used for local oscillators in receivers. Another drawback is the low value of their negative resistance that can give rise to matching difficulties.

2.8.1 Doping Profiles

IMPATT diodes are manufactured with many different doping profiles. In all variations, however, avalanche and drift regions can be distinguished [3]. Single drift profiles occur in P^+NN^+ and N^+PP^+ structures, the former being preferentially manufactured, since N type substrate is the more common. In P^+NN^+, the P-N junction is biased near avalanche breakdown. The P region injects electrons into the NN^+ region, along which they drift at saturated velocity. The holes injected from the N region do not drift. Hence the name of single drift profile. Double drift devices provide a higher efficiency and a higher output power. One possible double drift structure is P^+PNN^+. The central P-N junction operates in avalanche breakdown, injecting electrons that drift along the NN^+ region, and holes that drift along the P^+P region. Other possible profiles include the Read configurations. For single drift devices Read proposed N^+PIP^+ and P^+NIN^+ structures. The former is analysed below.

Although IMPATT diodes are manufactured in both Si and GaAs, higher efficiencies are obtained with the latter. Mesa technology is used for their fabrication. In the case of a P^+NN^+ profile, for example, two layers (N and P^+) are diffused by a double epitaxial growth process over a highly doped N^+ layer. Ionic implantation is also possible. Finally, the ohmic contacts are realised through a metal deposit.

2.8.2 Principle of Operation

A device with an N^+PIP^+ profile (Figure 2.46) is analysed here to illustrate the IMPATT's principle of operation. Assume a reverse bias $V = V_b$ is applied to the IMPATT device, where V_b is the breakdown voltage. A sinusoidal waveform is then superimposed, resulting in a total device voltage $v(t) = V_b + V_1 \cos \omega t$. When $v(t) > V_b$, a strong ionisation process (breakdown) starts at the N^+P junction resulting in the generation of electron-hole pairs. The electrons move towards the positive terminal, while the holes drift at saturated velocity v_s along the depletion zone (i.e. intrinsic, or I zone) towards the P^+ region.

The drift time will be shorter for the electrons from P, due to the shorter distance to be traversed. Their effect on the external current will not be observed until the holes reach P^+, however, in good agreement with the principle of electric neutrality. The device length may be chosen to provide the necessary delay for a 180° phase-shift between the device voltage and current.

The applied voltage, consisting of a sinusoidal waveform, superimposed on the bias voltage V_b is illustrated in Figure 2.47(a).

As shown in Figure 2.47(b), as long as $v(t) > V_b$, the number of carriers increases, even beyond the voltage maximum. This is explained by the fact that, during breakdown, electron-hole generation depends on the total carrier number. Since the avalanche current is

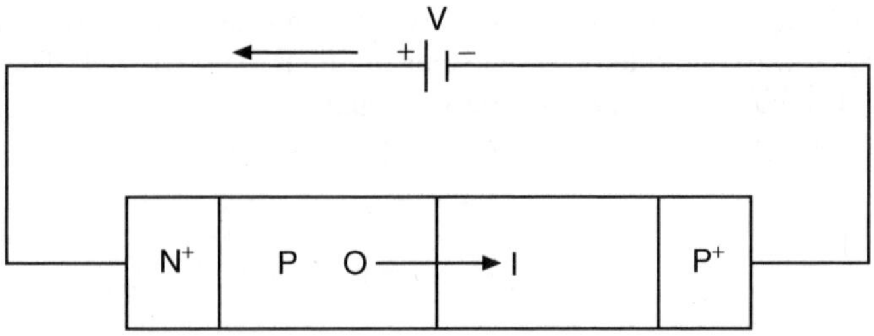

Figure 2.46 Schematic structure of IMPATT diode

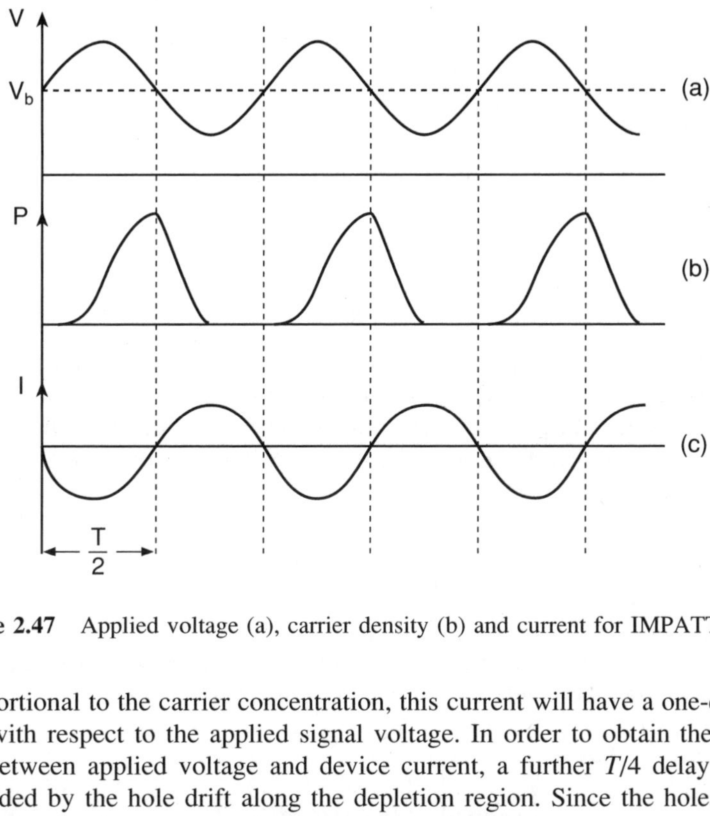

Figure 2.47 Applied voltage (a), carrier density (b) and current for IMPATT diode

directly proportional to the carrier concentration, this current will have a one-quarter period ($T/4$) delay with respect to the applied signal voltage. In order to obtain the desired 180° phase-shift between applied voltage and device current, a further $T/4$ delay is necessary. This is provided by the hole drift along the depletion region. Since the holes move at the constant velocity, v_s, the necessary device length is given by:

$$l = v_s \frac{T}{4} \tag{2.87}$$

Example 2.5

The carrier saturation velocity in silicon semiconductors is $v_s = 10^5$ m/s. What should be the length of a silicon IMPATT in order to obtain a negative resistance at about 12 GHz?

In order to obtain a negative resistance, the carrier drifting through the device must give rise to a phase shift of $\pi/4$, to be added to the avalanche phase shift of $\pi/4$. The drift time must therefore be $T/4$. Taking Equation (2.85) into account:

$$l = v_s \frac{T}{4}$$

where:

$$T = \frac{1}{12 \times 10^9} = 8.33 \times 10^{-11} \text{ s}$$

Therefore:

$$l = 2 \ \mu\text{m}$$

2.8.3 Device Equations

Ionisation rate is the probability of an electron-hole generation per carrier unit length. It can be calculated from the following formula [3]:

$$\alpha = C \exp\left(-\left(\frac{b}{E}\right)^m\right) \tag{2.88}$$

where m, b and C are parameters which depend on the material and type of particle. Electrons and holes therefore have different ionisation rates, α_n and α_p.

Since electric field is, in general, a function of distance x along the device, so are the ionisation rates. The electric field dependence can be clarified by remembering Poisson's equation, i.e.:

$$\frac{\partial E}{\partial x} = \frac{e}{\varepsilon_s}[N_D - N_A + p - n] \tag{2.89}$$

Integrating Equation (2.89) gives the electric field variation, $E(x)$, along the device and knowing this the ionisation rate, α, as a function of the x can be determined from Equation (2.88).

It can be shown that the condition for avalanche breakdown is given by [1]:

$$\begin{aligned} &\int_0^w \alpha_p \exp\left[-\int_0^x (\alpha_p - \alpha_n)dx'\right]dx = 1 \\ &\int_0^w \alpha_n \exp\left[-\int_x^w (\alpha_n - \alpha_p)dx'\right]dx = 1 \end{aligned} \tag{2.90}$$

where w is the width of the avalanche region.

Recall the carrier continuation Equations (2.20) and (2.21). For generation by impact avalanche, the coefficients, g_p and g_n, are given by:

$$g_p = \frac{\alpha_n J_n + \alpha_p J_p}{e}$$
$$g_n = \frac{\alpha_n J_n + \alpha_p J_p}{e} \tag{2.91}$$

The continuity equations for electrons and holes are given by:

$$\frac{\partial n}{\partial t} = \frac{1}{e}\left[\frac{\partial J_n}{\partial x} + \alpha_n J_n + \alpha_p J_p\right]$$
$$\frac{\partial p}{\partial t} = \frac{1}{e}\left[\frac{\partial J_p}{\partial x} + \alpha_n J_n + \alpha_p J_p\right] \tag{2.92}$$

where diffusion current in the drift region has been neglected.

The total current density J is:

$$J = J_n + J_p \tag{2.93}$$

J may also be written:

$$J = env_n + epv_p + \varepsilon_s \frac{\partial E}{\partial t} \tag{2.94}$$

where v_n and v_p are, respectively, the electron and hole-saturated velocities.

Equations (2.90), (2.92), (2.93) and (2.94) comprise a system of five equations (Equations (2.93) and (2.94) counting as one equation) in five unknowns (E, p, n, J_n and J_p) that may be solved numerically [3]. Once E and J are obtained the non-linear current-voltage relationship of the IMPATT diode can be found using:

$$V = \int_x E dx \tag{2.95}$$

2.8.4 *Equivalent Circuit*

In order to obtain a small signal model for the IMPATT diode, the electric field E and current density J are assumed to be given by:

$$E = E_0 + E_1 e^{j\omega t}$$
$$J = J_0 + J_1 e^{j\omega t} \tag{2.96}$$

where E_0 and J_0 are the DC components and E_1 and J_1 are the amplitudes of the AC components (assumed to be much smaller than the DC components).

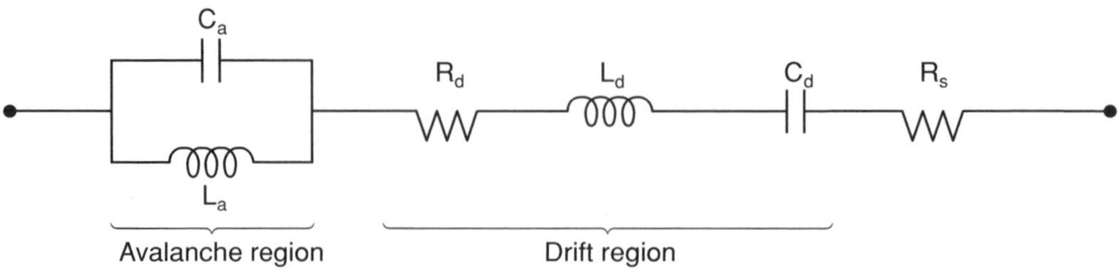

Figure 2.48 IMPATT diode equivalent circuit model

Solving the device Equations (2.90) to (2.95), it is possible to obtain the linear model of Figure 2.48.

It is composed of two sub-circuits that account, respectively, for the avalanche and drift zones [3], and a separate loss resistance R_s. The equivalent circuit for the avalanche region consists of a resonant circuit, with an avalanche inductance L_a and a capacitance C_a. The capacitance is given by:

$$C_a = \frac{\varepsilon_{sc} S}{w_a} \tag{2.97}$$

where w_a is the width of the avalanche region.

It can be shown that the device exhibits a negative resistance [1–3] for frequencies above the avalanche resonant frequency, $f_a = 1/\sqrt{(L_a C_a)}$.

The equivalent circuit for the drift region is a series resonant circuit, with capacitance C_d given by:

$$C_d = \frac{\varepsilon_{sc} S}{w - w_a} \tag{2.98}$$

where w is total device width.

For $f > f_a$ the resistance R_d is negative. At frequencies above f_a the avalanche sub-circuit has a capacitive behaviour. The drift region equivalent circuit, together with the avalanche capacitance, account for the phase shift.

Self-assessment Problems

2.20 What are the two mechanisms involved in the observation of negative-resistance in IMPATT diodes?

2.21 What is the reason for the $T/4$ delay between the avalanche current and the applied sinusoidal voltage?

2.22 Why are IMPATT diodes noisy?

2.9 Transistors

For the microwave circuit designer transistors are the key active elements most often used to achieve signal generation (oscillators), signal amplification and a wide range of other, more complex, switching and signal processing functions. (Photonic technology may come to challenge the supremacy of transistors in the future but this is not the case yet.) Traditional microwave transistors, including MESFETs and HEMTs, are normally constructed from III-V group compounds. HBTs are based on both III–V and SiGe compounds.

In the remainder of this chapter we will address basic transistor modelling and review the popular equivalent circuits used in microwave design. The principal focus will be on the different models used in different frequency ranges and excitation (i.e. large or small signal) amplitudes.

2.9.1 Some Preliminary Comments on Transistor Modelling

A model is a structure (physical, mathematical, circuit, etc.) that permits the behaviour of a device to be simulated, usually under a restricted set of conditions. Such models allow both circuit and device engineers to improve the performance of their designs.

In the context of modern microwave applications, the importance of semiconductor modelling lies in the fact that MMIC (monolithic microwave integrated circuit) technology does not offer the possibility of tuning once the fabrication has been completed. In this sense, modelling is a requirement for, rather than an aid to, design.

2.9.1.1 Model types

Models can be classified in a variety of different ways. Here we will classify them according to how they are obtained, i.e., by their extraction process.

Empirical models are obtained by describing (using mathematical functions) the macroscopic characteristics of devices, such as terminal currents and voltages, without regard to the physical processes that result in these characteristics. In contrast, physical models are derived from the known laws governing the microscopic physical process on which a device's behaviour ultimately depends.

Empirical models can offer the designer a level of accuracy in device behaviour prediction approaching that of measurements. In order to obtain an empirical model, a detailed characterisation of the device is required involving, for example, DC measurements, small signal S-parameters, pulsed current/voltage measurements, etc.

2.9.1.2 Small and large signal behaviour

One method of classifying models is on the basis of their ability to correctly predict device behaviour when the device is subjected to large input signals. Models restricted in application to small input signals are called small signal models while those which can be applied when large input signals are present are called large signal models.

2.9.1.2.1 Small signal models

Small signal models are widely used in microwave applications and often provide a simple, first order, method of assessing whether a given transistor might be suitable for a particular application of interest.

A small signal model is valid only when the applied signal voltages consist of small fluctuations about the quiescent point. It consists of a circuit that would have (to some required level of approximation) the same S-parameters, at least over the measured frequency range, as the device being modelled. Depending on the extraction process, the model might not only predict the small signal behaviour of the device over the measurement frequency band but also, to a greater or lesser extent, above and below the ends of this band. A good model, capable of such extrapolation, is very useful when the equipment available to make measurements does not cover the entire frequency range of interest. Individual components in this 'equivalent circuit' model can often be identified with different physical processes taking place in the device.

2.9.1.2.2 Large signal models

Large signal models are used when the assumption of small signals is invalid. They are analytical and are able to describe the large signal properties of the device. Like the small signal models, they consist of an equivalent circuit including non-linear components, e.g. non-linear current sources, voltage-controlled capacitances, etc. Non-linear components, along with appropriate characterisation procedures, allow the model user to simulate device behaviour under different excitation conditions (transient large signal conditions, harmonic balance conditions, etc). Small signal simulations can also be performed using this kind of model, because linearisation of a large signal model provides the same results as a small signal model.

The principal differences between large signal models lie in the mathematical laws chosen to represent the non-linearities. We therefore find models due to Curtice, Materka [6] etc. for MESFETs and HEMTs, or Gummel-Poon [8] for HBT devices, depending on the approach that each author suggests to match the behaviour of a measured non-linearity (e.g. $I_{ds}(v_{gs}, v_{ds})$ non-linear current source for MESFETs).

The distinguishing feature of a large signal model is its ability to correctly predict the magnitude and shape of the signal at the device output, irrespective (within certain limits) of the input signal's amplitude or frequency content. The model will therefore be valid even when the input signal exhibits large excursions about the bias, or quiescent, point.

Self-assessment Problems

2.23 What type of model would be appropriate for the design of a low noise amplifier?

2.24 What type of model would be appropriate for the design of a high power amplifier?

2.9.2 GaAs MESFETs

The GaAs MESFET (metal semiconductor field effect transistor) is widely used at microwave frequencies in the realisation of oscillators, amplifiers, mixers, etc. A simplified cross-section of a GaAs MESFET, is shown in Figure 2.49.

Three contacts can be seen on the surface of the MESFET in Figure 2.49. These contacts are called the source, gate and drain. The source and drain contacts are ohmic while the gate contact constitutes a Schottky diode junction. Figure 2.49 is drawn for the usual bias

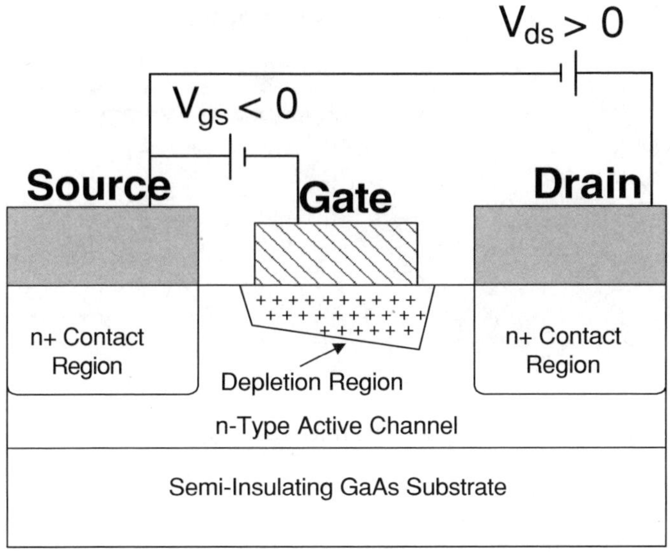

Figure 2.49 GaAs MESFET cross-section

configuration, i.e. negative gate to source voltage and positive voltage drop between drain and source terminals.

Under normal operating conditions, the drain terminal is positively biased with respect to the source terminal (which is often grounded), while the gate is negatively biased with respect to both drain and source. (This implies that no significant power is sunk at the gate terminal.) The volt drop along the conducting channel due to a current flowing from drain to source results in the gate reverse bias being progressively greater moving from source to drain. This creates a depletion region beneath the gate that is thicker at the drain end of the device than at the source end.

MESFETs can also be implemented in Si technology, the principal difference between GaAs and Si devices being operational frequency range. GaAs MESFETs can operate at higher frequency as a result of higher carrier mobility. This makes them suitable in applications where high operating frequency degrades the different merit figures of the transistor (e.g. gain and noise figure). Typically, 1 GHz (often taken to mark the lower edge of the microwave band) is the frequency above which GaAs transistors might be used with advantage.

To work properly at microwave frequencies, Gate length, L_g (see Figure 2.50), must be less than about 1 μm in order to avoid transversal propagation effects [6, 7]. These effects must be taken into account when wavelength becomes comparable to the physical length of the transistor contacts. High frequency operation therefore implies small device dimensions.

Figure 2.50 shows the most important MESFET dimensions. For microwave operation, the critical dimension is the gate length, L_g, since this determines the maximum operating frequency. Typically, for microwave operation, this gate length is in the range 0.1–1 μm.

Another important dimension in GaAs MESFET devices is gate width. For low noise applications, small gate widths must be employed. Large gate widths are usually reserved for high power applications.

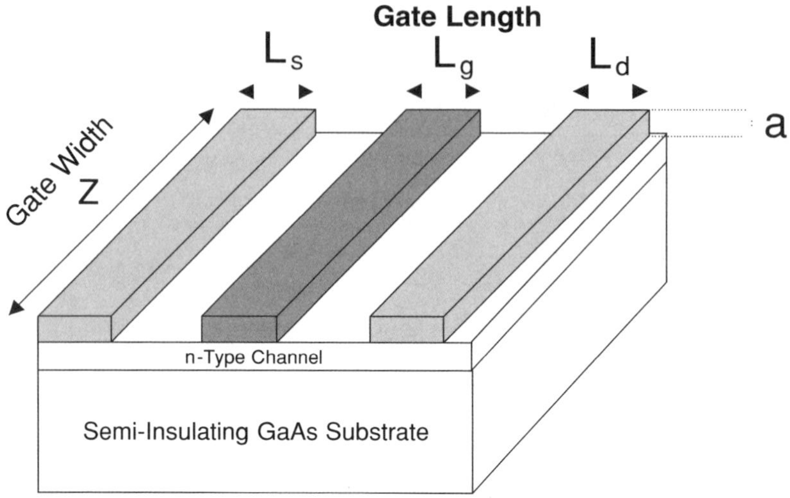

Figure 2.50 Longitudinal view of a GaAs MESFET

Self-assessment Problem

2.25 Small gate widths are necessary in devices intended for low noise applications. Why do you think this so?

2.9.2.1 Current-voltage characteristics

We can distinguish between two different kinds of microwave MESFET. In depletion mode devices (which are usually used in microwave applications) any negative gate-source voltage (greater than the pinch-off voltage) or zero gate-source voltage causes current flow from the drain to the source. In enhancement mode devices a positive volt drop from gate to source is necessary to allow current flow from the drain to the source. Depletion MESFETs are known as 'normally-off' devices while enhancement MESFETs are known as 'normally-on' devices.

The voltage applied to the gate terminal acts as a control of the current flowing from source to drain. In depletion devices, when a sufficiently negative voltage (measured with respect to source) is applied to the gate, the depletion region extends across the entire channel reducing drain-source current to zero. The voltage for which this just occurs is called the pinch-off voltage. When the gate is tied to the source (i.e. gate and source voltage are equal), the depletion region does not extend across the channel allowing drain-source current to flow, hence its alternative description as a normally-on device.

In enhancement devices, when the gate is tied to the source, the depletion region extends across the channel causing pinch-off of the drain-source current. In this case a positive gate voltage (measured with respect to source) is required to shrink the depletion layer sufficiently to open the channel for drain-source current conduction. These are called normally-off devices.

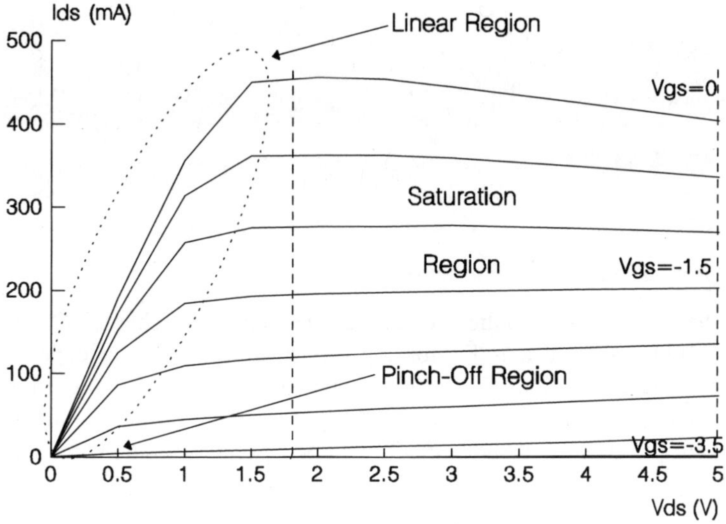

Figure 2.51 Typical I/V curves for a 10 × 140 μm GaAs MESFET device

Figure 2.51 shows a typical plot of drain-source current, I_{ds}, versus drain-source voltage, V_{ds}, for several different values of gate-source voltage, V_{gs}. The value of I_{ds} for $V_{gs} = 0$ is known as the saturated current level. Three characteristic operating regions for GaAs MESFETs are shown in Figure 2.51. The saturation region starts at the V_{ds} value where the slope of the I_{ds} curve (for a constant V_{gs} value) is zero, while the pinch-off region refers to the area below the V_{gs} voltage where no significant I_{ds} current flows for any value of V_{ds}.

It is sometimes desirable to know the behaviour of the derivatives of I_{ds} with respect to V_{gs} and V_{ds}, for example, when the minimisation of intermodulation distortion in a device is important. The output conductance of the MESFET is defined as the derivative of I_{ds} with respect to V_{ds}, when V_{gs} is kept constant, i.e.:

$$g_{ds} = \left(\frac{\partial I_{ds}}{\partial v_{ds}}\right)_{V_{gs}=constant} \tag{2.99}$$

The inverse of Equation (2.99) is the MESFET output resistance.

The transconductance of the MESFET is defined as:

$$g_m = \left(\frac{\partial I_{ds}}{\partial v_{gs}}\right)_{v_{ds}=constant} \tag{2.100}$$

The value of transconductance, and its dependence on frequency relate closely to gain that can be realised from a circuit incorporating the MESFET and the dependence of this gain on the frequency. Both I_{ds} and transconductance are greatly influenced by the device's geometrical dimensions. The larger the gate length and gate width, the larger both drain-source current and transconductance become.

Self-assessment Problem

2.26 What do you think is the relationship between device geometry (i.e. physical dimensions) and the available levels of drain current and transconductance for high power devices?

2.9.2.2 Capacitance-voltage characteristics

On changing the bias voltage applied to the gate terminal, the charge (Q_g) in the channel beneath this terminal suffers a redistribution. As a consequence, a capacitance appears between gate and source terminals (C_{gs}) and also between gate and drain terminals (C_{gd}). The value of these capacitances depends on the gate-source and gate-drain voltages respectively, thus making them (by definition) non-linear. The gate-source capacitance is defined by:

$$C_{gs} = \left(\frac{\partial Q_g}{\partial V_{gs}}\right)_{v_{gd}=constant} \tag{2.101}$$

and the gate-drain capacitance is defined by:

$$C_{gd} = \left(\frac{\partial Q_g}{\partial V_{gd}}\right)_{v_{gs}=constant} \tag{2.102}$$

The most important MESFET capacitance is C_{gs}. As would be expected, the value of C_{gs} is proportional to gate area. Other things being equal, larger gate width devices therefore have higher gate-source capacitance than the smaller gate width devices. If C_{gs} is high, the MESFET gate input impedance will become low as frequency increases, and at microwave frequencies the gate impedance may effectively become a short-circuit. Since gate terminal voltage acts as a device gain control, a MESFET operating in this regime will no longer be useful as an amplifying device. Small values of the C_{gs} are therefore necessary for MESFET operation as a microwave amplifier.

The term unilateral, in the context of transistors, relates to the existence of output (drain) to input (gate) feedback. A first estimate of how unilateral a device is can be deduced from the magnitude of the S_{12} scattering parameter (see Chapter 3). The smaller the magnitude of S_{12}, the more unilateral the device.

The value of C_{gd} indicates whether or not a device may be unilateral. High values mean that a device will not be unilateral while low values indicate that the device may or may not be unilateral depending on other device parameters. Unilateral devices are required for high frequency applications.

As for C_{gs}, the value of C_{gd} for a given gate length is directly proportional to the gate width (Golio, 1991).

Self-assessment Problem

2.27 One way of implementing a diode consists of shortening the drain and source terminal of a MESFET. Can you imagine a way to implement a voltage controlled capacitance?

2.9.2.3 Small signal equivalent circuit

The most widely used circuit model for the GaAs MESFET is shown in Figure 2.52.

The parasitic inductances, L_g, L_d and L_s, originate in the metallic contacts deposited on the device surface and any associated bonding wires. The value of these elements depends on the layout of the device, as well as the properties of its constituent materials. In most cases L_g is the largest inductance and L_s the smallest.

The origin of the parasitic resistances R_s and R_d, is the ohmic contact of drain and source terminals as well as a small contribution of the resistance due to the active channel. The resistance R_g represents the metalisation resistance resulting from the Schottky gate contact. The capacitances C_{pgi} and C_{pgi} arise from device packaging and may be important when the operating frequency of the MESFET is high. The capacitance C_{ds} accounts for coupling between drain and source terminals and is assumed to be bias independent. The equivalent circuit contains five non-linear components. These are:

C_{gs} used to model gate charge dependence on V_{gs}.
C_{gd} used to model gate charge dependence on V_{gd}.
I_{gs} used to model the gate-source diode.
I_{gd} used to model the gate-drain diode.
I_{ds} governed by V_{gs} and V_{ds} and used to model the primary non-linear element from which the amplification properties of the device are derived.

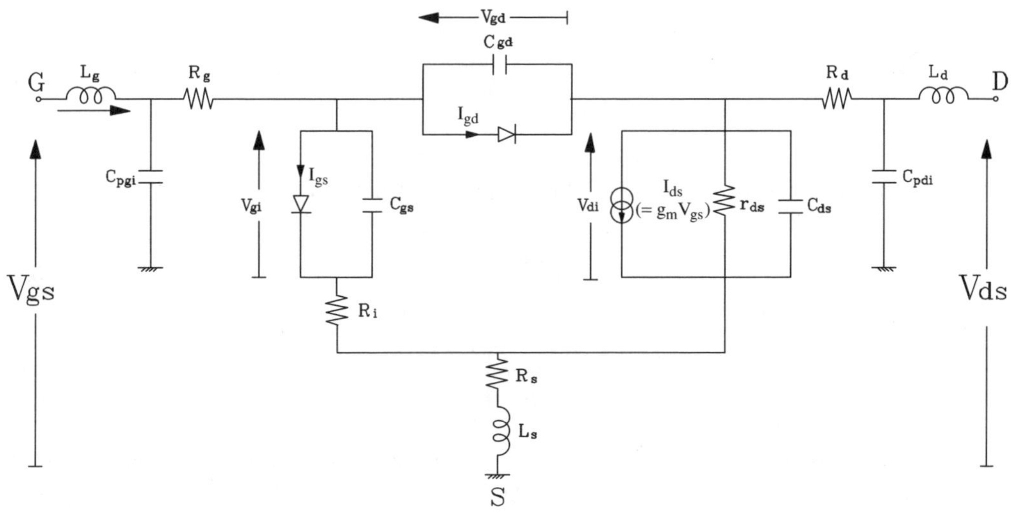

Figure 2.52 Equivalent circuit model for GaAs MESFET

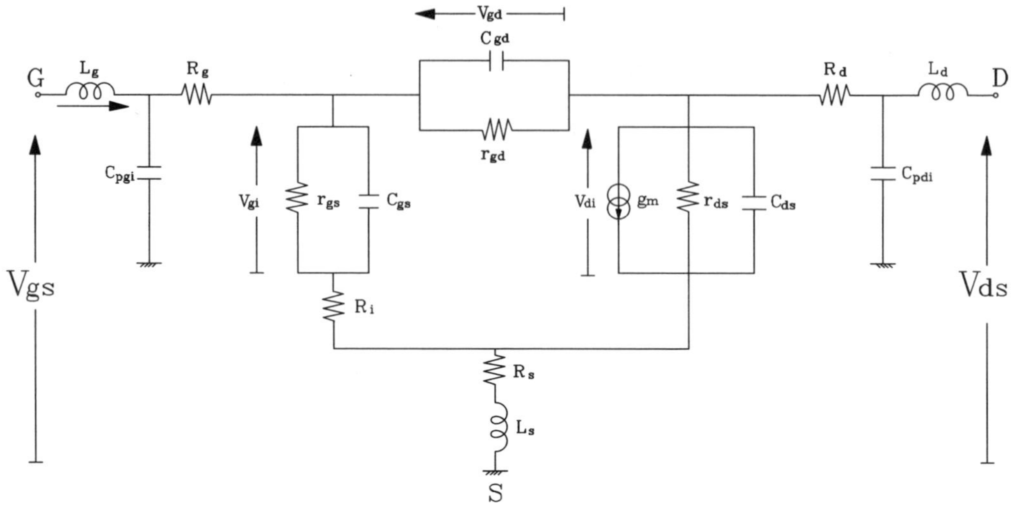

Figure 2.53 Small signal equivalent circuit for GaAs MESFET

For small signals the equivalent circuit of Figure 2.52 can be transformed, for a given bias point, to the linearised model shown in Figure 2.53. Here, C_{gs} and C_{gd} have been fixed at their value appropriate to the chosen bias point, and the non-linear gate-source and gate-drain diodes have been replaced with linear resistors r_{gs} and r_{gd}. The values of these resistors are, of course, those appropriate to the bias conditions, i.e.:

$$r_{gd} = \frac{1}{\left(\dfrac{\partial I_{gd}}{\partial v_{gd}}\right)} \quad (2.103)$$

$$r_{gs} = \frac{1}{\left(\dfrac{\partial I_{gs}}{\partial v_{gs}}\right)} \quad (2.104)$$

The small signal transconductance, g_m, and output conductance, g_{ds}, are given by:

$$g_m = \left(\frac{\partial I_{ds}}{\partial v_{gs}}\right)_{V_{ds}=constant} \quad g(2.105)$$

$$g_{ds} = \left(\frac{\partial I_{ds}}{\partial v_{ds}}\right)_{V_{gs}=constant} \quad g(2.106)$$

The value of g_m is related to the small signal gain at any given bias point. Knowledge of the dependence of g_m on frequency allows the prediction of small signal device gain as a function of operating frequency. Furthermore, the change in g_m with V_{gs} and V_{ds} (i.e. changes in bias point) can be identified.

g_{ds} depends (similarly to g_m) on both frequency and bias point. It is related to the DC output resistance, r_0, of the device by:

$$r_0 = \frac{1}{g_{ds}} \tag{2.107}$$

Another important device parameter is *transconductance time delay*, sometimes referred to more simply as *time delay*. Due to the time it takes for charge to redistribute itself when a voltage gate change occurs, the transconductance is not able to instantaneously follow fast voltage changes. A delay time is therefore added to the transconductance behaviour in order to obtain a more complete representation of the 'gain process'. This delay has to be taken into account at high operating frequencies. A typical value of this delay time for GaAs MESFET devices is between 1 ps and 3 ps.

For microwave operation, the equivalent circuit shown in Figure 2.53 can be obtained from S-parameter measurements, along with linear model extraction techniques developed by several authors [6, 7]. The complexity of these extraction methods is due to the difficulty of separating effects and identifying the value of the different elements when the operating frequency becomes high. As an additional handicap, g_m and g_{ds} have both been found to be frequency dependent [6]. To give an idea of the complexity of the extraction process, consider how the decreasing impedance of C_{ds} can obscure the variation of the transconductance with frequency.

Some second-order effects, e.g. trapping [6, 7, 8] cause a frequency dispersion to appear around tens of kilohertz. This dispersion can be observed, from a macroscopic point of view, as a sudden change in both g_m and g_{ds} from their DC values to their RF values. Several authors propose ways of modelling these effects. These second-order phenomena will not be treated here. It is important to be aware, however, that to correctly design circuits for operation at, or close to, the frequency where the dispersion appears these effects must generally be considered. When designing small signal broadband amplifiers, for example, the amplifier gain is proportional to g_m. If the frequency band of the amplifier to be designed covers the region of tens of kilohertz, care must be taken in the design process because the transconductance may exhibit variation at these frequencies, the gain of the amplifier being modified as a consequence.

In general, all the element values in the equivalent circuit model of a MESFET display a dependence on the geometric dimensions of the device. The relationships between device size and the value of the model elements are known as scaling rules. These rules are useful in estimating the value of the element values of the small signal model when direct measurement of them is difficult or impossible.

Self-assessment Problem

2.28 Calculate the small signal low frequency voltage gain of a GaAs MESFET (using the circuit in Figure 2.53) by ignoring all the parasitic resistors, inductor and capacitors.

2.9.2.4 Large signal equivalent circuit

The circuit shown in Figure 2.52 can be used to model the behaviour of a GaAS MESFET under any practical operating condition. It is therefore valid for DC, RF small signal and RF large signal conditions.

A large signal model consists of the values of the linear elements in the equivalent circuit along with the equations that define the non-linear elements. For the model shown in Figure 2.52 the latter comprise the current of the non-linear source, I_{ds}, as a function of v_{gs} and v_{ds} and the capacitance of the non-linear elements C_{gs} and C_{gd} as a function of their charge control voltages. (The equations defining the non-linear gate-source and gate-drain current sources are implicit in their representation as diodes.)

A variety of models can be found in the literature [6]. The optimum model choice depends on several different criteria such as accuracy, complexity and CPU time required to run the model etc. One of the most widely applied GaAs MESFET models, however, is that due to Curtice [6]. This incorporates an analytical expression for I_{ds} that closely approaches the measured behaviour of this non-linear element.

The general process of extracting a non-linear large signal model consists of fitting the appropriate measurements to a set of expressions, by optimising the expression parameters.

Measuring the gate-source and gate-drain junction currents for a set of forward bias conditions it is possible to characterise the non-linear current sources that characterise these junctions. The forward currents I_{gs} and I_{gd} are usually modelled by the expressions:

$$I_{gs} = I_{nss}(e^{\alpha V_{gs}} - 1) \tag{2.108}$$

$$I_{gd} = I_{nsd}(e^{\alpha V_{gd}} - 1) \tag{2.109}$$

In order to model reverse current breakdown, a diode-like current source is used. The breakdown current, $I_{d,br}$, (present when $v_d < -V_{br}$) is given by:

$$I_{gd,br} = \begin{cases} I_s(e^{-(v_{gs}-V_{br})\alpha}) & \text{for } v_{gd} < -V_{br} \\ 0 & \text{for } v_{gd} > -V_{br} \end{cases} \tag{2.110}$$

where V_{br} is the breakdown voltage.

2.9.2.5 Curtice model

The GaAs MESFET model that is currently most widely known was proposed by Curtice [6] and focuses on the non-linear elements C_{gs}, C_{gd} and I_{ds}.

In the expressions that follow the independent variables are intrinsic (or internal) voltages V_{gi} and V_{di}. To find the dependence of the non-linear elements on V_{gi} and V_{di} the value of the source, drain and gate resistances (R_s, R_d and R_g) must be known. With these resistance values it is possible to evaluate the intrinsic voltages using:

$$V_{gs} = I_g R_g + V_{gi} + (I_g + I_d)R_s \tag{2.111}$$

which gives V_{gi}, and:

$$V_{ds} = I_d R_d + V_{di} + (I_g + I_d)R_s \tag{2.112}$$

which gives V_{di}. (Here we have assumed there is no breakdown effect and only forward gate-source and drain-source currents are present.)

The dependence of I_{ds} on the intrinsic voltages V_{gi} and V_{di} is given by:

$$I_{ds}(V_{gi}, V_{di}) = \beta(V_{gi} - V_{TO})^2(1 + \lambda V_{di}) \tanh(\alpha V_{di}) \tag{2.113}$$

where α, β, λ and V_{TO} are the model parameters. The modelling process for I_{ds} consists of finding the values for the parameters of Equation (2.113) such that the equation best represents the real behaviour (i.e. the measured I/V characteristic) of the device.

The dependence of capacitances C_{gs} and C_{gd} on v_{gi} and v_{di} in the Curtice model, obtained from the classical theory of Schottky junctions, is given by:

$$C_{gs} = C_{gso}\left(1 - \frac{V_{gi}}{V_{bi}}\right)^{-1/2} \tag{2.114}$$

$$C_{gd} = C_{gdo}\left(1 - \frac{V_{di}}{V_{bi}}\right)^{-1/2} \tag{2.115}$$

where C_{gso}, C_{gdo} are the non-linear capacitances at zero bias voltage, V_{bi} is the built-in voltage, and V_{gi}, V_{di} are the intrinsic voltages.

From Equation (2.110) it is possible to obtain analytical expressions for g_m and g_{ds} as a function of the intrinsic voltages. Using Equations (2.105) and (2.106) these expressions are:

$$g_m = \frac{\partial I_{ds}}{\partial V_{gi}} = I_{ds}[2/(V_{gi} - V_{TO})] \tag{2.116}$$

$$g_{ds} = \frac{\partial I_{ds}}{\partial V_i} = \beta(V_{gi} - V_{TO})^2(1 + \lambda V_{di})\{\alpha/[\cosh^2(\alpha V_{di})]\} + \\ + \beta(V_{gi} - V_{TO})^2\lambda \tanh(\alpha V_{di}) \tag{2.117}$$

Many more models for GaAs MESFET devices are described in the literature.

Self-assessment Problem

2.29 How can a MESFET large signal model provide information about harmonic content when injecting a signal of frequency f_o through the gate terminal? (Use the Curtice model and consider the principal non-linearity to be drain current.)

2.9.3 HEMTs

The high electron mobility transistor (HEMT) is a type of field effect transistor. Unlike GaAs MESFETs, however, HEMT devices are heterostructures and as such are able to operate at higher frequencies. Figure 2.54 shows a cross-section of a typical HEMT device. It has three metal contacts at the gate, drain and source terminals. The source and drain

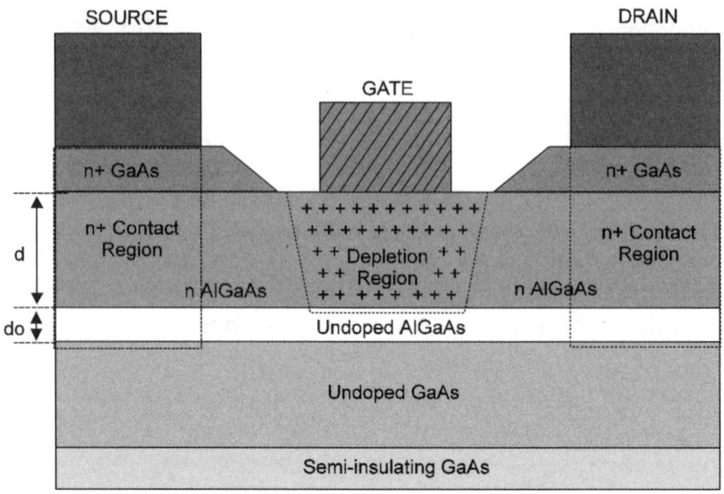

Figure 2.54 Cross-section of a HEMT device

terminals are ohmic contacts. The gate contact constitutes a Schottky junction. In these respects it is identical to a GaAs MESFET.

The improved performance of HEMTs over MESFETs comes about due to the higher value of the electron mobility (and therefore velocity) in these devices. This makes HEMTs suitable for applications where low noise and/or high frequency of operation are required.

Several different HEMT physical structures are possible [6, 7]. Each topology, or structure, has advantages in terms of some specific HEMT feature such as power dissipation capability, maximum operating frequency, noise performance, etc. and many commercial foundries offer an advice service to circuit designers on which HEMT topology will be most appropriate to a give device application.

The most important geometric dimension in an HEMT device (as for the GaAs MESFET) is gate length, which limits the maximum operating frequency of the device. Typical gate lengths for HEMTs are similar to those found in MESFETs. A further feature determining the general behaviour of the HEMT, however, is the thickness of the undoped, and the *n*-type, AlGaAs layers (see Figure 2.54). Typically, the thickness of the *n*-type AlGaAs layer may be between 0.03 and 0.2 μm. A typical thickness for the undoped layer (also known as spacer) might be 0.005 μm.

2.9.3.1 Current-voltage characteristics

From a macroscopic point of view, I/V characteristics of MESFETs and HEMTs are quite similar. The drain-source current (I_{ds}) dependence on gate width observed in MESFETs can also be observed in HEMTs.

A particular HEMT feature can be observed in the I/V output characteristics illustrated in Figure 2.55. As gate-source voltage (for constant v_{ds}) increases beyond a certain point, the corresponding increases in I_{ds} become smaller. This represents a degradation, or compression, of the device's transconductance, g_m (see Figure 2.56). Transconductance compression is an HEMT feature not found in MESFETs.

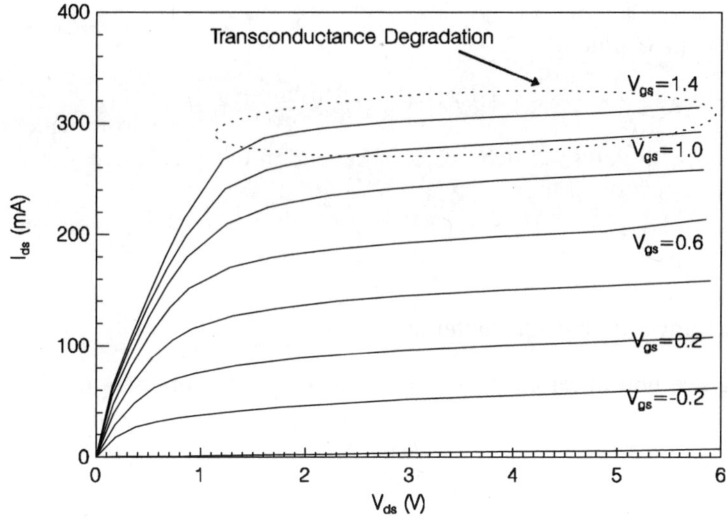

Figure 2.55 Typical I/V curves of a 6 × 150 μm HEMT device

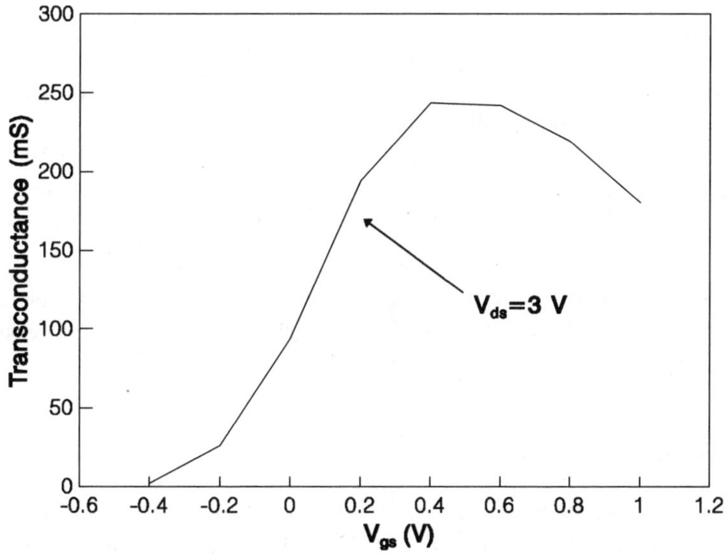

Figure 2.56 Typical g_m versus v_{gs} curves for a 6 × 150 μm HEMT

The same definitions of transconductance and output resistance as used for MESFETs can be applied to HEMTs. Due to its physical structure, however, the output conductances of HEMTs are higher than those observed for MESFET devices. Similar to MESFET devices, HEMT output resistance is inversely proportional to device gate width and directly proportional to gate length. Finally, transconductance is proportional to gate width, and inversely proportional to gate length.

Self-assessment Problem

2.30 Suppose you are using a HEMT device to implement a small signal LNA and, due to the low level of signal to be received you have to provide high gain. Looking at Figures 2.55 and 2.56, comment on the region you would choose in which to bias the device.

2.9.3.2 Capacitance-voltage characteristics

The following two non-linear capacitances, C_{gs} and C_{gd}, are important in HEMT devices:

$$C_{gs} = \left(\frac{\partial Q_g}{\partial V_{gs}} \right)_{v_{gd}=constant} \tag{2.118}$$

$$C_{gd} = \left(\frac{\partial Q_g}{\partial V_{gd}} \right)_{v_{gs}=constant} \tag{2.119}$$

(These definitions are the same as those applied to MESFET devices.) A study of the dependence of C_{gs} and C_{gd} on HEMT dimensions suggests that C_{gs} is proportional to gate area, while C_{gd} is proportional to gate width.

2.9.3.3 Small signal equivalent circuit

The same equivalent circuit model presented for MESFETs can be used for the HEMT (see Figure 2.53). The significance and interpretation of the equivalent circuit elements are the same as for MESFET devices although the element values, and the expressions used to model the non-linear elements, naturally differ.

2.9.3.4 Large signal equivalent circuit

As in the case of the small signal model, the circuit topology for the HEMT large signal model is identical to the corresponding equivalent circuit of a GaAs MESFET (see Figure 2.52). The element values and the expressions used to model the non-linear current sources and capacitances are, naturally, different, however.

As for MESFETs, the process of extracting a non-linear large signal model consists in fitting a set of expressions to a set of measurements by optimising the expressions' coefficients. The forward diode currents are well modelled by:

$$I_{gs} = I_{nss}(e^{\alpha V_{gs}} - 1) \tag{2.120}$$

$$I_{gd} = I_{nsd}(e^{\alpha V_{gd}} - 1) \tag{2.121}$$

and the breakdown current is given by:

$$I_{gd,br} = \begin{cases} I_s(e^{-(v_{gs}-v_{br})}) & \text{for } v_{gd} < -V_{br} \\ 0 & \text{for } v_{gd} > -V_{br} \end{cases} \tag{2.122}$$

where V_{br} is the breakdown voltage.

HEMT transconductance begins to degrade [6, 7], when a particular value of v_{gs}, often denoted by v_{pf}, is reached. The large signal model must take this degradation into account to accurately predict the behaviour of the device. To accurately model this effect, many models have been proposed [6]. As in the case of the MESFET, here we present one model as an example. To establish a point of comparison with the MESFET the example chosen is the HEMT Curtice model proposed by [6].

It is possible to represent the non-linear current source, I_{ds}, for HEMT devices in terms of the MESFET current source model, I_{dsFET}, i.e.:

$$I_{ds} = \begin{cases} I_{dsFET} & \text{for } V_{gi} \leq V_{pf} \\ I_{dsFET} - \dfrac{\xi}{\psi + 1}(V_{gi} - V_{pf})^{\psi+1} \tan(\alpha V_{di})(1 + \lambda V_{di}) & \text{for } V_{gi} > V_{pf} \end{cases} \tag{2.123}$$

$$I_{dsFET} = \begin{cases} 0 & \text{for } V_{gi} \leq V_{p0} \\ \beta(V_{gi} - V_{p0})^2 \tanh(\alpha V_{di})(1 + \lambda V_{di}) & \text{for } V_{gi} > V_{p0} \end{cases} \tag{2.124}$$

and α, β, λ, ξ, ψ, V_{pf} and V_{to} are the non-linear model parameters, V_{pf} being the gate-source voltage for which the transconductance degradation appears.

The intrinsic voltages V_{gi} and V_{di} (see Figure 2.52) may be calculated using:

$$V_{gi} = V_{gs} - I_g R_g - (I_g + I_d)R_s \tag{2.125}$$

$$V_{di} = V_{ds} - I_d R_d - (I_g + I_d)R_s \tag{2.126}$$

The non-linear capacitances, C_{gs} and C_{gs}, are quite different from those observed in MESFET devices. [6], for example, adapts capacitance expressions taken from MEYER's empirical MOSFET model [10] resulting in the following expressions:

$$C_{gs} = \begin{cases} \dfrac{2}{3}\left[1 - \dfrac{(V_{dss} - V_{di})^2}{(2V_{dss} - V_{di})^2}\right] C_G + C_{GS0} & \text{for } V_{di} < V_{dss} \\ \dfrac{2}{3} C_G + C_{GS0} & \text{for } V_{di} \geq V_{dss} \end{cases} \tag{2.127}$$

$$C_{gd} = \begin{cases} \dfrac{2}{3}\left[1 - \dfrac{V_{dss}^2}{(2V_{dss} - V_{di})^2}\right] C_G + C_{GS0} & \text{for } V_{di} < V_{dss} \\ C_{GS0} & \text{for } V_{di} \geq V_{dss} \end{cases} \tag{2.128}$$

where:

$$C_G = \begin{cases} C_{m0}(V_{gs} - V_{TO})^{1/X} & \text{for } V_{gi} > V_{TO} \\ 0 & \text{for } V_{gi} \leq V_{TO} \end{cases} \tag{2.129}$$

and:

$$V_{dss} = \begin{matrix} V_{ds0}\left(1 - \dfrac{V_{gi}}{V_{TO}}\right) & \text{for } V_{gi} > V_{TO} \\ 0 & \text{for } V_{gi} \leq V_{TO} \end{matrix} \tag{2.130}$$

Other approaches to modelling both the drain-source non-linear current source and the non-linear capacitances can be found in the literature, e.g. [6, 7]. The choice of a particular model depends on the application (and often, therefore, the circuit topology), the trade-off between model complexity and accuracy, and the effort required to obtain the model parameters.

Self-assessment Problem

2.31 Do you think the parasitic resistor R_g will affect the use of both MESFETs and HEMTs as low noise amplifiers? Explain your answer.

2.9.4 HBTs

The heterojunction bipolar transistor (HBT) has high transconductance and output resistance, high power handling capability and high breakdown voltage. The use of bipolar transistors in microwave applications is not common due to the limitation on their transition frequency. It can be shown that the transition frequency of a bipolar device depends on the base resistance; the higher the base resistance, the lower the transition frequency. A critical advantage of HBTs compared with BJTs is that they have very high base doping and hole injection into the emitter is suppressed. In this way it is possible to obtain low base resistance combined with wide emitter terminal dimensions and, as a consequence, very high operating frequency can be achieved. This makes these devices attractive for high power microwave and millimetre-wave applications. The principal advantages of HBTs over traditional BJTs, GaAs MESFETs and HEMTs can be summarised as follows:

1. Base resistance can be reduced as a consequence of higher base doping.
2. Base-emitter capacitance is reduced as a result of lower emitter doping.
3. Low transit time values due to the use of AlGaAs/GaAs material.
4. Low output conductance value.
5. High transconductance values.
6. High gain.
7. High breakdown voltages.
8. Low $1/f$ noise due to the absence of second-order effects appearing in GaAs MESFETs and HEMTs.

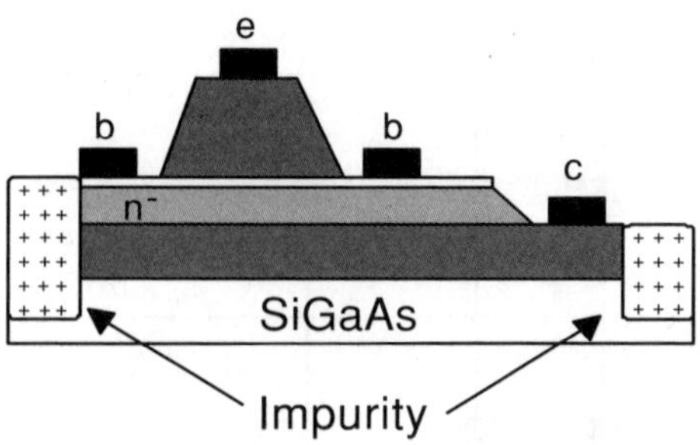

Figure 2.57 Cross-section of a AlGaAs/GaAs device

9. High transition frequencies due to the value of base resistance achieved.
10. High current/power handling capability.

The dominant factors in the rapid progress of HBT devices are improvements in the quality of materials and improvements in epitaxial layer growth techniques (e.g. molecular beam epitaxy and metallic organic chemical vapour deposition). Recently a new technology based on SiGe material has allowed HBT devices to be realised with the same RF and DC properties as III–V group based devices, decreasing the manufacturing cost as well as the technological process complexity.

We now outline a generic HBT model that, at least for first-order analysis, can be applied to both traditional III–V HBTs or to SiGe HBTs, only the element values varying for the different HBT materials.

A cross-section of an AlGaAs/GaAs HBT is shown in Figure 2.57. (In this particular device an InGaAs cap is used to obtain low contact resistance values. Berilium, Be, has been used as material for the base terminal. Base thickness can be reduced using carbon-doping techniques, this being a good way to reduce base resistance, thus achieving higher transition frequency.

Self-assessment Problem

2.32 From a manufacturing point of view do you think SiGe HBT devices present any advantages over MESFETs and HEMTs? (Hint: Think in terms of the integration of analogue and digital systems.)

For the HBT, in contrast to MESFETs and HEMTs, there are not several different equivalent circuit models to predict behaviour under different operating conditions (small signal, large signal, etc.). We will therefore discuss the basic operation of the HBT device, presenting at the same time the Gummel-Poon [8] model that may be extended to represent the main non-linear elements. This model is not, in itself, able to predict some second-order effects such as I/V-curve dependence on device self-heating for high values of base-current bias (important for high-power HBTs) and collector junction breakdown. The Gummel-Poon model is both the original and an interesting approach to HBT performance prediction and is

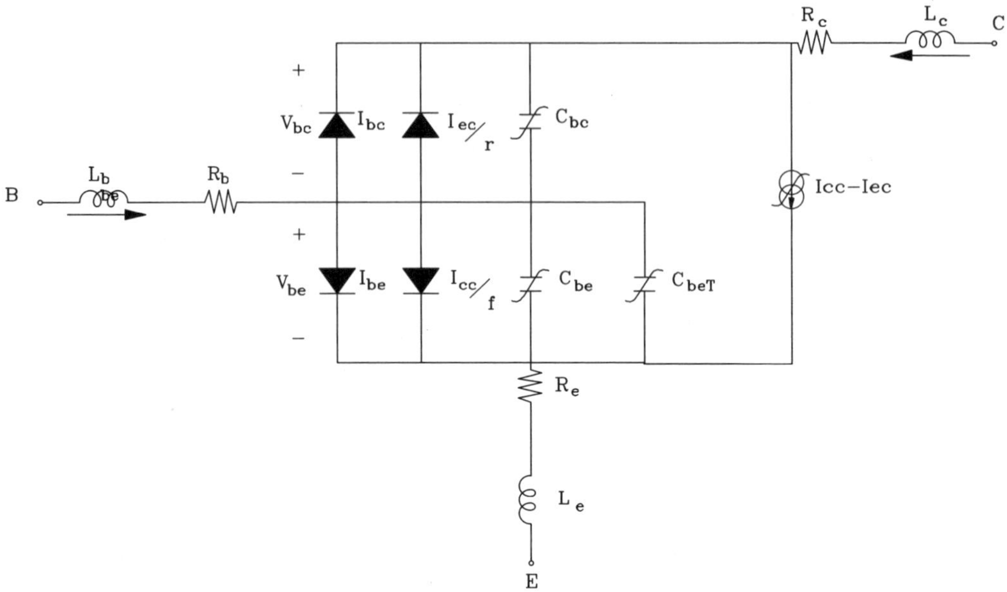

Figure 2.58 Large signal equivalent circuit for HBT

also useful as a starting point for understanding the device's physical behaviour. Other HBT models that attempt to account for effects not predicted by the Gummel-Poon model are discussed in [7].

A complete non-linear equivalent circuit model of an HBT is shown in Figure 2.58. Taking this model as a starting point, it is possible to explain the different operating conditions that can be observed in an HBT device. The study that we are going to carry out, under DC operation, is the same as proposed for BJT devices by other authors [8]. This approach can be justified taking since the important differences between BJT and HBT devices become apparent only at high frequencies (where the superior performance of the HBT over BJT allows its use in microwave applications), the DC behaviour being nearly the same for both devices.

Under forward bias conditions ($V_{be} \geq 0$ and $V_{bc} < 0$) device behaviour is well predicted by the simplified equivalent circuit shown in Figure 2.59.

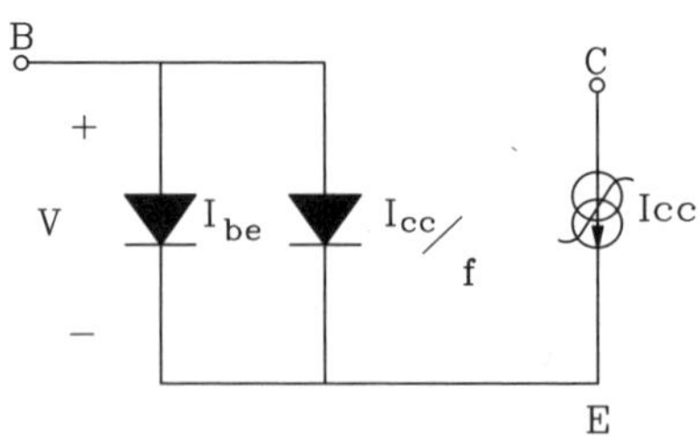

Figure 2.59 DC HBT equivalent circuit under forward bias

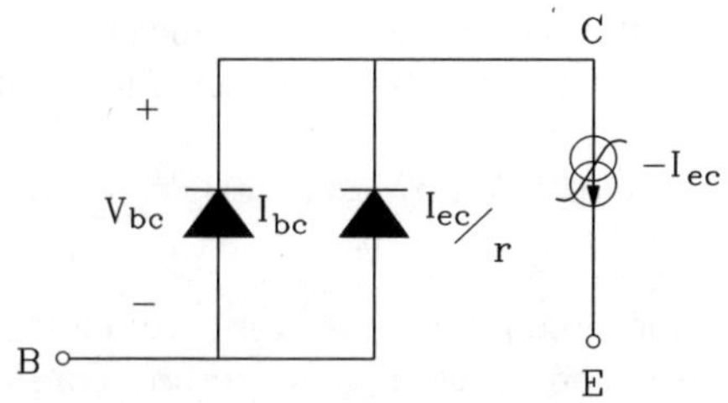

Figure 2.60 DC HBT equivalent circuit under reverse bias

The forward bias DC current sources proposed by Gummel-Poon [7, 8] are given, respectively by:

$$I_{cc} = I_{sf}\left(\exp\left(\frac{V_{be}}{n_f KT}\right) - 1\right) \quad (2.131)$$

and:

$$I_{be} = I_{se}\left(\exp\left(\frac{V_{be}}{n_e KT}\right) - 1\right) \quad (2.132)$$

The I_{cc} current source represents the collector current while base current is represented by two sources, one of them depending on β_f, the forward DC gain of the device.

Under reverse bias conditions device behaviour is predicted by the equivalent circuit shown in Figure 2.60.

The reverse bias DC current sources proposed by Gummel-Poon [8] are, respectively, given by:

$$I_{ec} = I_{sr}\left(\exp\left(\frac{V_{bc}}{n_r KT}\right) - 1\right) \quad (2.133)$$

$$I_{bc} = I_{sc}\left(\exp\left(\frac{V_{bc}}{n_c KT}\right) - 1\right) \quad (2.134)$$

As for MESFETs and HEMTs, the resistances R_b, R_c and R_e play an important role in the HBT modelling process. This is because when we measure the I/V curves of an HBT device we represent the measured collector current versus the external, or applied, voltages V_c and V_b. However, the voltages appearing in Equations (123)–(126) are intrinsic, or internal, voltages. Knowing R_b, R_c and R_e the intrinsic voltages can be calculated as follows:

$$V_{be} = V_b - I_e R_e - I_b R_b \quad (2.135)$$

$$V_{bc} = V_b - V_b + I_c R_c - I_b R_b \quad (2.136)$$

These resistances can be estimated from geometric and material considerations [7, 8] or from device measurements [8]. (The former method requires knowledge of some fabrication process parameters while the latter requires a complex measurement set-up, not available in most microwave laboratories, to carry out the device characterisation.)

Self-assessment Problem

2.33 What type of transistor would you select for a communication system in which minimisation of power consumption was a prime consideration? Explain your answer.

2.9.4.1 Current-voltage characteristics

Figure 2.61 shows a typical (grounded emitter) HBT I/V curve.

The currents in Figure 2.61 are governed by Equations (2.131)–(2.134).

More complex expressions for the HBT currents, taking into account several second-order effects such as self-heating, breakdown, etc., can be found in the literature, e.g. [7]. In most cases, however, the Gummel-Poon model is adequate. (The accuracy of this model will depend, of course, on how the different parameters have been extracted.)

2.9.4.2 Capacitance-voltage characteristics

Two different kinds of capacitances can be distinguished in HBT devices; depletion capacitances and diffusion capacitances.

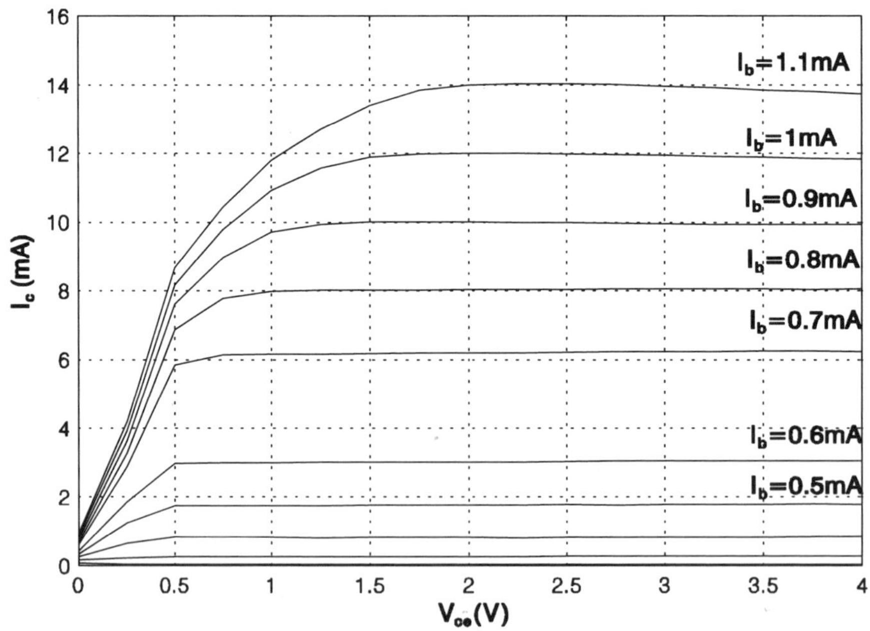

Figure 2.61 Typical I/V curve of a HBT device
Note: DEVICE: H67 HBT from Daimler-Benz I_b = 0.1 mA to 1.1 mA, increment 0.1 mA

2.9.4.2.1 Depletion capacitance

In HBT devices there are two intrinsic depletion, or junction, capacitances; the base-collector capacitance, C_{bc}, and the base-emitter capacitance, C_{be}. As in the case of MESFETs, the origin of these capacitances is in the electron and hole concentration changes as a result of base-emitter and/or base-collector voltage changes. C_{be}, which lowers the maximum operating frequency of the HBT, can be reduced by decreasing either the collector thickness or the base-collector area. The junction capacitances are given by the following expressions:

$$C_{be} = \frac{C_{beo}}{\sqrt{1 - \frac{V_{be}}{V_{built}}}} \tag{2.137}$$

$$C_{be} = \frac{C_{beo}}{\sqrt{1 - \frac{V_{be}}{V_{built}}}} \tag{2.138}$$

where C_{beo} and C_{bco} are the intrinsic capacitances when the applied voltages are zero, and V_{built} is the barrier height voltage.

Self-assessment Problem

2.34 Explain briefly why the capacitance C_{be} may limit HBT device performance depending on the operating frequency.

2.9.4.2.2 Diffusion capacitances

Electrons present in the base are due to diffusion current. The total base electron concentration near the collector defines the collector current. A direct relation between changes in electron concentration and changes in collector current therefore exists which means that a diffusion capacitance can be defined at the base terminal. For large forward bias, diffusion capacitance (in addition to depletion capacitance) must be considered.

A base diffusion time, τ_b, can be calculated using:

$$\tau_b = \frac{W_b^2}{2D_n} \tag{2.139}$$

where W_b is base thickness and D_n is the diffusion constant for electrons in the base. For the collector, we can define a transit time given by:

$$\tau_c = \frac{X}{2v_{sat}} \tag{2.140}$$

where X is the collector thickness and v_{sat} is the electron saturate velocity. The transition frequency, f_t, of the HBT is then given [7] by:

$$f_t = \frac{1}{2\pi\tau_{ec}} \tag{2.141}$$

where:

$$\tau_{ec} = r_e + (C_{be} + C_{bc}) + \tau_b + \tau_c + C_{bc}(R_c + R_e) \qquad (2.142)$$

and the emitter junction resistance, r_e, in Equation (2.142) is:

$$r_e = \frac{\partial V_{be}}{\partial I_e} \approx \frac{n_f KT}{I_e} \qquad (2.143)$$

The base-emitter diffusion capacitance can be calculated from:

$$C_{bet} = \frac{\tau_c + \tau_b}{r_e} \qquad (2.144)$$

where:

$$\tau_b = \frac{W_b^2}{2D_n} + \tau_{bX}\left(1 - \frac{I_c^*}{I_c}\right) \qquad (2.145)$$

and I_c^* is the value of collector current corresponding to the minimum of the τ_{ec} versus $1/I_c$ plot. (N.B. the (correction) term containing τ_{bx} is only applied when $I_c > I_c^*$.)

The above equations express an increase of τ_{ec} for low values of $1/I_c$ due to the hole and electron injection into the collector. This is known as the *base widening effect* and can be modelled by replacing the base width by an effective width equal to $W_b + X$ where X is collector thickness. This results in an effective diffusion capacitance that can be calculated from Equation (2.144).

2.9.4.3 Small signal equivalent circuit

Figure 2.62 shows the small signal equivalent circuit of an HBT based on a one-dimensional transistor model [7].

Other equivalent circuit topologies that model the small signal behaviour of the transistor are possible and some of these, extracted from geometrical parameters and manufacturing information, are quite different from those applied to BJTs. Such models may approximate the device performance very well including second-order effects not predicted by the equivalent circuit of Figure 2.62.

The admittance parameters [Y] of the small signal equivalent circuit of Figure 2.62 can be calculated analytically and, once known, these can be transformed into scattering (s-) parameters (see Section 3.8.2, Chapter 3). The value of the different elements of the equivalent small signal circuit can then be found using linear extraction techniques [7]. The values obtained can be refined by applying an optimisation process to improve the model's accuracy (i.e. make the model's S-parameters match those of the device which would be measured more closely).

A wide variety of small signal HBT models can be found in the literature, e.g. [7, 8]. The appropriate choice of model depends on the application and on the modelling facilities available by the user. The trade-off between the simplicity of the extraction process and the simplicity with which the final model can be used may also be a consideration.

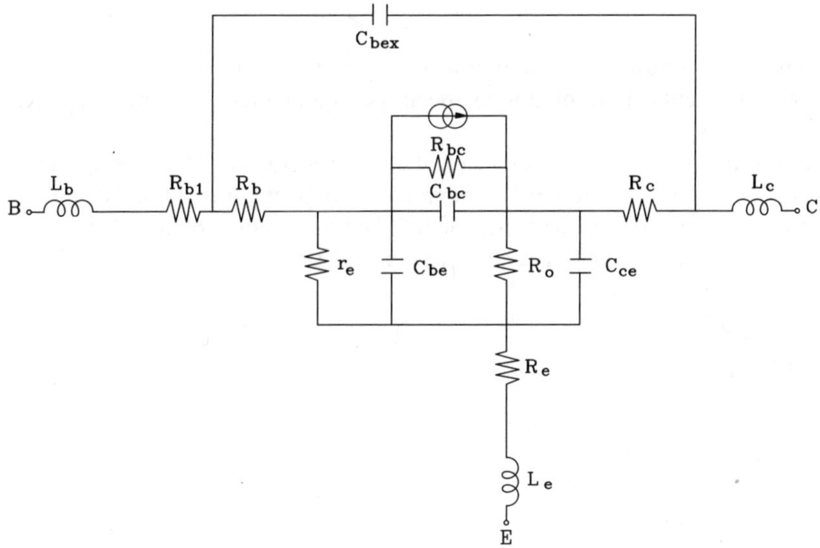

Figure 2.62 Small signal HBT equivalent circuit model

2.9.4.4 Large signal equivalent circuit

The most widely used large signal circuit model for the HBT is that shown in Figure 2.58. This model is implemented in most of the non-linear circuit simulation packages (e.g. MDS, LIBRA, PSPICE). Other model topologies have been proposed in the literature, e.g. [7]. The procedure to extract the large signal model is the same as for MESFETs and HEMTs. Forward and reverse bias current source expressions, found from a set of forward and reverse bias DC measurements [7, 8] are given by Equations (2.125)–(2.128).

Since the final DC model must predict the collector current I/V curves, a parameter refinement procedure involving optimisation methods [7] may be necessary. In this case a set of I_c versus V_c curves (emitter grounded) with I_b as parameter are measured experimentally, along with V_b. From the circuit in Figure 2.58 and Equations (2.125)–(2.128) the following current balance expressions can be found:

$$I_c = I_{sf}\left[\exp\left(\frac{qV_{be}}{n_f KT}\right) - 1\right] - I_{sr}\left[1 + \frac{1}{\beta_r}\right]\left[\exp\left(\frac{qV_{bc}}{n_r KT}\right) - 1\right] - I_{sc}\left[\exp\left(\frac{qV_{bc}}{n_c KT}\right) - 1\right] \quad (2.146)$$

$$I_b = \frac{I_{sf}}{\beta_f}\left[\exp\left(\frac{qV_{be}}{n_f KT}\right) - 1\right] + I_{se}\left[\exp\left(\frac{qV_{be}}{n_e KT}\right) - 1\right] + \frac{I_{sr}}{\beta_r}\left[\exp\left(\frac{qV_{bc}}{n_c KT}\right) - 1\right] + I_{sc}\left[\exp\left(\frac{qV_{bc}}{n_c KT}\right) - 1\right] \quad (2.147)$$

The internal voltages V_{be} and V_{bc} are calculated from the external V_b and V_c voltages using:

$$V_{be} = V_b - (I_b + I_c)R_e - I_b R_b \quad (2.148)$$

$$V_{bc} = V_b - V_c - I_b R_b \tag{2.149}$$

It is then possible to optimise the value of the parameters in Equations (2.146) and (2.147) to fit the measurements, thus obtaining the large signal model of the non-linear current source, I_c.

In the case of the non-linear capacitances, the process consists of choosing the parameters of Equations (2.137)–(2.145) to get good agreement between measurements and model predictions. In this process we take as experimental data the results obtained from the extraction of the HBT small signal model at each different bias point, as described for MESFET's in Section 2.9.4.3.

2.10 Problems

2.1 What are the phenomena giving rise to the two parallel capacitors in the equivalent circuit model of the PN diode?

2.2 Calculate the conversion losses in a varactor multiplier with input frequency 4 GHz. The diode parameters are $I_0 = 10^{-12}$ A, $\alpha = 40$ V^{-1} and $\gamma = 0.33$.

2.3 What are the phenomena giving rise to the formation of the potential barrier in the Schottky diode?

2.4 A Schottky diode, connected in parallel across a transmission line, is used for the detection of a microwave signal at 9 GHz. The diode characteristics are $I_0 = 10^{-10}$ A, $\alpha = 40$ V^{-1}. The amplitude of the microwave signal is 7 mV. Calculate the amplitude of the detected DC current.

2.5 What are the three main effects resulting from the presence of a long intrinsic region embedded in a PIN diode?

2.6 Obtain the possible phase-shift values between input and output provided by the circuit of Figure 2.36, when three PIN diodes are used with transmission line sections of length: $l_x = \lambda g/10$, $l_1 = 3\lambda g/20$, $l_3 = \lambda g/4$.

2.7 What should be the length of a Gunn diode operating in dipole layer mode at 10 GHz?

2.8 A switch is manufactured by shunting a PIN diode across a 50 Ω transmission line. The diode has a length $d = 100$ μm, an average carrier lifetime $\tau = 6$ μs and a mobility $\mu = 600$ cm^2 V^{-1} s^{-1}. The simplified model includes a variable resistor, along with an intrinsic capacitor, $C_I = 0.1$ pF, and a loss resistance, $R_s = 0.5$ Ω. Calculate: (a) the insertion losses for a bias current $I_d = 0$ mA; and (b) the isolation for a bias current $I_d = 12$ mA.

2.9 A Gunn diode has the simplified non-linear model given by a non-linear conductance $G(V) = -15 + 2V$ mS and a parallel capacitance value $C = 2$ pF. Design: (a) a stable negative resistance amplifier at 9 GHz; and (b) a stable free-running oscillator with maximum output power at 9 GHz.

2.10 Sometimes, when designing mixers for high frequency communication systems, GaAs MESFETs are employed as resistive elements. Given that the most common mixer structure includes diodes as non-linear elements, explain how to bias a GaAs MESFET to be used in the linear region as a resistor. (NB Gate-source and gate-drain junctions are Schottky junctions. In the I/V GaAs MESFET curves three, well-differentiated, operating regions can be distinguished.)

2.11 Suppose you are designing a single stage small-signal amplifier based on a GaAs MESFET device. Using the small-signal equivalent circuit MESFET model and considering $R_i = R_s = 0\ \Omega$, $L_s = 0$ H, discuss qualitatively the elements of the equivalent circuit responsible for the gain degradation of the amplifier with increasing the frequency. What is the influence of R_s and L_s in the particular case of $R_s \neq 0$ and/or $L_s \neq 0$ (maintaining $R_i = 0$), on the behaviour of the device as an amplifier?

2.12 Consider the MESFET equivalent circuit model. The I_{ds} parameters for the Curtice model presented for a GaAs MESFET device (Equation (2.113)) are $\beta = 0.12$ mA, $V_{TO} = -2.5$ V, $\lambda = 0.4 \times 10^{-2}$ V^{-1} and $\alpha = 2.34$ V^{-1}. The C_{gs} and C_{gd} non-linear capacitances (Equations (2.114) and (2.115)) are known with $C_{gso} = 2.3 \times 10^{-11}$ F, $C_{gdo} = 0.02 \times 10^{-12}$ F and $V_{bi} = 0.7$ V. The rest of the elements can be assumed to be open circuits (capacitances) or short circuits (resistances and inductances). For a gate-source bias of $v_{gs} = -0.25$ V and a drain-source bias of $v_{ds} = 3$ V, calculate the small-signal equivalent circuit model of the device. (Note that neither the forward gate-source current nor the breakdown effect need be considered.)

2.13 When designing oscillators for digital communications systems, a critical figure of merit is the oscillator phase noise. Analogue oscillators are based on transistors as the active devices responsible for oscillation. The phase noise can be viewed as a low frequency noise, very near to the carrier, that degrades the device behaviour. Which of the three transistor types discussed in this chapter would be most suitable for application in a reference oscillator? Explain your answer. (Hint: think about the flicker, i.e. $1/f$, noise.)

References

[1] S.M. Sze, *Physics of Semiconductor Devices*, John Wiley & Sons, New York, 1981.

[2] A. Vapaille and R. Castagne, *Dispositifs et circuits intégrés semiconducteurs*, Dunod Université, Bordas, Paris, 1990.

[3] K. Chang, *Microwave Solid-State Circuits and Applications*, Wiley Series in Microwave and Optical Engineering, John Wiley & Sons, New York, 1994.

[4] P.F. Combes, J. Graffeuil and J.F. Sautereau, *Microwave Components, Devices and Active Circuits*, John Wiley & Sons, New York, 1988.

[5] E.A. Wolff and R. Kaul, *Microwave Engineering and System Applications*, John Wiley & Sons, New York, 1988.

[6] J.M. Golio, *Microwave MESFETs and HEMTs*, Artech House, Norwood, MA, 1991.

[7] R. Anholt, *Electrical and Thermal Characterization of MESFET'S, HEMT's and BT's*, Norwood, MA, Artech House, 1995.

[8] R.S. Muller and T.I. Kamins, *Device Electronics for Integrated Circuits*, John Wiley & Sons, Inc., New York, 1982.

[9] K.V. Shalimova, *Semiconductors Physics*, MIR, 1975.

[10] J. Meyer, 'MOS models and circuit simulation', *RCA Rev*, vol. 32, March 1971, pp. 42–63.

3

Signal Transmission, Network Methods and Impedance Matching

N. J. McEwan, T. C. Edwards, D. Dernikas and I. A. Glover

3.1 Introduction

Signal transmission, network methods and impedance matching are all fundamental topics in RF and microwave engineering. Signals must be transmitted between devices such as mixers, amplifiers, filters and antennas, and at frequencies where wavelength is comparable to, or shorter than, the separation of these devices, transmission line theory is required to design the connecting conductors properly. There are several technological implementations of transmission lines. The most familiar, in the RF context, is coaxial cable and this technology is still important (with others) where long and/or flexible lines are required. The most important transmission line for shorter distances, where rigidity is not a disadvantage, is microstrip and this technology is, therefore, given particular attention in this chapter.

Network methods refer to the collection of mathematical models that relate the electrical quantities at the ports (inputs and outputs) of a device. The device may be passive (such as a piece of transmission line) or active (such as an amplifier). Usually, but not always, devices have two ports: an input and an output. Some two-port descriptions may be familiar from other applications, e.g. h-parameters for lower frequency electronics and ABCD-parameters for power systems. The two-port description used at RF and microwave frequencies employs s-parameters and it is these on which this chapter therefore concentrates. There are good practical reasons for adopting different parameter sets for different applications (to do with ease of measurement and/or ease of use) but all such sets contain equivalent information about the device and translation between sets is straightforward. The utility of network parameters is that they characterise the systems aspects of a device or subsystem completely – but free from the potentially distracting details of how the device or subsystem works.

Impedance matching (or impedance compensation) is important because it affects the way devices interact when they are connected together. It consists of altering the input or output impedance of a device to make it more compatible in some way with another device to which it is to be connected. It may be designed to realise one of several quite different objectives either at a single frequency or over a band of frequencies (e.g. minimising reflected

Microwave Devices, Circuits and Subsystems for Communications Engineering Edited by I. A. Glover, S. R. Pennock and P. R. Shepherd

power, maximising power transfer, maximising gain or minimising noise figure). Alternatively, matching may be designed to achieve some satisfactory compromise between more than one of these objectives.

Loosely speaking, signal transmission is the process by which signals are transferred by transmission lines from one device another, network methods describe the change in a signal as it enters, traverses and leaves a device and impedance matching is the technique used to optimise the overall desired characteristics of the device and its associated input and output transmission lines. This collection of technologies and techniques represents a basic tool kit for the RF and microwave design engineer.

3.2 Transmission Lines: General Considerations

A transmission line is a structure that is used to guide electromagnetic waves along its length. The most obvious practical purpose is either to transport power from place to place, as, for example, in power cables, such as overhead lines on pylons, or to transport information – but to transmit information, some energy must be transported in any case.

In RF engineering, a third and extremely important function is as circuit elements, for example, in impedance transforming networks, or in filters. This function is based on the ability of transmission lines to store energy, as more familiar circuit elements such as inductors and capacitors do.

3.2.1 Structural Classification

Transmission lines have an immense variety of forms, and it is important to realise that they need not conform to elementary ideas of an electric circuit. Here we are restricting ourselves to a specialised subset, which can be at least partially understood from a circuit point of view. Before focusing on this, we need to set this group within the context of more general structures, and to see the features that distinguish it.

A transmission line does not need to have a uniform cross-sectional structure along its length. There are structures, for example, that use periodic corrugations to guide waves along their surface. Our first stage of specialisation, therefore, is to consider only structures that are longitudinally homogeneous. This still leaves, however, a great variety.

Consider Figure 3.1 that sets out examples of important transmission lines classified according to the number of conductors they contain, and according to the general class of electromagnetic wave or propagation 'mode' that they support.

The first example (Figure 3.1(1)) has no conductors at all – it is just a rod of dielectric, but it can still trap and guide an electromagnetic wave. This is extremely important practically in the form of an optical fibre. It can also be used at 'high' radio frequencies, i.e. in microwave or millimetre-wave bands, when it would be referred to as a 'dielectric waveguide'. There is no very obvious way we could apply concepts like voltage and current to this structure.

In Figure 3.1(2) we have a transmission line with only one conductor – a conventional rectangular waveguide. In Figure 3.1(3) we see a more modern 'finline' or 'E-plane' structure. Here there is a central section with a printed conductor pattern, lending itself to the production of a microwave integrated circuit. This is considered an attractive structure for work at millimetric frequencies. Notice that these still do not much resemble the simple idea of

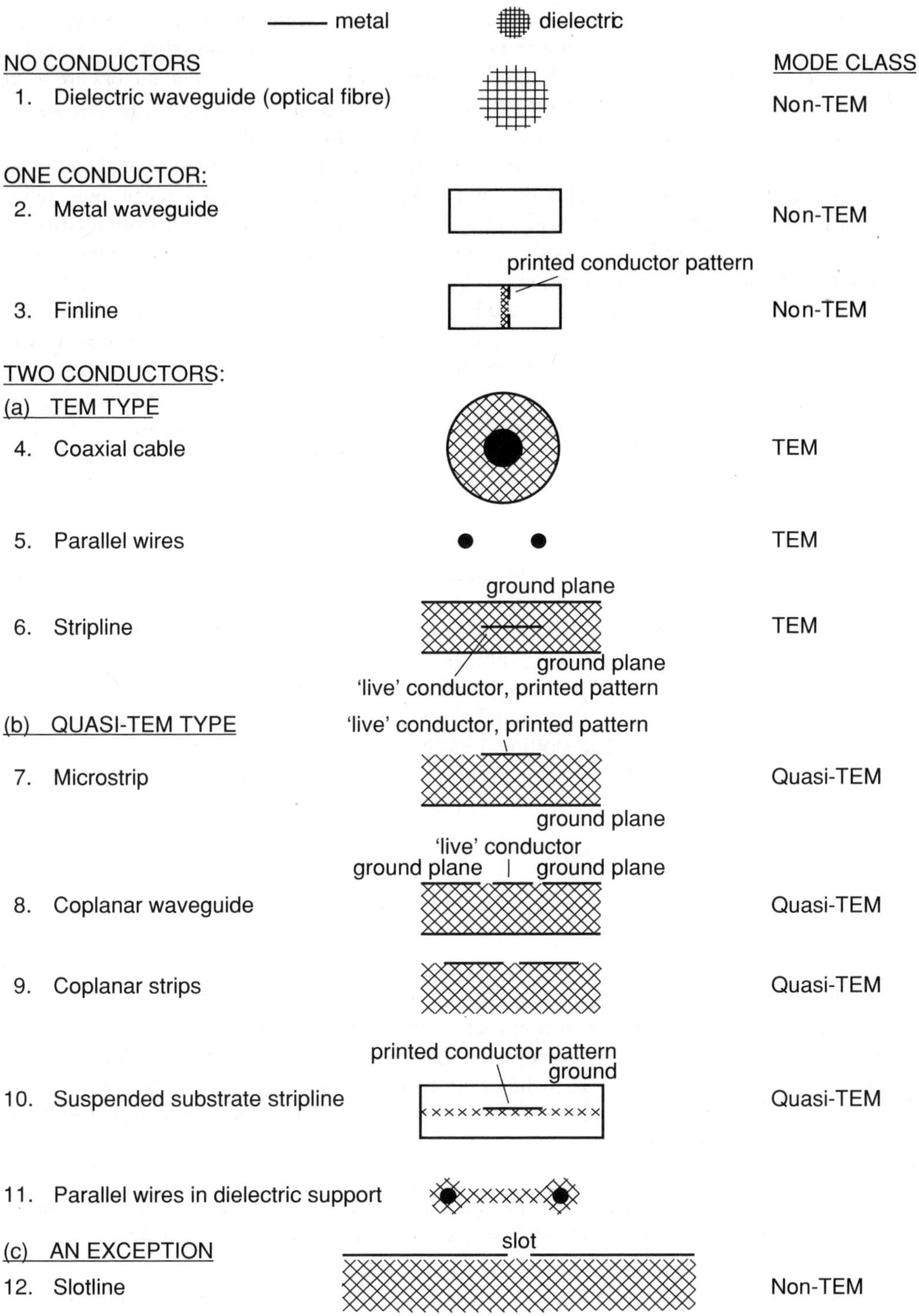

Figure 3.1 Classes of transmission line

a circuit – they are both *short-circuit* at DC – which is connected with their description as non-TEM (transverse electromagnetic) structures. All the remaining transmission line examples have been classified as two conductor lines. (Sometimes more than two conductors are used but they still fall into the same general type.)

All transmission lines have some tendency to radiate into space the power they are meant to be guiding, especially at bends and discontinuities. The enclosed structure of the coaxial cable, Figure 3.1(4), largely prevents this and makes it suitable as a general-purpose radio frequency line. The parallel wire line, Figure 3.1(5), may be seen in old-fashioned open telephone lines, overhead power lines, and sometimes as lines connecting high-power, low and medium frequency radio transmitters to their antennas.

In Figure 3.1(6) we have a structure suitable for microwave integrated circuits, where the 'live' conductor may be given a complex pattern by printed circuit methods. However, it is mechanically awkward to include other electronic components in it and to assemble.

The structures in Figures 3.1(7), (8), (9) and (10) retain the suitability for 'printed' production methods and microwave integrated circuits (MICs) while avoiding the mechanical drawbacks of Figure 3.1(6). Microstrip, Figure 3.1(7), is by far the most widely used, while coplanar waveguide, Figure 3.1(8), is gaining in popularity. The line in Figure 3.1(10) is especially useful in low-loss applications such as filters. The structure shown in Figure 3.1(9) is used only for a few special purposes.

Note that, in the microstrip form of line, it is easy to break the 'live' conductor in order to insert a component in series with it, but if we want to connect a component in shunt between the live conductor and ground, we have to cut or drill the dielectric. Coplanar waveguide and coplanar strips do not suffer from this problem.

The line in Figure 3.1(11) is a variant of the parallel wire line where the mechanical support is built in. It is mainly used for relatively short runs linking radio equipment and antennas at VHF frequencies. The slotline, Figure 3.1(12), can be, and is, used for complex MICs but it remains rather specialised and is not particularly easy to use.

3.2.2 Mode Classes

The following important points can be made about the classification of transmission lines:

1. All the two-conductor lines (except slotline), and only these, are classified as transverse electromagnetic (TEM), or quasi-TEM mode, lines.
2. The lines in this class can be recognised as those that could carry DC excitation and which conform to the idea of a complete circuit with 'go' and 'return' conductors.
3. In the two-conductor family, TEM lines can be recognised as those in which the dielectric constant is *uniform* over the cross-section of the line, while those with a *non-uniform* dielectric are quasi-TEM lines.

(In a few special cases, magnetic materials may also be involved, and here the permeability also has to be uniform for a true TEM line.) A further important point is that:

4. All the TEM and quasi-TEM lines can treated, to a good first approximation at least, by *distributed circuit theory*.

Slotline, Figure 3.1(12), looks as though it should be classed as a quasi-TEM line, and it would support DC excitation. It turns out, however, to be a special case that is not adequately described by quasi-TEM mode theory. (This is because the conductors are nominally infinite in extent. Anticipating something to be discussed later, the magnetic field lines in the slotline mode cannot form complete loops in a transverse plane, because they would have to penetrate the conductors to do so.) This and the other non-TEM lines need more difficult field theory treatments that are beyond the scope of this text.

3.3 The Two-Conductor Transmission Line: Revision of Distributed Circuit Theory

The physical reality of the waves on a transmission line involves fields continuously distributed in the space between the conductors, and a current density continuously distributed over the conductors. Distributed circuit theory is a simpler way of analysing the line. It has the advantage of avoiding field theory, and using circuit concepts, that are easier to understand. It is also useful in providing insight into the quasi-TEM lines for which the exact theory is rather difficult.

Distributed circuit theory makes a number of assumptions, namely:

1. The voltage, v, across the line is a well-defined quantity at any transverse plane along its length
2. The distributed parameters:

   Inductance L
   Capacitance C
   Resistance R
   Conductance G

   are all specified *per unit length* and assumed to be well defined and to make sense physically.
3. The current density can be integrated over a conductor to find the total, or 'lumped', longitudinal current, i, at any point on the line. It is assumed that v and i, together with the distributed parameters, provide a sufficient basis for analysis without worrying about the detailed structure of the fields.

In treating junctions between different lines, or between lines and other components, we usually also assume that:

4. Kirchhoff's laws are obeyed at junctions; this is a good approximation if the transverse line dimensions are electrically small, i.e. much less than a wavelength. (Longitudinal dimensions can be as large as one wants.)

It is important to realise that these are only assumptions, even though they may at first sight seem self-evident. (1), (2) and (3) can, in fact, be shown to be exactly justifiable for ideal TEM lines, while for quasi-TEM lines they are only approximations, valid for moderate frequencies, but nevertheless very useful.

Parameter C, which has units of farad per metre (F/m), can be worked out as just the capacitance found when DC is applied across the line. L, R, and G have units of henry per metre (H/m), ohm per metre (Ω/m) and siemen per metre (S/m), respectively.

3.3.1 The Differential Equations and Wave Solutions

Having made the assumptions stated in Section 3.3, we now take an infinitesimal section of the line with length δx, draw an equivalent circuit for it as shown in Figure 3.2, and analyse it using normal circuit concepts.

The voltage, v, and current, i, on the line are functions of both position and time:

$$v = v(x,t) \tag{3.1}$$

$$i = i(x,t) \tag{3.2}$$

The voltage across the infinitesimal section of line (with positive on the left) is $-(\partial v/\partial x)\delta x$ and this can be equated to the voltage developed across the series resistance and inductance. The current leaving the section on the right is smaller than that entering on the left by an amount $-(\partial i/\partial x)\delta x$, and this must be equated to the current flowing through the shunt conductance and shunt capacitance. The following two equations can thus be deduced:

$$\frac{\partial v}{\partial x}\delta x = -(R\delta x)i - (L\delta x)\frac{\partial i}{\partial t} \tag{3.3}$$

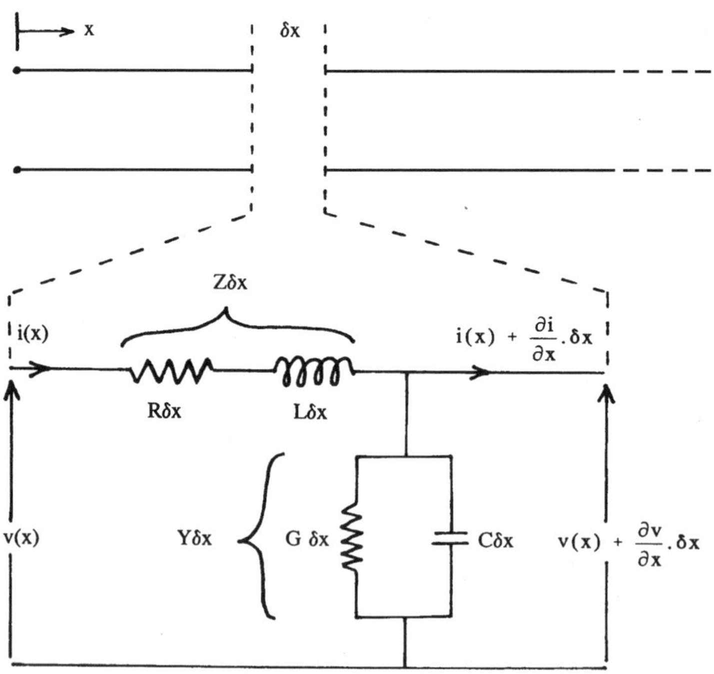

Figure 3.2 Lumped parameter model of an elemental length of transmission line

$$\frac{\partial i}{\partial x}\delta x = -(G\delta x)v - (C\delta x)\frac{\partial v}{\partial t} \tag{3.4}$$

Dividing through by δx:

$$\frac{\partial v}{\partial x} = -Ri - L\frac{\partial i}{\partial t} \tag{3.5}$$

$$\frac{\partial i}{\partial x} = -Gv - C\frac{\partial v}{\partial t} \tag{3.6}$$

To progress further with these equations, we have to consider a single harmonic (i.e. sinusoidally time varying) component of i and v. For such a *single frequency* component we can write:

$$v = Ve^{j\omega t} \tag{3.7}$$

$$i = Ie^{j\omega t} \tag{3.8}$$

Note that we are now using the exponential phasor notational convention of dropping the 'real part' sign. V and I are now complex but are dependent on position only, i.e.:

$$V = V(x) \tag{3.9}$$

$$I = I(x) \tag{3.10}$$

Differentiating with respect to x and t:

$$\frac{\partial v}{\partial x} = \frac{dV(x)}{dx}e^{j\omega t} \tag{3.11}$$

$$\frac{\partial i}{\partial x} = \frac{dI(x)}{dx}e^{j\omega t} \tag{3.12}$$

$$\frac{\partial v}{\partial t} = j\omega V(x)e^{j\omega t} \tag{3.13}$$

$$\frac{\partial i}{\partial t} = j\omega I(x)e^{j\omega t} \tag{3.14}$$

Substituting Equations (3.11)–(3.14) into Equations (3.5)–(3.8) we have:

$$\frac{dV}{dx}e^{j\omega t} = -RIe^{j\omega t} - Lj\omega Ie^{j\omega t} \tag{3.15}$$

$$\frac{dI}{dx}e^{j\omega t} = -GVe^{j\omega t} - Cj\omega Ve^{j\omega t} \tag{3.16}$$

Finally, we can divide through by $e^{j\omega t}$ to obtain two equations that are vital to understanding wave propagation on the line:

$$\frac{dV}{dx} = -(R + j\omega L)I = -ZI \tag{3.17}$$

$$\frac{dI}{dx} = -(G + j\omega C)V = -YV \tag{3.18}$$

Z is the series impedance per unit length, and Y the shunt admittance per unit length. Differentiating with respect to x:

$$\frac{d^2V}{dx^2} = -(R + j\omega L)\frac{dI}{dx} = (R + j\omega L)(G + j\omega C)V \tag{3.19}$$

$$\frac{d^2I}{dx^2} = -(G + j\omega C)\frac{dV}{dx} = (R + j\omega L)(G + j\omega C)I \tag{3.20}$$

Each of these is a standard differential equation, the solution for V being of the form:

$$V(x) = V_1 e^{-\gamma x} + V_2 e^{\gamma x} \tag{3.21}$$

where V_1, V_2 and γ are suitable constants. It can be easily verified by direct substitution that:

$$\gamma = \sqrt{(R + j\omega L)(G + j\omega C)} = \sqrt{ZY} \tag{3.22}$$

γ is the *complex propagation constant*, and is commonly written as $\gamma = \alpha + j\beta$ where α is the *attenuation constant* (in neper m^{-1}) and β is the *phase constant* (in radian m^{-1}).

Self-assessment Problems

3.1 What is d/dx of the function $Ve^{\gamma x}$, where V, γ are constants? Now differentiate again and substitute to show that a function of this form can satisfy the equation in d^2V/dx^2, given the stated value for γ. Why does the solution still work if we replace γ by $-\gamma$?

3.2 If a quantity changes by 1.0 neper (–1.0 neper), it has increased by a factor of e (decreased by the factor $1/e$). Prove that a 1.0 neper change of voltage is equivalent to 8.686 decibels (to 3 decimal places).

3.3.2 Characteristic Impedance

Differentiating Equation (3.21) with respect to x and equating to Equation (3.17):

$$\frac{dV}{dx} = -\gamma V_1 e^{-\gamma x} + \gamma V_2 e^{\gamma x} = -ZI \tag{3.23}$$

Therefore:

$$I(x) = \frac{\gamma}{Z} V_1 e^{-\gamma x} - \frac{\gamma}{Z} V_2 e^{\gamma x} \tag{3.24}$$

Now, from Equation (3.22):

$$\frac{\gamma}{Z} = \frac{\sqrt{ZY}}{Z} = \sqrt{\frac{Y}{Z}} \tag{3.25}$$

and writing out Y and Z in terms of the parameters G, C, R, and L we have:

$$\frac{\gamma}{Z} = \sqrt{\frac{G + j\omega C}{R + j\omega L}} \tag{3.26}$$

The reciprocal of this quantity is called the *characteristic impedance* of the line, usually written Z_0, i.e.:

$$I(x) = \frac{V_1 e^{-\gamma x} - V_2 e^{\gamma x}}{Z_0} \tag{3.27}$$

V_1 and V_2 are constants (independent of x) determined by boundary conditions, i.e. by the terminating impedance and voltage source at the end(s) of the line. They have dimensions of volts.

V_1 is the complex coefficient of a *forward wave*, referenced to the specified origin $x = 0$, and V_2 is the coefficient of a *reverse wave*. The characteristic impedance Z_0 is *not* the ratio of total voltage over total current on the line at any point. The reason for this is the minus sign that appears in Equation (3.27). This equation shows, as we would expect, that the current is reversed for the reverse travelling wave. The characteristic impedance can, therefore, be defined as the ratio of voltage over current for *either the forward wave or reverse wave considered alone* (remembering, of course, the change of current sign for the reverse wave). Thus we have:

$$Z_0 = \sqrt{\frac{R + j\omega L}{G + j\omega C}} \tag{3.28}$$

3.4 Loss, Dispersion, Phase and Group Velocity

We have mathematical expressions for the forward and reverse travelling waves, but it is important to be able to visualise what these mean in terms of waveforms in space and time. We have:

$$v(x,t) = V_1 e^{j\omega t} e^{-\gamma x} + V_2 e^{j\omega t} e^{\gamma x} \tag{3.29}$$

$$i(x,t) = \frac{V_1 e^{j\omega t} e^{-\gamma x} - V_2 e^{j\omega t} e^{\gamma x}}{Z_0} \tag{3.30}$$

Alternatively, rearranging exponential factors:

$$v(x,t) = V_1 e^{-\alpha x} e^{j(\omega t-\beta x)} + V_2 e^{\alpha x} e^{j(\omega t+\beta x)} \tag{3.31}$$

$$i(x,t) = \frac{V_1 e^{-\alpha x} e^{j(\omega t-\beta x)} - V_2 e^{\alpha x} e^{j(\omega t+\beta x)}}{Z_0} \tag{3.32}$$

$e^{j(\omega t \mp \beta x)}$ in Equations (3.31) and (3.32) represents a wave travelling in the $\pm x$ direction and $e^{\mp \alpha x}$ represents decreasing (increasing) amplitude with increasing x.

To examine the behaviour of the space-time function $\text{Re}(e^{j\omega t}e^{-j\beta x})$ let us first put $\alpha = 0$ and look at the expression for the wave with the negative sign before β. The voltage is $V = \text{Re}(V_1 e^{j\omega t}e^{-j\beta x})$. To keep it simple we can assume V_1 is real and take it outside the 'Re' sign. (If it isn't real, the effect is just a simple phase shift anyway.)

Then the voltage is proportional to $\text{Re}(e^{j\omega t}e^{-j\beta x})$ which can of course also be written, using the basic properties of exponentials, as $\text{Re}(e^{j(\omega t-\beta x)})$. We know that this is just $\cos(\omega t - \beta x)$.

Imagine a snapshot of this function at time zero – we would just get $\cos(-\beta x)$ and the graph would be a cosine function drawn along the x-axis.

Now imagine that time advances a little. To get (say) any peak on the cosine function, we must increase x a bit to get the same value of the argument $(\omega t - \beta x)$. In other words, our graph of the function in *space* is moving down the x-axis as time advances.

3.4.1 Phase Velocity

The rate at which the oscillatory pattern appears to move down the x-axis is obviously ω/β. This is what we call the *phase velocity*, the velocity at which any single frequency component appears to be moving in any kind of wave-propagating system. (It will be written v_{phase} or just v_p from now on.)

If, however, we sit at any point in space, i.e. a particular value of x, we just see the function oscillating sinusoidally in time but of course, the *phase* of the oscillation is more and more retarded as we go further down the x-axis. (This is why β is called the phase constant and is measured in rad m^{-1}.)

Self-assessment Problem

3.3 Prove that ω/β gives the phase velocity: if time increases by δt, by how much must x change to keep the argument of the cosine function the same?

3.4.2 Loss

Let us now allow α to be non-zero, so that the voltage will be proportional to $\text{Re}(e^{-\alpha x}e^{j\omega t}e^{-j\beta x})$. The term $e^{-\alpha x}$ is just a real number and we can just take it outside the real-part sign, so this expression is in fact equal to $e^{-\alpha x}\cos(\omega t - \beta x)$.

The term $e^{-\alpha x}$ is an exponentially decaying function of x, and it just multiplies a term that is the same as the sinusoidal wave travelling to the right which we discussed before.

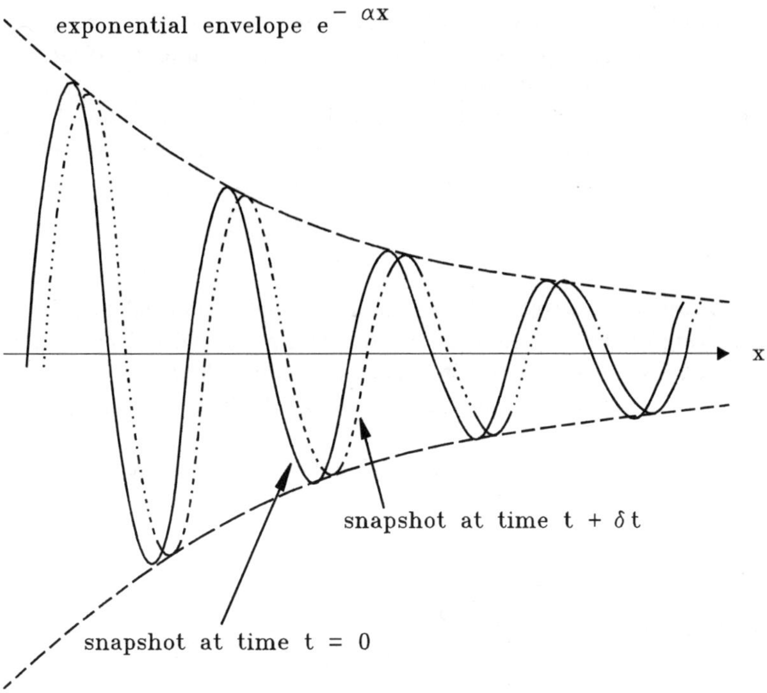

Figure 3.3 Snapshots of a travelling wave with attenuation

We should therefore picture an exponentially decaying envelope, *fixed in space*, describing the amplitude of an oscillation in space and time which travels to the right at the phase velocity, as pictured previously. The sinusoidal variation is graphed 'within' the fixed envelope. This completes our visualisation of the wave, Figure 3.3.

The reduction of the amplitude of the wave as it travels is called *loss* or *attenuation* and it is obviously due to the absorption of energy in the line. It is clearly the line parameters R and G that give rise to this loss, and a line that only had inductance and capacitance would be *lossless*. In some circumstances, discussed later, it is a good approximation to ignore the loss.

3.4.3 Dispersion

So far only waves at a single frequency have been considered. If a more general waveform were to be transmitted, it could be decomposed into its frequency components by Fourier analysis. The waveform could then be observed at a point further down the line, found by adding up the frequency components again, after their amplitudes and phases have been modified by the propagation constant of the line.

If the line has a phase velocity that is frequency independent, or at least frequency independent over some specified frequency band within which we construct our signal, then it is said to be *dispersionless*.

If the line is also lossless over the given band, then all the frequency components travel down the line 'in step' and with no change in amplitude. In this case we would see a time-delayed replica of the original waveform at a point further down the line. The term *dispersion* refers to frequency-dependent phase velocity of a transmission structure, or to the distortion of a transmitted waveform that this produces.

If the line is not lossless and α varies noticeably over the bandwidth of the transmitted waveform, there is an additional distortion (*amplitude distortion*) of the waveform due to the change in relative amplitudes of its frequency components.

3.4.4 Group Velocity

By modulating a continuous wave (CW), i.e. an unmodulated sinusoidal carrier, with a much more slowly varying modulating signal, we can obtain a narrowband signal (i.e. one whose bandwidth is much less than its centre frequency). A simple description can be given of the propagation of such a signal down a dispersive line, which leads to the useful concept of *group velocity*.

The simplest example is obtained by multiplying the carrier wave by a sinusoidal modulating signal at a much lower frequency:

$$v(t) = \cos(\omega_m t)\cos(\omega_c t) \tag{3.33}$$

If you have previously studied modulation theory, you may recognise this waveform as that generated if the modulating signal cos $\omega_m t$ is modulated onto the carrier using double-sideband, suppressed carrier, amplitude modulation. You should be able to visualise the waveform, which looks like the carrier wave cos $\omega_c t$ lying within the amplitude envelope function $|\cos \omega_m t|$, and with 180° phase changes of the carrier at the points where cos $\omega_m t$ changes sign. You should also know that the spectrum just consists of equal amplitude sidebands at the carrier frequency plus and minus the modulating frequency.

We shall write this out explicitly using the trigonometric identity:

$$\cos(a)\cos(b) = \frac{1}{2}\cos(a - b) + \frac{1}{2}\cos(a + b) \tag{3.34}$$

which holds for any angles a, b. Equation (3.33) now reads:

$$\begin{aligned} v(t) &= \frac{1}{2}\cos(\omega_c t - \omega_m t) + \frac{1}{2}\cos(\omega_c t + \omega_m t) \\ &= \frac{1}{2}\cos(\omega_1 t) + \frac{1}{2}\cos(\omega_2 t) \end{aligned} \tag{3.35}$$

where we have written $\omega_c - \omega_m = \omega_1$ for the lower sideband frequency, and $\omega_c + \omega_m = \omega_2$ for the upper sideband frequency.

Now we have two ways of looking at $v(t)$: as a carrier with amplitude modulation, or simply as the sum of two pure frequencies of equal amplitudes. In the first form we cannot easily work out how the waveform would travel down a line, but in the second form it is easy because we know how signals at a single frequency are transmitted.

To keep it simple we shall assume that the voltage is applied to a line that is lossless but may be dispersive, so that the phase constant will be some known function of frequency, $\beta = \beta(\omega)$, but the phase velocity ω/β is not necessarily constant.

Let us write $\beta_2 = \beta(\omega_2)$, $\beta_1 = \beta(\omega_1)$ for the phase constants at the upper and lower sideband frequencies. It will be assumed that the voltage is forced to be equal to $v(t)$ at the input end of the line, which we can assume to extend to infinity, or to end in a matched termination, so that no reflections are present and only waves travelling in the positive direction away from the input point are excited. At a general point x on the line, the voltages in the lower and upper frequency travelling waves on the line can now be written respectively as:

$$\frac{1}{2}\cos(\omega_1 t - \beta_1 t)$$

and:

$$\frac{1}{2}\cos(\omega_2 t - \beta_2 t)$$

We just have two single frequencies superimposed, and each travelling at its appropriate phase velocity. Hence the total voltage at a general point on the line is:

$$v(x,t) = \frac{1}{2}\cos(\omega_1 t - \beta_1 x) + \frac{1}{2}\cos(\omega_2 t - \beta_2 x) \tag{3.36}$$

We can rearrange this again using the same trigonometric identity, in its inverse form:

$$\cos(c) + \cos(d) = 2\cos\left[\frac{1}{2}(c+d)\right]\cos\left[\frac{1}{2}(c-d)\right] \tag{3.37}$$

for any angles c, d if we just let: $c = a + b$, $d = a - b$. The result is:

$$\begin{aligned} v(x,t) &= \cos\left[\frac{1}{2}(\omega_1 + \omega_2)t - \frac{1}{2}(\beta_1 + \beta_2)x\right]\cos\left[\frac{1}{2}(\omega_2 - \omega_1)t - \frac{1}{2}(\beta_2 - \beta_1)x\right] \\ &= \cos\left[\omega_c t - \frac{1}{2}(\beta_1 + \beta_2)x\right]\cos\left[\omega_m t - \frac{1}{2}(\beta_2 - \beta_1)x\right] \end{aligned} \tag{3.38}$$

We can now see that the total wave on the line is the product of two terms, *each of which looks like a waveform travelling down the line*. One term is the rapidly oscillating carrier wave at frequency ω_c, and the other is the modulation envelope at frequency ω_m. If we sit at one point x on the line and watch the time variations, we again see a single-sideband suppressed carrier AM signal, but both the carrier oscillations and the modulation envelope have been shifted in phase, and not necessarily in a way which corresponds to a simple time delay of the waveform $v(t)$ we started off with at the line input.

On the other hand, we could take, as discussed before, a snapshot of the two terms at a given time, in which case, each would look sinusoidal in space. At a slightly later time, a snapshot would show the two sinusoids to have moved in the positive x direction, but not necessarily by the same amount. The rapidly oscillating carrier wave appears to be moving at a velocity:

$$\frac{\omega_c}{\frac{1}{2}(\beta_1 + \beta_2)}$$

Even if β is varying non-linearly with ω, this becomes asymptotically equal to the phase velocity $\omega_c/\beta(\omega_c)$ evaluated at the carrier frequency when the modulating frequency tends to zero. On the other hand, the second term, the modulation envelope, appears to be travelling at a velocity:

$$\frac{\omega_m}{\frac{1}{2}(\beta_2 - \beta_1)}$$

which is equal to $\delta\omega/\delta\beta$ if we write $\delta\omega = 2\omega_m = \omega_2 - \omega_1$ and $\delta\beta = \beta_2 - \beta_1$.

In the limit where ω_m tends to zero, the velocity of the modulation envelope becomes equal to the derivative $d\omega/d\beta$, or $1/(d\beta/d\omega)$, evaluated at the carrier frequency ω_c. This quantity is called the *group velocity*, v_g:

$$v_g = \frac{d\omega}{d\beta} = \frac{1}{\frac{d\beta}{d\omega}} \tag{3.39}$$

Although for simplicity a signal with only two frequency components was considered, it is clear that the same result would hold for a more general waveform of the form:

$$\begin{aligned} v(t) &= \text{Re}[m(t)e^{j\omega t}] \\ &= \text{Re}[m(t)] \cos \omega_c t - \text{Im}[m(t)] \sin \omega_c t \end{aligned} \tag{3.40}$$

where $m(t)$ is an arbitrary modulating waveform. (By allowing m to be complex, frequency, phase or mixed phase/amplitude modulation can be included.) The spectrum of the whole signal is a shifted image of the spectrum of m, centred on the carrier frequency. If the group velocity is (to a good enough approximation) constant over the whole spectrum of the modulated signal, then the modulating signal is transmitted without distortion and travels down the line at the group velocity.

This condition tells us that a graph of β versus ω is a straight line over the frequency range of interest. If this straight line passes through the origin, we also have v_{phase} constant and equal to v_{group} over the band in question and the entire signal is transmitted without distortion.

Where the straight line does not pass through the origin, v_{phase} is not constant and not equal to v_{group}. Nevertheless, the modulating waveform is not distorted and the only distortion of the signal is a progressively increasing phase shift between the modulating signal and the oscillations of the carrier as we move down the line. Usually this will not matter for a communications system, or it can easily be corrected at the receiver. A line that is dispersive in the strict sense, therefore, need not necessarily cause signal distortion that is practically significant. A more useful criterion may be that dispersion of the modulating signal will only occur if group velocity is not constant.

A rough criterion is that dispersion will start to cause noticeable distortion of the modulation when:

$$x\Delta\omega\left[\frac{d^2\beta}{d\omega^2}\right]\frac{1}{8} \approx \frac{1}{2}\,\text{radian} \tag{3.41}$$

where x is the distance from the starting point and $\Delta\omega$ is the full bandwidth (in rad/s not Hz) of the signal being transmitted. The left-hand side of this expression is the approximate phase error of frequency components at the edges of the band, relative to those near the carrier frequency, caused by the variation in group velocity.

(You may have come across the almost identical concepts of group delay, group delay or phase distortion, amplitude distortion and intercept distortion in connection with the response of linear networks such as filters. The main difference in the present context is that all these effects increase in proportion to how far we go down the line when observing the signal. Group delay distortion corresponds to dispersion; intercept distortion corresponds to the carrier appearing to travel at a phase velocity different from the group velocity of the modulation envelope.)

3.4.5 Frequency Dependence of Line Parameters

The distributed parameters R, L and G of a transmission line are not in fact constant, but turn out to be frequency dependent. The variations in R and L are due to the *skin effect*. Electromagnetic waves and their associated currents decay exponentially with depth in a highly conducting medium, and are greatest at the surface. The *skin depth*, usually written δ, is the distance over which a $1/e$ decay of the E, H fields and current occurs. Where conductivity is high, as for example in metals, the skin depth is given to a very good approximation by:

$$\delta = \sqrt{\frac{2}{\omega\mu\sigma}} \tag{3.42}$$

where σ is the conductivity of the material and μ is its permeability. Notice that skin depth is *inversely proportional to the square root of the frequency*. The implication is that at high frequencies, current is only being carried in a thin layer at the surface of conductors. Since skin depth is the effective depth over which current is being carried, *the resistance of the conductor increases as the square root of frequency*.

Where the transverse dimensions of the conductors in a transmission line are substantially smaller than the skin depth, the current flow becomes uniform over the conductor cross-section and the resistance R takes its familiar DC value, which is inversely proportional to the conductor cross-sectional area. At high frequencies, where the skin depth is small compared with the conductor dimensions, the resistance increases as the square root of the frequency, and (for a given line geometry) decreases only in inverse proportion to the linear size of the conductors, rather than their area (because the current is only flowing in a thin surface layer). A transition between the two behaviours occurs at frequencies where the skin depth is comparable with the conductor dimension.

Self-assessment Problems

3.4 For copper, taking $\sigma = 56 \times 10^6$ S/m and $\mu = \mu_0 = 4\pi \times 10^{-7}$ H/m, show that the skin depth is 9.5 mm at mains frequency (50 Hz) and 0.95 mm at 5 kHz. At what frequency is the skin depth equal to 1 micron (10^{-6} m)?

3.5 Show that the DC resistance measured between opposite edges of any square sheet of a conducting material of conductivity σ and thickness d is $1/\sigma d$. Why is this independent of the size of the sheet? (Hence the term ‘ohms per square’ when specifying surface resistances.)

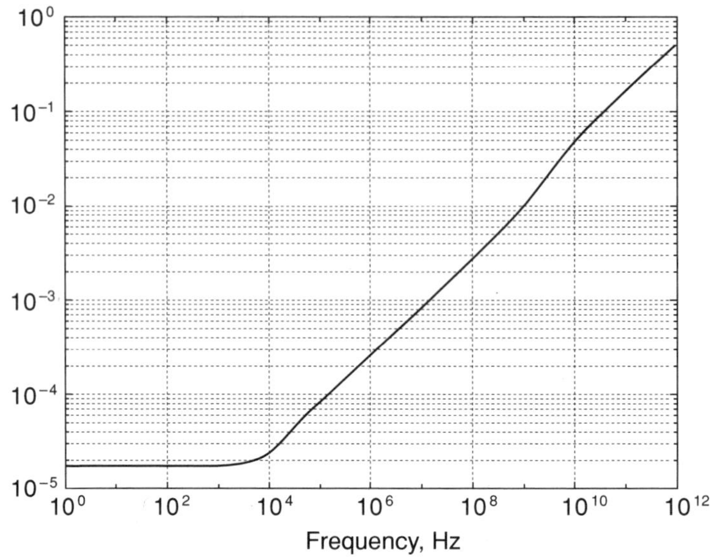

Figure 3.4 Surface resistance of a 1 mm thick copper conductor

Figure 3.4 shows how the surface resistance (measured in ohms per square) of a copper conductor 1 mm thick varies with frequency. The skin depth becomes equal to the thickness at 4.5 kHz. Resistance is very close to the value $\sqrt{(\omega\mu/2\sigma)}$ over the range 10 kHz to 1 GHz. At extremely high frequencies, the resistance rises above this value because of the effect of surface roughness. When the skin depth becomes comparable with the scale size of the microscopic irregularities on the conductor surface, the current is flowing in a convoluted (and hence longer) path over these irregularities, and the resistance is increased. In the example shown, the rms roughness was taken as 1 micron. Because of this effect, care is often taken to provide a good surface finish on conductors for microwave applications.

The skin effect also causes some frequency dependence of the inductance. Consider, for example, a total current I flowing in the centre conductor of a coaxial cable. At a radius r from the centre line, the magnetic field strength H is given by $i/2\pi r$, where i is the total current flowing through a circle of radius r. At zero frequency the current is uniformly distributed over the conductor cross-section; i, and therefore H, is non-zero right down to $r = 0$. Within the space between the conductors, the field is $I/2\pi r$.

If, however, the same total current I were flowing near the surface of the inner conductor, the field outside the conductor would remain as $I/2\pi r$ but inside the conductor it would fall to zero within a short distance of the surface. For the same current both the total stored magnetic field energy, and the total magnetic flux (per unit length of line) between the centre line and infinity, would be reduced. This argument indicates that, with the skin effect operating, inductance will be greater at low frequencies and will fall noticeably at the transition frequency where skin depth becomes comparable with conductor dimensions. At high frequencies, the magnetic flux within the metal becomes negligible, but the flux in the space between the conductors remains unchanged; the inductance is therefore expected to become constant at a rather lower value. Figure 3.5 shows the typical behaviour for a

TEM line with a central conductor diameter of about 2 mm. For typical line geometries L may be 20–30% greater at low frequencies.

This effect can also be described in terms of the surface impedance of a metal, which can be shown to be given (to a very good approximation for high conductivity) by:

$$Z_{\text{surface}} = \frac{(1+j)}{\sigma\delta} \tag{3.43}$$

when the conductor is at least two or three skin depths thick. The real part of this expression is the resistance rising as $\sqrt{\omega}$ (already discussed). We see that there is a *positive* reactive part of the same magnitude, corresponding to the inductance contributed by the magnetic flux within the conductor. However, a reactance rising like $\sqrt{\omega}$ corresponds to an inductance proportional to $1/\sqrt{\omega}$.

Self-assessment Problem

3.6 Explain why this is.

In Figure 3.5 we can see the *excess* inductance falling with frequency in this manner, above the transition frequency.

In nearly all everyday RF and microwave engineering, we are operating well above the skin depth transition frequency of the conductors. Exceptions occur where very thin conductors are used, for example, in some monolithic microwave integrated circuits (MMICs).

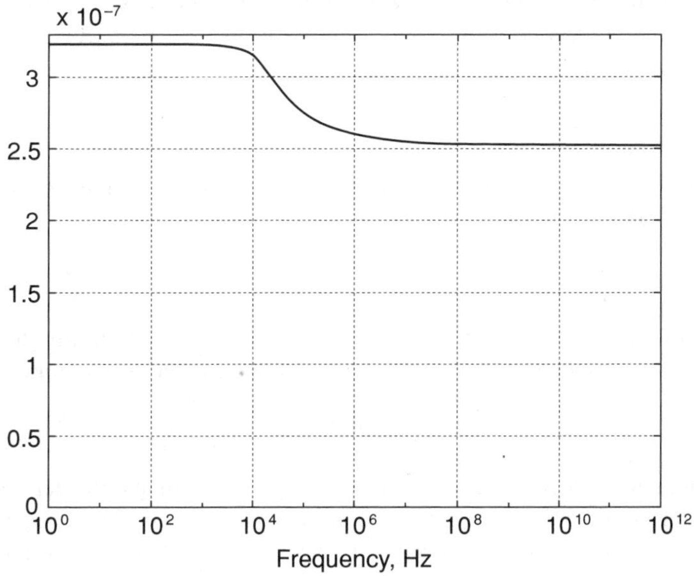

Figure 3.5 Typical variation of line inductance with frequency

For TEM lines, C is constant, and L becomes constant at the high frequencies where the skin depth is small. In quasi-TEM lines, the distributed circuit method is not an accurate description at high frequencies; however, constant values of L and C can be assumed in the frequency range above the skin depth transition frequency but below the frequencies where the inherently dispersive property of quasi-TEM modes become apparent. (This feature of the quasi-TEM lines is discussed later.)

3.4.5.1 Frequency dependence of *G*

Transmission line sections that are deliberately made very lossy may sometimes be used as attenuators and dummy loads. Loss might be introduced by filling the space between the conductors with a material that has substantial ohmic conductivity. However, except for these special cases, our transmission lines are normally filled with dielectrics, such as modern plastics, that are extremely good insulators. In this case true conduction makes a completely negligible contribution to G.

However, the parameter G also describes a quite different mechanism of energy absorption, namely *dielectric loss*. If an electric field is applied to a dielectric and then removed, not all the energy put into the dielectric is released again; some is converted into heat (cf. hysteresis in magnetic or elastic materials). Where the electric field is cycling sinusoidally, there is a component of the dielectric displacement current (of which more later) in phase with the applied field, therefore causing power to be dissipated in the dielectric.

The effect is usually quantified by assigning a complex value to the dielectric constant (relative permittivity) ε_r of the material; we write:

$$\varepsilon_r = \varepsilon_r' - j\varepsilon_r'' \tag{3.44}$$

The real part is the dielectric constant in its usual (low frequency) sense, and the ratio of the imaginary to the real part is conventionally referred to as the *loss tangent*, written:

$$\tan\delta = \frac{\varepsilon_r''}{\varepsilon_r'} \tag{3.45}$$

(This δ is not to be confused with the δ used for skin depth.) The current flowing into the line capacitance is $j\omega CV$ (per unit length) and:

$$C = (\varepsilon_r' - j\varepsilon_r'')C_{\text{empty}} \tag{3.46}$$

where C_{empty} is the (conceptual) capacitance of the same line without any dielectric. So we equate the in-phase component of above current to the term GV, obtaining:

$$G = \omega\varepsilon_r'' C_{\text{empty}} \tag{3.47}$$

ε_r is always frequency dependent to some extent, but for many dielectrics its real part is nearly constant, and tan δ is often fairly constant over quite a wide range of frequency. In a frequency range where ε_r'' is constant, we have G *directly proportional to frequency*, quite unlike a normal resistor. Materials with very low loss tangents are available for radio frequency and microwave use. Some examples are given in Table 3.1.

Table 3.1 Lowloss dielectric materials

		1 kHz	1 MHz	100 MHz	3 GHz	25 GHz
Alumina	ε_r'	8.83	8.80	8.80	8.79	?
	$\tan\delta$	0.00057	0.00033	0.00030	0.0010	?
PTFE	ε_r'	2.1	2.1	2.1	2.1	2.08
	$\tan\delta$	< 0.0003	< 0.0002	< 0.0002	< 0.00015	0.0006

3.4.6 High Frequency Operation

Above a certain frequency, a transmission line can be considered to be operating in a *high frequency* regime. Under these conditions:

(a) the line has, in a certain sense, less tendency to distort transmitted pulses;
(b) the characteristic impedance becomes real and constant;
(c) a lossless approximation can be made for many purposes.

In the high frequency regime we can also make convenient analytical approximations for α and β and use them to analyse point (a).

To be operating in the high frequency region, we need the two conditions:

(i) $\omega L \gg R$
(ii) $\omega C \gg G$

Even if the skin effect is operating and R is increasing with frequency, it will only increase as $\sqrt{\omega}$ and the term ωL increases more rapidly with ω. Hence the condition (i) can always be reached in normal metallic lines. For any given line there is a characteristic transition frequency at which $\omega L = R$. (Typically this is rather lower than the skin depth transition frequency, but not by a very large factor.) Condition (ii) looks as though it is also a consequence of making ω large, but this is not true in the commonest case where G is dominated by dielectric loss and may also be increasing as fast as ω. Rather the condition follows from using a low-loss dielectric where $\tan\delta$ is very small.

Self-assessment Problem

3.7 Convince yourself that condition (ii) follows from small $\tan\delta$.

If dielectric loss is neglected, the propagation constant can be written as:

$$\gamma = \sqrt{(j\omega L + R)j\omega C} = j\omega\sqrt{LC}\sqrt{1 + \frac{R}{j\omega L}} \tag{3.48}$$

Now assuming $R/j\omega L \ll 1$, the second square root can be approximated using the binomial expansion (or Taylor series) as:

$$\sqrt{1+X} \approx 1 + \frac{X}{2} - \frac{X^2}{8} + \frac{X^3}{16} - \frac{5X^4}{128} + \ldots \tag{3.49}$$

Self-assessment Problem

3.8 Convince yourself of this.

The following approximations can be found for the attenuation and phase constants:

$$\alpha \approx \frac{R}{2}\sqrt{\frac{C}{L}} + \text{higher order terms} \tag{3.50}$$

$$\beta \approx \omega\sqrt{LC}\left[1 + \frac{1}{8}\left(\frac{R}{\omega L}\right)^2 - \frac{5}{128}\left(\frac{R}{\omega L}\right)^4\right] + \text{higher order terms} \tag{3.51}$$

Self-assessment Problem

3.9 Set $X = R/j\omega L$ to obtain the approximations for α and β given in Equations (3.50) and (3.51).

Equation (3.50) shows that we expect to find the attenuation of transmission lines (in dB/metre) increasing as the square root of frequency (since R does) most of the time in RF and microwave work. Although the loss increases with frequency, the rate of increase of the square root function decreases like $1/\sqrt{\omega}$, and this implies that the *change* in attenuation over a given small band of frequencies falls as the centre frequency rises. Hence there will tend to be less amplitude distortion of a modulated signal of fixed bandwidth as the carrier frequency is raised.

The second equation shows that the effect of line resistance on phase velocity is very small at high frequencies. Also, when R is varying as $\sqrt{\omega}$, the second term in the expression for β is frequency independent, giving no group delay distortion even though the phase velocity is not quite constant. Very small group delay distortion arises from the third term.

For an ideal TEM line with $R = G = 0$ and constant L and C, we would have:

$$\gamma = \sqrt{j\omega L j\omega C} = j\omega\sqrt{LC} \tag{3.52}$$

Hence:

$$\beta = \omega\sqrt{LC} \tag{3.53(a)}$$

and:

$$v_{phase} = \frac{1}{\sqrt{LC}} \tag{3.53(b)}$$

This line would propagate a wave of arbitrary shape with no distortion or loss – we could break up our arbitrary pulse into its frequency components, which would all travel at the same velocity (and with no attenuation) and hence would preserve the same pulse shape.

3.4.6.1 Lossless Approximation

Under high frequency conditions, it can be seen from the approximate expressions for α and β that (even with skin effect operating) lines tend to a 'lossless' condition. This does not mean that the attenuation α itself becomes small, but only the *attenuation per wavelength* becomes very small, i.e. the ratio α/β becomes small. Radio frequency circuits such as filters, matching networks, etc. are generally concerned with wavelength-scale structures and the lossless approximation is very good when designing them. Lines in this condition also have Z_0 *real*, to a very good approximation.

Self-assessment Problems

3.10 Ignoring G, show that α/β tends to zero at high frequencies, even with R increasing as $\sqrt{\omega}$.

3.11 Using Equation (3.28), explain why Z_0 becomes real (i.e. purely resistive) when conditions (i) and (ii) hold. Show that Z_0 is then approximately equal to $\sqrt{(L/C)}$.

3.12 Ignoring R and instead letting G be non-zero, and assuming that G is purely due to dielectric loss, show that α/β is approximately one-half the tan β of the dielectric.

In almost all RF and microwave work, the conditions (i) and (ii) will be found to hold to a very good approximation. A real, frequency-independent Z_0 can be assumed for TEM lines, which greatly simplifies design work and also makes it easy to provide a broadband, low reflection termination (in the form of a resistor) for the transmission lines. For quasi-TEM lines, Z_0 shows weak frequency dependence at frequencies so high that the distributed circuit model becomes inadequate, but it can be shown to remain resistive.

3.4.6.2 The Telegrapher's Equation and the Wave Equation

The Telegrapher's Equation is:

$$\frac{\partial^2 v}{\partial x^2} = RGv + (RC + LG)\frac{\partial v}{\partial t} + LC\frac{\partial^2 v}{\partial t^2} \tag{3.54}$$

The derivation of this equation implicitly assumes that all the parameters R, G, C and L are frequency-independent constants (which they may be over wide ranges of frequency). Even then the equation can only be solved directly for the special case $R = G = 0$ for which it reduces to:

$$\frac{\partial^2 v}{\partial x^2} = LC \frac{\partial^2 v}{\partial t^2} \tag{3.55}$$

This is called the *one-dimensional wave equation.* The general solution of the 1-D wave equation is:

$$v = f(x - ct) + g(x + ct) \tag{3.56}$$

where $c = 1/\sqrt{(LC)}$. Note that f and g in this solution are *arbitrary functions of a single variable.* (The only requirement is that they are assumed to be 'smooth' enough, i.e. to be twice differentiable.)

The first term in f represents a *forward travelling wave*, and the second term in g a *reverse travelling wave.* c is clearly the velocity of these waves. The implication of this is that *a system obeying the 1-D wave equation can propagate a pulse of arbitrary shape without any distortion.*

When R and G are non-zero, the equation can only be solved by transform techniques, which means in effect going into the frequency domain and solving at an arbitrary single frequency. In this case, we can also handle the frequency dependence of R, G, and L, which of course are not truly constant for real lines.

Self-assessment Problems

3.13 (a) Take $\partial/\partial x$ of Equation (3.5) to show that:

$$\frac{\partial^2 v}{\partial x^2} = -R\frac{\partial i}{\partial x} - L\frac{\partial^2 i}{\partial x \partial t}$$

(b) Substitute Equation (3.6) into the second term on the right-hand side of the above equation.

(c) Now take $\partial/\partial t$ of Equation (3.6) and substitute this into the result of step (b). (Remember that second derivatives are symmetrical, i.e.: $\partial^2 i/\partial x\partial t = \partial^2 i/\partial t\partial x$, etc.)

You should now have derived the Telegrapher's Equation.

3.14 Show that for $R = G = 0$, the Telegrapher's Equation reduces to Equation (3.55).

3.15 If $f(x)$ is an arbitrary smooth function of a single variable x, and $f'(x)$ and $f''(x)$ represent its first and second derivatives, we can turn f into a function of two variables x and t by replacing its argument with $(x - ct)$, where c is a constant.

(Note that $\partial/\partial x$ of $f(x - ct)$ is $f'(x - ct)$, and $\partial^2/\partial x^2$ of $f(x - ct)$ is $f''(x - ct)$.) Hence:

(a) Show that $\partial/\partial t$ of $f(x - ct)$ is $-cf'(x - ct)$. (Hint: if not immediately obvious, this is a simple case of the function of a function rule – or just think about how much change in the argument $(x - ct)$ is made by changes δx in x and δt in t, respectively.)
(b) Find $\partial^2/\partial t^2$ of $f(x - ct)$, and hence show that $f(x - ct)$ is a solution of the 1-D wave equation if $c = 1/\sqrt{(LC)}$.
(c) By the same methods show that a function $g(x + ct)$ is also a solution, where g is a second arbitrary function.

In Self-assessment Problem 3.14 you have proved directly in the time domain, what was already proved less directly in the frequency domain, that a line with $R = G = 0$, and frequency independent constants L, C, can propagate an arbitrary pulse shape without loss or distortion, at velocity $1\sqrt{(LC)}$.

3.5 Field Theory Method for Ideal TEM Case

When we first start learning electronics, we become used to thinking in circuit terms where everything can be analysed in terms of voltages and currents, and Kirchhoff's laws can be applied. In RF engineering, and especially at microwave and millimetric frequencies, we have to start facing the fact that these are only approximations and that a rigorous description involves us in analysing distributed fields.

Nevertheless we try to go on using circuit theory as far as possible, because it is easier to understand, is more suitable for analysis and especially synthesis, and can be used in CAD with limited computing power. Most people find field theory difficult, and in any case many field problems can only be solved exactly by numerical techniques.

Fortunately most day-to-day RF engineering can still be done on the basis of circuit theory. In the distributed circuit theory of the line we have done just this, and it works well when the transverse dimensions of the line are small. Sometimes, for more difficult structures, we rely on field theory specialists to tackle the field solution and provide us with an equivalent circuit model which makes it possible to go on using circuit analyses.

A good example occurs in the treatment of discontinuities such as a junction between two sections of transmission line of different impedance. The circuit view gives us a simple expression for the reflection. The reality is that there is a complex electromagnetic field structure set up around the discontinuity, and the reflection coefficient given by the expression is only an approximation that works well when the transverse dimensions of the structure remain much less than one wavelength. The situation can be 'patched up' by converting the field solution into an equivalent circuit model of the discontinuity, containing some additional fictitious circuit elements.

It is not the intention here to give a comprehensive treatment of field theory but it is useful to give a proper treatment of one special problem where the solution is exact and fairly simple. This is the TEM wave which, as pointed out in the previous discussion of line classification, exists on a two-conductor line (making the approximation that the conductors are perfect), *where the medium filling the line has properties that are uniform over the line cross-section.* This will give us, at least, some physical insight into the nature of transmission line fields, and also into how the distributed circuit model can be related to the field picture.

3.5.1 Principles of Electromagnetism: Revision

It is assumed here that the reader has a general knowledge of basic electromagnetics and is familiar, in particular, with the following principles:

1. Where there is a changing magnetic field linking a loop, there must be an electromotive force (EMF) induced around the loop. (The law of induction.)
2. Flow of electric current through a loop must be associated with a magnetomotive force (MMF) acting round the loop. (Ampère's law.)
3. The flux of electric field out of a closed surface is proportional to the charge contained within it. (Gauss's Law.)
4. A changing electric field, even in a vacuum, can act like a current, to be included in law 2.

The third and fourth ideas introduce the idea of *displacement current*, which may be less familiar than the first two laws. In a vacuum there is an apparent current density given by $\varepsilon_0 \partial \underline{E}/\partial t$, where ε_0 is a fundamental constant of nature called the *permittivity of free space* (equal, to four significant figures, to 8.854 picofarad per metre).

The three fundamental laws of electromagnetism, shown above, can be written mathematically as follows:

$$\oint_C \underline{E} \cdot \underline{dl} = -\int_S \frac{\partial \underline{B}}{\partial t} \cdot \underline{n} dS \tag{3.57}$$

$$\oint_C \underline{H} \cdot \underline{dl} = \int_S (\underline{J} + \frac{\partial \underline{D}}{\partial t}) \cdot \underline{n} dS \tag{3.58}$$

$$\oint_{S'} \underline{D} \cdot \underline{n} dS = \int_V \rho dV = Q \tag{3.59}$$

In the first two expressions, C denotes *any* closed loop in space (which need not coincide with any physical structure), and S denotes *any* surface that spans that loop. The small circle on the integral sign emphasises that we are integrating round a closed loop and returning to the same point at which we started. The dot symbol is for the scalar product of two vectors and $\underline{dl}$ denotes an infinitesimal element of C ($\underline{dl}$ has a direction, so is a vector). dS denotes an infinitesimal element of area making up S and $\underline{n}$ denotes the unit normal to the surface S at any point on it. The requirement that $\underline{n}$ is normal to the surface does not specify the sign of $\underline{n}$. The right-hand screw rule, with which you may already be familiar, can define the sign convention. (If the curled fingers of the right hand indicate the sense in which we traverse C, then the thumb indicates the approximate direction in which $\underline{n}$ should be pointing.)

Equation (3.57) is usually expressed in words by the statement that the EMF induced round a loop is equal to minus the rate of change of magnetic flux linking it. Likewise, Equation (3.58) says that the MMF round a loop is equal to the total current, including displacement current, linking it. In Equation (3.59), S' is any closed surface and V is the volume it contains; ρ is the charge density, and Q the total charge, within V.

We are using *two* quantities, $\underline{E}$ and $\underline{D}$, for electric fields as a convenient way of dealing with the effect of a dielectric. An electric field acting on a dielectric induces a continuous distribution of atomic-scale electric dipoles throughout it. We use the symbol $\underline{P}$ to denote the total induced dipole moment per unit volume, and define the *electric flux density* $\underline{D}$ by the fundamental relationship:

$$\underline{D} = \varepsilon_0 \underline{E} + \underline{P} \tag{3.60}$$

It can then be shown that the charge Q or charge density ρ only needs to include the contribution from unbalanced total numbers of positive and negative charge carriers, averaged over the volume of a few molecules, at the point in question. The effect of the atomic dipoles in the dielectric is automatically taken into account.

When the electric field in the dielectric changes, the movement of charges in the atomic dipoles represents a real physical current. This current $\partial \underline{P}/\partial t$ is added to the vacuum displacement current to make up the total displacement current $\partial \underline{D}/\partial t$ in Equation (3.58). Then the current density $\underline{J}$ only needs to include the conduction current due to the movement of free charge carriers.

Likewise in magnetic materials there is a distribution of atomic-scale magnetic dipoles through the material. These are mainly due to electron spin and they behave like currents circulating in atomic-sized loops. They are conveniently taken care of by using the relation:

$$\underline{B} = \mu_0 \underline{H} + \underline{M} \tag{3.61}$$

where $\underline{M}$ is the average magnetic dipole moment of the atomic magnets per unit volume, averaged over the volume of a few molecules, in the material. With this definition the effect of the atomic magnets is automatically included, and the current $\underline{J}$, as stated before, only has to include the conduction current. μ_0 is a fundamental constant defined exactly as $4\pi \times 10^{-7}$ H/m (in the S.I. system of units).

Self-assessment Problem

3.16 Calculate $1/\sqrt{(\mu_0 \varepsilon_0)}$ using the values given. What is this quantity?

In the RF context we are usually only dealing with fairly weak fields for which it can be assumed that $\underline{P} \propto \underline{E}$ and $\underline{M} \propto \underline{B}$ in most materials. We then characterise the medium by two constants:

$$\underline{B} = \mu \underline{H} \tag{3.62}$$

$$\underline{D} = \varepsilon \underline{E} \tag{3.63}$$

where μ and ε are called the permeability and permittivity of the medium, respectively. It is usually convenient to write

$$\varepsilon = \varepsilon_r \varepsilon_0 \tag{3.64}$$

$$\mu = \mu_r \mu_0 \tag{3.65}$$

where ε_r, μ_r are pure numbers called the relative permittivity and relative permeability. ε_r is also called the 'dielectric constant'.

For the following treatment of the TEM wave, we need to write the first two electromagnetic laws in their *differential* form, viz:

$$\nabla \times \underline{E} = -\frac{\partial \underline{B}}{\partial t} \tag{3.66}$$

$$\nabla \times \underline{H} = \underline{J} + \frac{\partial \underline{D}}{\partial t} \tag{3.67}$$

If you are not familiar with vector analysis, the left-hand side of Equation (3.66) is called the *curl* of the field $\underline{E}$, and it is a new vector whose (x, y, z) components are given by:

$$\left(\frac{\partial E_z}{\partial y} - \frac{\partial E_y}{\partial z}, \quad \frac{\partial E_x}{\partial z} - \frac{\partial E_z}{\partial x}, \quad \frac{\partial E_y}{\partial x} - \frac{\partial E_x}{\partial y} \right) \tag{3.68}$$

or equivalently in determinant form as:

$$\nabla \times \underline{E} = \begin{vmatrix} \underline{x} & \underline{y} & \underline{z} \\ \partial/\partial \boldsymbol{x} & \partial/\partial \boldsymbol{y} & \partial/\partial \boldsymbol{z} \\ E_x & E_y & E_z \end{vmatrix} \tag{3.69}$$

where $\underline{x}$, $\underline{y}$, $\underline{z}$ are unit vectors along the directions of the x, y, z axes.

Self-assessment Problem

3.17 Expand the determinant in Equation (3.69) and show that it gives the same result as Equation (3.68).

The differential forms of the law are in fact precisely equivalent to the integral forms. The differential forms can quite easily be deduced from Equations (3.57) and (3.58) by considering the case where C has been shrunk to an infinitesimal rectangle. (This deduction represents the proof of Stokes's theorem.)

For the rest of Section 3.5, we shall assume sinusoidal time variation at a single frequency, using $e^{j\omega t}$ notation for this. We also assume the line contains a uniform linear medium with constants ε and μ. The equations we have to satisfy then reduce to the following (which have to be satisfied throughout the space between the conductors):

$$\nabla \times \underline{E} = -\mathrm{j}\omega\mu\underline{H} \tag{3.70}$$

$$\nabla \times \underline{H} = \mathrm{j}\omega\varepsilon\underline{E} \tag{3.71}$$

3.5.2 The TEM Line

The main point that will now be demonstrated is that the *time-varying fields in the TEM wave can be constructed simply from a knowledge of the static electric field.* If $\underline{E}_t$ denotes the static electric field, all we have to do is multiply it by $e^{j\omega t} \cdot e^{-j\beta x}$, where x denotes distance along the line and β is a suitably chosen phase constant, to obtain a valid solution of the Maxwell equations:

$$\underline{E} = \underline{E}_t e^{j\omega t} e^{-j\beta x} \tag{3.72}$$

The geometry that will be used is shown in Figure 3.6, for an arbitrary pair of parallel conductors making up a TEM line. The x-axis is along the line and the y-, and z-axes can be any pair of axes which, together with x, make up a right-handed coordinate system.

3.5.3 The Static Solution

If we apply a DC voltage to our line, a static electric field is set up on the line. Assuming an infinitely long line, it is obvious by symmetry that this field is purely transverse, i.e. it has no x-component. It is equally obvious that the field cannot be dependent on x either, so we shall write: $\underline{E}_{static} = \underline{E}_t(y, z)$, where the suffix t means that the field is transverse (no x components) and the arguments (y, z) are included as a reminder that the field components are functions of y and z, but not of x or time. We can write $\underline{E}_t$ in its components as:

$$(0,\ E_{t,y}(y, z),\ E_{t,z}(y, z))$$

If we can solve the static problem, i.e. find the functions $E_{t,y}(y, z)$, $E_{t,z}(y, z)$, then we can calculate the capacitance C per unit length of the line.

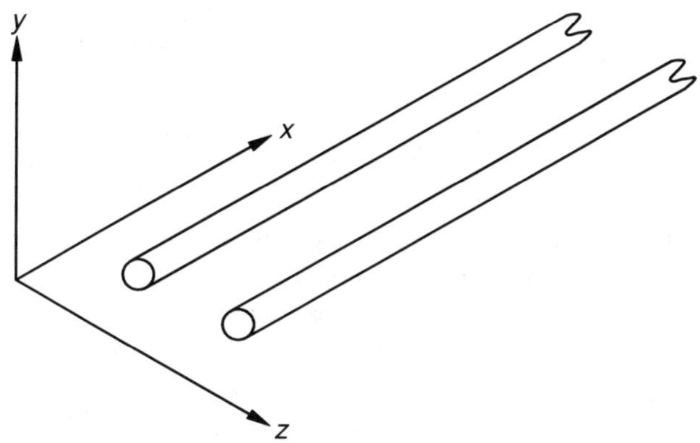

Figure 3.6 Coordinate system for TEM line analysis

What equations would have to be satisfied for the static field? Because the field is static, the Maxwell Equation (3.66) tells us that:

$$\nabla \times \underline{E}_t = \underline{0} \tag{3.73}$$

This equation is in fact the condition that, if a test charge is moved from a point A to point B, the work done is independent of the path, so the potential difference between the two points is well defined.

We must also assume that any dielectric in the line is perfectly insulating, so that when we apply the static voltage there is no build-up of charge at any point in the dielectric, and hence the density ρ of free charges in Equation (3.59) is zero.

Since there is no z field component and the other field components only depend on x and y, it can be shown from Equation (3.73) that the static field satisfies:

$$\frac{\partial E_{t,z}}{\partial y} - \frac{\partial E_{t,y}}{\partial z} = 0 \tag{3.74}$$

Self-assessment Problem

3.18 Show that Equation (3.74) is correct.

Now consider Equation (3.59) with its right-hand side set to zero. If the volume V is shrunk to an infinitesimal cuboid, it is quite easy to show that the differential form of the equation becomes:

$$\frac{\partial D_x}{\partial x} + \frac{\partial D_y}{\partial y} + \frac{\partial D_z}{\partial z} = 0 \tag{3.75}$$

In this case $\underline{D} = \varepsilon \underline{E}$ with a *constant* ε, so we can write:

$$\frac{\partial E_x}{\partial x} + \frac{\partial E_y}{\partial y} + \frac{\partial E_z}{\partial z} = 0 \tag{3.76}$$

And finally, $\underline{E}_t$ has no x component (and no dependence on x either!), so the first term in this expression is zero, leading to:

$$\frac{\partial E_{t,y}}{\partial y} + \frac{\partial E_{t,z}}{\partial z} = 0 \tag{3.77}$$

In a static field, at any point on a conductor surface, there must be no component of electric field acting tangentially to the surface. (This holds even if the conductivity is finite.) If any such field component were present, there would be current flowing along the surface, charges would be redistributing and the field would not be static. This is called the *boundary condition* obeyed by the electric field at a conductor surface.

The same boundary condition is obeyed in a time-varying field if the conductors are perfect, i.e. of infinite conductivity. Because the conductivity of metals is so high, this can actually be used as a very good approximation when solving time-varying field problems.

Equations (3.74) and (3.77) together with the boundary condition can be regarded as the defining equations for the electric field in the static problem. Of course they only define the form of the field and not its absolute magnitude, which would only be known when the voltage applied between the conductors has been specified.

Even these equations are not simple to solve for a general conductor configuration, and a numerical solution has to be used for the general case. Only two cases can be solved with negligible effort. These are:

(a) Where the lines are wide parallel plates, $\underline{E}_t$ is approximately a uniform, parallel field normal to the plates – except that at the edges there is a fringing field which is harder to calculate.
(b) For a pair of coaxial cylinders, the exact solution is that $\underline{E}_t$ is everywhere in a radial direction, and its intensity is proportional to $1/r$ where r is the distance from the centre line.

Analytical solutions can be found, however, by the method of conformal transformations for a number of useful geometries. The most important of these geometries are a pair of parallel cylinders, a cylinder parallel to a plane, and a strip conductor parallel to one ground plane (like a microstrip with no dielectric) or parallel to two ground planes (like a stripline).

(You may have come across a graphical method known as *curvilinear squares* that can be used to find approximately the field lines and equipotential curves for an arbitrary pair of conductors. With some experience, remembering that the field lines have to enter the surface normally, you can usually make a fairly good sketch by hand of the approximate shape of the field lines.)

3.5.4 Validity of the Time Varying Solution

Step 1. Finding the associated magnetic field

Starting from an arbitrary explicit expression for an electric field $\underline{E}$, we can work out what $\nabla \times \underline{E}$ is by differentiating it with respect to position. Then if we look at the first of Maxwell's Equations, Equation (3.70), we can see that it will tell us what the magnetic field associated with this electric field would have to be.

This procedure will be used taking the Equation (3.72) as a starting point. This gives:

$$\nabla \times \underline{E} = \begin{vmatrix} \underline{x} & \underline{y} & \underline{z} \\ \partial/\partial x & \partial/\partial y & \partial/\partial z \\ 0 & E_{t,y}(y, z)e^{-j\beta x} & E_{t,z}(y, z)e^{-j\beta x} \end{vmatrix} \tag{3.78}$$

(The electric field is assumed to have no x-component.) The first term of the expanded determinant reads: $\underline{x}$ times $[(\partial/\partial y\, E_{t,z} \cdot e^{-j\beta x}) - (\partial/\partial z\, E_{t,y} \cdot e^{-j\beta x})]$. The term $e^{-j\beta x}$ does not vary with y or z so we can take it outside the expression, which reduces to: $e^{-j\beta x} \cdot [\partial/\partial y\, E_{t,z} - \partial/\partial z\, E_{t,y}]\underline{x}$. Now we have already seen that $[\partial/\partial y\, E_{t,z} - \partial/\partial z\, E_{t,y}] = 0$ is one of the defining equations of the static field, so this first term in the determinant is actually zero.

The second term reads: $-\underline{y}$ times $[(\partial/\partial x\, E_{t,z} \cdot e^{-j\beta x}) - (\partial/\partial z$ of zero)]. Since E_t does not depend on x, the only contribution to this whole term arises from differentiating $e^{-j\beta x}$ with respect to x, which just gives $-j\beta \cdot e^{-j\beta x}$, so this whole term in the end reads: $(j\beta \cdot e^{-j\beta x} \cdot E_{t,z})\underline{y}$.

The third term reads: $\underline{z}$ times $[(\partial/\partial x\, E_{t,y} \cdot e^{-j\beta x}) - (\partial/\partial y$ of zero)] and this becomes: $(-j\beta e^{-j\beta x} E_{t,y})\underline{z}$ in exactly the same way. So we have finally shown that:

$$\nabla \times \underline{E} = (j\beta e^{-j\beta x} E_{t,z})\underline{y} - (j\beta e^{-j\beta x} E_{t,y})\underline{z} \tag{3.79}$$

if $\underline{E}$ is derived from the static electric field in the way we assumed. Now if this is compared with the Maxwell Equation (3.70), we can deduce that the associated magnetic field must be given by:

$$\underline{H} = \frac{\beta}{\omega\mu}[(-e^{-j\beta x} E_{t,z})\underline{y} + (e^{-j\beta x} E_{t,y})\underline{z}] \tag{3.80}$$

We can now see that the magnetic field has the same components as the electric field, except that the y and z components are interchanged, the y component is given a minus sign, and there is an overall coefficient of proportionality $\beta/\omega\mu$. A simple way of expressing this is that *the magnetic field $\underline{H}$ is also transverse, and it is proportional to the electric field and at right angles to it.* This tells us that the magnetic field lines are at right angles to the electric field lines!

Another simple way of writing this expression is:

$$\underline{H} = \frac{\beta}{\omega\mu}\underline{x} \times \underline{E} \tag{3.81}$$

You will be aware that the cross-product is at right angles to both of the vectors making it up, so this again shows us that the $\underline{H}$ field is transverse to the line and is at right angles to $\underline{E}$.

The operation we just did looks very abstract, but we shall see later that it is fairly easy to picture physically in terms of the law of induction.

So far we have simply assumed a form for the electric field, and worked out what the associated magnetic field would have to be. To show that we have a proper field solution, we need to show that the other Maxwell Equation (3.71):

$$\nabla \times \underline{H} = j\omega\varepsilon\underline{E} \tag{3.82}$$

can be satisfied. In the process we find what the so far unspecified value for β has to be. We therefore take the curl of the expression (3.81) to obtain:

$$\nabla \times H = \frac{\beta}{\omega\mu}\begin{vmatrix} \underline{x} & \underline{y} & \underline{z} \\ \partial/\partial x & \partial/\partial y & \partial/\partial z \\ 0 & -E_{t,z}(y, z)e^{-j\beta x} & E_{t,y}(y, z)e^{-j\beta x} \end{vmatrix} \tag{3.83}$$

Remembering that the $\partial/\partial x$ and $\partial/\partial y$ operate only on the E_t terms, and the $\partial/\partial z$ only on the $e^{-j\beta x}$ we obtain:

$$\nabla \times \underline{H} = j\frac{\beta^2}{\omega\mu}\left[\left(e^{-j\beta x}E_{t,y}\right)\underline{y} + \left(e^{-j\beta x}E_{t,z}\right)\underline{z}\right] + \frac{\beta}{\omega\mu}e^{-j\beta x}\left[\frac{\partial E_{t,y}}{\partial y} + \frac{\partial E_{t,z}}{\partial z}\right]\underline{x} \tag{3.84}$$

Again we saw that $(\partial/\partial y E_{t,y} + \partial/\partial z E_{t,z}) = 0$ is one of the defining equations of the static field, while the expression $[(e^{-j\beta x} \cdot E_{t,y})\underline{y} + (e^{-j\beta x} \cdot E_{t,z})\underline{z}]$ is just the time-varying electric field $\underline{E}$ again. So we have shown that:

$$\nabla \times \underline{H} = j\frac{\beta^2}{\omega\mu}\underline{E} \tag{3.85}$$

Then, if we compare this with the Maxwell Equation:

$$\nabla \times \underline{H} = j\omega\varepsilon\underline{E} \tag{3.86}$$

we see that we have a solution provided that:

$$\beta^2 = \omega^2\mu\varepsilon \tag{3.87}$$

So we have now shown that, for a transmission line with a uniform dielectric and perfect conductors, we can generate time-varying electric and magnetic fields that satisfy Maxwell's field equations. Both of the fields are purely transverse to the line, and the electric field is identical in form to the static electric field. The time-varying, travelling wave is consequently known as the TEM (Transverse Electromagnetic) mode for the line.

3.5.5 Features of the TEM Mode

The TEM mode is convenient in having several features that make its behaviour easy to analyse, although these are not of great importance to a line's practical performance:

1. The TEM time varying fields are easy to calculate – it is only necessary to solve for the static electric field on the line.
2. The voltage across the line is well defined. Because of Equation (3.73), the integral of electric field from one conductor to the other (keeping within a given transverse plane) is independent of the path chosen.
3. Because the voltage is well defined, the characteristic impedance defined as voltage over current in the wave is also well defined. It also agrees with other definitions, e.g. using the relation that I^2Z_0 or V^2/Z_0 gives the average power transported in the wave, if I and V are the rms current and voltage in the wave. The Z_0 also turns out to be inherently frequency independent (see below). For quasi-TEM lines such as microstrip, the various definitions of Z_0 show different, though fairly minor, variations with frequency, while agreeing at low frequency.
4. Currents on the line are purely longitudinal. On quasi-TEM lines, small transverse components of current exist.

Potentially much more significant in performance terms is the following property:

5. *The TEM mode is inherently non-dispersive*, i.e. in the ideal case the phase velocity is independent of frequency, so that a pulse of arbitrary shape can be sent down the line without distortion. The phase velocity is ω/β which we can now see is given by:

$$v_{phase} = \frac{1}{\sqrt{\mu\varepsilon}} = \frac{c}{\sqrt{\mu_r\varepsilon_r}} \tag{3.88}$$

Two effects can still cause the TEM wave to be slightly dispersive in practice:

(i) The finite conductivity of the conductors, neglected in the ideal TEM theory, has a slight effect on the phase velocity. As discussed in Section 3.5.3, this becomes negligible at high frequencies, and more important is the frequency-dependent attenuation that the resistance produces. (It also produces small components of electric field along the line.)
(ii) For certain materials, ε and μ may be noticeably frequency-dependent. The most common case is where a non-magnetic (μ_r almost 1) and low-loss dielectric is used, and these materials usually have dielectric constants whose real parts are nearly constant over very wide ranges of frequency.

However, the dispersion is not an inherent property of the mode itself. This may be contrasted with non-TEM modes such as those in a metallic waveguide, which are dispersive even if it is filled with vacuum and made of perfect conductors. Likewise it will be seen later that the 'quasi-TEM' mode of a microstrip, and surface wave modes on a microstrip substrate, are inherently dispersive even given perfect conductors and frequency-independent ε_r of the dielectric substrate.

The TEM mode is therefore the best-behaved type for propagating wide-band signals over long transmission lines with minimum waveform distortion. The quasi-TEM lines are potentially more distorting, but this is not usually a major issue in RF and microwave subsystems where line lengths and signal bandwidths are moderate.

3.5.5.1 A useful relationship

An important and useful relationship can easily be derived from the field theory of the TEM mode, and even more easily by comparing the field and circuit descriptions. According to the distributed circuit theory, the velocity of waves on an ideal (lossless) line is $1/\sqrt{(LC)}$, while the field theory gives it as $1/\sqrt{(\mu\varepsilon)}$. We can therefore deduce:

$$LC = \mu\varepsilon = \mu_r\varepsilon_r\mu_0\varepsilon_0 = \frac{\mu_r\varepsilon_r}{c^2} \tag{3.89}$$

Because perfect conductors have been assumed, there is no field penetration into the conductors and clearly L must be understood as the high frequency limiting value in Figure 3.5, which is the value that would be calculated using perfect conductors. The relationship is worth committing to memory, as it is valuable for calculating line inductances, and in the analysis of quasi-TEM lines.

Self-assessment Problems

3.19 For lossless lines, or lines working at high frequency, use the distributed circuit theory to deduce the following useful relationships:

$$Z_0 = v_p L$$
$$Z_0 = 1/(v_p C)$$
$$Z_0 = Z_{0,empty}\surd(\mu_r/\varepsilon_r)$$

(Subscript 'empty' refers to the value that a parameter would take if we could remove the dielectric from the line while leaving the conductor geometry unchanged. For the third relation you also need Equations (3.58) and (3.89).)

3.20 Standard coaxial cable has $Z_0 = 50\ \Omega$ (at high frequency). (a) For a cable filled with polythene dielectric having $\varepsilon_r = 2.26$, calculate L, C and the phase velocity. (b) What would be the 'empty' values of Z_0, C and L for this cable?

3.5.6 Picturing the Wave Physically

The time varying field was derived using the rather abstract-looking differential forms of the Maxwell Equations, so it is useful to finish this section of the theory with a physical picture of the wave which shows how the electromagnetic laws are satisfied in terms of the more familiar circuital laws, and how the theory links up with the distributed circuit description.

Consider Figure 3.7. Looking at loop A in the transverse plane, there is no EMF round this loop because of the curl-free property of the static $\underline{E}$ field, and the law of induction is satisfied because there is no longitudinal component of $\underline{H}$ and therefore no magnetic flux through the loop.

If we draw a second longitudinal loop B orientated so that its short edges are normal to the $\underline{E}$ field, this loop is normal to the $\underline{E}$ field at all points so there cannot be any EMF round it. If it is spanned by a cylindrical surface, the $\underline{H}$ field, being normal to $\underline{E}$, is everywhere parallel to this surface. The $\underline{H}$ field lines do not pass through the surface and there is no magnetic flux through it. Again the induction law is satisfied.

The interesting case is loop C whose edges bc, da lie in the conductor surfaces. There is no EMF along either edge, because there is (as required by the boundary condition for perfect conductors) no $\underline{E}$ field component tangential to the surface. However, the line integral of $\underline{E}$ from a to b can be identified as the voltage $V(x)$ across the line at the plane x, (taking the inner conductor as positive) and that from c to d as minus the voltage at $x + \delta x$. Hence the EMF round loop C is $V(x) - V(x + \delta x)$ which is taken as $(-\partial V/\partial x) \times \delta x$ when δx is small.

You should now be able to see that the magnetic field is flowing through the loop C, and is in fact at right angles to it. If we took any small sub-loop within C, you should be able to visualise how the law of induction, Equation (3.57), is satisfied in the TEM wave by equating the EMF round the sub-loop, which is equal to the area of the sub-loop times the rate of change *with x* of the radial electric field, to the rate of change *with time* of the

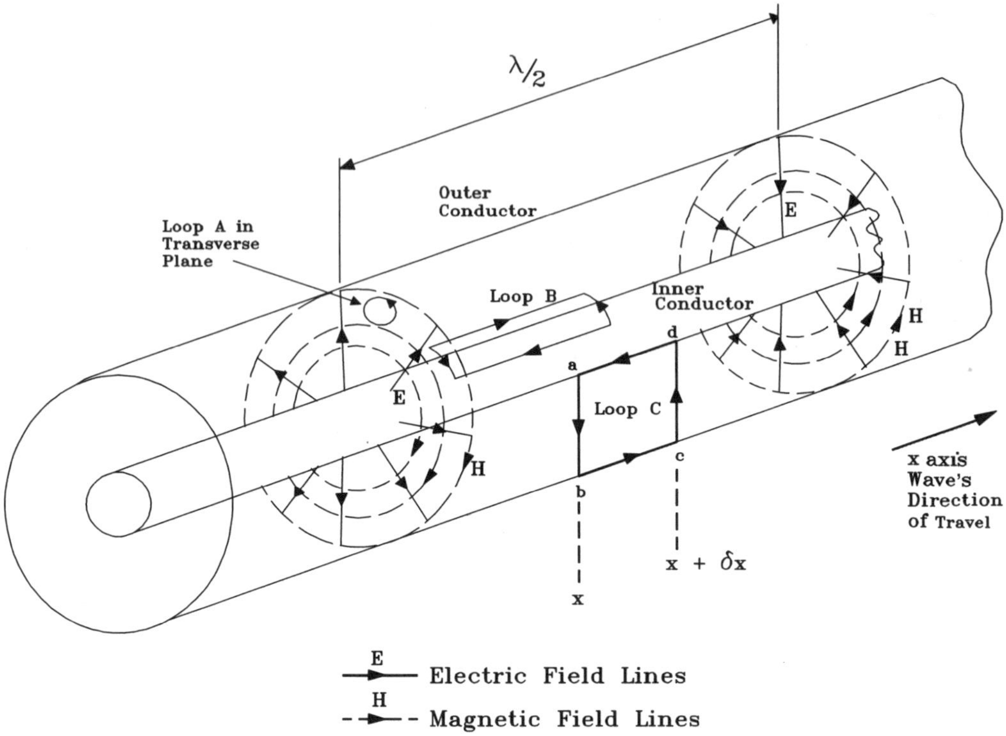

Figure 3.7 Form of the TEM wave in a coaxial line

magnetic flux through it. For a sinusoidal travelling wave of the type discussed before, all the field quantities are proportional to the function $\cos(\omega t - \beta x)$. $\partial/\partial x$ and $\partial/\partial t$ of this function are both proportional to $\sin(\omega t - \beta x)$ so you should be able to picture how the law of induction can be satisfied in the TEM travelling wave at all times and positions.

Self-assessment Problem

3.21 Prove that $\partial/\partial x$ and $\partial/\partial t$ of $\cos(\omega t - \beta x)$ are both proportional to $\sin(\omega t - \beta x)$ and try visualising it in terms of the snapshots of the cosine function.

Applying the law of induction to the whole of loop C, one of the distributed circuit theory equations can be deduced. Inductance is *defined* as the ratio of magnetic flux linking a circuit to the current flowing in it. The flux Φ linking the loop C is, by definition, $L\ \delta x \times I$ where I is the local value of the current flowing in the centre conductor. (For an ideal TEM wave we know that the magnetic field is what would be calculated for a static current – but assuming perfect conductors and therefore no flux penetrating them – and L is frequency independent and equal to the static value.) If I flows in the positive x direction, the right-hand screw rule says that the magnetic field is circulating in a clockwise sense when looking in the positive x direction. However, the right-hand rule applied to loop C says that, in working out the EMF round C, the flux passing through C has to be defined in the opposite sense when applying the law of induction. So we can write:

$$\text{Flux linking } C \text{ is } \Phi = -L\delta x I$$

Therefore:

$$\text{EMF round } C = -d\Phi/dt = L\delta x\, \partial I/\partial t$$

But:

$$\text{EMF round } C = (-\partial V/\partial x) \times \delta x$$

as discussed before.

Therefore, cancelling δx, $\partial V/\partial x = -L\partial I/\partial t$, which was one of the distributed circuit equations for an ideal lossless line.

Self-assessment Problem

3.22 Of the three loops shown in Figure 3.7, only one has a non-zero magnetomotive force round it, and a non-zero electric flux passing through it. Which one is it? From this starting point, give a similar physical explanation of how the other circuital law, Equation (3.58), is satisfied at all positions and times.

The other equation of the distributed circuit theory can be deduced by applying Equation (3.58) to a loop similar to loop B placed close to the surface of the centre conductor, but it is quicker to do it just using the law of charge conservation. (This law can be shown to be implicit in Equation (3.58).) Consider a short section of conductor between two transverse planes at x, $x + \delta x$. There is a net current into this piece of conductor given by $I(x) - I(x + \delta x)$, and the charge conservation principle says that this net inflow must be equated to a rate of build up of charge in that section. Hence we can write:

$$I(x) - I(x + \delta x) = \frac{\partial I}{\partial x}\delta x = \frac{d}{dt}(Q\delta x) \tag{3.90}$$

where Q is the total charge per unit length on the conductor, and of course δx can be cancelled.

At any point on one of the conductors the flux of electric field out of that conductor, which is ε times the local value of the normal component of $\underline{E}$, must be equated to the surface charge density ρ on the conductor. Knowing the functional form $\underline{E}(x, y)$ enables us to integrate $\underline{E}$ from one conductor to the other to obtain a voltage V, and to integrate ρ round the conductor circumference in a fixed transverse plane to obtain Q per unit length. The capacitance per unit length C is defined as the ratio Q/V. The circuit approach reduces the complexity of the field function to a 'lumped' pair of variables Q and V and the relation $Q = CV$. We now have that $\partial I/\partial x = -dQ/dt = -C\partial V/\partial t$, which is the second distributed circuit equation. In the TEM time varying wave the $\underline{E}$ field has the same form as the static field, and so the capacitance C is frequency independent and takes its static value.

Again picturing the sinusoidally varying travelling wave $\cos(\omega t - \beta x)$, you should be able to visualise the charge density growing with time at the points on the centre conductor where the spatial gradient of I is negative, the electric field correspondingly growing with time at those points, and the space and time rates of change both proportional to $\sin(\omega t - \beta x)$ as before.

The time varying field was found as a solution of Maxwell's Equations in the space between the conductors, together with the boundary condition at the conductor. In this approach the fields tell us what the charges have to be, rather than the other way round. This is fine because, in assuming that boundary condition, we are assuming that the conductors are perfect and that the charges can redistribute themselves instantaneously to produce the given fields.

At the end of all this work we have used the rigorous field theory to justify the distributed circuit view of the line as an exact description when the line is ideal (lossless) and of true TEM type.

It is easy to see that the pure TEM mode cannot exist on one of the lines where the dielectric constant is not uniform over the cross-section, as the wave would have to take two different velocities in the dielectric and air regions. However, it can be shown that, at low frequencies, a 'quasi-static' approximation can be made. In this regime, the longitudinal components of the $\underline{E}$ and $\underline{H}$ fields are small, and the transverse parts approximate closely to their static form. (The static form of $\underline{H}$ would have to be calculated assuming perfect conductors, i.e. no penetration of $\underline{H}$ into the conductors and no normal component of $\underline{H}$ at the surface.) This field pattern is called the 'quasi-TEM' mode of the line. The distributed circuit description of the line continues to work well in this regime.

3.6 Microstrip

This transmission technology has been important for at least two and a half decades already, and microstrip technology remains at the forefront of the options for RF, microwave and high-speed systems implementation. Its significance is actually increasing because of the expanding applications for RF and microwave technology as well as high-speed digital electronics. Currently many companies are appreciating that microstrip technology is the best answer to problems associated with maintaining the operating integrity of digital systems clocked at frequencies around and above 200 MHz. Examples of systems using microstrip include:

1. Satellite (DBS, GPS, Intelsat, VSATs, LEO systems, etc.).
2. Wireless (cellular, PCN, WLAN, etc.).
3. EW (ECM, ECCM), radars and communications for defence.
4. High-speed digital processors.
5. Terminals for fibre optic transmission.

Many other examples could be cited.

Microstrip as a concept begins with the requirement for increased integration at RF and microwave frequencies, leading to microwave integrated circuits (MICs). This requirement was first recognised in the late 1960s, mainly driven by military applications and hence leading to designs such as broadband EW amplifiers, etc. Although this class of applications is still growing, in spite of overall defence reductions, it is the other applications sectors listed above that drive most current interests in microstrip technology and design.

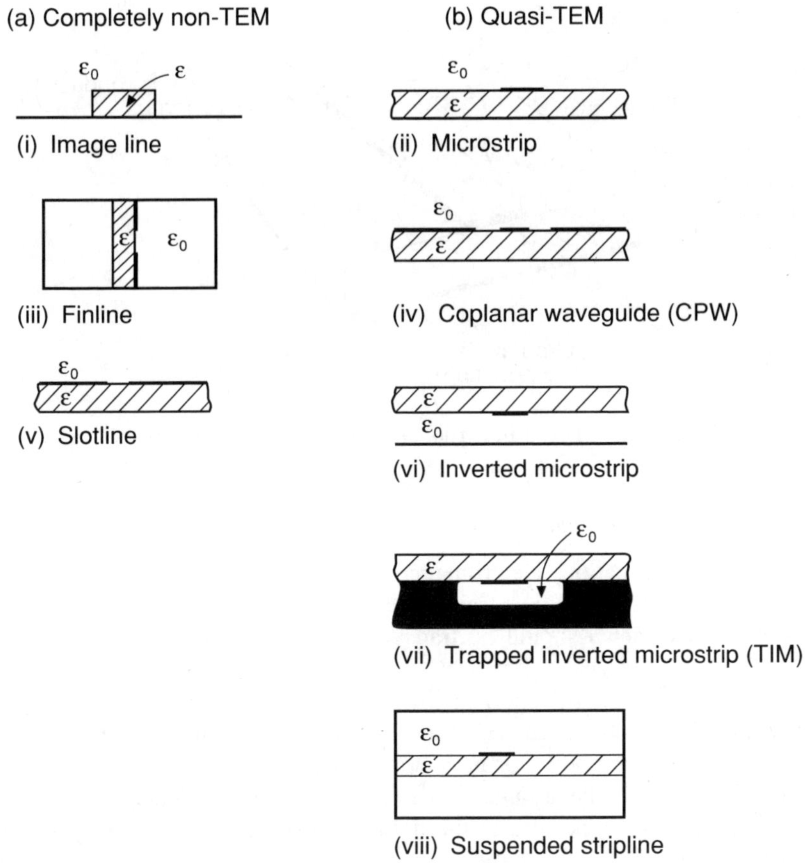

Figure 3.8 A range of planar and semi-planar transmission line structures suitable as candidates for microwave circuit integration. (Substrates are shown shaded in each case and metal is shown in dense black.)

Until the late 1960s the great majority of RF and microwave transmission used these types of structures – and they are still used extensively. However, in order to accommodate both *active* and *passive chip* insertion, some form of planar transmission structure is required. Conventional PCBs (even single-layer ones) are unsatisfactory, mainly because of their radiation losses and cross-talk problems. Multilayer PCBs are worse in these respects.

Microstrip is totally grounded (earthed) on one side of the dielectric support, which provides a better electromagnetic environment than open structures (such as PCB conductor tracks).

A range of possible MIC-oriented transmission line structures is shown (as cross-sections) in Figure 3.8. In each case ε_0 indicates 'air' (strictly vacuum) and ε_r is the relative permittivity of the supporting dielectric. Here we always refer to this supporting dielectric as the '*substrate*' (with thickness denoted by h).

Of the structures shown in Figure 3.8 the most important are: microstrip, CPW, finline and suspended stripline. The rest of this chapter, however, is entirely concerned with microstrip.

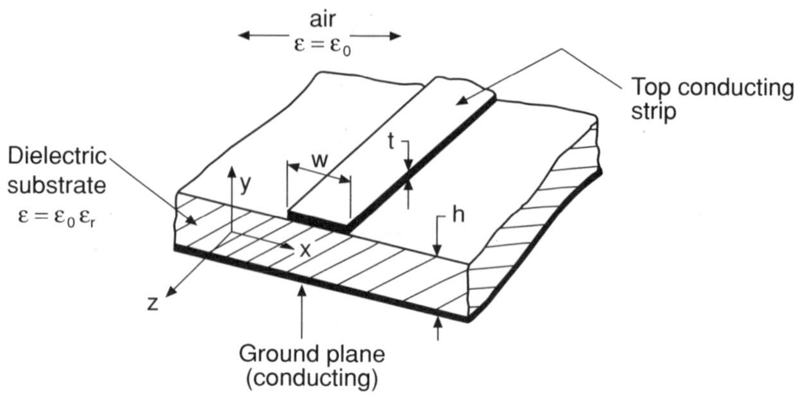

Figure 3.9 Three-dimensional view of microstrip geometry

3.6.1 Quasi-TEM Mode and Quasi-Static Parameters

As a rough first approximation, electromagnetic wave propagation in microstrip can be likened to that in coaxial lines, i.e. transverse electromagnetic (TEM). This, of course, implies that all electric fields should be transverse to, and orthogonal with, the magnetic fields.

However, even a cursory inspection of the structure (see Figure 3.9) reveals a distinct dielectric discontinuity – between the substrate and the air – and any mixed medium like this cannot support a true TEM mode – or indeed any 'pure' mode (TE, TM, etc.).

Because the presence of the ground plane (the grounded 'underside' of the structure) together with the substrate concentrates the field between the strip and the ground there is some reasonable resemblance to a 'flattened-out' coaxial ('co-planar') line. It is this feature that gives rise to the 'quasi-TEM' concept and suites of design approaches (curves and formulas) have been developed that provide engineering design level accuracy, at relatively low frequencies at least.

Self-assessment Problems

3.23 Why cannot microstrip support any pure (TEM, TE, TM) mode?

3.24 Give two reasons why microstrip is preferred above other media for MICs.

3.6.1.1 Fields and static TEM design parameters

A simplified cross-section showing only the electric field is shown in Figure 3.10, with magnetic and electric fields shown in three-dimensional detail in Figure 3.11(a) and (b). The presence of longitudinal components of fields is clear from these diagrams, and in the next subsection we show the importance of quantifying the effects of these for accurate high-frequency design.

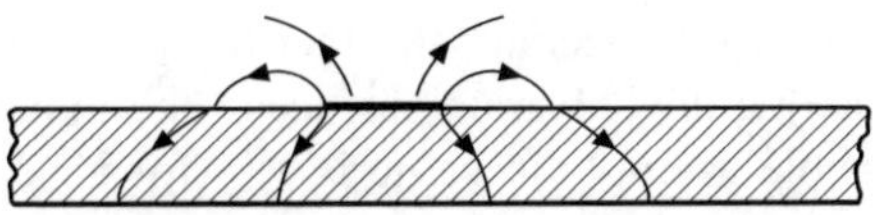

Figure 3.10 Simplified cross-section, showing electric field only

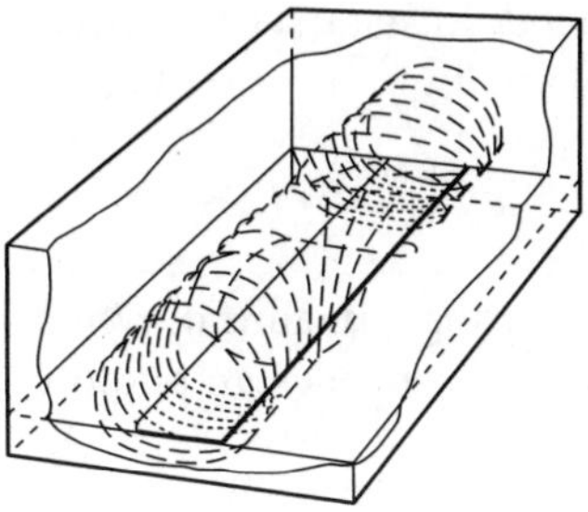

(a) Magnetic field distribution

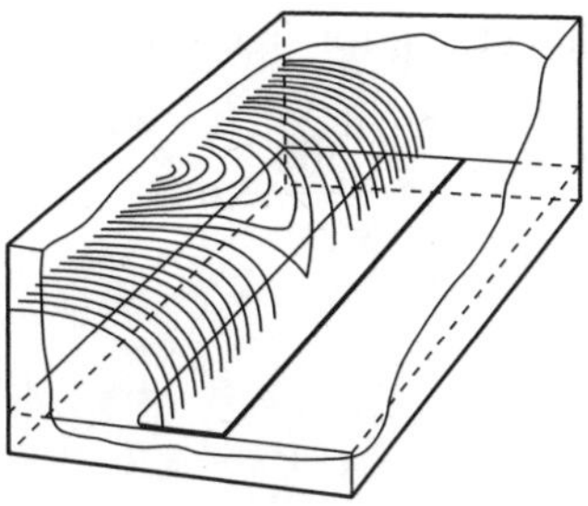

(b) Electric field distribution (partial view)

Figure 3.11 Three-dimensional views showing (a) magnetic field alone; (b) electric field alone (in air, only for simplicity)

3.6.1.2 Design aims

Comprehensive passive networks can be designed and built using interconnections of microstrips. Given the relative permittivity ε_r and the thickness h (usually in mm) of the substrate, as well as the desired characteristic impedance Z_0 (in ohms) and the frequency of operation f (in GHz), the design aims are to determine:

1. the physical width w of the strip, and
2. the physical length l of the microstrip line.

Initially, the static TEM parameters can be used to find w but first we need to define two special quantities relating to microstrip, namely, the effective microstrip relative permittivity, ε_{eff}, and the filling factor, q.

The quantity ε_{eff} simply takes into account the fact that propagation is partly in the substrate and partly in the air above the substrate surfaces. The maximum asymptotic value of ε_{eff} is ε_r and the lowest possible asymptotic value is 1.0 (for air).

The filling factor, q, also takes into account the fact that propagation is partly in the substrate and partly in the air. However, q just tells us the proportionate filling effect due to the substrate, and its maximum possible asymptotic value is 1.0 (representing fully filled), with a lowest possible asymptotic value of 0.5 (implying an exactly evenly filled space). Both ε_{eff} and q must be used together in static TEM design synthesis.

It can be shown that the following useful relation holds:

$$\varepsilon_{eff} = 1 + q(\varepsilon_r - 1) \tag{3.91}$$

Also the characteristic impedance of the structure in free-space Z_{01} (i.e. entirely air-filled), which is a useful normalising parameter for design purposes, is given by:

$$Z_{01} = Z_0\sqrt{\varepsilon_{eff}} \tag{3.92}$$

Neglecting dispersion (see Section 3.6.2), the wavelength (λ_g) in a line intended to be one wavelength long is given (in millimetres), for frequency F (in GHz), by:

$$\lambda_g = \frac{300}{F\sqrt{\varepsilon_{eff}}} \text{ mm} \tag{3.93}$$

Self-assessment Problem

3.25 Calculate the physical length of a three-quarter-wavelength microstrip when ε_r is 9.8, the filling factor is 0.76 and $F = 10$ GHz.

3.6.1.3 Calculation of microstrip physical width

In practice, microstrip physical width, w, is almost always determined within computer CAD (CAE) routines. We can use graphical approximations for w (as well as other parameters), however, and one approach is outlined here. Such methods provide excellent checks on computed results.

Presser (following the fundamental work of Wheeler) developed the data for the curves shown in Figure 3.12.

We now run through the sequence of steps required to use these curves, with the following warning:

> It will frequently be found that designs require extrapolation of the curves beyond their useful accuracy. These curves should only be used when initial calculations indicate that mid-ranges are applicable.

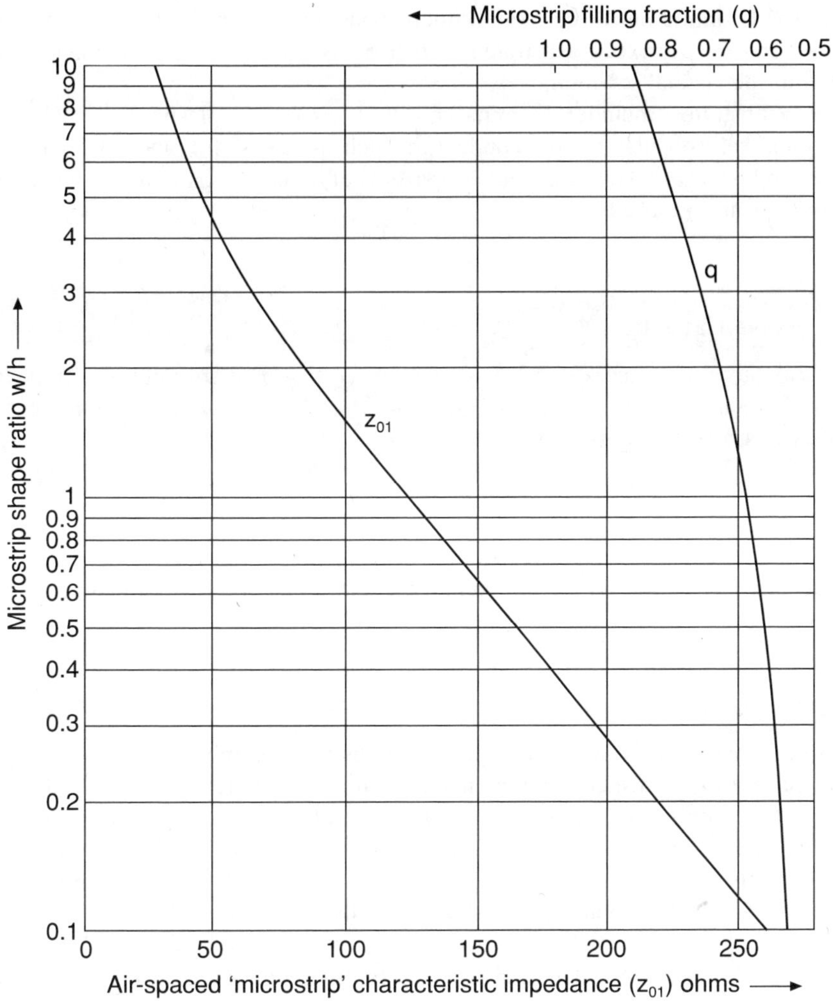

Figure 3.12 Generalised curves facilitating approximate analysis or synthesis of microstrip

The calculation sequence is:

1. Make the initial (very approximate) assumption the $\varepsilon_{eff} = \varepsilon_r$. This only provides starting values. (To accelerate the process you can take ε_{eff} just below the value of ε_r.)
2. Calculate the air-spaced characteristic impedance under the approximation of step 1.
3. Use Figure 3.13 to find the width to height ratio, w/h, applicable to this value of airspaced characteristic impedance. Also note the corresponding value of q.
4. Calculate the updated value of q using Equation (3.91).

This completes one iteration of the sequence. In most cases at least three iterations should be completed before final values are established.

Here we shall confine our discussion to microstrips on uniform, isotropic, and non-ferrite substrates. (Occasionally exotic substrate microstrips are required, design guidance for which is available in the specialist literature.)

Important substrates include: alumina, co-fired ceramics, plastic (e.g. PTFE, glass-reinforced) and semi-insulating semiconductors such as GaAs, InP and, of course, silicon. At moderate to high microwave frequencies GaAs is often used – particularly where medium to high volumes are required.

Self-assessment Problem

3.26 Using the same substrate as in SAP 3.25, and desiring a characteristic impedance of 31.3 ohms, determine, using graphical synthesis, the width of the microstrip if the substrate thickness is 0.5 mm.

Relatively accurate formulas for analysing and synthesising microstrip are available. Design-orientated texts provide several suitable expressions and give guidance on limits of applicability.

3.6.2 Dispersion and its Accommodation in Design Approaches

In any system, non-linearity in the frequency f versus wave number β (or phase coefficient) results in what is termed dispersion. One manifestation of the presence of such dispersion is group-delay distortion of signals transmitted through such a system.

Dispersion can be chromatic, 'waveguide', or modal. Chromatic dispersion is observed in many materials and extends through optical as well as microwave and millimetre-wave frequencies. In fact, all the varieties of dispersion listed are important in optical fibres to varying extents.

All of the planar and semi-planar microwave transmission structures are dispersive and microstrip is no exception. In this case the dispersion is modal and arises from variations in coupling between longitudinal section electric (LSE) and longitudinal section magnetic (LSM) modes. That these types of modes must exist should be clear by re-visiting Figure 3.11. As the frequency is increased so the strength of coupling between modes also increases and, because the currents associated with most of the modes are concentrated beneath the strip (adjacent to the dielectric), the fields overall are proportionately contained more within the substrate. Hence, the effective microstrip permittivity increases. For design purposes the questions that must be answered are:

1. Precisely how much is this increase?
2. How does this vary with other microstrip parameters?

Since the effective microstrip permittivity is known to be frequency (f) dependent, we acknowledge this fact writing it as $\varepsilon_{eff}(f)$. This represents a very important microstrip design parameter – especially at high frequencies.

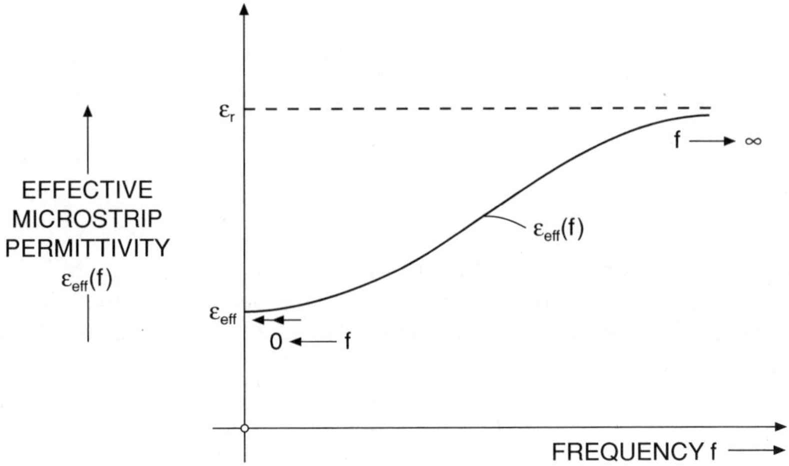

Figure 3.13 Microstrip dispersion: $\varepsilon_{eff}(f)$ versus frequency f

Calculations of wavelength in microstrip must now be based upon $\varepsilon_{eff}(f)$ and *not* (at high frequencies at least) using the low frequency parameter ε_{eff}. The formula for wavelength therefore becomes:

$$\lambda_g = \frac{c}{f\sqrt{\varepsilon_{eff}(f)}} \tag{3.94}$$

Calculations of w are substantially unaffected by this dispersion (although strictly the characteristic impedance is also frequency dependent). In most cases w can be computed as was shown earlier.

Dispersion is therefore generally viewed in terms of varying $\varepsilon_{eff}(f)$ the limits being clearly defined in Figure 3.13.

As might be expected, early attempts to fully quantify this dispersion were based upon various frequency-dependent field solutions (founded ultimately upon Maxwell's Equations). One of the earliest offerings was the work of Itoh and Mittra who employed a spectral domain method to solve this problem.

A clear difficulty with all these approaches, however accurate, is the extensive computational time required to find even $\varepsilon_{eff}(f)$ and hence wavelength in the microstrip. Also, even for relatively narrow bands of operation, the wavelengths are generally required over a range of frequencies and for as many microstrips as there are in the final design. Consequently, a search has been conducted over many years, for practical closed form expressions (or families of expressions) that would enable microstrip dispersion to be quantified rapidly and efficiently. A major breakthrough was achieved in 1972 by Getsinger, who formulated and published his 'microstrip dispersion model'. This was proven to work accurately, within around 3% or so, for microstrips on alumina-type substrates operating up to about 12 GHz. Getsinger modelled microstrip line as separated structures for which approximate analysis was more straightforward and provided the following expressions:

$$\varepsilon_{eff}(f) = \frac{\varepsilon_r - \varepsilon_{eff}}{1 - G(f/f_p)^2} \tag{3.95}$$

where:

$$f_p = \frac{Z_0}{2\mu_0 h} \tag{3.96}$$

and μ_0 is the free-space permeability ($4\pi \times 10^{-7}$ H/m). The parameter G is purely empirical, thereby giving some flexibility to the formula. G is dependent mainly upon Z_0 but also to a lesser extent upon h. Getsinger deduced from measurements of microstrip ring resonators on alumina that:

$$G = 0.6 + 0.009Z_0 \tag{3.97}$$

These expressions are accurate (typically to within about 3%) where alumina substrates having thickness between about 0.5 and 1 mm are used and frequencies remain below 12 GHz. For other substrates, including plastics with much lower permittivities, the formulas are still reasonably accurate up to this frequency – although Z_0 should remain above 20 ohms. Where substrates differing from plastics or alumina are used, and particularly when the frequency rises above 12 GHz or so, different expressions are necessary.

Since Getsinger's original work, many variations have been devised and reported. One example is that due to Kobayashi, who developed the following linked relationships:

$$\varepsilon_{eff}(f) = \varepsilon_r - \frac{\varepsilon_r - \varepsilon_{eff}}{1 + (f/f_{50})^m} \tag{3.98}$$

where:

$$f_{50} = \frac{f_{k,TM_0}}{0.75 + (0.75 - (0.332/\varepsilon_r^{1.73}))w/h} \tag{3.99}$$

$$f_{k,TM_0} = \frac{c \tan^{-1}\left(\varepsilon_r \sqrt{\dfrac{\varepsilon_{eff} - 1}{\varepsilon_r - \varepsilon_{eff}}}\right)}{2\pi h \sqrt{\varepsilon_r - \varepsilon_{eff}}} \tag{3.100}$$

$$m = m_0 m_c \tag{3.101}$$

$$m_0 = 1 + \frac{1}{1 + \sqrt{w/h}} + 0.32\left(\frac{1}{1 + \sqrt{w/h}}\right)^3 \tag{3.102}$$

$$m_c = \begin{cases} 1 + \dfrac{1.4}{1 + w/h}\left\{0.15 - 0.235\exp\left(\dfrac{-0.45f}{f_{50}}\right)\right\}, & \text{where } w/h \le 0.7 \\ 1, & \text{where } w/h > 0.7 \end{cases} \tag{3.103}$$

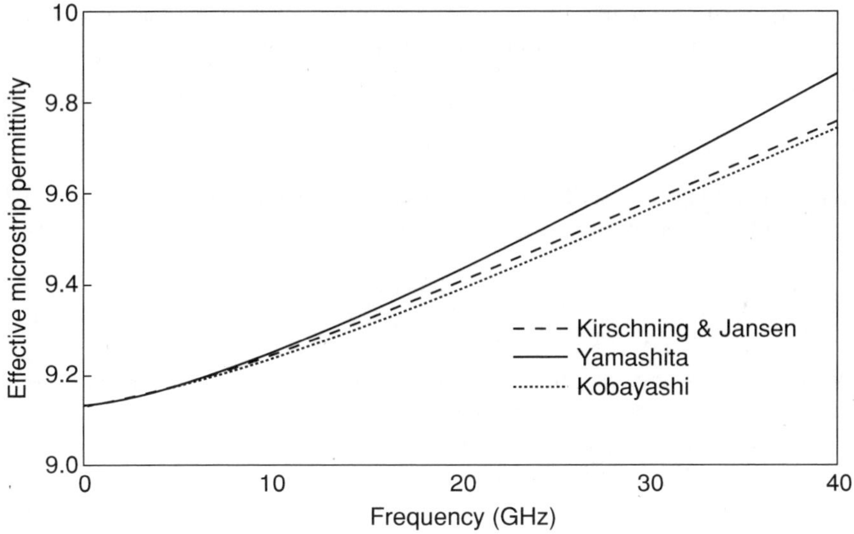

Figure 3.14 Dispersion curves for a microstrip on a GaAs substrate (comparison of predictions)

The consistency of results computed using various closed form dispersion expressions is indicated by the curves in Figure 3.14, for microstrip on a 0.127 mm thick GaAs substrate ($\varepsilon_r = 13$) and microstrip width (w) of 0.254 mm.

It is noteworthy that the closest agreement is between the curves of Kirschning and Jensen and Kobayashi and that the curves extend to 40 GHz.

Self-assessment Problem

3.27 Using Getsinger's expressions, calculate, for the same microstrip line on alumina as applies to previous SAPs, the wavelength at 11.5 GHz. Re-calculate the wavelength for a microstrip with an impedance of 80 ohms (all other parameters identical). What does this tell us about the effect of impedance on the 'degree of dispersion'?

3.6.3 Frequency Limitations: Surface Waves and Transverse Resonance

The generation of surface waves, comprising electromagnetic energy trapped close to the surface of the substrate, and also a transverse resonance effect, both set limits to the maximum operating frequency associated with a particular microstrip.

The lowest-order TM *surface wave* mode is the first limiting phenomenon in this category and analysis leading to the corresponding frequency limit starts by considering the substrate as a dielectric slab. An eigenvalue expression is set down and the net phase coefficient is calculated, from which the relationship for the TM surface wave frequency is determined as:

$$f_{TEM1} = \frac{c \tan^{-1}\varepsilon_r}{\sqrt{2}\pi h \sqrt{\varepsilon_r - 1}} \tag{3.104}$$

For narrow microstrips on reasonably high permittivity substrates (usually greater than $\varepsilon_r = 10$), which is typically the most critical situation, this reduces to:

$$f_{TEM1} = \frac{106}{h\sqrt{\varepsilon_r - 1}} \tag{3.105}$$

leading to the following formula for maximum substrate thickness allowable while avoiding the generation of this wave:

$$h = \frac{0.354\lambda_0}{\sqrt{\varepsilon_r - 1}} \tag{3.106}$$

Clearly, keeping the substrate as thin as possible (consistent with other engineering constraints) ensures that the shortest possible wavelength, i.e. the highest frequency, is supported while avoiding the generation of the lowest order TM surface wave.

We next consider the lowest order *transverse microstrip resonance*. A resonant mode can become set up transversely across the width of a microstrip, which can couple strongly with the dominant quasi-TEM mode. It is therefore important to ensure that the maximum operating frequency for a given MIC remains below the frequency associated with the generation of this resonance, for the widest strip in the MIC. This resonance will be manifested as a half-wave with its node in the centre and antinodes beyond the edges of the microstrip (due to side-fringing) as shown in Figure 3.15.

The side-fringing equivalent distance is denoted by d (= $0.2h$ for most microstrips), and the equation dictating the resonance is therefore:

$$\frac{\lambda_{CT}}{2} = w + 2d$$

$$= w + 0.4h \tag{3.107(a)}$$

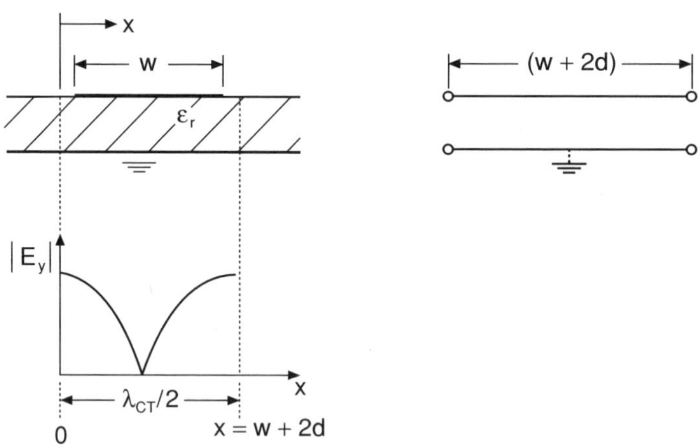

Figure 3.15 Transverse microstrip resonance; standing voltage wave and equivalent transverse electrical 'line'

or, equivalently:

$$f_{CT} = \frac{c}{\sqrt{\varepsilon_r}(2w + 0.8h)} \qquad (3.107(b))$$

Slots in the microstrip, usually down the centre to cut into the longitudinal current, can suppress this transverse resonance.

Self-assessment Problem

3.28 For the two microstrips in the previous problems (including the 80 ohm one); calculate the frequencies of onset for the lowest order TM mode and for transverse resonance. Compare these results and comment on the practical operating implications.

3.6.4 Loss Mechanisms

In common with all types of electrical circuit elements, microstrips dissipate power. The following three loss mechanisms are the most important:

1. Conductor
2. Dielectric
3. Radiation.

The first two, conductor and dielectric losses, are more or less continuous along the direction of propagation of the electromagnetic wave in the microstrip transmission line. Radiation loss, on the other hand, tends to occur from discrete apertures such as nominally open conductor ends. There are also parasitic losses due to surface wave propagation, when this occurs.

Both conductor and dielectric losses are described in terms of contributions to the loss coefficient α of the microstrip transmission line. Conductor loss is influenced strongly by the skin effect and hence also by surface roughness (on the underside of the strip). Denoted by α_c, it is given to a good approximation by:

$$\alpha_c' = 0.072\frac{\sqrt{f}}{wZ_0}\lambda_g\left[1 + \frac{2}{\pi}\tan^{-1}\left\{1.4\left(\frac{\Delta}{\delta_s}\right)^2\right\}\right] \qquad (3.108)$$

Self-assessment Problem

3.29 What approaches to design tend to lead to low α_c?

The dielectric loss is a consequence of the dissipation mechanism within the dielectric of the substrate, usually characterised by the loss tangent, tan δ. This loss is reduced by the fact that some of the energy is transported in air. The resulting loss, in decibels per microstrip wavelength, is:

$$\alpha'_d = 27.3\frac{\varepsilon_r(\varepsilon_{eff} - 1)\tan\delta}{\varepsilon_{eff}(\varepsilon_r - 1)} \quad \text{[dB/microstrip wavelength]} \tag{3.109}$$

Clearly, the only practical way to maintain low dielectric loss is to ensure that substrate materials have low tan δ.

Typically microstrip structures have physically open sections and other *discontinuities* (see Section 3.6.5). In general energy is radiated by such structures – although this energy will be associated with near field components only, since the conducting top and side-walls of the metallic box normally used to enclose the subsystem reflect any radiation and set up standing waves or multiple reflections. Little of this energy thus escapes from the enclosed module, but it does contribute to overall losses and should be minimised.

An open-ended microstrip can be treated as having an associated shunt admittance Y given by:

$$Y = G_r + G_s + jB \tag{3.110}$$

where G_r is the radiation conductance (modelling the radiation loss), G_s is the surface wave conductance (modelling the surface wave loss) and B is the shunt susceptance. The radiation conductance may be approximated by:

$$G_r Z_0 \approx \frac{4\pi h w_{eff}}{3\lambda_0^2\sqrt{\varepsilon_{eff}}} \tag{3.111}$$

where w_{eff} is the effective microstrip width, defined by:

$$w_{eff}^2 = c^2/4f_p^2\varepsilon_{eff} \tag{3.112}$$

In this expression f_p is the quantity defined by Getsinger, see Section 3.6.2. (This is slightly non-linearly frequency dependent, i.e. dispersive.)

Self-assessment Problem

3.30 Assuming the same 31.3 ohm microstrip as defined for previous SAPs:

1. Calculate the conductor and dielectric losses at a frequency of 14 GHz for a substrate with tan $\delta = 0.002$. Compare your results and comment on the practical implications. (Ignore surface roughness.)
2. Again, at a frequency of 14 GHz, calculate the equivalent radiation loss resistance and sketch the complete equivalent circuit for a length of line with one open end.

What would be the result of *encouraging* (instead of minimising) these radiation losses? Is this approach at all useful at relatively low frequencies? Explain.

3.6.5 Discontinuity Models

Continuous, uninterrupted, sections of microwave transmission lines are unrealistic in practice and networks must be created that involve graded or sudden changes in structure. These changes, called *discontinuities*, are very important in MIC (and MMIC) design. The following represent examples of microstrip discontinuities:

1. The foreshortened open end.
2. The series gap.
3. Vias (i.e. short circuits) through to the ground plane.
4. The right-angled corner or bend (unmitred or mitred).
5. The step change in width.
6. The transverse slit.
7. The T-junction.
8. Cross-junctions.

In order to show instances of several of these discontinuities a typical single-stage MIC transistor amplifier (GaAsFET) layout is shown in Figure 3.16.

Modelling and computation of the effects of all these discontinuities have been the subject of much research effort over the past four decades.

We shall restrict our considerations here to the following types of discontinuities:

1. The foreshortened open end.
2. Vias.
3. The *mitred* right-angled bend.
4. The T-junction.

3.6.5.1 The foreshortened open end

As already discussed, radiation and surface waves are generally launched from this type of discontinuity. There is also, however, susceptance (B in the admittance expression) at

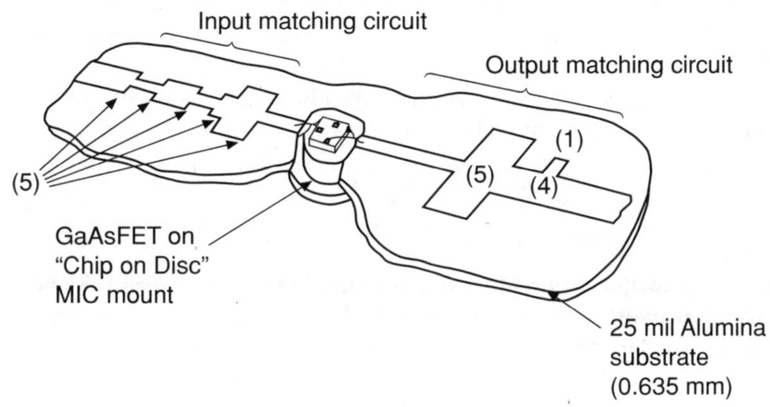

Figure 3.16 Single-circuit layout of a single-stage GaAsFET MIC amplifier

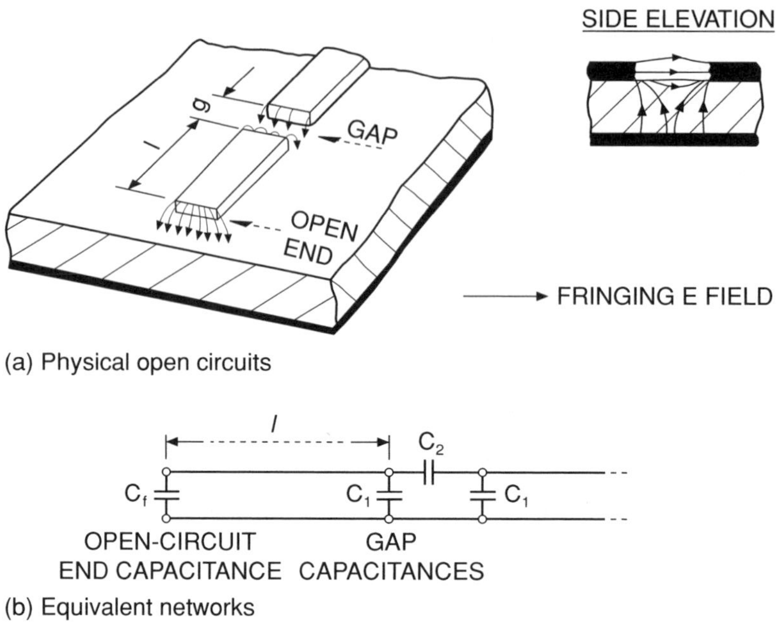

Figure 3.17 Microstrip open end and series gap: (a) physical circuits, (b) equivalent networks

this plane, due mainly to capacitance resulting from the local electric field fringing from the open end of the microstrip down to the ground plane. A practical example of this feature (and also a series gap) is shown in Figure 3.17(a) and the equivalent lumped capacitance associated with these structures is shown in Figure 3.17(b).

In *some* CAE software approaches the end-fringing capacitance, C_f, may be used directly within the design. However, it is important to observe that this particular discontinuity frequently occurs at the 'end' of an open circuit stub – which could be part of a filter or a matching network (see Section 3.9 and subsequent sub-sections). In this case it is the *complete effective electrical length of the stub* that is required in the design. This poses an initial problem, because we have mixed types of circuit elements here – one distributed (the microstrip itself) and the other lumped (C_f), Figure 3.18.

In terms of C_f directly it can be shown that this additional section of line, to be added in series beyond the microstrip open end, is given by:

$$l_{e0} \approx \frac{cZ_0C_f}{\sqrt{\varepsilon_{eff}}} \tag{3.113}$$

Equation (3.113) is adequate providing the designer knows the accurate value of C_f for all microstrips in the network. This is not particularly convenient, however, and an empirical expression (that embodies only microstrip parameters) such as:

$$l_{e0} = 0.412h\left(\frac{\varepsilon_{eff} + 0.3}{\varepsilon_{eff} - 0.258}\right)\left(\frac{w/h + 0.262}{w/h + 0.813}\right) \tag{3.114}$$

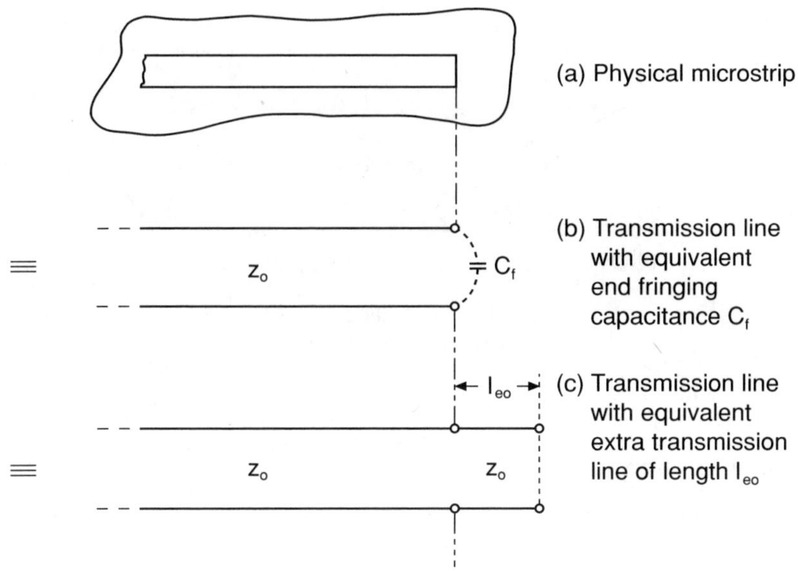

Figure 3.18 Equivalent elements to represent the microstrip open end fringing field

is useful. Error bounds of approximately 5% apply when using this expression over a wide range of types of microstrips on differing substrates. This degree of error should be acceptable in most cases.

3.6.5.2 Microstrip vias

The provision of short-circuiting holes between conductor patterns is a familiar one applying to circuits operating from DC to millimetre waves. Lower frequency circuits, including the important multi-chip modules (MCMs) require such vias. They are also highly significant in monolithic integrated circuits of all types – ICs and MMICs.

It is instructive to pose the question *when is a short circuit not actually a short circuit?* The answer is, when the frequency is sufficiently high.

Even for a conducting hole prepared in an extremely thin substrate, the term 'short circuit' is only ever near-perfect at DC. As the frequency increases so losses due to the skin effect mount and the inductive reactance becomes increasingly significant.

A short-circuiting conductive hole, manufactured at the otherwise 'open' end of a microstrip is shown in Figure 3.19.

A useful optimising condition that has been derived for this structure is:

$$\ln\left(\frac{w_{eff}}{\pi d_e}\right) \approx \left(\frac{\pi d_e}{w_{eff}}\right)^2 \tag{3.115}$$

where $d_e = 0.03 + 0.44d$, d is the actual (physical) hole diameter and w_{eff} is the effective microstrip width. Provided this condition is satisfied, the hole should remain an acceptable broadband short circuit over the frequency range 4 to 18 GHz. Calculations using Equation

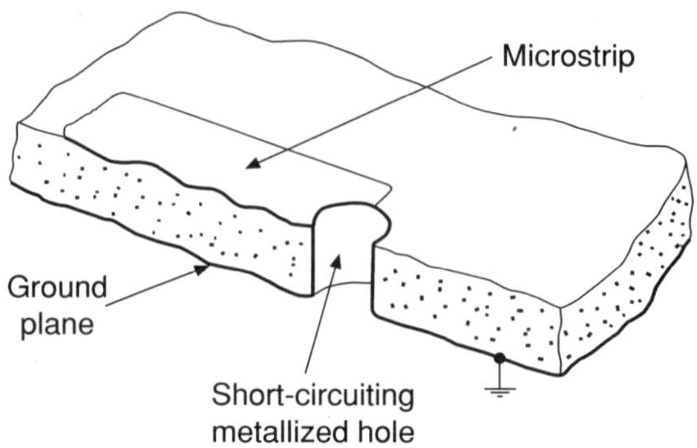

Figure 3.19 A shunt metallised hole in microstrip (at an otherwise 'open' end)

(3.115) require substitution and iteration, since the equation is transcendental in both d_e (and hence d) and w_{eff}.

3.6.5.3 Mitred bends

If a sudden bend is created in microstrip, sharp corners are inevitable on both the inside and the outside of the bend. These give rise to substantial current discontinuities because the major proportion of the total current flows in the outside edges. In turn, these current disturbances lead to excess inductances. Another, equally valid, way of viewing the situation is to appreciate that such sharp bends lead to significant mismatches, even when both ends of the complete microstrip line are terminated. It is evident, therefore, that an attempt must be made to reduce the effect of the bend discontinuity. The most important approach to this problem is to chamfer (or mitre) the 'long' outside edge, as shown in Figure 3.20.

On first consideration it might be thought that increasing advantage would be obtained the greater the degree of chamfer. However, in the limit this results in local microstrip-narrowing and current-crowding in the cornered region – so there is actually an optimum degree of chamfer. It has been shown that this optimum is, to a good approximation, given by:

$$b \sim 0.57w \tag{3.116}$$

3.6.5.4 The microstrip T-junction

In many instances it is necessary to branch from one microstrip into another. Examples include stubs and branched signal routing. Such branching is often achieved using a right-angled (T-shaped) junction, Figure 3.21. These junctions involve inherent discontinuities that need to be modelled.

In common with the microstrip bend, and for identical reasons, we must include series inductance to account for current disturbances, and electric field distortion means that capacitance has to be introduced locally at the junction plane. It is also necessary to account for the effects of general impedance loads introduced at some plane along the branching microstrip line. This is accommodated by means of an equivalent transformer element that transforms this impedance so that its effect is equivalently introduced into the main microstrip line.

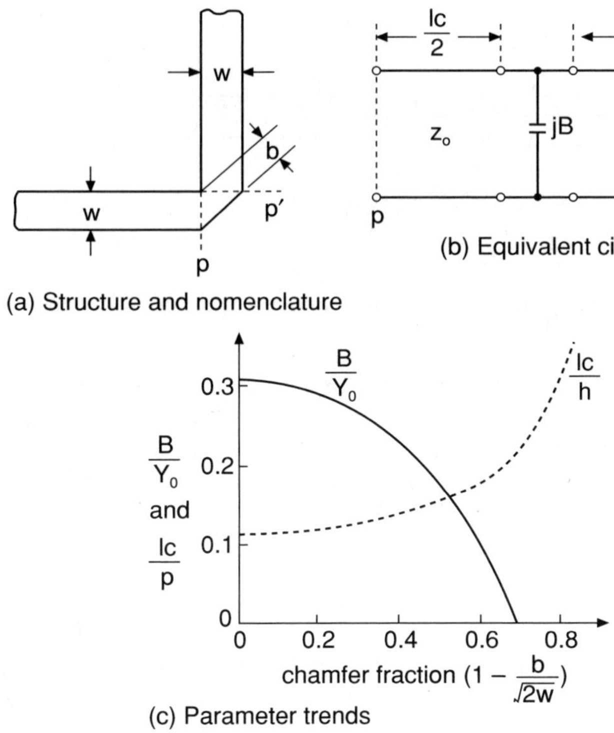

Figure 3.20 The chamfered (or mitred), right-angled, microstrip bend and relationships

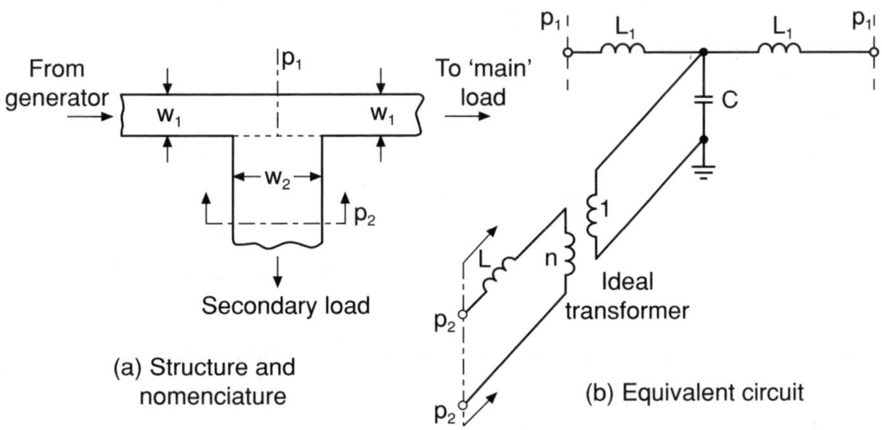

Figure 3.21 The microstrip T-junction and its elementary equivalent circuit

There have been some difficulties in using this equivalent circuit, especially in terms of accurately modelling the transformer (which requires a frequency-dependent turns ratio). Driven by these problems, research has yielded semi-empirical design expressions based upon including effective shifts in reference planes in this structure. The nomenclature for the reference planes is shown in Figure 3.22.

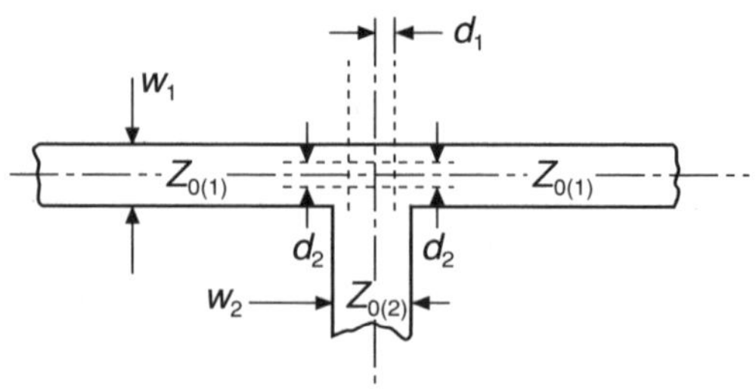

Figure 3.22 Reference planes and impedances for the T-junction

In all the following expressions, relating to the T-junction, the subscripts 1 and 2 denote microstrip number 1 (the main line) and microstrip 2 (the branch line), the impedances are characteristic impedances for the lines, and the capacitance C is the local equivalent junction capacitance taking account of the electric field disturbances. In this model effective microstrip widths (defined earlier) are also used, and their values for either line are given by:

$$w_{eff1,2} = \frac{h\eta}{Z_{0(1,2)}\sqrt{\varepsilon_{eff1,2}}} \tag{3.117}$$

where η is the impedance of free space (376.7 Ω). The turns ratio, n, reference plane displacements and capacitance are given by the following (complicated) algebraic relationships:

$$n^2 = \left[\frac{\sin\{\pi(w_{eff1}/\lambda_{g1})(Z_{0(1)}/Z_{0(2)})\}}{\pi(w_{eff1}/\lambda_{g1})(Z_{0(1)}/Z_{0(2)})}\right]^2 [1 - \{\pi(w_{eff1}/\lambda_{g1})(d_2/w_{eff1})\}^2] \tag{3.118}$$

The displacement of the reference plane for the primary line (1) is:

$$\frac{d_1}{w_{eff2}} = 0.05\frac{Z_{0(1)}}{Z_{0(2)}}n^2 \tag{3.119}$$

and the displacement of the reference plane for the secondary arm (2) is:

$$\frac{d_2}{w_{eff1}} = 0.5 - \left\{0.076 + 0.2\left(\frac{2w_{eff1}}{\lambda_{g1}}\right)^2 + 0.663\exp\left(-1.71\frac{Z_{0(1)}}{Z_{0(2)}}\right) - 0.172\ln\left(\frac{Z_{0(1)}}{Z_{0(2)}}\right)\right\}\frac{Z_{0(1)}}{Z_{0(2)}} \tag{3.120}$$

The shunt capacitance is determined by the following expression for the condition $Z_{0(1)}/Z_{0(2)} \leq 0.5$:

$$\frac{\omega C\lambda_{g1}}{Y_{0(1)}w_{eff1}} = \left(\frac{2w_{eff1}}{\lambda_{g1}} - 1\right)\frac{Z_{0(1)}}{Z_{0(2)}} \tag{3.121}$$

and, for the condition $Z_{0(1)}/Z_{0(2)} \geq 0.5$, by:

$$\frac{\omega C\lambda_{g1}}{Y_{0(1)}w_{eff1}} = \left(\frac{2w_{eff1}}{\lambda_{g1}} - 1\right)\left(2 - 3\frac{Z_{0(1)}}{Z_{0(2)}}\right) \tag{3.122}$$

Discrepancies resulting from the inherent approximations in these expressions increase when, for the main line, twice the effective width divided by the wavelength exceeds about 0.3. It has been found by measurements that the d_2 prediction by Equation (3.120) is typically too large at microwave frequencies – by as much as a factor of two or more. (At microwave frequencies, therefore, halving d_2 generally provides a better result.)

Self-assessment Problem

3.31 For the same microstrip line specified for earlier questions (the 31.3 ohm line), calculate:

(a) The equivalent open end effective length. What is the electrical length, in degrees, of this at a frequency of 20 GHz? How does this electrical length compare with that of a one-eighth wavelength of the microstrip line itself?

(b) The physical diameter of an optimum broadband short-circuiting via.

(c) The width of microstrip conductor remaining after chamfering a right-angled bend. (Also calculate this width for a 90 ohm microstrip – other parameters the same – and comment on any obvious practical problem arising.)

3.6.6 Introduction to Filter Construction Using Microstrip

Compared with coaxial lines, waveguides, or dielectric resonators, most planar structures, including microstrip, exhibit relatively low Q-factors (over 100 is often difficult to realise). This is a serious limitation in respect of several types of filter design. With specifications suitable for many purposes, however, a range of classic filters can be constructed in microstrip. (Where Q-factors much above 100 are absolutely essential, then alternative technologies must be selected, e.g. dielectric resonator 'pill' chips surface mounted on to otherwise microstrip MICs.)

We shall exclusively discuss low-pass filters at this point, because the design requirements for these encompass the microstrip techniques described previously.

3.6.6.1 Microstrip low-pass filters

The approach here begins by taking the model of lumped component low-pass filter (LPF) design comprising a cascade of C-L-C sections. The aim is to first determine the (often unrealisable) set of lumped component values, e.g. 1 pF, 3 nH, 2 pF, and then to transform these values into realisable microstrip sections. The general lumped network topology is shown in Figure 3.23(a) and the derived distributed (microstrip) configuration is shown in Figure 3.23(b).

The element values in the lumped network are obtained by conventional filter synthesis. This starts with low-pass filter synthesis determined by the required insertion loss characteristic and then proceeds to the microwave values by performing suitable frequency transformations (e.g. Butterworth, Chebychev). We assume here that the calculations have been performed and that the microwave-version lumped LPF topology is known (for a single

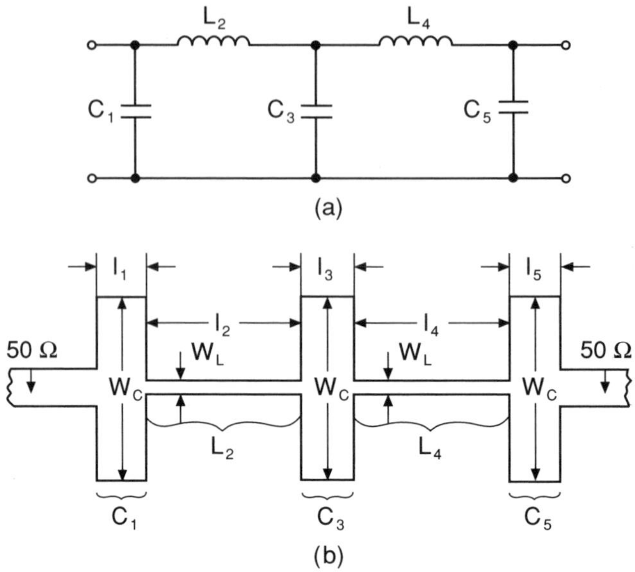

Figure 3.23 Lumped network topology (a) and microstrip configuration (b) for a low-pass filter

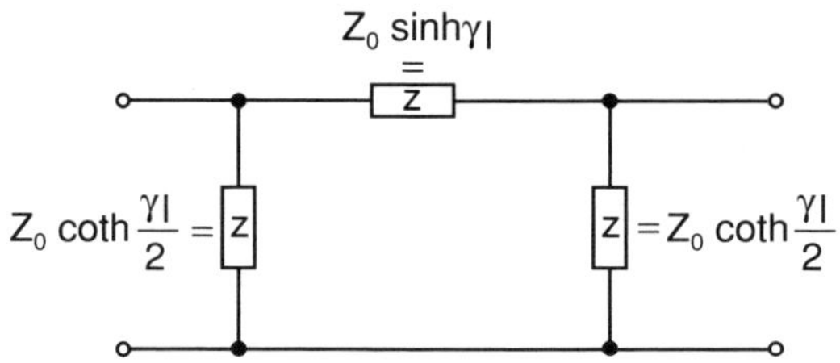

Figure 3.24 Equivalent π-network of impedance elements representing a length l of transmission line having propagation coefficient γ

or double-terminated filter, as desired). The next stage is to realise each inductive and capacitive element in microstrip form.

Fundamentally, we have a situation where inductive and capacitive lines are adjacent, which can be represented by the equivalent π-network shown in Figure 3.24.

In this approach to LPF design the effects of losses are neglected so that all elements are (as required by the lumped element model) reactive/susceptive. The basic lumped element π-network, its lumped-distributed equivalent and the final microstrip form of the structure are shown in Figure 3.25.

Fundamental transmission line theory shows that the input reactance of a line of length l is given by:

$$X_L = Z_0 \sin\left(\frac{2\pi l}{\lambda_g}\right) \tag{3.123}$$

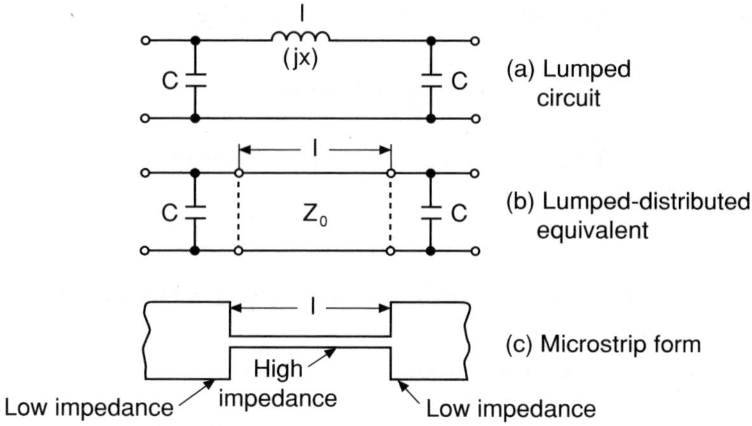

Figure 3.25 Inductive line with adjacent capacitive lines

which can be re-arranged to yield the length of the required section of line:

$$l = \frac{\lambda_g}{2\pi} \sin^{-1}\left(\frac{\omega L}{Z_0}\right) \tag{3.124}$$

When (and only when) the section of line is electrically short (much less than a quarter wavelength) it is possible to simplify Equation (3.124):

$$l \approx \frac{f\lambda_g L}{Z_0} \tag{3.125}$$

In most cases, however, the full expression given in Equation (3.124) will be required.

The shunt end susceptances are given by:

$$B_L = \frac{1}{Z_0} \tan\left(\frac{\pi l}{\lambda_g}\right) \tag{3.126}$$

which for short electrical lengths yields the following approximate expression for capacitance:

$$C_L \approx \frac{1}{2fZ_0\lambda_g} \tag{3.127}$$

In addition to these components, when realising the physical microstrip line there will also be the discontinuity elements due to the step in impedances. We did not consider this particular discontinuity, but where the step is large the main capacitive element of the discontinuity is approximately that of the wide adjacent capacitive microstrips.

A short length of microstrip line having relatively low characteristic impedance (i.e. a wide line) is predominantly capacitive and its shunt susceptance is given by:

$$B = \frac{1}{Z_0} \sin\left(\frac{2\pi l}{\lambda_g}\right) \tag{3.128}$$

which can be re-arranged to yield the length of this short line:

$$l = \frac{\lambda_g}{2\pi} \sin^{-1}(\omega C Z_0) \tag{3.129}$$

Again, since these lines are always electrically short the following approximation holds with reasonable accuracy:

$$l \approx f \lambda_g Z_0 C \tag{3.130}$$

The inductances associated with this predominantly capacitive line are usually negligible.

3.6.6.2 Example of low-pass filter design

We start with the basic specification, which is a 3 dB cut-off frequency of 2 GHz. (Note: cut-off frequencies are not always specified at the 3 dB level for filters – always check.) The design is based on a 5th order Butterworth response, and a *double-terminated 50 ohm system*.

We are assuming that the fundamental calculations have been completed as far as the microwave prototype lumped network. Following the nomenclature of Figure 3.23(a) the values are:

$$C_1 = 0.98 \text{ pF};\ L_2 = 6.4 \text{ nH};\ C_3 = 3.18 \text{ pF};\ L_4 = 6.4 \text{ nH};\ C_5 = 0.98 \text{ pF}$$

(and, to repeat, a 50 ohm system extends on both the signal input and output ends of this circuit).

All the design expression given above are now employed to develop the required microstrip line lengths, on an alumina substrate of relative permittivity 9.6 and thickness 0.635 mm. The final 2 GHz LPF layout is shown in Figure 3.26.

In practice this type of design is generally developed using proprietary RF/microwave CAE software.

There are many further possibilities with LPF (and other) filter designs. Notably, the low-impedance line sections can be replaced with 'quasi-lumped' capacitive pads – usually circular configurations. It is important to appreciate that these are indeed strictly 'quasi-lumped' because, as the frequency is increased, resonant modes can occur within the pad. Obviously this complicates the design.

3.7 Coupled Microstrip Lines

When any forms of transmission line, specifically TEM propagating (or closely related) forms, are located adjacently, then a fraction of the signal energy travelling on one line is coupled across to the other line. This is edge-coupling or *parallel coupling* and accounts for the unwanted cross-talk experienced in many circuits. The coupling process results in contra-directional travelling waves; the coupled wave travels in the opposite direction to that of the incident wave.

Such coupling is important in many instances, extending from *directional couplers* that are required in many applications, to parallel coupled resonators that are highly significant in parallel-coupled bandpass filters.

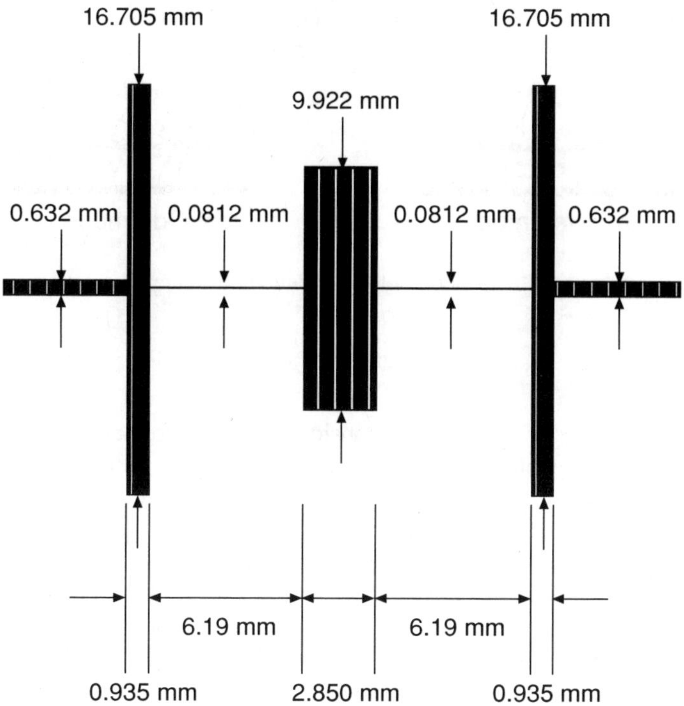

Figure 3.26 Layout of the 2 GHz Butterworth microstrip LPF

There is considerable flexibility in the configuration of this type of coupling – for instance, the lines do not have to be identical, they could have different geometries.

Here we shall concentrate upon the pair of identical parallel-coupled microstrips. The configuration is shown in Figure 3.27.

Throughout the following subsections it is important to remember that electromagnetic transmission on microstrips is *quasi-TEM* and not pure TEM. This has far-reaching consequences for the behaviour and achievable specifications of microstrip coupled lines. Analysis of coupled microstrips begins with the concept of superimposing the effects of two separately identifiable modes associated with this situation – the so-called *even* and *odd* modes. This concept is vital to understanding, and designing, any component using coupled microstrips.

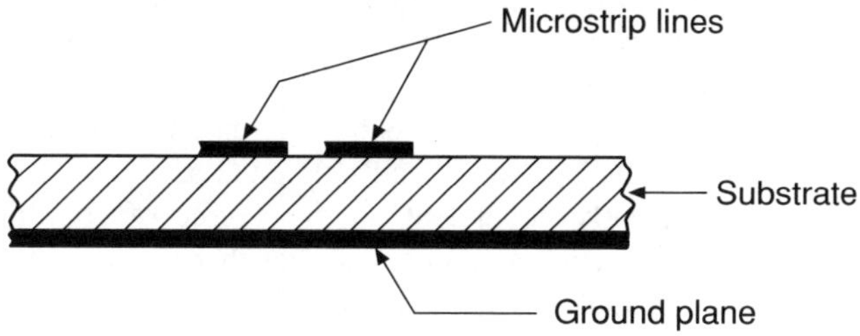

Figure 3.27 Cross-section showing a pair of parallel-coupled (or 'edge-coupled') microstrips

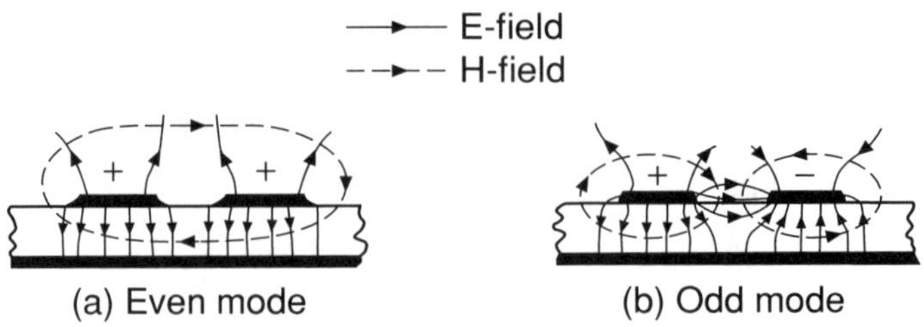

Figure 3.28 Even- and odd-mode field distributions for parallel-coupled microstrips

3.7.1 Theory Using Even and Odd Modes

The concept of these modes arises from considering the extremes of polarisation for the driving and coupled voltage on each microstrip. The aim is to eventually use the results from considering these extremes to develop design expressions for the coupled situation (such as degree of coupling, directivity, etc.).

The *even mode* is defined as the field situation produced when the voltage of each microstrip has *identical polarity* (either both positive or both negative). In contrast, the *odd mode* is defined as the field situation that exists when the voltages on each microstrip are of *opposite polarity*. These two extreme situations are shown in Figure 3.28 (in which only outlines of the electric and magnetic fields are considered).

Consequently, instead of a single characteristic impedance, phase velocity, capacitance, effective microstrip permittivity and electrical length (θ) being associated with a single microstrip as covered in some detail in Section 3.6 and subsequent subsections, we have two definitions for each of these quantities – one for each mode. The notation therefore becomes:

Z_{0e} and Z_{0o}	for characteristic impedances
C_{ie} and C_{io}	for capacitances
ε_{effe} and ε_{effo}	for effective relative permittivities

Extensive analysis has been undertaken in order to generate design data for these parameters under a wide variety of microstrip conditions. We shall not consider this analysis in detail here, but we shall need to be clearly aware of the results, consequences and limitations.

First let us consider the overall nomenclature associated with these coupled microstrips. This differs only by virtue of having one extra physical parameter – namely the strip separation, s. Otherwise the microstrips have widths, w, and are manufactured on a substrate of relative permittivity ε_r and thickness h as for single microstrip. The cross-section defining physical dimensions is shown in Figure 3.29.

Note that there is one further important dimensional parameter not shown in Figure 3.29, and this is the length of the coupled region, usually denoted as l.

All the parameters are inherently dependent on the degree of coupling required. This might seem obvious but it is important to appreciate that this fact extends to the widths of the microstrips as well as the separation s and the length l. Also, as with single microstrip, all the parameters of the coupled microstrips are strictly frequency-dependent and so dispersion exists in this structure – but with important differences compared to the single microstrip.

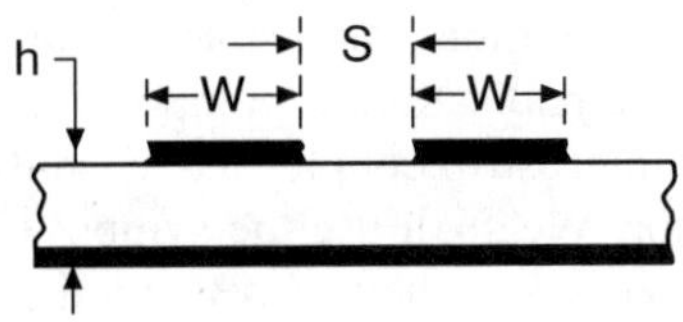

Figure 3.29 Nomenclature for the cross-sectional dimensions of parallel-coupled microstrips

Modelled on the results for single microstrip, the odd- and even-mode characteristic impedances are given by:

$$Z_{0o} = \frac{1}{cC_{1o}\sqrt{\varepsilon_{effo}}} \tag{3.131}$$

and:

$$Z_{0e} = \frac{1}{cC_{1e}\sqrt{\varepsilon_{effe}}} \tag{3.132}$$

To a limited extent, with some difficulties in terms of design use due to curve convergence, interpolation, and other approximations, the graphs of Figure 3.30 can be used for approximate designs. They are also useful for general guidance.

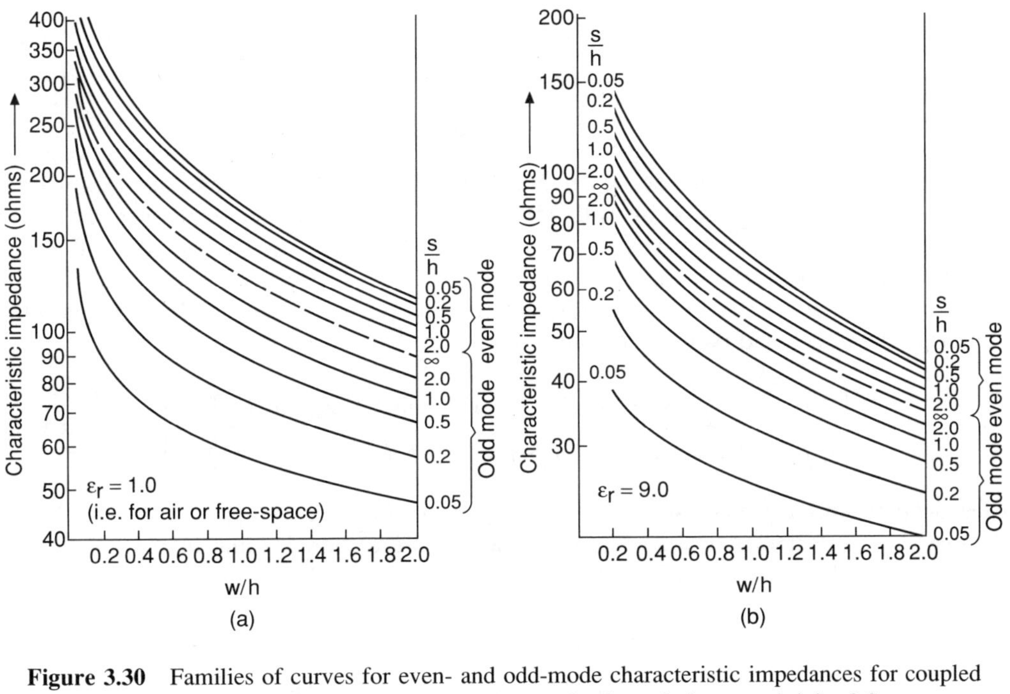

Figure 3.30 Families of curves for even- and odd-mode characteristic impedances for coupled microstrips in (a) air and (b) on a substrate having relative permittivity 9.0

We now consider some design equations applicable to a microstrip coupler and also to the use of coupled microstrips for the realisation of components such as band-pass filters.

If we are aiming to design a microstrip coupler based on this configuration, then we start with the required coupling factor. We shall use the symbol C for this parameter as a linear ratio, and the symbol C' when it is specified in dB (the usual practice). All the following simple and direct relationships are for mid-band.

The coupling factor is the given by:

$$C' = 20 \log_{10}\left|\frac{Z_{0e} - Z_{0o}}{Z_{0e} + Z_{0o}}\right| \quad \text{[dB]} \tag{3.133}$$

There also exists an approximate impedance interrelationship that is required to develop expressions for the even- and odd-mode characteristic impedances and which may be used as a check on relative values during a design process. This is:

$$Z_0^2 \approx Z_{0e}Z_{0o} \tag{3.134}$$

The error in this expression increases strongly as the degree of coupling increases.

Using Equations (3.133) and (3.134) the important even- and odd-mode characteristic impedances can readily be derived. These are:

$$Z_{0e} \approx Z_0\sqrt{\frac{1 + 10^{C'/20}}{1 - 10^{C'/20}}} \tag{3.135}$$

$$Z_{0o} \approx Z_0\sqrt{\frac{1 - 10^{C'/20}}{1 + 10^{C'/20}}} \tag{3.136}$$

Equations (3.135) and (3.136), are the essential starting expressions for proceeding with a microstrip coupler design. Almost always the designer will know the single microstrip characteristic impedance Z_0 and also the required mid-band coupling factor C' (the expressions as written here automatically convert the dB value of C' into its linear equivalent).

A completely accurate and universal relationship for Z_0 as a function of Z_{0o} and Z_{0e} is obtainable from fundamental analysis of the coupled microstrip system. This starts with the two-port ABCD matrix for the voltages and currents on each line, incident (denoted by subscript 1) and coupled (denoted by subscript 2):

$$\begin{bmatrix} V_1 \\ I_1 \end{bmatrix} = \begin{bmatrix} \cos\theta_e + \cos\theta_o & j(Z_{0e}\sin\theta_e + Z_{0o}\sin\theta_o) \\ j(Y_{0e}\sin\theta_e + Y_{0o}\sin\theta_o) & \cos\theta_e + \cos\theta_o \end{bmatrix} \begin{bmatrix} V_2 \\ I_2 \end{bmatrix} \tag{3.137}$$

(See Section 3.8.1 for a revision of matrix representations for two-port networks.) For a theoretically perfect match (also infinite isolation) the diagonal terms in the admittance matrix must be forced into equality. We need this condition and when invoked it leads to the following expressions:

$$\frac{Z_{0e}}{Z_0}\sin\theta_e + \frac{Z_{0o}}{Z_0}\sin\theta_o = \frac{Z_0}{Z_{0e}}\sin\theta_e + \frac{Z_0}{Z_{0o}}\sin\theta_o \tag{3.138}$$

or:

$$Z_0^2 = Z_{0e}Z_{0o}\frac{Z_{0e}\sin\theta_e + Z_{0o}\sin\theta_o}{Z_{0e}\sin\theta_o + Z_{0o}\sin\theta_e} \tag{3.139}$$

There are two highly significant features in these expressions, both important to understanding couplers (and coupled microstrips generally). These are:

1. Because of the differing field configurations in the even and odd modes, the electrical lengths θ_e and θ_o can never be exactly equal, including at mid-band.
2. These electrical lengths (θ_e and θ_o) are naturally (differing) functions of frequency.

These features tell us that mid-band design will never be precise and will require compensation in order to adjust the coupling factor and increase the directivity. They also tell us that we are dealing with a frequency-dependent device and therefore some broad-banding approach is desirable.

Returning to the mid-band design approach, for moderate to relatively loose couplings (typically 10 dB and looser), we can use expressions for the characteristic impedances as functions of the speed of light c and the capacitances of the individual microstrips both in air and with the substrate dielectric – for each mode. The effective microstrip permittivities are also the ratio of these capacitances – again for each mode:

$$Z_{0e} = \frac{1}{c\sqrt{C_e C_{e1}}} \tag{3.140}$$

and:

$$Z_{0o} = \frac{1}{c\sqrt{C_o C_{o1}}} \tag{3.141}$$

Also:

$$\varepsilon_{effe} = \frac{C_e}{C_{e1}} \tag{3.142}$$

and:

$$\varepsilon_{effo} = \frac{C_o}{C_{o1}} \tag{3.143}$$

There is a powerful *approximate synthesis* approach for coupled microstrips, which is useful in both coupler and band-pass filter design. This is presented next.

The aim here is to obtain first-cut approximate results for w and s. (More accurate figures can then subsequently be derived by analysis, re-calculation of the characteristic impedances, and iteration – straightforward on a computer.)

There are two main stages in this approximate synthesis:

1. Determine (intermediate step) shape ratios for equivalent single-modelled microstrips: $(w/s)_{se}$ and $(w/s)_{so}$. All subscripts 's' here refer to this single microstrip model.
2. Use these shape ratios to find the final coupled microstrips widths w and separation s.

For stage 1 it is necessary to use halved values of the characteristic impedances. For the single microstrip ratio $(w/s)_{se}$ therefore we have:

$$Z_{0se} = \frac{Z_{0e}}{2} \tag{3.144}$$

and, for the single microstrip shape ratio $(w/s)_{so}$ we must use:

$$Z_{0so} = \frac{Z_{0o}}{2} \tag{3.145}$$

Explicit equations are available for the single microstrip shape ratios and also for s/h:

$$\left(\frac{w}{h}\right)_{se} = \frac{2}{\pi}\cosh^{-1}\left(\frac{2d - g + 1}{g + 1}\right) \tag{3.146}$$

If $\varepsilon_r \leq 6$:

$$\left(\frac{w}{h}\right)_{so} = \frac{2}{\pi}\cosh^{-1}\left(\frac{2d - g - 1}{g - 1}\right) + \frac{4}{\pi\left(1 + \frac{\varepsilon_r}{2}\right)}\cosh^{-1}\left(1 + 2\frac{w/h}{s/h}\right) \tag{3.147(a)}$$

or if $\varepsilon_r \geq 6$:

$$\left(\frac{w}{h}\right)_{so} = \frac{2}{\pi}\cosh^{-1}\left(\frac{2d - g - 1}{g - 1}\right) + \frac{1}{\pi}\cosh^{-1}\left(1 + 2\frac{w/h}{s/h}\right) \tag{3.147(b)}$$

where:

$$g = \cosh\left(\frac{\pi s}{2h}\right) \tag{3.148}$$

and:

$$d = \cosh\left(\pi\frac{w}{h} + \frac{\pi s}{2h}\right) \tag{3.149}$$

In order to use these expressions as a design aid, Equation (3.146) must be solved simultaneously with *either* Equation (3.147(a)) or (3.147(b)) – depending on the permittivity of the substrate used. However, the second terms in Equations (3.147(a)) and (3.147(b)) may often be ignored (especially with tight coupling, since then w/h will generally be some factor greater than s/h). Under this condition a direct formula for s/h is obtained:

$$\frac{s}{h} = \frac{2}{\pi}\cosh^{-1}\left[\frac{\cosh\{(\pi/2)(w/h)_{se}\} + \cosh\{(\pi/2)(w/h)_{so}\} - 2}{\cosh\{(\pi/2)(w/h)_{so}\} - \cosh\{(\pi/2)(w/h)_{se}\}}\right] \tag{3.150}$$

Although errors exceeding 10% occur when this technique is used alone, it is possible to obtain much improved accuracy by starting with these approximate relations and then solving accurate *analysis* expressions, followed by iteration. This approach is summarised below:

1. Starting with the desired Z_{0e} and Z_{0o} (from C' or from filter design), find initial values $(w/h)_1$ and $(s/h)_1$ using the approximate synthesis approach.
2. Apply these first $(w/h)_1$ and $(s/h)_1$ quantities to *re-calculate* Z_{0e} and Z_{0o} using relatively accurate *analysis* formulas.
3. Compare the two sets of Z_{0e} and Z_{0o} values, noting the differentials, and iterate the above calculations until the two sets of values are within a specified error limit. (In the case of a coupler, also compare values of the coupling factor C required.)

The values of w and s at this final stage are the closest to those required in the design.

The above expressions can be combined and data plotted in the form of the graph shown in Figure 3.31.

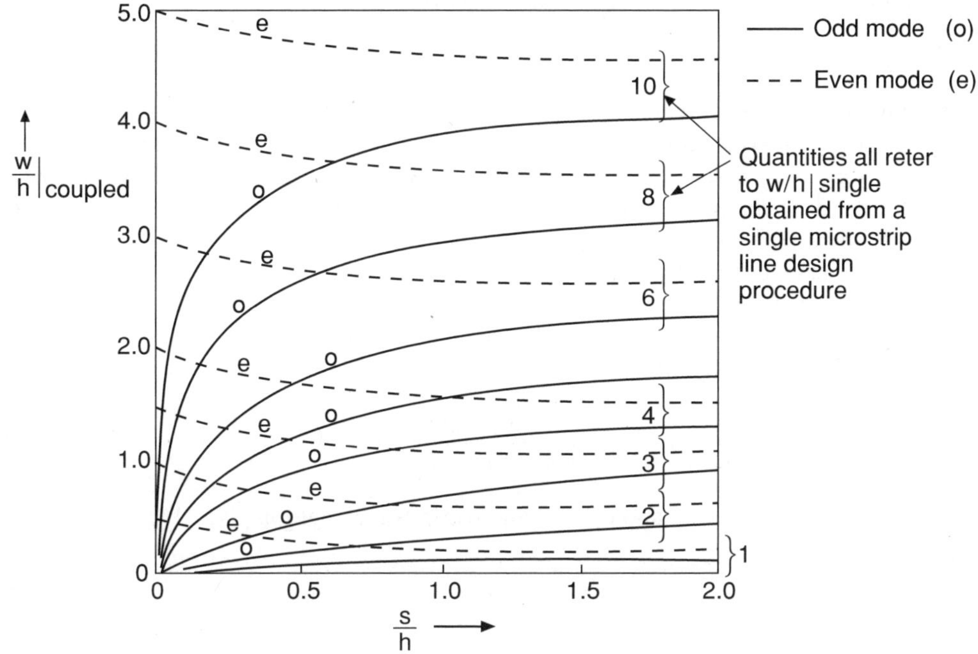

Figure 3.31 Family of curves for use in the approximate synthesis design (strictly applicable only for $\varepsilon_r = 6$)

This family of curves can often be particularly difficult to use because values are frequently within crowded regions of the graph.

Self-assessment Problem

3.32 A microstrip coupler, using parallel-coupled lines, is required to have a coupling factor of 10 dB. (Note that this means the voltage level coupled into the coupled line is −10 dB (i.e. 10 dB down) with respect to that on the incident line.) The single microstrip feeder connecting lines have terminated characteristic impedances of 50 ohms. The substrate has a relative permittivity of 9.0 and a thickness of 1 mm. Calculate the width and separation of the coupled lines using: (a) The curves shown in Figure 3.30, and (b) the approximate synthesis curves of Figure 3.31.

3.7.1.1 Determination of coupled region physical length

So far, in the design approaches for parallel-coupled microstrips, we have only concentrated on the determination of cross-sectional dimensions: w and h. A further quantity must be established to complete designs, namely the physical length of the coupled region, l.

In mid-band, i.e. for the maximum coupling condition, the electrical length of the coupled region should be an integral number of quarter wavelengths:

$$(n - 1)\lambda_g/4 \tag{3.151}$$

To avoid harmonic effects (repeated responses at higher frequencies) it is best if the region is simply one-half wavelength, i.e. $n = 2$, and the length is simply $\lambda_g/4$. However, at high frequencies this leads often to physically short structures that may be difficult to manufacture and $n = 3, 4$, etc. may be desired. It is pertinent to ask the question, here, *what is the value* λ_g? Remember, there are the two modes (even and odd) and they have different phase velocities and hence different wavelengths.

Various methods for dealing with this problem have been put forward. We shall describe the weighted mean approach here, in which the characteristic impedances are also invoked. Using this approach the weighted mean phase velocity is:

$$v_n = \frac{Z_{0e} + Z_{0o}}{(Z_{0e}/v_e) + (Z_{0o}/v_o)} \tag{3.152}$$

and (following weighting) the appropriate microstrip wavelength is given by:

$$\lambda_{gn} = \frac{v_n}{f_0} \tag{3.153}$$

This value is then used in Equation (3.151) to find the appropriate length. Some discontinuity will exist around the end regions of the coupler, where the single microstrips connect. This may be minimised by chamfering the entering corners (see Section 2.5.3).

Self-assessment Problem

3.34 Calculate the quarter-wave coupling region length for the coupler of the previous SAP, operating at a centre frequency of 5 GHz.

3.7.1.2 Frequency response of the coupled region

As remarked earlier in this section all the parameters of the coupled region are in fact frequency dependent due to two underlying phenomena:

1. The inherent frequency behaviour of any TEM parallel coupler.
2. The dispersion in the quasi-TEM microstrips differing between even and odd modes.

The first contribution to the frequency dependence, the TEM situation, will actually be interpreted to encompass this quasi-TEM structure and the relevant expression obtained from fundamental analysis is:

$$C(\theta) = C(j\omega) = \frac{jC\sin\theta}{\sqrt{1 - C^2}\cos\theta + j\sin\theta} \tag{3.154}$$

In this expression C is the mid-band value of the coupling factor already described in some detail. The phase angle θ is the general, frequency-dependent, quantity $\theta = \beta l = 2\pi l/\lambda_g$. While l will obviously be constant, β will vary since the wavelength λ_g will change with frequency.

For the ideal matched (i.e. terminated) coupler of this type the frequency response, neglecting dispersion, will closely follow that shown in Figure 3.32.

Unlike single microstrips, for coupled microstrip lines dispersion affects the even and odd modes in different ways. Referring back to Figure 3.28 it can seen that the field distributions

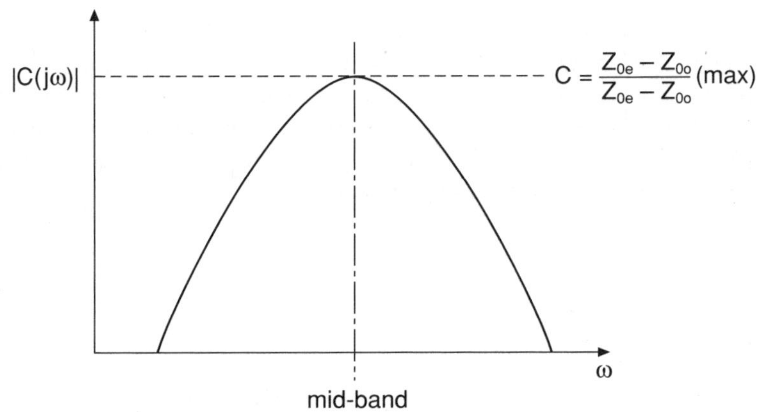

Figure 3.32 Frequency response for a loosely-coupled ideal matched coupler

are markedly different for each of these modes, including at low frequencies. It should therefore be clear that, as frequency increases and the coupling beneath the strips intensifies (see single microstrip dispersion in Section 3.6.2), the detailed changes in field distributions must differ between the even and odd modes.

As well as functions of frequency and substrate thickness, the effective microstrip permittivities for the even and odd modes are now also functions of the low frequency limiting values and the characteristic impedances (obtained by analysis).

Getsinger has shown how his (single microstrip) dispersion expressions may be adapted to approximately predict this dispersive behaviour. The key to this follows from studying the effects on characteristic impedance substitution, summarised thus:

1. For the even mode the two microstrips are at the same potential, but the total current flowing down the system is twice that of a single strip with otherwise identical structure (twice because of the paralleled conductors). This implies that the correctly substituted impedance should be one half of Z_{0e}. For the even mode case substitute, for Z_0, 0.5 Z_{0e}.
2. For the odd mode the two microstrips are at opposite potentials and the potential difference between the strips is therefore twice that of a single microstrip. However, the current is the same as for a single microstrip, and the correctly substituted impedance should therefore be twice Z_{0o}. For the odd mode case substitute, for Z_0, $2Z_{0e}$.

These two substitutions should be made in the formulas for Getsinger's single microstrip line case (see Section 3.6.2).

This approach is still approximate, including at relatively low microwave frequencies, and it should only be used with caution. More accurate expressions are available in the literature, but these remain elaborate algebraic functions with extended numerical coefficients. They are only really useful, therefore, for programming into CAE software routines, and not for 'hand calculations'.

Self-assessment Problem

3.36 Considering again the coupler structure designed in the previous SAP, calculate the length of the coupled region assuming the mid-band operating frequency to be 12 GHz. Use Getsinger's approach for dispersion. Note the differences between the effective relative permittivities and compare the wavelength calculated here with that calculated using the rough approximation: $(5/12)\lambda_g$, where λ_g is the value for 5 GHz determined in the earlier SAP.

3.7.1.3 Coupler directivity

Ideally, a coupler should only couple signal energy from its transmission port (i.e. directly through port 2) and its coupled port (port 3). However, in practice, 'stray' energy is also fed to the fourth port.

The ability of a coupler to reject unwanted signal energy from being coupled back from its fourth port is referred to as its directivity. Working in terms of voltages the definition of directivity is:

$$D = \left|\frac{V_4}{V_3}\right| \tag{3.155}$$

The analysis leading to expressions for this directivity, derived from the fundamental equations governing the coupled line behaviour, is quite complicated. A more amenable set of expressions is:

$$D = \left\{\frac{4|\zeta|}{\pi\Delta(1 - |\zeta|^2)}\right\}^{-2} \tag{3.156}$$

in which:

$$\zeta = \left(\frac{\rho_e}{1 + \rho_e^2}\right) - \left(\frac{\rho_o}{1 + \rho_o^2}\right) \tag{3.157}$$

where:

$$\rho_e = \frac{Z_{0e} - Z_0}{Z_{0e} + Z_0} \tag{3.158(a)}$$

$$\rho_o = \frac{Z_{0o} - Z_0}{Z_{0o} + Z_0} \tag{3.158(b)}$$

and:

$$\Delta = \frac{\lambda_{go}}{\lambda_{ge}} - 1 \tag{3.159}$$

Calculations using these expressions reveal that the best values of directivity achievable with the basic parallel-coupled structure are only around 12 or 14 dB, which is within the same region of value as the coupling factor in many instances, making it unacceptable for most applications. Techniques have therefore been developed for compensating for the coupler in order to improve performance.

3.7.1.4 Coupler compensation by means of lumped capacitors

This is arguably the most sophisticated method, in the engineering sense, for compensating for the behaviour of a microstrip coupled-line coupler. If lumped capacitors are introduced anywhere across the coupler they will only function in the odd mode, because only then is there a potential difference across the capacitors. Furthermore, if such capacitors are implemented across the end planes of the coupler they will not unduly disturb the main coupled region effects. The configuration is shown, in network outline, in Figure 3.33. (There are actually other advantageous performance features resulting from the introduction of such capacitors.)

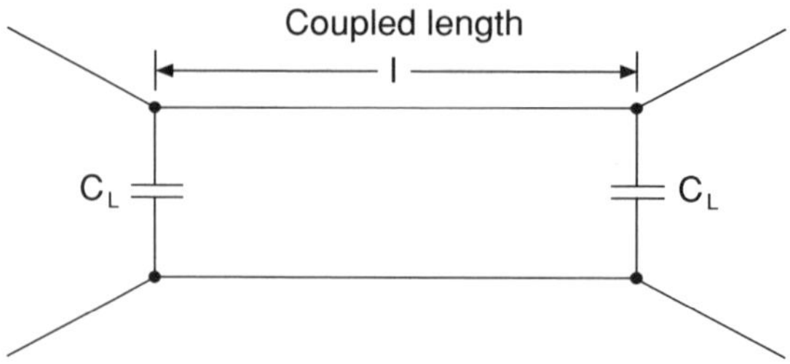

Figure 3.33 Schematic diagram of coupled microstrips, with compensating capacitors across the ends

The analysis of the circuit begins with expanding the ABCD matrix for this network (refer to Section 3.8.1 for a review of ABCD parameters). Again, this analysis is fairly lengthy and we shall only state the principal results here. The capacitance value is given by:

$$C_L = \frac{\cos\left(\frac{\pi}{2}\sqrt{\frac{\varepsilon_{effo}}{\varepsilon_{effei}}}\right)}{2Z_{0oi}\omega} \tag{3.160}$$

in which the final '*i*' subscripts for the even-mode relative permittivity and the odd-mode characteristic impedance refer to the ideal case (equal even- and odd-mode electrical lengths) and $\omega = 2\pi f$.

In practice, the capacitance values are usually much less than 1 pF (typically 0.05 pF or so) and these can be difficult to realise except by precision thin-film interdigital approaches – a technology that also leads to small gap dimensions.

For 10 dB couplers the directivity exceeds 20 dB over, typically, an 8 to 20 GHz band (which is very broadband). Input VSWR maximises at around 1.4. The broadbanding appears to be an essential unexpected advantage of this compensation method.

There are other ways in which these couplers can be compensated, notably using a dielectric overlay which serves to almost equalise the phase velocities. This overlay may be introduced as a dielectric layer co-fired during the manufacturing process (thin film or thick film), or it could be introduced in the form of a surface-mounted chip. Some excellent results have been achieved using this technique. A combined approach could also be explored in which both the compensating lumped capacitors and the dielectric overlay might be introduced together.

Given the limitations of practically any variation on the parallel-coupled microstrip coupler, especially where relatively tight coupling is required (e.g. 3 dB or greater), radically different configurations are available. One of these (not strictly microstrip) is the combined directional coupler. This uses a cascade of interdigital coupling elements (typically well over 10) along an otherwise microstrip-like arrangement. This enables coupling factors approaching 0 dB to be achieved and up to at least a three-octave bandwidth at low microwave frequencies. The viability of this approach is limited to low microwave frequencies because of the essentially quasi-lumped technology.

For fairly tight coupling, typically around 3 dB or tighter, over at least one octave of bandwidth, the Lange coupler remains an excellent choice. This is covered in the next section, which also deals with some so-called 'hybrid' couplers.

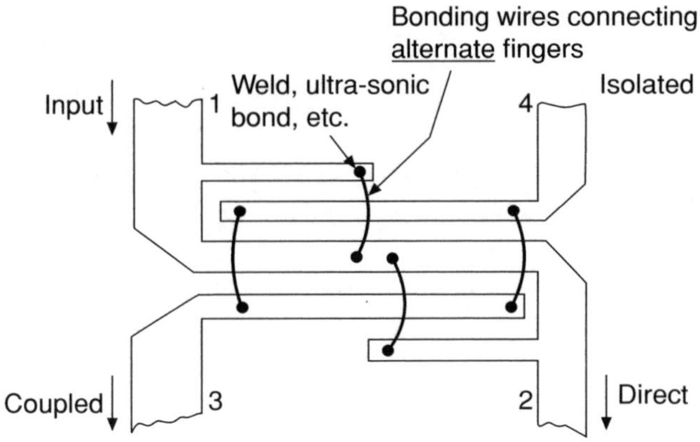

Figure 3.34 The basic form of the Lange coupler

3.7.2 Special Couplers: Lange Couplers, Hybrids and Branch-Line Directional Couplers

There have been many instances throughout the history of scientific and engineering inventions where an essentially heuristic or empirical effort has led to exciting and sustained patents. This is true of the Lange microstrip coupler, invented by Julius Lange while working at Texas Instruments in 1961. The Lange coupler was first reported at a microwave conference and initial results indicating its useful performance were shown – but for many years no formal design equations were available and the realisation of these types of couplers remained empirical. The general arrangement of the Lange coupler is shown in Figure 3.34.

It is immediately obvious that the Lange coupler structure differs greatly from the simple parallel-coupled microstrip structure. The principal differences are:

1. There are interleaved, quarter-wave, coupling sections. (Two are shown here – in practice there are many more.)
2. Bond wires straddle alternate coupling elements.

The first aspect (1) indicates that we are still principally interested in a quarter-wave coupling effect (i.e. a quadrature hybrid) at mid-band. The second aspect is arguably the most innovative, namely, the introduction of bond wires, although this complicates the manufacturability of the device.

The interdigital nature of the quarter-wave coupling sections ensures that substantial power transfer can occur over a broad band, and the wires serve to transfer further power – taking up what otherwise would be open ends of coupling microstrip fingers. It is essential to ensure that the bond wires remain much less than eighth-wavelengths at the highest design frequency, otherwise they will present serious design problems. In any case the inductance of these bond wires must be minimised.

Basic design expressions, enabling the 'standard' parallel-coupled microstrip design already covered to be used for the Lange coupler are:

$$Z_0^2 = \frac{Z_{0e}Z_{0o}(Z_{0e} + Z_{0o})^2}{\{Z_{0e} + (k-1)Z_{0o}\}\{Z_{0o} + (k-1)Z_{0e}\}} \qquad (3.161)$$

This equation provides the relationship defining all the impedances. The voltage coupling factor is given by:

$$C = 10^{-\{\text{voltage coupling ratio in dB/20}\}}$$

$$= \frac{(k-1)Z_{0e}^2 - (k-1)Z_{0o}^2}{(k-1)(Z_{0e}^2 + Z_{0o}^2) + 2Z_{0e}Z_{0o}} \tag{3.162}$$

and the expressions for the even- and odd-mode characteristic impedances are:

$$Z_{0e} = Z_{0o}\frac{C+q}{(k-1)(1-C)} \tag{3.163}$$

and:

$$Z_{0o} = Z_0\left(\frac{1-C}{1+C}\right)^{1/2}\frac{(k-1)(1+q)}{(C+q)+(k-1)(1-C)} \tag{3.164}$$

where q is related to the number of fingers in the design (k) and the voltage coupling ration (C) by:

$$q = \{C^2 + (1-C^2)(k-1)^2\} \tag{3.165}$$

The length of the entire coupler should be a quarter-wavelength at the lowest frequency in the desired band, whereas the length of the intermediate fingers should be made a quarter-wavelength at the highest frequency involved.

In the Lange coupler the even- and odd-mode phase velocities (and hence wavelengths, etc.) are very nearly equal, as required in the ideal case. Double octave bandwidths, e.g. 10 to 40 GHz, are achievable.

Self-assessment Problem

3.37 (a) Suggest, with an explanation, a method for reducing the inductance of the bond wires in Lange couplers.

(b) Determine the coupling conductor widths and separations for a Lange coupler having the following characteristics:

1. Coupling factor = −10 dB.
2. Centre frequency = 10 GHz.
3. Operating in a 50 ohm system.
4. Substrate relative permittivity 2.23 and thickness 0.234 mm.

Compare these values with those applying to a simple parallel-coupled microstrip coupler, and comment on the practical implications.

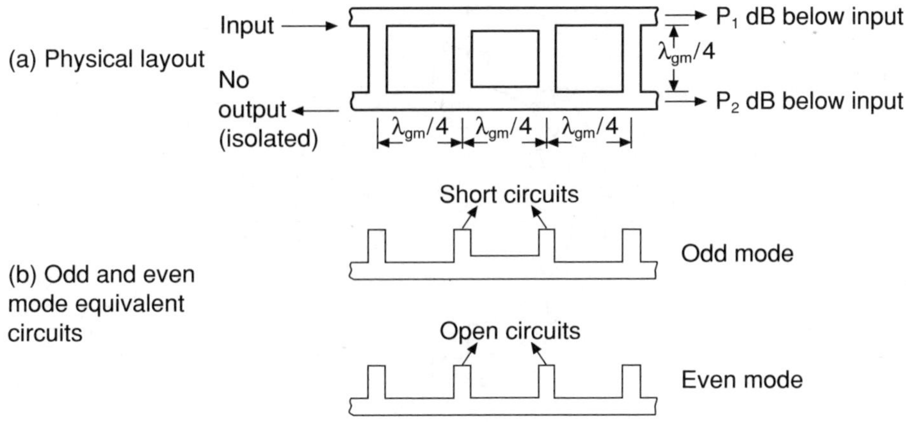

Figure 3.35 The orthogonal branch-line directional coupler

A wide and ever-growing variety of further coupler configurations exists. Amongst these are: branch-line directional couplers (orthogonal and ring types) and the 'rat-race' or ring hybrid. We indicate some examples of these configurations here, but shall not provide design routines or expressions.

The basic configuration of the (orthogonal) branch-line directional coupler is shown in Figure 3.35.

These types of couplers have the following properties:

1. Elements such as chokes or filters (particularly in-line band-stop) can be introduced into joining branches.
2. In many cases relatively high RF/microwave voltages can be handled (there is no risk of breakdown across a gap).
3. Tight coupling is easy to achieve – but only over limited bandwidths – typically 50%.
4. Impedance transforming is a further use of these designs.

Ring forms of couplers have been known for many years in waveguide, coaxial and strip-line configurations. The basic design requirements are similar for microstrip realisation and the 3 dB branch-line and the 'rat-race' or 'hybrid ring' are shown in Figures 3.36 and 3.37 respectively.

3.8 Network Methods

Network theory is a formidable aid to the electrical engineer. A thorough grasp of this subject enables a multitude of applications to be easily tackled. Examples of the wide applicability of network theory are its use in CAE packages to simulate highly complex and large-scale circuits, to the analysis of single semiconductor devices.

In the following subsections we concentrate on describing a parameter set, the scattering or 's' parameters, that is of particular relevance to the microwave circuit designer. The requirement for s-parameters in high frequency applications and the techniques for converting *s*-parameter data into other parameter sets (z, y, h, ABCD, etc.) are also described.

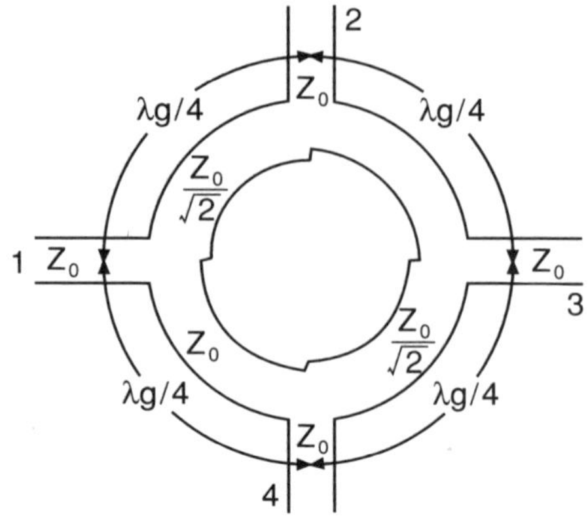

Figure 3.36 The 3 dB ring-form branch-line directional coupler

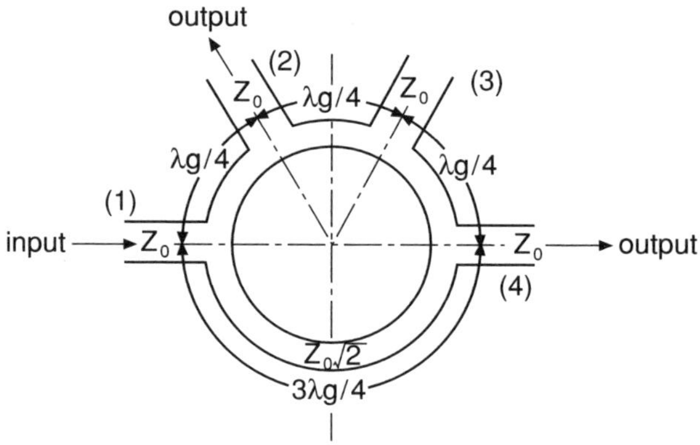

Figure 3.37 The 'rat-race' or 'hybrid ring' coupler

3.8.1 Revision of z, y, h and ABCD Matrices

A general two-port network is shown in Figure 3.38.

The most common parameter sets used to describe such a network are:

(i) z-parameters

$$V_1 = z_{11}I_1 + z_{12}I_2 \tag{3.166(a)}$$

$$V_2 = z_{21}I_1 + z_{22}I_2 \tag{3.166(b)}$$

(ii) y-parameters

$$I_1 = y_{11}V_1 + y_{12}V_2 \tag{3.167(a)}$$

$$I_2 = y_{21}V_1 + y_{22}V_2 \tag{3.167(b)}$$

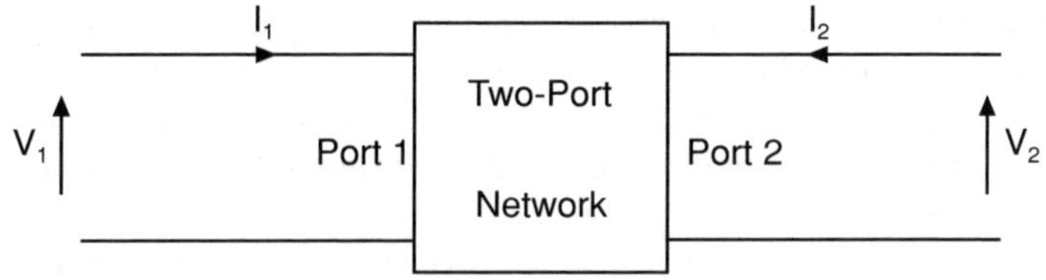

Figure 3.38 Two-port network

(iii) h-parameters

$$V_1 = h_{11}I_1 + h_{12}V_2 \tag{3.168(a)}$$

$$I_1 = h_{21}I_1 + h_{22}V_2 \tag{3.168(b)}$$

(iv) ABCD parameters

$$V_1 = AV_2 - BI_2 \tag{3.169(a)}$$

$$I_1 = CV_2 - DI_2 \tag{3.169(b)}$$

The selection of a particular set is very much influenced by the application. For example, manufacturers of bipolar transistors usually provide users with the h-parameters of the device. This set yields the input and output impedance of the transistor (h_{11}, $1/h_{22}$), its current gain (h_{21}) and its reverse voltage gain (h_{12}), which are key parameters in the selection of a device. In many circuit simulation packages, however, the admittance (or y-) parameters are used because of the ease with which large circuits can be described and their matrices manipulated. The choice of a particular set is also influenced by the ease with which the parameters can be measured in the particular situation of interest. Very often conversion between one parameter set and another is required to ease the task in hand. This is readily performed by simple substitution. For example, to determine z_{11} from the y-parameters, the measurement condition for z_{11} is first determined:

$$z_{11} = \left.\frac{V_1}{I_1}\right|_{I_2 = 0} \tag{3.170}$$

Then expressions for V_1 and I_1 are found from the Y-matrix:

$$V_1 = \frac{I_2 - y_{22}V_2}{y_{21}} \tag{3.171(a)}$$

$$I_1 = \frac{y_{11}(I_2 - y_{22}V_2)}{y_{21}} + y_{12}V_2 \tag{3.171(b)}$$

Finally, terms containing I_2 can be cancelled to give the required parameter transformation:

$$z_{11} = \left.\frac{V_1}{I_1}\right|_{I_2 = 0} = \frac{y_{22}}{y_{11}y_{22} - y_{12}y_{21}} \tag{3.172}$$

A similar process can be used to convert between any of the parameter sets. Notice that none of these parameter sets require knowledge of the source or load impedances connected to the network. Notice also, however, that these parameter sets do require a voltage or a current to be set to zero. From a measurement point of view, this requires a short or open circuit termination. As will be seen later, at microwave frequencies these two conditions are difficult to satisfy. An alternative parameter set that avoids this requirement, and is therefore particularly useful for microwave applications, is now described.

3.8.2 Definition of Scattering Parameters

Consider a sinusoidal voltage source of internal impedance Z_S applying a voltage to a transmission line of characteristic impedance Z_0 terminated in a load impedance Z_L, as shown in Figure 3.39.

The application of such a sinusoidal voltage causes a forward voltage and current wave to propagate towards the load. These have the complex form:

$$V(x, t) = A \exp(j\omega t - \psi_x) \quad (3.173(a))$$

$$I(x, t) = \frac{A}{Z_0} \exp(j\omega t - \psi_x) \quad (3.173(b))$$

If the load Z_L is perfectly matched to the line impedance Z_0, then all of the power incident on the load will be absorbed by the load. If Z_L is not matched to Z_0, however, then a wave will be reflected from the load. The reflected wave, which travels along the line towards the source, has the complex form:

$$V(x, t) = B \exp(j\omega t + \psi_x) \quad (3.174(a))$$

$$I(x, t) = -\frac{B}{Z_0} \exp(j\omega t + \psi_x) \quad (3.174(b))$$

If impedance Z_S is equal to Z_0, then the source is matched to the line.

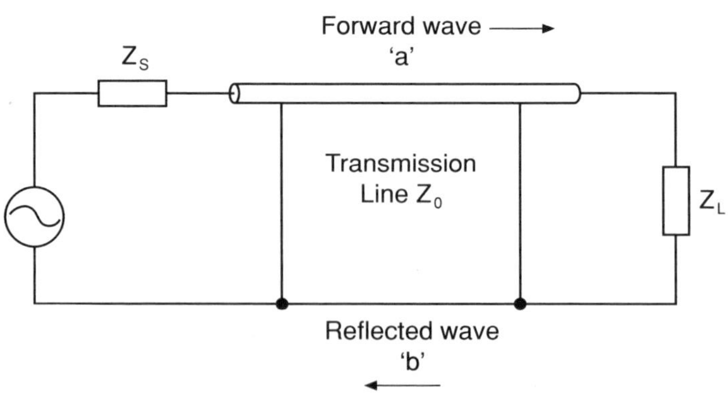

Figure 3.39 Transmission line as a network

The forward and backward travelling waves can be normalised (such that the amplitude of each term represents the square root of wave power) and combined, i.e.:

$$\frac{V(x,t)}{\sqrt{Z_0}} = \frac{A}{\sqrt{Z_0}}\exp(j\omega t - \psi_x) + \frac{B}{\sqrt{Z_0}}\exp(j\omega t + \psi_x) \qquad (3.175(a))$$

$$I(x,t)\sqrt{Z_0} = \frac{A}{\sqrt{Z_0}}\exp(j\omega t - \psi_x) - \frac{B}{\sqrt{Z_0}}\exp(j\omega t + \psi_x) \qquad (3.175(b))$$

The forward wave 'a' (see Figure 3.39) can therefore be defined as:

$$a = \frac{A}{\sqrt{Z_0}}\exp(j\omega t - \psi_x) \qquad (3.176(a))$$

and the reflected wave 'b' can be defined as:

$$b = \frac{B}{\sqrt{Z_0}}\exp(j\omega t + \psi_x) \qquad (3.176(b))$$

hence:

$$\frac{V(x,t)}{\sqrt{Z_0}} = a + b \qquad (3.177(a))$$

and:

$$I(x,t)\sqrt{Z_0} = a - b \qquad (3.177(b))$$

Alternatively (but equivalently), a and b can be defined as:

$$a = \frac{V(x,t) + I(x,t)Z_0}{2\sqrt{Z_0}} \qquad (3.178(a))$$

and:

$$b = \frac{V(x,t) - I(x,t)Z_0}{2\sqrt{Z_0}} \qquad (3.178(b))$$

If the ratio $s = b/a$ (referred to as the reflection coefficient) is used as a figure of merit, then:

$$s = \frac{b}{a} = \frac{V(x,t) - I(x,t)Z_0}{V(x,t) + I(x,t)Z_0} = \frac{Z_L - Z_0}{Z_L + Z_0} \qquad (3.179)$$

When Z_L is equal to Z_0, then s will be zero and we have a matched line. When Z_L is a short circuit ($Z_L = 0$), then s will be unity with a phase angle of $-180°$. When Z_L is infinity (open circuit), s will again be unity, but its phase angle is 0°. Between these extreme values of Z_L, the forward and reflected waves interact to form a standing wave. s therefore indicates the amount of mismatch present in the system. Maximum power is injected into the line when Z_S

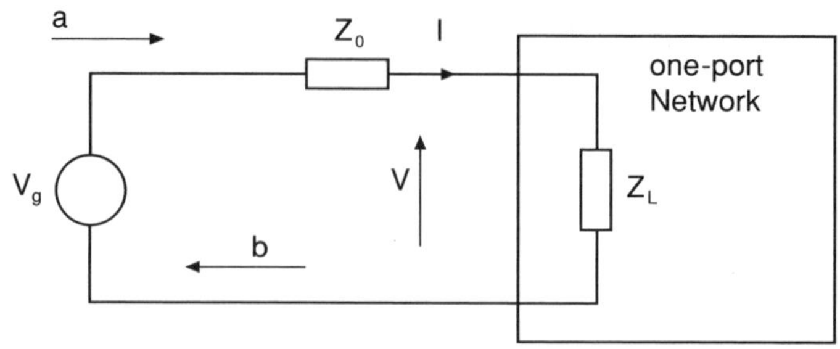

Figure 3.40 One-port network

and Z_L are equal to the impedance Z_0. In this fully matched state, the voltage/current ratio will be constant along the line and equal to Z_0. The scattering parameters are based on s, i.e. the interaction between the forward and reflected waves.

3.8.3 S-Parameters for One- and Two-Port Networks

The one-port problem has been described in the context of a transmission line in the previous section. Here we relate the solutions to the more general one-port device shown in Figure 3.40, where Z_0 is the output impedance of the signal source (generating voltage V_g) and Z_L is the device (or network) input impedance, and extend the s-parameter description to two-port networks.

The forward wave (a) and the reflected wave (b) are given by:

$$a = \frac{V + IZ_0}{2\sqrt{Z_0}} \tag{3.180(a)}$$

and:

$$b = \frac{V - IZ_0}{2\sqrt{Z_0}} \tag{3.180(b)}$$

(Note that the explicit argument notation x, t has been dropped but still applies.) The one-port reflection coefficient or s-parameter is given by Equation (3.179), repeated here:

$$s = \frac{b}{a} = \frac{Z_L - Z_0}{Z_L + Z_0} \tag{3.181}$$

For a two-port network, the same idea can be applied. For this case, the 'a' and 'b' variables are defined for port 1 as:

$$a_1 = \frac{V_1 + Z_0 I_1}{2\sqrt{Z_0}} \tag{3.182(a)}$$

$$b_1 = \frac{V_1 - Z_0 I_1}{2\sqrt{Z_0}} \tag{3.182(b)}$$

and for port 2 as:

$$a_2 = \frac{V_2 + Z_0 I_2}{2\sqrt{Z_0}} \tag{3.183(a)}$$

$$b_2 = \frac{V_2 - Z_0 I_2}{2\sqrt{Z_0}} \tag{3.183(b)}$$

The wave (b_1) travelling away from port 1 towards the source is given by the wave incident on port 1 times the reflection coefficient for port 1 (s_{11}) plus the wave incident on port 2 times the transmission factor from port 2 to port 1 (s_{12}). (The two-port parameter s_{11} is therefore the equivalent of s for the one-port network.) Similarly, the wave (b_2) travelling away from port 2 towards the load is given by the wave incident on port 2 times the reflection coefficient for port 2 (s_{22}) plus the wave incident on port 1 times the transmission factor from port 1 to port 2 (s_{21}). The full two-port s-parameter equations therefore have the form:

$$b_1 = s_{11}a_1 + s_{12}a_2 \tag{3.184(a)}$$

$$b_2 = s_{21}a_1 + s_{22}a_2 \tag{3.184(b)}$$

The input (port 1) reflection coefficient is therefore given by:

$$s_{11} = \left.\frac{b_1}{a_1}\right|_{a_2 = 0}$$

$$= \frac{(V_1 - Z_0 I_1)/2Z_0}{(V_1 + Z_0 I_1)/2Z_0} = \frac{\dfrac{V_1}{I_1} - Z_0}{\dfrac{V_1}{I_1} + Z_0}$$

$$= \frac{Z_{in} - Z_0}{Z_{in} - Z_0} \tag{3.185}$$

Note the requirement for $a_2 = 0$ to yield s_{11} explicitly. This condition is satisfied when port 2 of the network is terminated in Z_0 as shown in Figure 3.41.

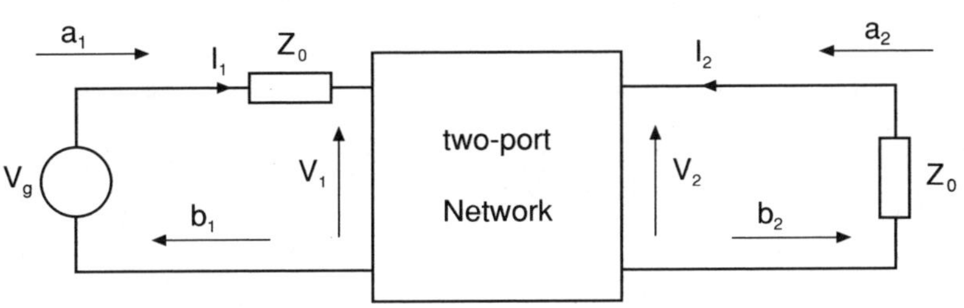

Figure 3.41 Two-port network terminated at port 2 in Z_0

The same condition ($a_2 = 0$) is also needed to yield s_{21} explicitly:

$$s_{21} = \left.\frac{b_2}{a_1}\right|_{a_2 = 0}$$

$$= \frac{(V_2 - Z_0 I_2)/2Z_0}{(V_1 + Z_0 I_1)/2Z_0} = \frac{V_2 - Z_0 I_2}{V_1 + Z_0 I_1}$$

$$= \frac{2V_2}{V_g} \tag{3.186(a)}$$

Therefore:

$$|s_{21}|^2 = \frac{4|V_2|^2}{|V_g|^2}$$

$$= \frac{\text{Power delivered to port 2}}{\text{Power delivered by } V_g} \tag{3.186(b)}$$

s_{21} therefore represents the forward transmission coefficient of the network.

s_{22} and s_{12} are determined in a similar manner but now the signal generator is applied as port 2 and port 1 is terminated in Z_0 as shown in Figure 3.42.

In this mode a_1 is set to zero since port 1 is terminated in Z_0. For this case, s_{22} and s_{12} are determined as:

$$s_{22} = \left.\frac{b_2}{a_2}\right|_{a_1 = 0}$$

$$= \frac{V_2 - Z_0 I_2}{V_2 + Z_0 I_2} = \frac{Z_{out} - Z_0}{Z_{out} + Z_0} \tag{3.187}$$

$$s_{12} = \left.\frac{b_1}{a_2}\right|_{a_1 = 0}$$

$$= \frac{V_1 - Z_0 I_1}{V_2 + Z_0 I_2} = \frac{2V_1}{V_g} \tag{3.188(a)}$$

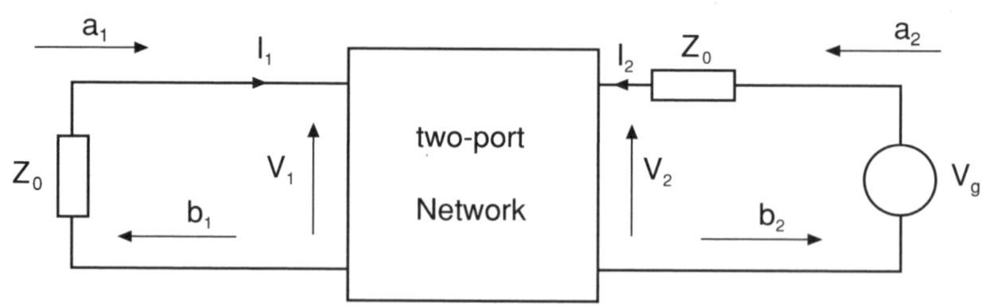

Figure 3.42 Two-port network terminated at port 1 in Z_0

Therefore:

$$|s_{12}|^2 = \frac{4|V_1|^2}{|V_g|^2} = \frac{\text{Power delivered to port 1}}{\text{Power delivered from } V_g} \quad (3.188(b))$$

s_{22} therefore represents the reflection coefficient of port 2 and s_{12} is the reverse transmission coefficient. s_{11} and s_{22} relate to the input and output impedances of the network, and s_{21} and s_{12} relate to the forward and reverse gains of the network.

It is clear that all the s-parameters can be determined by terminating the input or output ports of the network in Z_0. In modern network analysers, the reference impedance (usually 50 ohm) can be specified very accurately over a broad frequency band. As will be seen in Section 3.10.2, however, the calibration (error correcting techniques) needed to improve the accuracy of the measured data requires the use of open and short circuit terminations. These can also be specified quite accurately over a broad frequency band. If this is the case, why are the s-parameters preferred over other parameter sets for microwave work? The answer is considered in the following section.

3.8.4 Advantages of S-Parameters

The acquisition of z, y, h and ABCD parameters requires either a short or open circuit termination. The difficulties of achieving this at microwave frequencies is often quoted as being their main disadvantage. It should be stressed, however, that this difficulty is in ensuring that the *terminals* of the device or circuit are open or short circuit. Broadband open- and short-circuit reference terminations for calibration purposes are available in a variety of connector styles and are used for s-parameter calibration purposes (see Section 3.10.2). The use of such reference terminations would not, however, satisfy the required criteria for the above parameter sets because of the stray parasitic effects introduced by the terminals (probably a microstrip arrangement) of the circuit under test.

Even if a true short or open condition could be established, such a termination can in many cases cause instability. This is true of microwave semiconductor devices having a high bandwidth. Such devices need to be terminated in some (known) intermediate impedance (usually 50 ohm) for stability. A further measurement disadvantage of the above parameters is that the voltages and currents on which they depend vary along a transmission line. It is also true that the parasitic components of the probe used to measure the voltage and current can affect the device or network behaviour.

The s-parameter approach avoids all these measurement problems. The word measurement needs to be stressed here, since all of the disadvantages quoted for z, h, y and ABCD matrices are measurement-related. It is often the case that the s-parameters data are converted into z, y, h or ABCD form for analysis purposes. The converted information can be trusted since the accuracy of the original measured data will be high. The procedure for the conversion of s-parameter data into z-parameter format is discussed in the next section.

3.8.5 Conversion of S-Parameters into Z-Parameters

The conversion between s-parameters and other parameter sets and vice versa is performed using the same basic approach as illustrated previously. For this case, however, the load and source impedance (both assumed to be Z_0) must be taken into account.

The terminal currents and voltages are:

$$I_1 = \frac{a_1 - b_1}{\sqrt{Z_0}} \tag{3.189(a)}$$

$$V_1 = \sqrt{Z_0}\,(a_1 + b_1) \tag{3.189(b)}$$

where subscripts refer to port number, a refers to incident waves and b refers to reflected waves.

$$I_2 = \frac{a_2 - b_2}{\sqrt{Z_0}} \tag{3.190(a)}$$

$$V_2 = \sqrt{Z_0}\,(a_2 + b_2) \tag{3.190(b)}$$

For the z-parameter set, the two-port equations are:

$$V_1 = z_{11}I_1 + z_{12}I_2 \tag{3.191(a)}$$

$$V_2 = z_{21}I_1 + z_{22}I_2 \tag{3.191(b)}$$

which can be rewritten as:

$$Z_0(a_1 + b_1) = z_{11}(a_1 - b_1) + z_{12}(a_2 - b_2) \tag{3.192(a)}$$

and:

$$Z_0(a_2 + b_2) = z_{21}(a_1 - b_1) + z_{22}(a_2 - b_2) \tag{3.192(b)}$$

Solving for b_1 and b_2 gives:

$$b_1 = \frac{z_{12}(a_2 - b_2) + a_1(z_{11} - Z_0)}{Z_0 + z_{11}} \tag{3.193(a)}$$

$$b_2 = \frac{z_{21}(a_1 - b_1) + a_2(z_{22} - Z_0)}{Z_0 + z_{22}}$$

$$= \frac{2a_1Z_0z_{21} + a_2[(Z_0 + z_{11})(z_{22} + Z_0) - z_{21}z_{12}]}{(Z_0 + z_{11})(z_{22} + Z_0) - z_{21}z_{12}} \tag{3.193(b)}$$

Therefore:

$$s_{21} = \left.\frac{b_2}{a_1}\right|_{a_2 = 0}$$

$$= \frac{2Z_0z_{21}}{(Z_0 + z_{11})(Z_0 + z_{22}) - z_{21}z_{12}} \tag{3.194(a)}$$

and:

$$s_{22} = \left.\frac{b_2}{a_2}\right|_{a_1 = 0}$$

$$= \frac{(Z_0 + z_{11})(z_{22} + Z_0) - z_{21}z_{12}}{(Z_0 + z_{11})(Z_0 + z_{22}) - z_{21}z_{12}} \tag{3.194(b)}$$

Similarly:

$$b_1 = \frac{2a_2Z_0z_{12} + a_1[(Z_0 + z_{22})(z_{11} - Z_0) - z_{21}z_{12}]}{(Z_0 + z_{11})(z_{22} + Z_0) - z_{21}z_{12}} \tag{3.195}$$

Therefore:

$$s_{12} = \left.\frac{b_1}{a_2}\right|_{a_1 = 0} \tag{3.196(a)}$$

$$= \frac{2Z_0z_{12}}{(Z_0 + z_{11})(Z_0 + z_{22}) - z_{21}z_{12}} \tag{3.196(a)}$$

and:

$$s_{11} = \left.\frac{b_1}{a_1}\right|_{a_2 = 0}$$

$$= \frac{(Z_0 + z_{22})(z_{11} - Z_0) - z_{21}z_{12}}{(Z_0 + z_{11})(Z_0 + z_{22}) - z_{21}z_{12}} \tag{3.196(b)}$$

Clearly, this procedure is reversed to provide the z-parameters from the s-parameters:

$$b_1 = s_{11}a_1 + s_{12}a_2 \tag{3.197(a)}$$

$$b_2 = s_{21}a_1 + s_{22}a_2 \tag{3.197(b)}$$

Using the definition for a_1, a_2, b_1, and b_2, these equations can be rewritten as:

$$V_1 - Z_0I_1 = s_{11}(V_1 + Z_0I_1) + s_{12}(V_2 + Z_0I_2) \tag{3.198(a)}$$

and:

$$V_2 - Z_0I_2 = s_{21}(V_1 + Z_0I_1) + s_{22}(V_2 + Z_0I_2) \tag{3.198(b)}$$

Solving for V_2:

$$V_2 = \frac{Z_0\{2s_{21}I_1 + I_2[(1 - s_{11})(1 + s_{22}) + s_{12}s_{21}]\}}{(1 - s_{22})(1 - s_{11}) - s_{12}s_{21}} \tag{3.199}$$

Therefore:

$$z_{22} = \left.\frac{V_2}{I_2}\right|_{I_1 = 0}$$

$$= \frac{Z_0[(1 - s_{11})(1 + s_{22}) + s_{12}s_{21}]}{(1 - s_{22})(1 - s_{11}) - s_{12}s_{21}} \tag{3.200(a)}$$

and:

$$z_{21} = \left.\frac{V_2}{I_1}\right|_{I_2 = 0}$$

$$= \frac{2Z_0 s_{21}}{(1 - s_{22})(1 - s_{11}) - s_{12}s_{21}} \qquad (3.200(b))$$

Similarly,

$$V_1 = \frac{Z_0\{I_1[(1 - s_{22})(1 + s_{11}) + s_{12}s_{21}] + 2s_{12}I_2\}}{(1 - s_{22})(1 - s_{11}) - s_{12}s_{21}} \qquad (3.201)$$

Therefore:

$$z_{11} = \left.\frac{V_1}{I_1}\right|_{I_2 = 0}$$

$$= \frac{Z_0[(1 - s_{22})(1 + s_{11}) + s_{12}s_{21}]}{(1 - s_{22})(1 - s_{11}) - s_{12}s_{21}} \qquad (3.202(a))$$

and:

$$z_{12} = \left.\frac{V_1}{I_2}\right|_{I_1 = 0}$$

$$= \frac{2Z_0 s_{12}}{(1 - s_{22})(1 - s_{11}) - s_{12}s_{21}} \qquad (3.202(b))$$

The same procedure can be adopted to convert the s-parameters into y, h or ABCD parameters and vice versa.

3.8.6 Non-Equal Complex Source and Load Impedances

In the above treatment of scattering parameters, it has been assumed that the impedance at ports 1 and 2 are equal (Z_0). This is because network analysers usually have both ports terminated in the same impedance. However, in many circuit applications, this condition may not apply with different impedances at port 1 (Z_{01}) and at port 2 (Z_{02}). It is also true, in general, that the terminating impedances will be complex. Although the s-parameter approach is still valid under these conditions, the expressions defining the forward and reflected waves need to be modified. For port 1 they become:

$$a_1 = \frac{V_1 + Z_{01}I_1}{[2(Z_{01} + Z_{01}^*)]^{1/2}} \qquad (3.203(a))$$

$$b_1 = \frac{V_1 - Z_{01}^* I_1}{[2(Z_{01} + Z_{01}^*)]^{1/2}} \qquad (3.203(b))$$

and for port 2 they become:

$$a_2 = \frac{V_2 + Z_{02}I_2}{[2(Z_{02} + Z_{02}^*)]^{1/2}} \tag{3.204(a)}$$

$$b_2 = \frac{V_2 - Z_{02}^*I_2}{[2(Z_{02} + Z_{02}^*)]^{1/2}} \tag{3.204(b)}$$

where * denotes the complex conjugate. The conversion between s- and z- (or y, h or ABCD) parameters clearly needs to consider these modified expressions. If the previously outlined approach is followed, then the following equations result for the z-parameters in terms of the s-parameters:

$$z_{11} = \frac{(Z_{01}^* + s_{11}Z_{01})(1 - s_{22}) + s_{12}s_{21}Z_{01}}{(1 - s_{22})(1 - s_{11}) - s_{12}s_{21}} \tag{3.205(a)}$$

$$z_{12} = \frac{2s_{12}(R_{01}R_{02})^{1/2}}{(1 - s_{22})(1 - s_{11}) - s_{12}s_{21}} \tag{3.205(b)}$$

$$z_{21} = \frac{2s_{21}(R_{01}R_{02})^{1/2}}{(1 - s_{22})(1 - s_{11}) - s_{12}s_{21}} \tag{3.205(c)}$$

$$z_{22} = \frac{(Z_{02}^* + s_{22}Z_{02})(1 - s_{11}) + s_{12}s_{21}Z_{02}}{(1 - s_{22})(1 - s_{11}) - s_{12}s_{21}} \tag{3.205(d)}$$

In these expressions, R_{01} and R_{02} represent the real component of Z_{01} and Z_{02} respectively.

The expressions defining the s-parameters in terms of the z-parameters are:

$$s_{11} = \frac{(z_{11} - Z_{01}^*)(z_{22} - Z_{02}) + z_{12}z_{21}}{(z_{11} + Z_{01})(z_{22} + Z_{02}) - z_{12}z_{21}} \tag{3.206(a)}$$

$$s_{12} = \frac{2z_{12}(R_{01}R_{02})^{1/2}}{(z_{11} + Z_{01})(z_{22} + Z_{02}) - z_{12}z_{21}} \tag{3.206(b)}$$

$$s_{21} = \frac{2z_{21}(R_{01}R_{02})^{1/2}}{(z_{11} + Z_{01})(z_{22} + Z_{02}) - z_{12}z_{21}} \tag{3.206(c)}$$

and:

$$s_{22} = \frac{(z_{11} - Z_{01})(z_{22} - Z_{02}^*) - z_{12}z_{21}}{(z_{11} + Z_{01})(z_{22} + Z_{02}) - z_{12}z_{21}} \tag{3.206(d)}$$

Self-assessment Problem

3.38 Verify these results by following the procedure described previously. Repeat for the other parameter sets.

3.9 Impedance Matching

An important task for the RF engineer is impedance compensation or, as is more commonly known, matching. This refers to consideration, in the design process, of the impedance properties of the particular subsystem in order to ensure it is compatible with its environment. Impedance matching arises as a consequence of the power transfer theorem, which points out that a badly mismatched design will result in large power losses.

The need for close control of impedance is more general, however, than simply considerations of power transfer. For several applications, rather than perfect matching (for maximum power transfer), the requirement is for controlled mismatch. When an amplifier is designed, for example, the selection of the input and output matching networks not only provides for good power transfer, but also determines other characteristics of the design such as gain and noise figure.

Impedance matching may be relatively straightforward for narrow-band designs but the complexity increases with increasing bandwidth. The difficulty arises from variation with frequency in the network properties. Complex circuits are therefore required when broadband designs are implemented.

3.9.1 The Smith Chart

Tedious mathematical calculations are required if the basic transmission line equations are used to provide solutions to even simple problems. In the early days of RF engineering replacement of these calculations with a faster, graphical, method in order to both accelerate and simplify the design process was therefore very desirable. With the advent of CAD software the requirement for graphical methods has become less essential from a practical point of view. The insight that a geometrical representation of impedance problems (and their solution) yields, however, has meant that the most popular of these methods, a chart devised by Philip H. Smith, has become part of the standard language of the RF designer.

The Smith chart provides a graphical tool that allows quick and accurate manipulation of impedance data. One of its most useful applications is the design of matching networks although it can also be applied to solve many other problems associated with transmission lines.

The chart is developed via a transformation of a Cartesian coordinate representation of normalised impedance to a polar coordinate representation. Lines of constant normalised resistance, r, and normalised reactance, x, on the Cartesian plane are transformed to circles on the polar plane.

To gain an insight into the development of the Smith chart, consider the equation governing the reflection coefficient at the load:

$$\Gamma_\ell = \frac{Z_\ell - Z_0}{Z_\ell + Z_0} = |\Gamma_\ell| e^{j\vartheta_\ell} = \Gamma_x + j\Gamma_y \tag{3.207}$$

where Z_ℓ is the load impedance. Assuming that the load reflection coefficient is $|\Gamma_\ell| \leq 1$, indicative of a passive load, then Γ_ℓ must lie within or on the unit radius circle. Furthermore, the reflection coefficient at a distance d from the end of the line (having a characteristic impedance of Z_0) will also lie within the unity circle since:

$$\Gamma_d = |\Gamma_\ell| e^{-2\alpha d} e^{j(\vartheta_\ell - 2\beta d)} = |\Gamma_d| e^{j(\vartheta_\ell - 2\beta d)} \tag{3.208}$$

where α and β are attenuation and phase constants respectively.

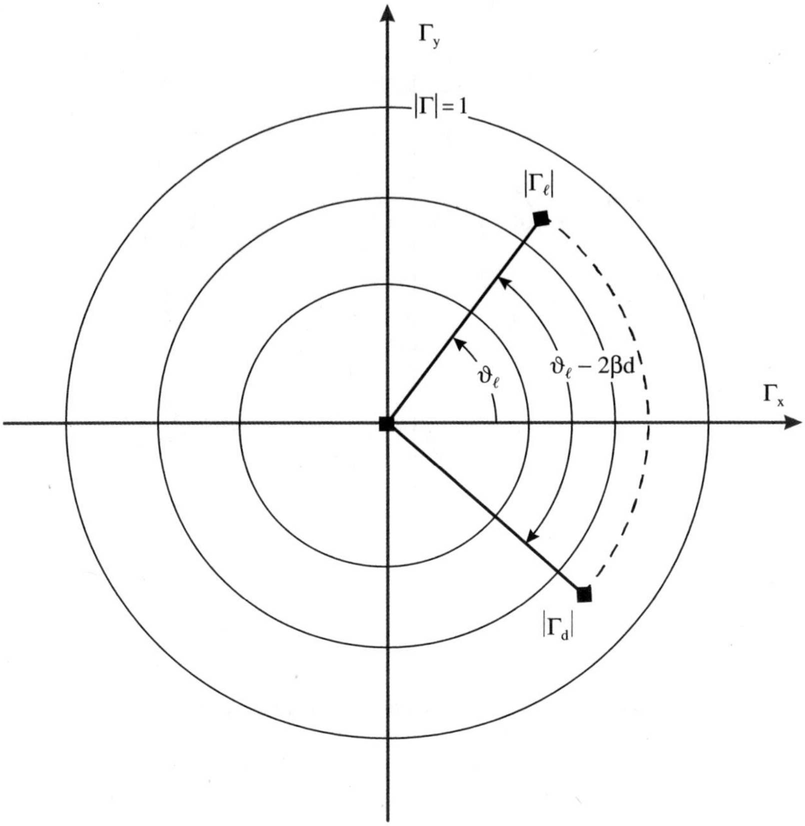

Figure 3.43 Circles of constant reflection coefficient, $|\Gamma|$

In simple terms Equation (3.208) means that for a length d the (complex) reflection coefficient will be rotated by an angle of $-2\beta d$ radians (Figure 3.43).

From Equation (3.207) we can derive the normalised impedance in terms of the reflection coefficient:

$$z = \frac{Z}{Z_0} = \frac{1 + \Gamma_\ell e^{-2\gamma d}}{1 - \Gamma_\ell e^{-2\gamma d}} \tag{3.209}$$

where $\gamma = \alpha + j\beta$ is the propagation constant. For the purpose of derivation we can assume that the impedance is given at the load, i.e. $d = 0$. This does not reduce the generality of the argument, however, since any given point of the line can be simulated by a different physical load at the same position:

$$z = \frac{1 + \Gamma_\ell}{1 - \Gamma_\ell} = \frac{R_\ell + jX_\ell}{Z_0} = r + jx \tag{3.210(a)}$$

also:

$$\Gamma_\ell = \frac{z - 1}{z + 1} = \Gamma_x + j\Gamma_y \tag{3.210(b)}$$

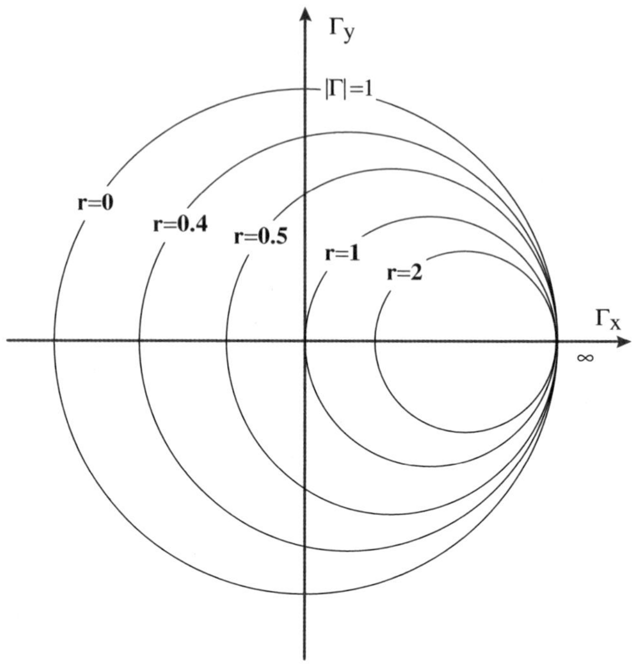

Figure 3.44 Circles of constant normalised resistance, *r*

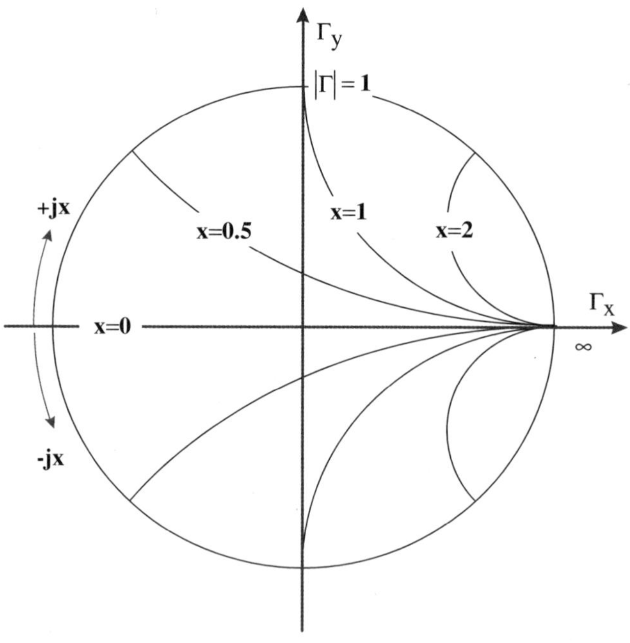

Figure 3.45 Circles of constant normalised reactance, *x*

From manipulation of these two equations we have:

$$r = \frac{1 - \Gamma_x^2 - \Gamma_y^2}{(1 - \Gamma_x)^2 + \Gamma_y^2} \tag{3.211(a)}$$

and:

$$x = \frac{2\Gamma_y}{(1 - \Gamma_x)^2 + \Gamma_y^2} \tag{3.211(b)}$$

Rearranging the previous pair of equations we arrive at the following:

$$\left(\Gamma_x - \frac{r}{1+r}\right)^2 + \Gamma_y^2 = \left(\frac{1}{1+r}\right)^2 \tag{3.212(a)}$$

$$(\Gamma_x - 1)^2 + \left(\Gamma_y - \frac{1}{x}\right)^2 = \left(\frac{1}{x}\right)^2 \tag{3.212(b)}$$

The curves described by Equations (3.212(a)) and (3.212(b)) on the Γ-plane represent a family of circles. Equation (3.212(a)) describes the locus of points having constant r whereas Equation (3.212(b)) describes the locus of points having a constant x. These two sets of circles are shown in Figures 3.44 and 3.45.

The geometrical parameters of the curves are given explicitly in Table 3.2:

Table 3.2 Centre coordinates and radius of circles of constant r and x

	Centre	Radius
Circles of constant r	$\left(\frac{r}{1+r}, 0\right)$	$\frac{1}{1+r}$
Circles of constant x	$\left(1, \frac{1}{x}\right)$	$\frac{1}{x}$

The superposition of these two families of curves results in the Smith chart shown in Figure 3.46.

Although the constant Γ circles (Figure 3.43) are not usually explicitly shown on the Smith chart, they are implied, i.e. the reflection coefficient plane in radial co-ordinates is implicitly superimposed on the normalised impedance plane. Thus any point on the chart can be read as an impedance or as a reflection coefficient.

The important properties of the Smith chart (see Figure 3.47) can be summarised as follows:

1. It represents all passive impedances on a grid of constant r and x circles.
2. It contains the corresponding reflection coefficients in polar co-ordinates; the angle being read on the peripheral scale and the magnitude being calculated using:

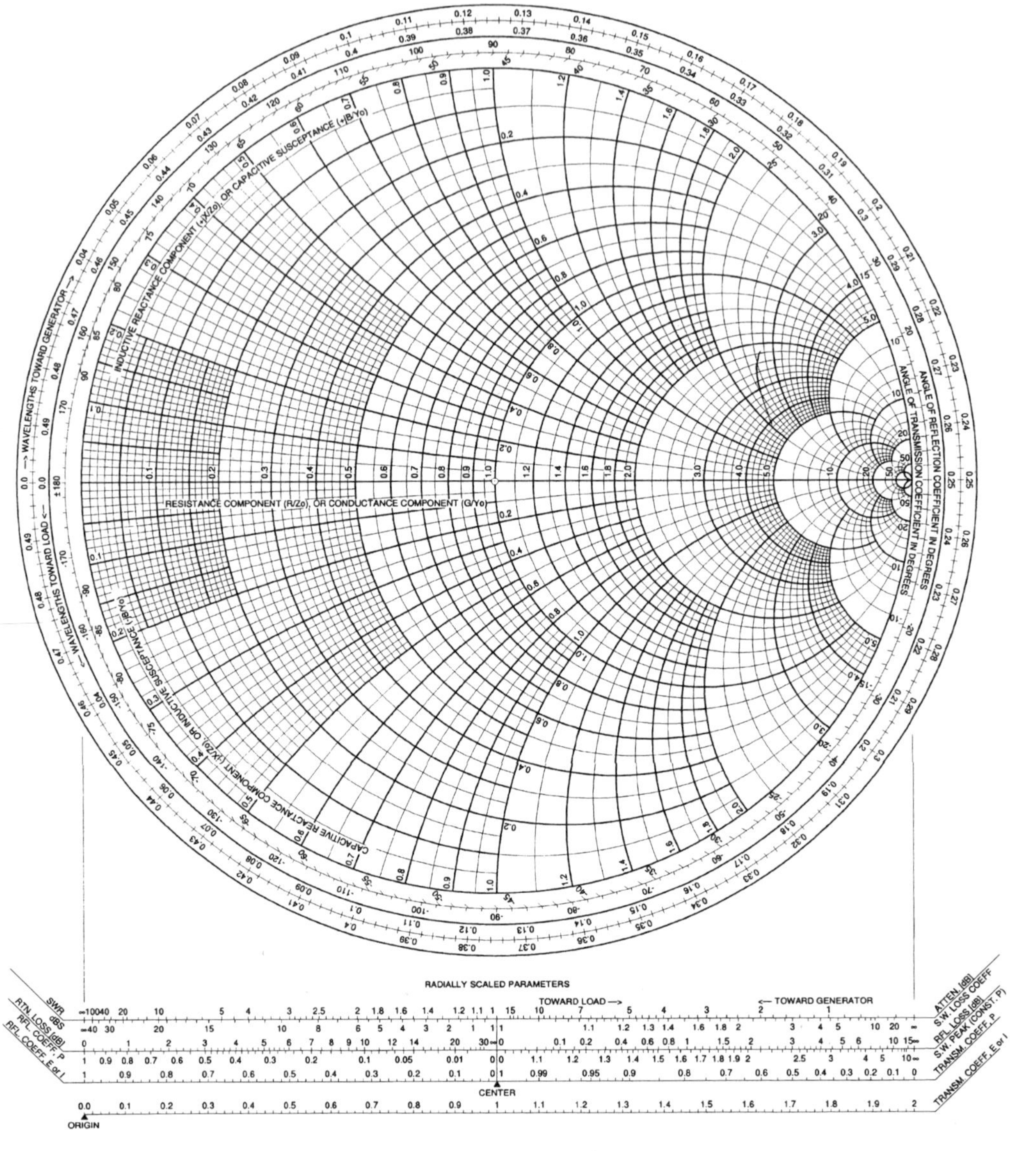

Figure 3.46 Smith chart

$$|\Gamma| = \frac{\text{radius of circle on which point is located}}{\text{radius of the chart}}$$

3. The upper half of the diagram represents positive reactance values (inductive elements).
4. The lower half of the diagram represents negative reactance values (capacitive elements).

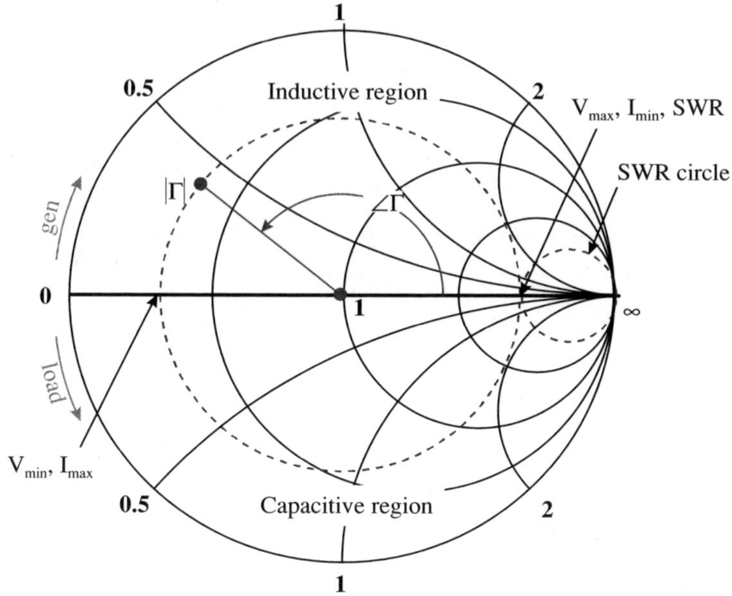

Figure 3.47 Properties of the Smith chart

5. The arc around the Smith chart corresponds to a movement of half a wavelength ($\lambda/2$); any other movement across an arc between any two points can be read on the chart periphery in terms of wavelengths.
6. The horizontal radius to the left of the centre corresponds to voltage minimum and current maximum (V_{min}, I_{max}).
7. The horizontal radius to the right of the centre corresponds to the standing wave ratio (SWR), the voltage maximum and the current minimum (V_{max}, I_{min}).
8. By rotating the reflection coefficient plane by 180° the corresponding quantities are read as admittances. The transformation is $\Gamma \to -\Gamma$, in which case:

$$-\Gamma = -\left(\frac{z-1}{z+1}\right) = \frac{1-z}{1+z} = \frac{\dfrac{1}{z}-1}{\dfrac{1}{z}+1} = \frac{y-1}{y+1} \quad (3.213)$$

i.e., when $\Gamma \to -\Gamma$ then $z \to y$. An admittance chart can therefore be created by simply rotating the impedance chart through 180°. This transformation is shown in Figure 3.48.

A variation on the Smith chart called the immittance chart (*im*pedance ad*mittance*) is generated if we superimpose the impedance and the admittance charts (Figure 3.49). The advantage is that we can now read each point as admittance or impedance simply by reading values from the different grids, thus making the impedance to admittance transformation easier.

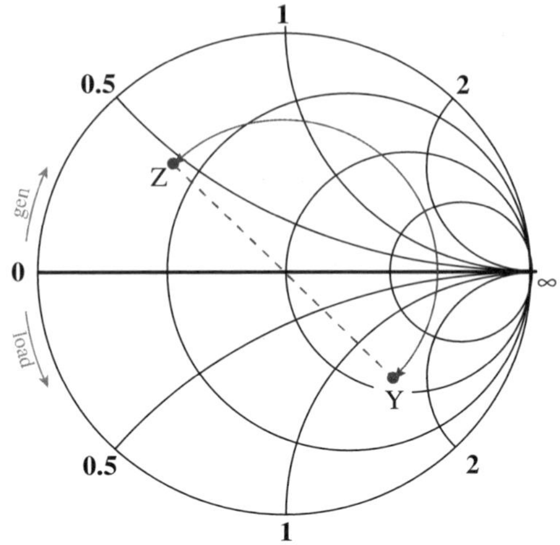

Figure 3.48 Impedance – admittance transformation

Finally, we can generalise the Smith chart to include active elements, by permitting values for the reflection coefficient that are larger than unity. The result, called the compressed Smith chart, is shown, schematically, in Figure 3.50. This form is especially useful for the design of oscillators, where negative resistances are common. In the usual format it allows gain of 10 dB, which corresponds to reflection coefficient magnitude less than 3.16. All the properties of the normal Smith chart hold for the compressed Smith chart.

3.9.2 Matching Using the Smith Chart

The single most common use of the Smith chart is matching a load-impedance to a source-impedance.

3.9.2.1 Lumped element matching

There are several networks that provide impedance transformation but the simplest approach is to connect a series or shunt lumped component to the load thereby changing the input impedance. To evaluate the effect of a lumped element connected to the load we can examine the following cases.

1. Series connection of ideal inductor or capacitor

The connection of a series element, Z_{el}, to an arbitrary load, Z_L, (Figure 3.51) will alter the overall input impedance, Z_{in}, of the network.

Since the component is introduced in series the input impedance is given by:

$$Z_{in} = Z_{el} + Z_L \tag{3.214}$$

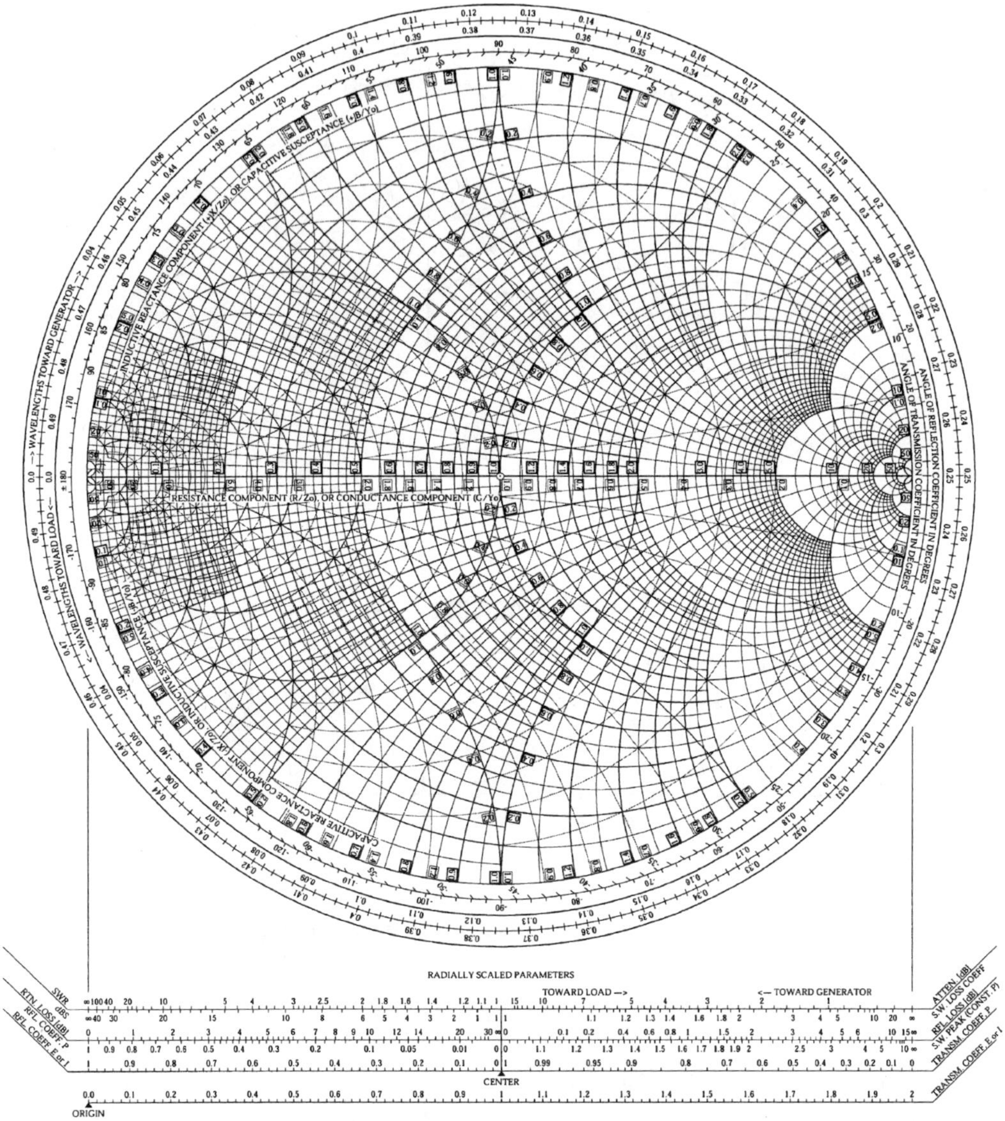

Figure 3.49 Immitance chart

If the component introduced is purely imaginary, i.e. $Z_{el} = jX_{el}$, then the Smith chart description of the effect is movement along a circle of constant r. The sense of the movement is determined by the sign of the reactance introduced. An inductor (positive reactance) changes the overall impedance towards the upper half of the chart, whereas a capacitor (negative reactance) changes it towards the lower half of the chart, Figure 3.52.

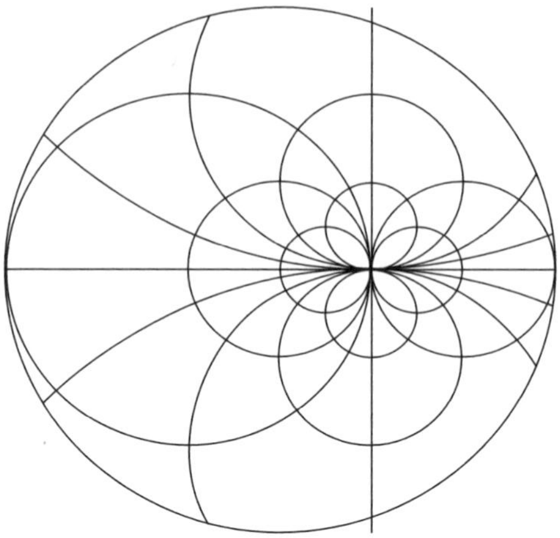

Figure 3.50 Compressed Smith chart

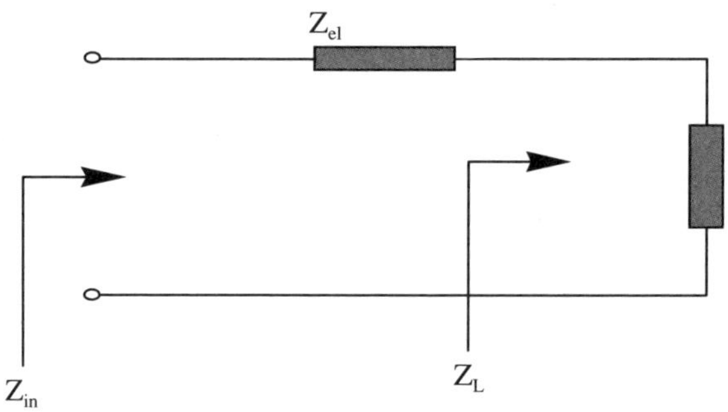

Figure 3.51 Series connection of a lumped element

2. Shunt connection of ideal inductor or capacitor

It is easier to follow the effect of a shunt element, Figure 3.53, using the admittance chart.
The input admittance for the parallel network is given by:

$$Y_{in} = Y_{el} + Y_L \tag{3.215}$$

If the component introduced is purely imaginary, i.e. $Y_{el} = jB_{el}$, then the Smith chart description is movement along a circle of constant (normalised conductance) g. The sense of the movement is determined by the sign of the susceptance introduced. A capacitor (positive susceptance) changes the overall admittance towards the lower half of the (admittance) chart, whereas an inductor (negative suseptance) changes it towards the upper half of the chart, Figure 3.52.

A combination of series and shunt elements can be used to match an arbitrary load to the source (at a single frequency). The most convenient tool for multi-element matching is the

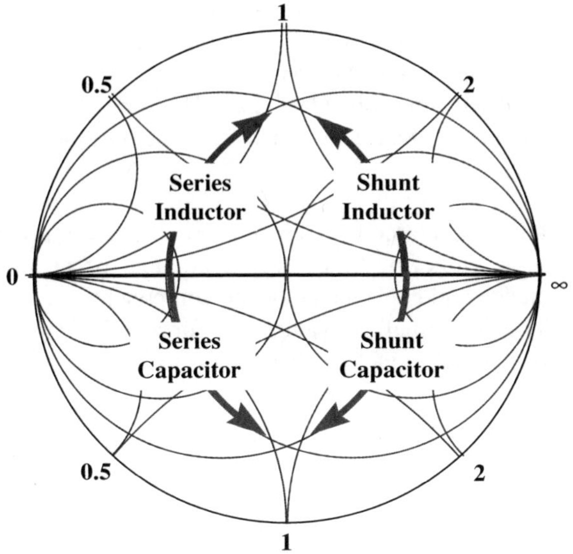

Figure 3.52 Effect of series element on impedance and shunt elements on admittance

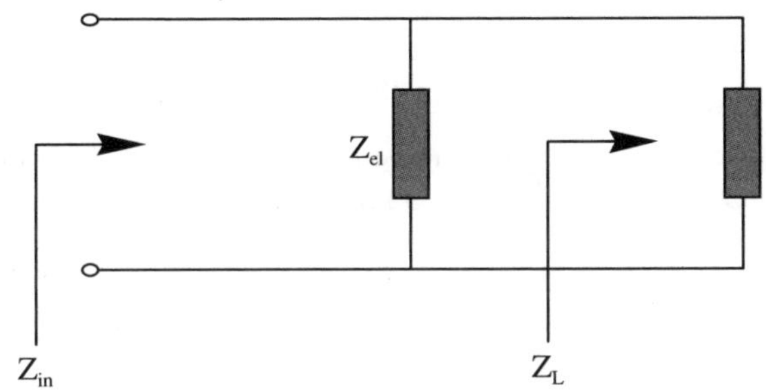

Figure 3.53 Shunt connection of lumped element

immittance chart, since it allows rapid transformation between the admittance and impedance. The general technique (illustrated in Figure 3.53) is summarised in the following steps:

1. Normalise the load impedance, Z_L, to $z_L = Z_L/Z_0$ and plot this on the immittance chart
2. Determine the admittance of the matching element using (a) or (b) below:

   (a) If the last element of the matching network is a parallel component, then the admittance chart is used, in which case movement from z_L is along a circle of constant g (adding susceptance) until the unit normalised resistance ($r = 1$) circle is intersected. This point gives the required normalised input admittance y_{in}. The normalised admittance, y_{el}, of the required element is then determined from the normalised version of Equation (3.215).

(b) If the last element of the matching network is a series component, then the impedance chart is used, in which case movement is along a circle of constant r (adding reactance) until the unit normalised conductance ($g = 1$) circle is intersected. This point gives the required normalised input impedance, z_{in}. The normalised impedance, z_{el}, of the required series element is then determined from the normalised version of Equation (3.214).

3. Determine the admittance of the second element from the new impedance value (on either of the $g = 1$ or the $r = 1$ circle) by following the previous procedure.
4. Determine the inductance and capacitance of the components from their impedance and admittance values. The de-normalised values are, for a series capacitor, series inductor, parallel capacitor and parallel inductor, given respectively by:

$$C = \frac{1}{\omega x Z_0} \tag{3.216(a)}$$

$$L = \frac{x Z_0}{\omega} \tag{3.216(b)}$$

$$C = \frac{b}{\omega Z_0} \tag{3.216(c)}$$

$$L = \frac{Z_0}{\omega b} \tag{3.216(d)}$$

where Z_0 is the impedance used to normalise the Smith chart (the value corresponding to the centre of the chart), x is the normalised reactance and b is the normalised susceptance.

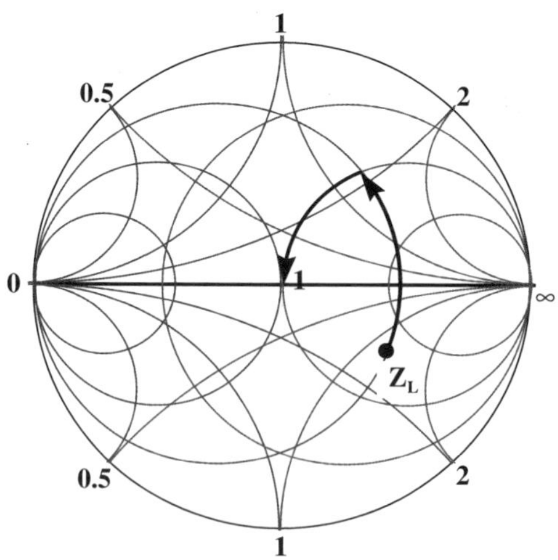

Figure 3.54 Matching using lumped elements

The value selected for Z_0 can be different from the source impedance. When lumped elements are used to provide matching we normally choose a convenient value that brings all the impedances used towards the centre of the chart. (Theoretically this value can be selected to be complex but the difficulty involved in manipulating complex numbers usually means that a real value is chosen.)

3.9.2.2 Distributed element matching

As the operational frequency increases, it becomes correspondingly difficult to employ lumped components. The alternative is to incorporate distributed elements in the form of transmission line segments. A transmission line can exhibit inductive or capacitive behaviour depending on its length, characteristic impedance and terminal load. This property can be exploited in order to design matching networks. There are several techniques for the design of such distributed matching networks. The most common ones are presented next.

3.9.2.3 Single stub matching

The distributed element equivalent of the lumped element L-network consists of a length of transmission line connected directly to the load (corresponding to the series component) and a transmission line stub connected in parallel to the line (Figure 3.55). The principle behind this configuration is that any length of line moves the admittance of the load along an arc of a circle of constant reflection coefficient. After a certain rotation (corresponding to the required transmission line length), the real part of the admittance will be equal to that of the source. At that point the introduction of a susceptive load will provide the required matching. This can be provided by a lumped inductor or capacitor but it is more common in practice to incorporate the susceptive properties of a short circuited or open circuited line.

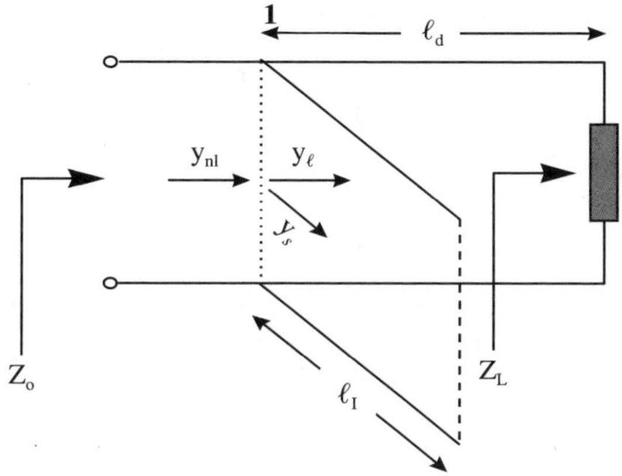

Figure 3.55 Single stub matching network

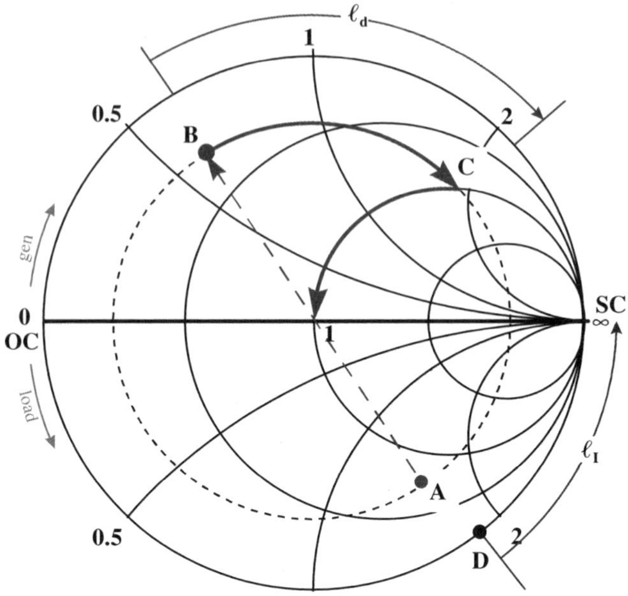

Figure 3.56 Single stub matching

The design steps, illustrated in Figure 3.56, are summarised as follows.

1. Normalise the impedance of the load to the characteristic impedance of the line and plot this on the Smith chart (point A).
2. Translate the impedance to admittance (point B), after which the Smith chart should be read as an admittance chart.
3. Move the reference plane towards the generator along a constant reflection coefficient circle until the unit conductance circle ($g = 1$) is intercepted (point C). The arc between B and C determines the length (in wavelengths) of the series line.
4. The susceptance of the stub introduced at point C should cancel the existing susceptance at that position. At the connection point therefore:

$$y_{n1} = y_\ell + y_S \tag{3.217(a)}$$

i.e.:

$$y_S = y_{n1} - y_\ell \tag{3.217(b)}$$

where y_S is the normalised admittance of the stub, y_{n1} is the normalised admittance after the stub has been added (usually equal to the normalised source admittance), and y_ℓ is the admittance looking towards the load at the point of the introduction of the series line, see Figure 3.55.

5. Determine the admittance y_S (which is purely imaginary) on the perimeter of the Smith chart (point D). The required length of the stub (ℓ_I) is provided by the arc between point D and the short or open circuit termination (SC or OC on the Smith chart) in the direction of the load.

Admittance has been employed since it is easier to use for parallel elements, distributed or lumped. A series stub can, in principal, provide the same effect although this solution is rarely used in practice. Nevertheless the design process remains the same apart from the fact that the impedance chart is used instead of the admittance chart.

The selection of an open-circuit or short-circuit stub depends on the kind of transmission line. When coaxial lines are employed, for example, it is easier to implement a short circuit whereas in microstrip it is easier to implement an open circuit.

The source impedance here, as in most practical cases, is assumed to be equal to the characteristic impedance of the line. In the case where the source and the characteristic impedance are different, the normalising impedance is still that of the line (Z_0). The design process varies, but only slightly, since the source impedance does not coincide with the centre of the chart. The basic steps, however, are the same.

3.9.2.4 Double stub matching

Although any impedance can be matched with a single stub, it is necessary to vary (i.e. select freely) both the stub's position and length. This is not always easy to achieve and in some applications the position of the stub is fixed. In this case two or three stubs in fixed positions from the load can be used to provide the matching.

The double stub configuration is shown in Figure 3.57. It consists of two short-circuit stubs connected in parallel at fixed positions on the line. (The stubs may also be open circuit or a combination of open and short circuit.) The distance between the stubs (ℓ_{d2}) is usually selected to be 1/8, 3/8, 5/8 of a wavelength, whereas the position of the first stub from the load (ℓ_{d1}) depends on the region (of the Smith chart) in which the load is most likely to be found.

The design process, illustrated in Figure 3.58, proceeds as follows:

1. Normalise the load impedance to the characteristic impedance of the line (Z_O) and plot it on the Smith chart.
2. Transfer the normalised impedance to the admittance chart (because the stubs are connected in parallel and it is, therefore, easier to manipulate admittances).

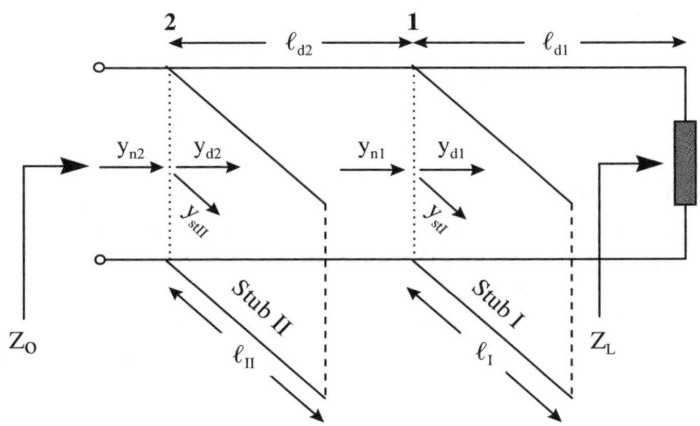

Figure 3.57 Double stub matching network

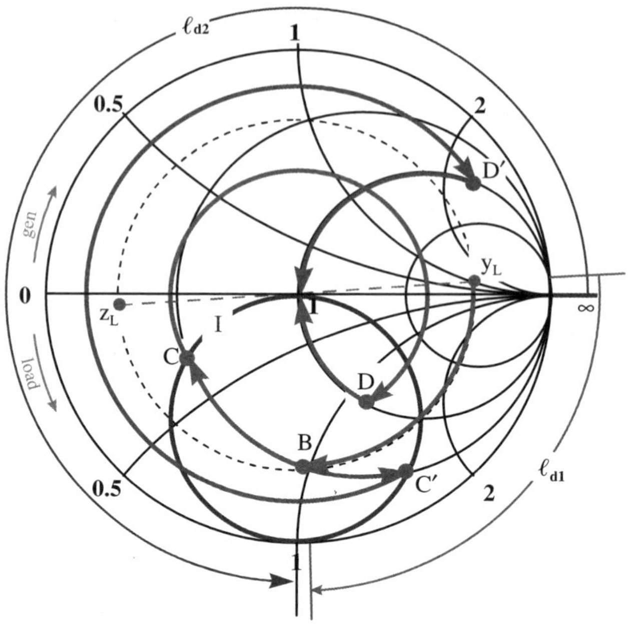

Figure 3.58 Double stub matching

3. Rotate the unity conductance circle ($g = 1$) towards the load by an amount corresponding to the length ℓ_{d2}, thus determining the spacing circle (circle I).
4. Move the normalised load admittance (y_L) along a constant reflection coefficient circle towards the generator by an amount corresponding to ℓ_{d1} and read the admittance y_{d1} at point B.
5. Determine the intersections of the constant conductance circle that contains point B and the spacing circle (points C and C'), and read the admittances y_{n1} and y'_{n1}.
6. Calculate the susceptance to be introduced by the stub I from:

$$y_{stI} = y_{n1} - y_{d1} \tag{3.218(a)}$$

$$y'_{stI} = y'_{n1} - y_{d1} \tag{3.218(b)}$$

7. Determine the lengths ℓ_I, ℓ'_I from the chart's circumferential arc in the direction of the load between the admittance y_{stI} and y'_{stI} and the short circuit (or open circuit if appropriate) termination of the stub.
8. Rotate points C and C' towards the generator, along a constant reflection coefficient circle, determining points D and D' on the $g = 1$ circle. These points correspond to the admittances y_{d2} and y'_{d2} just before stub II.
9. Calculate the susceptance to be introduced by stub II from:

$$y_{stII} = y_{n2} - y_{d2} \tag{3.219(a)}$$

$$y'_{stII} = y_{n2} - y'_{d2} \tag{3.219(b)}$$

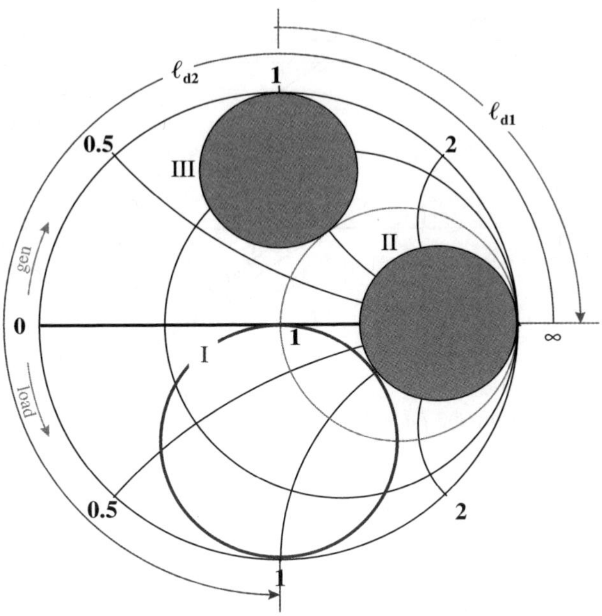

Figure 3.59 Load admittancies that cannot be matched using double stub

10. Finally determine lengths ℓ_{II}, ℓ'_{II} from the chart's circumferential arc in the direction of the load between the admittance y_{stII} and y'_{stII} and the short circuit (or open circuit if appropriate) termination of the stub.

In general, there are two solutions for a given load. Admittances exist, however, that cannot be matched with a given stub position. No y_{n1} admittances falling within circle II (tangent to the spacing circle), see Figure 3.59, can be matched. For those admittances that lie on the circumference of circle II, only one solution exists. The loads for which matching cannot be achieved are those lying within circle III, determined by moving circle II by ℓ_{d2} towards the load.

To solve the 'no solution' problem an additional stub can be employed at the load (Figure 3.60) to move the (composite) load admittance outside the critical area.

3.9.3 Introduction to Broadband Matching

The term broadband matching loosely implies that impedance compensation should be achieved over frequency ranges larger than 50%. The design difficulty in this case arises from the large variations of the load impedance across the matching frequency band. When such frequency-dependent loads (e.g. transistors or broadband antennas) require matching, the design aims usually shift from achieving perfect matching to improving the broadband performance. This is usually possible to achieve when some maximum permissible SWR is acceptable, thus enabling acceptable reflection coefficient over the whole frequency range while allowing some frequency ripple. In general, the larger the required bandwidth, the

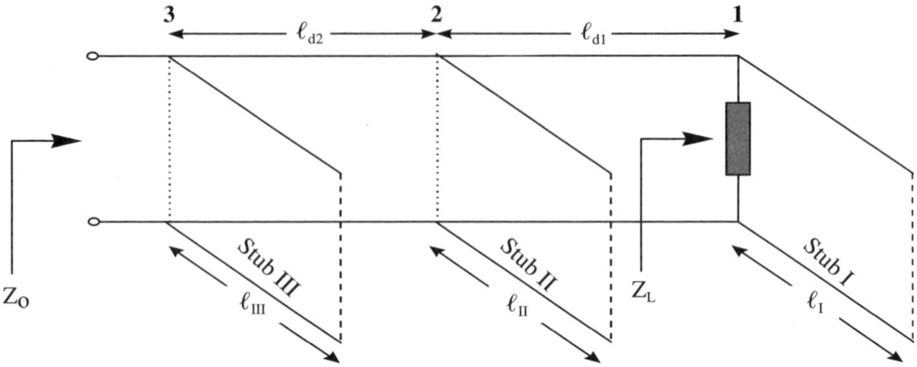

Figure 3.60 Triple stub matching network

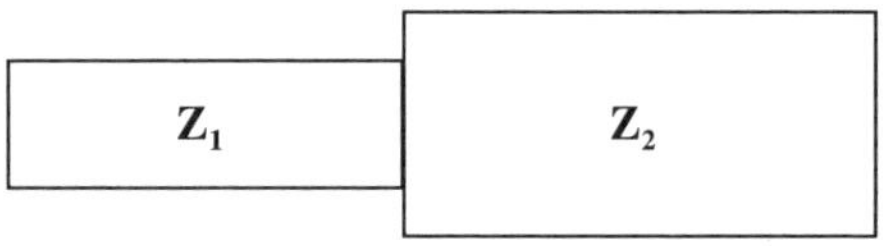

Figure 3.61 Layout of a line transformer

greater must be the allowed matching ripple and the more modest the guaranteed matching performance at a particular frequency.

Distributed elements usually have poor broadband performance due to their fixed geometric characteristics although some distributed networks exhibit better broadband performance than others. A line transformer (Figure 3.61), for example, allows better broadband matching than a single stub, which in turn is more broadband than a double stub. Generally speaking, the larger and the more often, abrupt changes in characteristic impedance are encountered, the smaller the operational bandwidth of the design.

In practice when broadband performance is needed, it is usual to use multi- (more than three) element designs. The advantage of these designs is that they can determine the overall quality factor (Q) of the network, the lower the Q, the more broadband the design. The Smith chart is a valuable tool in this case as well. The Q of a series impedance circuit is the ratio of the reactance to the series resistance. Each point on the Smith chart therefore has a Q associated with it. The locus of impedances on the chart with equal Q is a circular arc that passes through the open circuit (O/C) and short circuit (S/C) load points. Several Q circles are shown in Figure 3.62.

Constant Q circles can be used to provide limits within which the matching network should remain in order to provide a required operational bandwidth. The design process, illustrated in Figure 3.63, is as follows.

1. Normalise the source and load impedances with a convenient value (Z_0) and plot them on the Smith chart.
2. Plot the Q circle that corresponds to the overall quality factor required (a low value, usually less than 5 is normally selected).

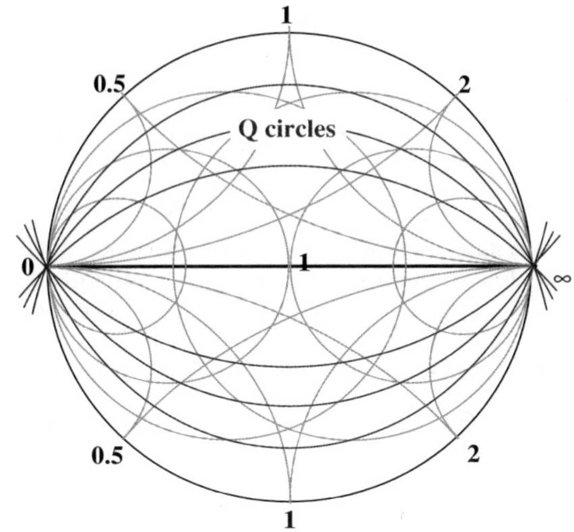

Figure 3.62 Constant Q circles

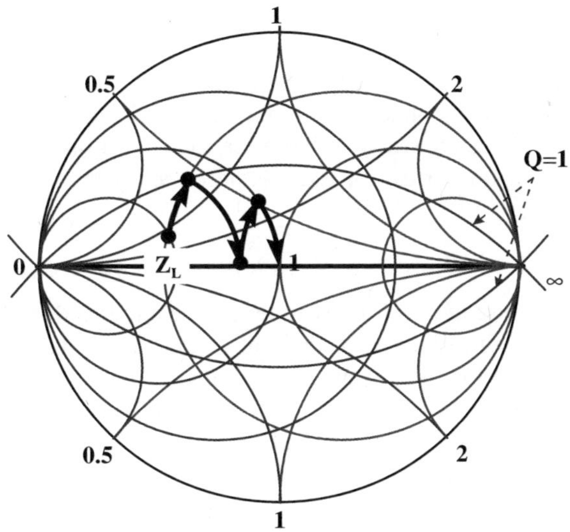

Figure 3.63 Broadband matching using lumped components

3. If the matching network is a π-network (e.g. the last element is a parallel component) move from the load along a constant g-circle on the immittance chart until the Q-circle is reached. If the matching network is a T-network (e.g. the last element is a series component) move from the load along a constant r-circle until the Q-circle is reached.
4. Introduce an element of the opposite sign from the first one until the real axis is reached.
5. Repeat the previous process introducing as many elements as required until the source is reached.

When a, lossless, two-element L-section is used for matching, the Q is automatically limited by the source and load impedance:

$$Q_S = Q_P = \sqrt{\frac{R_P}{R_S} - 1} \qquad (3.220)$$

where R_P is the parallel resistance and R_S is the series resistance of the L-section. A section of at least three elements should, therefore, be used when broadband matching is required.

3.9.4 Matching Using the Quarter Wavelength Line Transformer

The input impedance, Z_g, of a $\lambda/4$ line with characteristic impedance Z_T terminated in load Z_L is given by:

$$Z_g = \frac{Z_T^2}{Z_L}$$

i.e.:

$$Z_T = \sqrt{Z_g \cdot Z_L} \qquad (3.221)$$

When both the input and load impedance of the transformer are purely resistive (real), then a quarter wavelength line having characteristic impedance Z_T can be used to match Z_L to Z_g. When one or both impedances are reactive, however, then a $\lambda/4$ line is not sufficient. In such cases a single section transformer can be used as described below.

3.9.5 Matching Using the Single Section Transformer

A technique similar to the use of a $\lambda/4$ transformer can be employed to match any impedance point that lies within the unit resistance or the unit conductance circle. A single section line is incorporated between the load and source impedances (Figure 3.64).

The following design process (illustrated in Figure 3.65) can be used to determine the required length and characteristic impedance of the line:

1. Normalise Z_L by Z_0.
2. Connect point A corresponding to z_L with the centre of the Smith chart.
3. Find the perpendicular bisector and determine Point C at the intercept with the real axis.

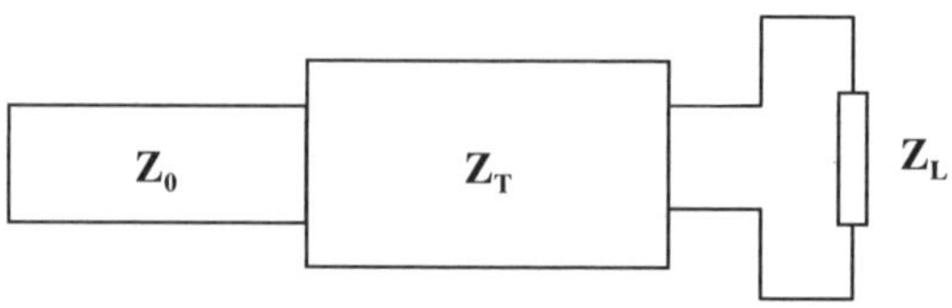

Figure 3.64 Single section transformer

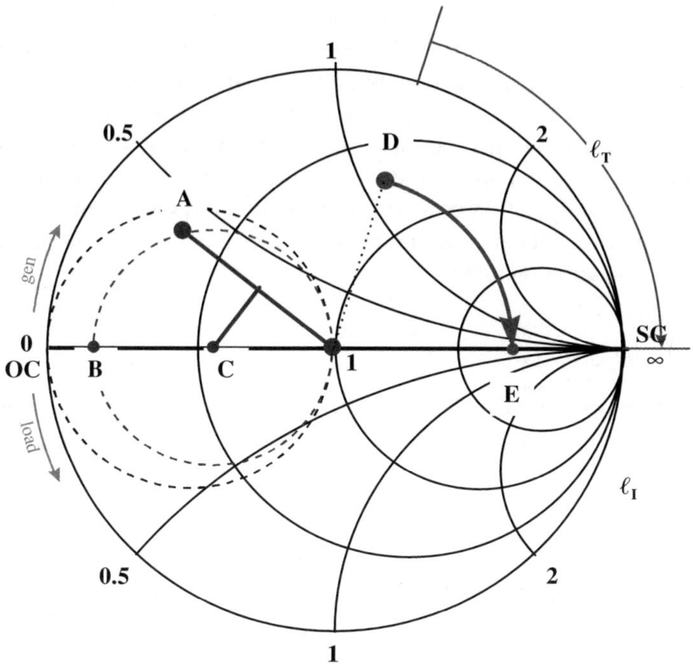

Figure 3.65 Matching using a single section line

4. Draw a circle around point C with radius equal to the distance from the centre thus arriving at point B, corresponding to a de-normalised impedance Z_B.
5. Calculate the characteristic impedance of the single section line (Z_T) using:

$$Z_T = \sqrt{Z_B \cdot Z_0} \tag{3.222}$$

6. Re-normalise Z_L to the calculated value of Z_T determining point D.
7. Move towards the generator along a constant reflection coefficient circle until the real axis is intersected at point E. The length of the line in wavelengths is determined from the arc l_T along the chart circumference.

A quick investigation of the design process identifies the locus of the points that can be matched using a single section. Point B will lie outside the Smith chart for any load impedance outside the unit resistance and conductance circles. To extend the method to cover these impedance values a second section of line is needed to move the load, when required, into the permitted region.

3.10 Network Analysers

The development of network analysers able to measure accurately and quickly the s-parameters of a device or circuit over a broad frequency range has made a significant contribution to the advancement of microwave engineering. Measurements that in the past took days to perform can now be completed in a fraction of the time. In addition, measurements which in the past would not have been feasible are now common. Examples of this are measurements on production lines to monitor the performance of circuits or devices. Computer-controlled instruments enable the information gathered to be stored for statistical purposes. Test programs can also be generated during the product design cycle, thereby integrating the design and production stages. In the following subsections we describe the principles of operation of a typical network analyser, the models and calibration procedures employed to reduce measurement errors, and the test jigs and probes required for the characterisation of semiconductor devices.

3.10.1 Principle of Operation

A simplified diagram of a typical vector network analyser is shown in Figure 3.66. The synthesised signal source generates either a continuous wave or a swept RF signal. The frequency range of the source depends on the user's needs. For example, in the characterisation of gallium-arsnide devices, a 40 GHz source would be required because of the high frequency (f_t > 25 GHz) capability of these devices. The characterisation of circuits for mobile communications applications operating at 1.8 GHz, however, would not require such a high frequency source. The power delivered by the source to the test set is usually levelled using automatic control. Phase locking is achieved by routing a portion of the RF signal through the test set to the R input of the receiver. Here, the signal is sampled by a phase detection circuit and fed back to the source.

The RF signal from the source is applied to the circuit/device under test through the test set. The signal transmitted through the device or reflected from its input is fed through the A and B inputs to the receiver. Here the transmitted and reflected signals are compared with the incident or reference signal at input R.

In the receiver the R, A and B RF inputs are converted or sampled to form a low frequency intermediate frequency (IF). The sampling process retains the magnitude and phase information of the RF signal. The IF data is usually converted into digital signals using an analog-to-digital converter (ADC) for further processing.

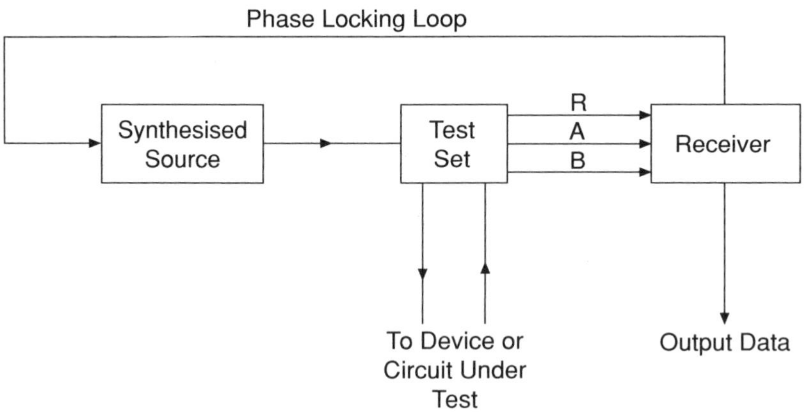

Figure 3.66 Main circuit blocks of a vector network analyser

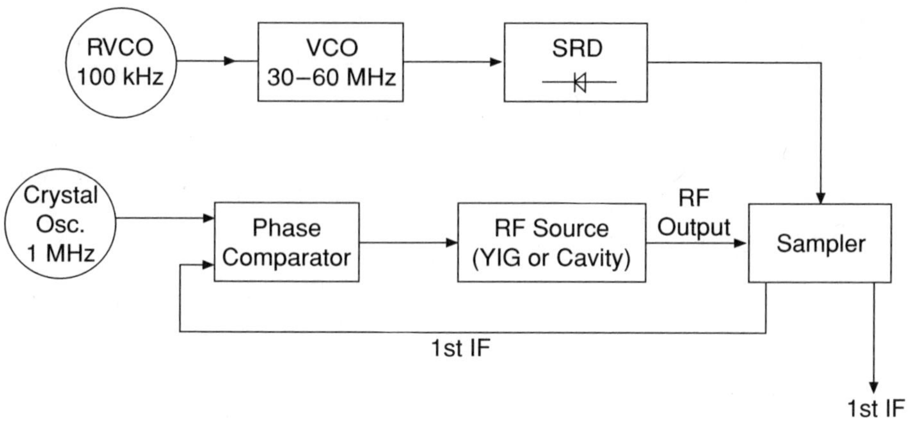

Figure 3.67 Simplified schematic of a signal source

In the following subsections each of the network analyser components is described in more detail.

3.10.1.1 The signal source

A simplified schematic diagram of a possible signal source for a 3 GHz network analyser is shown in Figure 3.67. The reference voltage controlled oscillator (RVCO) generates a 100 kHz reference signal. This is fed to the VCO which generates a signal in the 30 to 60 MHz frequency range. This signal can be a continuous wave (CW) or swept, and is synthesised and phase locked to the 100 kHz reference signal.

The signal from the VCO is fed into the step recovery diode (SRD), which multiplies the incoming fundamental signal into a comb of harmonic frequencies. The harmonics are used as the first local oscillator signal to the sampler.

The crystal oscillator generates a 1 MHz reference signal, which is fed to the phase comparitor. This in turn sets the RF source to a first approximation of the RF signal frequency. The signal is fed to the sampler and combined with the SRD signal to generate a difference frequency. This difference frequency is filtered and fed back to the phase comparitor. The difference between the filtered frequency difference and the 1 MHz reference frequency generates a voltage that adjusts the RF source frequency closer to the required frequency. This iterative process continues until the difference frequency is equal to the 1 MHz signal.

3.10.1.2 The two-port test set

Figure 3.68 shows a simplified diagram of a test set for two-port circuit or device testing. The power splitter diverts a portion of the incoming RF signal to the R input of the receiver and this acts as the reference signal. The remaining RF input signal is routed through a switch to one of two bi-directional bridges. The switch enables either forward (S_{11}, S_{21}) or reverse (S_{22}, S_{12}) measurements to be performed on a two-port network. The position of the switch is controlled by the network analyser and depends on the measurement specified by the user. A bias tee is normally included for each port to enable the device under test to be biased.

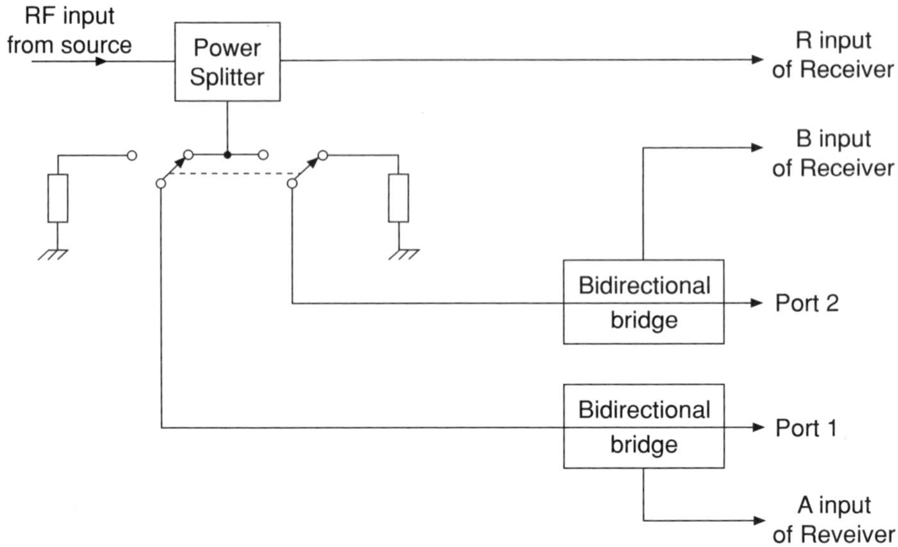

Figure 3.68 Simplified schematic of a two-port test set

3.10.1.3 The receiver

A simplified schematic of a receiver is shown in Figure 3.69. The three sampler and mixed circuits are the same. The sampler and mixer essentially down-convert the incoming RF signal to a fixed low (4 kHz) 2nd IF. The amplitude and phase of this IF signal correspond to those of the incoming RF signal. The 2nd IF produced by the mixer is essentially the difference between the 1st IF (see source description) and the 2nd reference frequency.

The 2nd IF signal from each sampler/mixer circuit is fed to the multiplexer which directs each of the signals to the analog-to-digital converter. The resulting digital information is then processed.

A typical automated or computer-controlled measurement system is shown in Figure 3.70. The PC controls the DC bias supply (PSU1 and PSU2), the network analyser and the printer/plotter through the IEEE-488 or GPIB interface bus. The PC enables rapid and complex measurements to be made on circuits or devices. Since the power supplies are also under computer control, the s-parameter measurements can be performed as a function of bias. This would be important in device characterisation, where the non-linearity of the component is an aspect of interest to the circuit designer.

3.10.2 Calibration Kits and Principles of Error Correction

Prior to any device or circuit measurement, a calibration procedure should be followed to minimise the effects of the sources of errors inherent in the system. Here the system includes the network analyser, connecting cables, connectors, test jigs used to accommodate the device, etc. Although the calibration procedures are built into the network analyser itself, it is important to understand how they operate so that their limitations are understood.

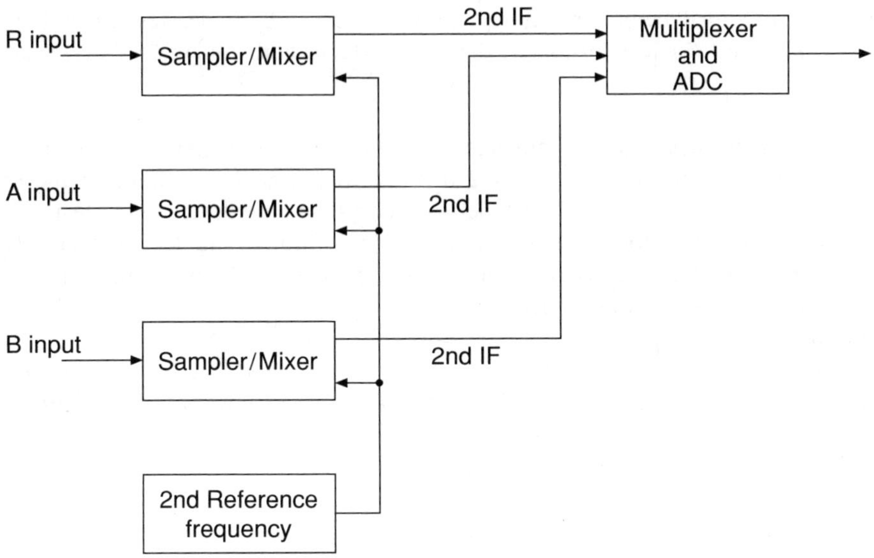

Figure 3.69 Receiver configuration

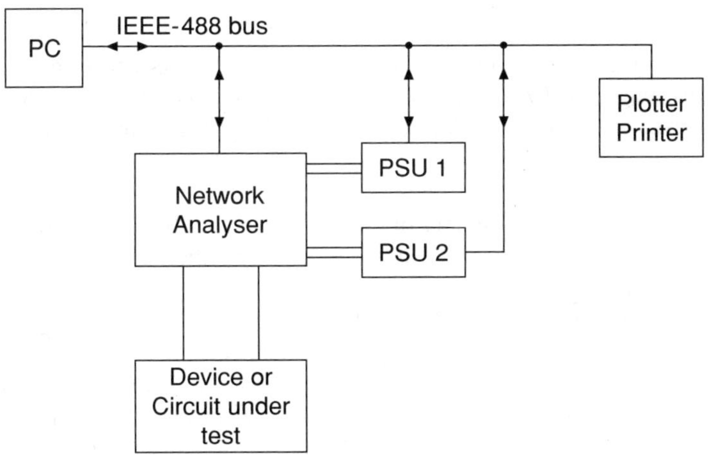

Figure 3.70 Typical equipment configuration for s-parameter measurements

The errors present in the system can be divided into two groups: systematic and random. Random errors cannot be accounted for by the calibration procedures but systematic errors are predictable and can therefore be compensated for. The principal systematic errors are:

1. *Source mismatch*
   This is the error caused by the impedance mismatch between port 1 of the network analyser (the source) and the input of the device under test. Some of the incident signal will be reflected back and forth between the source and the device under test, a fraction

of the 'error' signal being transmitted through the device under test to port 2 of the network analyser. Such a mismatch therefore affects both reflection and transmission measurements.

2. *Load mismatch*
   This is the error caused by the mismatch between the output impedance of the device under test and the impedance of port 2 of the network analyser. Some of the incident signal will be reflected back and forth between port 2 and the device under test, a fraction of the reflected signal from port 2 being transmitted through the device to port 1. If the device has a low insertion loss, then the signal reflected back and forth between port 1 and port 2 can cause significant errors.
3. *Cross-talk*
   The coupling of energy between the signal paths of the network analyser affects the quality of transmission measurements. In modern network analysers, however, the isolation is good and the largest uncertainty arises from the device under test. The calibration procedures described later eliminate much of this error.
4. *Directivity*
   Ideally, the directional bridge or coupler in the test set should isolate the forward and reverse travelling waves completely. In reality, a small amount of incident signal appears at the coupled output due to leakage. Directivity specifies how well the coupler can separate the two signals. This error is usually small in modern network analysers, the largest uncertainty arising from the device under test.
5. *Tracking*
   This is the vector sum of all the previously described sources of error as a function of frequency, the magnitude and phase of the errors being taken into account. It is usual for the tracking error to be split into two components: the transmission and reflection tracking error.

As can be seen, there are a large number of errors that need to be accounted for, the majority of which are addressed by the error models employed by the network analyser. These are discussed next, starting with the simplest, one-port, case.

For one-port devices, the measured reflection coefficient (s_{11M}) differs from the actual (i.e. true) reflection coefficient (s_{11A}) of the device under test because of the errors discussed above. The source mismatch error (E_S), directivity error (E_D) and tracking error (E_T) are most important. Knowledge of these errors allows the actual reflection coefficient to be determined using:

$$s_{11M} = E_D - \frac{s_{11A}E_T}{1 - E_S s_{11A}} \tag{3.223}$$

To evaluate the three sources of error, three standards are employed. A 50 ohm reference load effectively measures the directivity error E_D since $s_{11A} = 0$. A reference open and short are used to evaluate E_S and E_T. Each of these reference terminations should be measured over the same frequency range that will be employed to measure the device or circuit under test, and the results stored. Once this is done, the device under test can be measured and its actual reflection coefficient (s_{11A}) determined.

For the two-port case a similar procedure is employed but the device or circuit under test must be terminated in the system characteristic impedance. This should be as good as the reference 50 ohm load used to determine the directivity value for the one-port situation. For the two-port case, the most significant errors are source-mismatch, load-mismatch, isolation and tracking. For this case, the actual s-parameters (s_{ijA}) of the device under test are determined from the measured s-parameters (s_{ijM}) using:

$$s_{11A} = \frac{A(1 + BE_{SR}) - CDE_{LF}}{(1 + AE_{SF})(1 + BE_{SR}) - CDE_{LF}E_{LR}} \quad (3.224(a))$$

$$s_{21A} = \frac{C[1 + B(E_{SR} - E_{LF})]}{(1 + AE_{SF})(1 + BE_{SR}) - CDE_{LF}E_{LR}} \quad (3.224(b))$$

$$s_{12A} = \frac{D[1 + A(E_{SF} - E_{LR})]}{(1 + AE_{SF})(1 + BE_{SR}) - CDE_{LF}E_{LR}} \quad (3.224(c))$$

$$s_{22A} = \frac{B(1 + AE_{SF}) - CDE_{LR}}{(1 + AE_{SF})(1 + BE_{SR}) - CDE_{LF}E_{LR}} \quad (3.224(d))$$

where:

$$A = \frac{s_{11M} - E_{DF}}{E_{RF}} \quad (3.225(a))$$

$$B = \frac{s_{22M} - E_{DR}}{E_{RR}} \quad (3.225(b))$$

$$C = \frac{s_{21M} - E_{XF}}{E_{TF}} \quad (3.225(c))$$

$$D = \frac{s_{12M} - E_{XR}}{E_{TR}} \quad (3.225(d))$$

The error quantities in Equations (3.225) and (3.226) are identified in Table 3.3.

Table 3.3 Network analyser error terms

Symbol	Error
E_{DF}	Forward directivity error
E_{DR}	Reverse directivity error
E_{XF}	Forward isolation error
E_{XR}	Reverse isolation error
E_{SF}	Forward source-mismatch error
E_{SR}	Reverse source-mismatch error
E_{LF}	Forward load-mismatch error
E_{LR}	Reverse load-mismatch error
E_{TF}	Forward transmission tracking error
E_{TR}	Reverse transmission tracking error
E_{RF}	Forward reflection tracking error
E_{RR}	Reverse reflection tracking error

The above approach takes into account 12 error terms. As previously discussed, the error terms are determined with 50 ohm, open and short reference terminations. The error value should be determined over the same frequency range to be employed in the circuit testing. The signal strength, cabling, connectors, etc. should be the same for the calibration and circuit testing.

A variety of kits is available to perform the calibration procedure. These include kits for connectors using 7 mm, type N, 3.5 mm, SMA, etc. Each kit includes a 50 ohm, short and open termination. Reference adaptors are available to convert from one connector family to another.

The effects of frequency and temperature drifts can be minimised by performing the calibration process at regular intervals. Good RF cables and connectors are an integral part of the system and good care should be taken of them. Errors such as those introduced by connector repeatability and noise cannot be accounted for.

3.10.3 Transistor Mountings

The measurement of the s-parameters of a semiconductor device needs to accommodate two situations, i.e. the measurement of a packaged or naked discrete device, and the measurement of a discrete on-wafer device. These cases are considered in the following subsections.

1. S-parameter measurements of packaged devices

For a packaged device such as a bipolar transistor, a suitable device holder or jig is shown in Figure 3.71. This consists of a brass body, usually gold-plated to minimise losses, which supports two alumina substrates, each with a 50 ohm microstrip line. These lines are attached to suitable high-frequency 50 ohm connectors (such as SMA) and via high-frequency 50 ohm cables to the network analyser ports. To eliminate the effects of any air gaps between the substrate and the body of the jig, a film of conductive paste is applied to the underside of the substrate. Sandwiched between the two halves of the jig is a brass spacer, which is used to hold the device. The jig is designed so that different spacers can be utilised depending on the size of the device package.

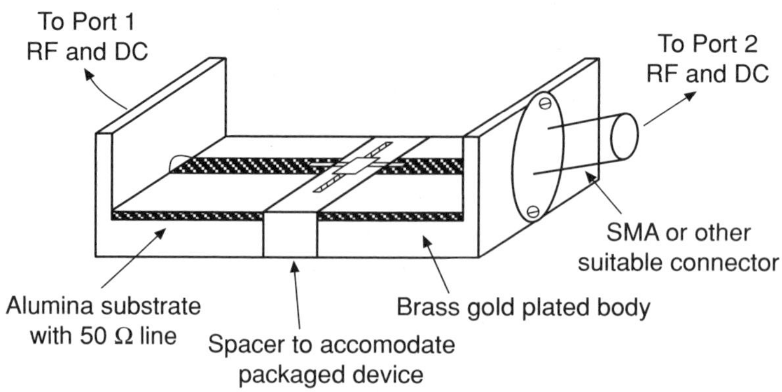

Figure 3.71 Microwave jig for packaged device measurement

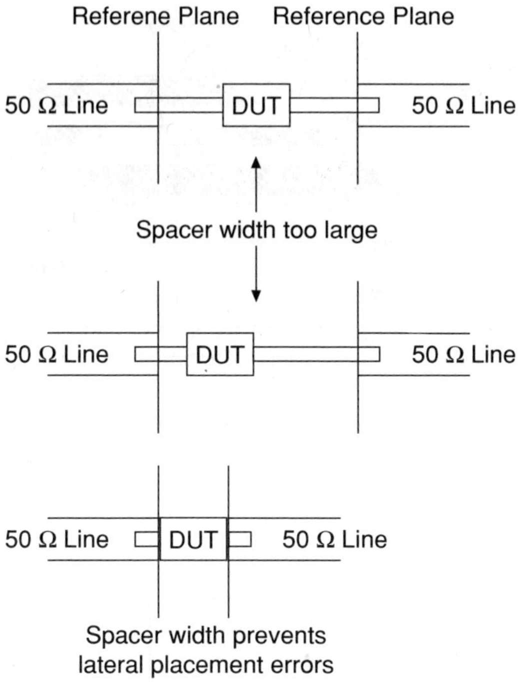

Figure 3.72 Spacer design

The spacer should be designed so that placement of the device on the jig is repeatable. Placement differences between one device and another identical device will affect the measured data, as shown in Figure 3.72.

The device itself is normally configured in common-emitter mode with the emitter terminal connected to the body of the jig. The DC base current to the device is applied via the network analyser RF port 1, while the collector-emitter voltage is applied via port 2. There is therefore no need to have separate cables for biasing the device, and this improves the RF performance of the jig. The leads of the device are not normally soldered to the 50 ohm lines of the jig. A reasonable contact can be achieved with the use of a plastic clamp which does not affect the accuracy of the RF measurements.

For a FET device the same jig can be used. In this case, the device would be configured in common-source mode. Port 1 would be used to carry the gate-source DC bias, while port 2 would be used to apply the drain-source voltage.

Open-circuit calibration of the structure can be performed by leaving the 50 ohm jig lines open. The through line calibration can be performed by bridging across the two 50 ohm lines on the jig with a separate substrate containing a 50 ohm line. Alternatively, special substrates can be designed for the jig to facilitate the calibration process.

2. S-parameter measurements of discrete naked devices

Due to their small size and lack of suitable ground connection points, the easiest way of measuring such devices mechanically is to mount them onto an artificial package. The device

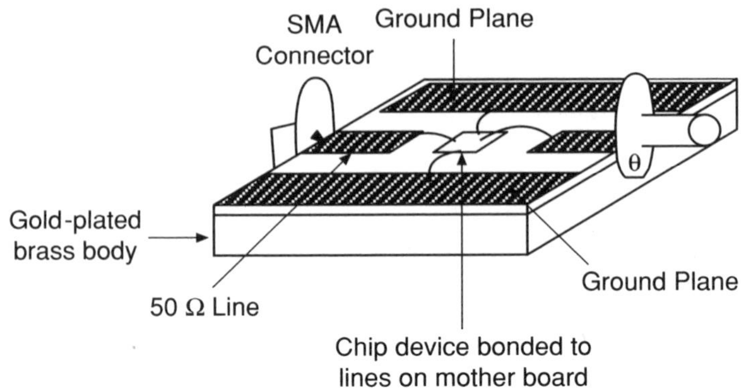

Figure 3.73 Probing a chip device

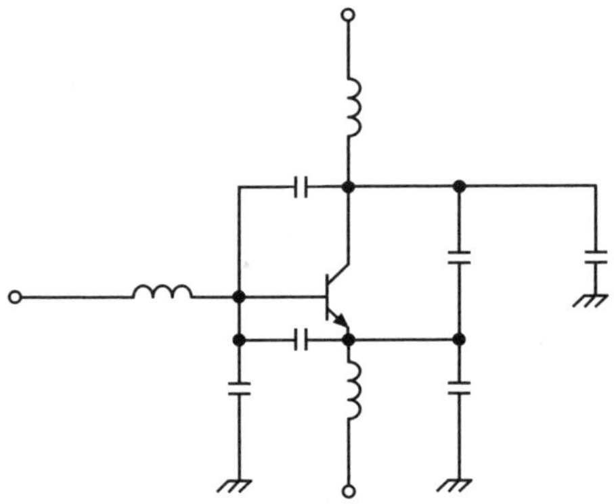

Figure 3.74 Electrical equivalent circuit of mounting shown in Figure 3.73

connection pads can then be bonded to the artificial package or motherboard. From here, SMA or other suitable connectors can be used to connect to the network analyser. A possible scenario is shown in Figure 3.73.

The principal problem with such an approach is the de-embedding or calibration of the system. In this respect the removal from the measured data of the parasitic effects introduced by the bond wires and mounting of the device would present problems. This could be overcome by developing an electrical model for the package, which can then be added to the intrinsic model of the device. For the scheme shown in Figure 3.73, a possible model would be as shown in Figure 3.74.

Assuming that the motherboard, 50 ohm lines, cables, connectors, etc. have been correctly calibrated, the measured s-parameters would correspond to the model shown in Figure 3.74. Notice that in the measurement of a packaged device, the same situation exists, i.e. the s-parameters measured correspond to the intrinsic device and its package.

The estimation of the parasitic elements of the package model is difficult to do from a measurement point of view. However, transmission line graphs can be employed to calculate the series inductances and parallel capacitances. In essence, this approach compares the package to a 50 ohm air line.

3. S-parameter measurements of on-wafer devices

The measurements of the s-parameters of a device on a wafer requires specialised probes in order to maintain a 50 ohm environment to the device contact pads. Many foundry houses lay out their devices as shown in Figure 3.75. The device is configured in common-emitter mode and a co-planar design (ground–signal–ground) is employed for the device pads or contact points. Each pad is typically 100 μm × 100 μm with a pitch between pads of 150 μm. To make contact with the device, a probe such as the one shown in Figure 3.76 is employed.

The gold-plated brass body supports the connector and substrate. The substrate itself uses a co-planar design for the lines, and the contact pads at the tip of the substrate are of the same dimensions as those of the device on the wafer. To measure a transistor, two such probes would be required. Commercially available probes have a frequency range in excess of 40 GHz.

The probe would usually be attached to a micromanipulator to enable its position to be finely controlled in the *x*, *y* or *z* direction. The device or the wafer sits on a prober as shown

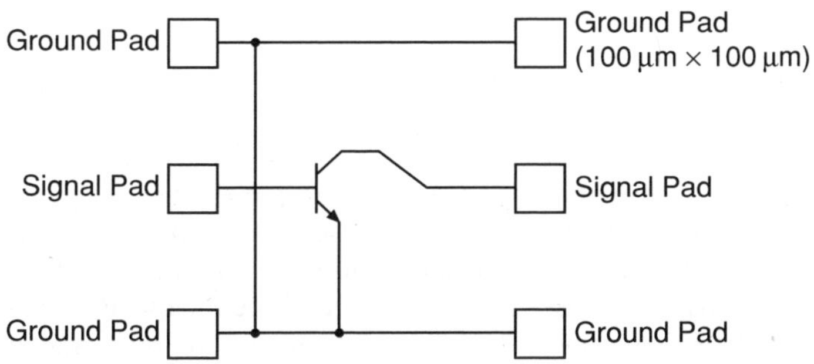

Figure 3.75 Device layout on a wafer

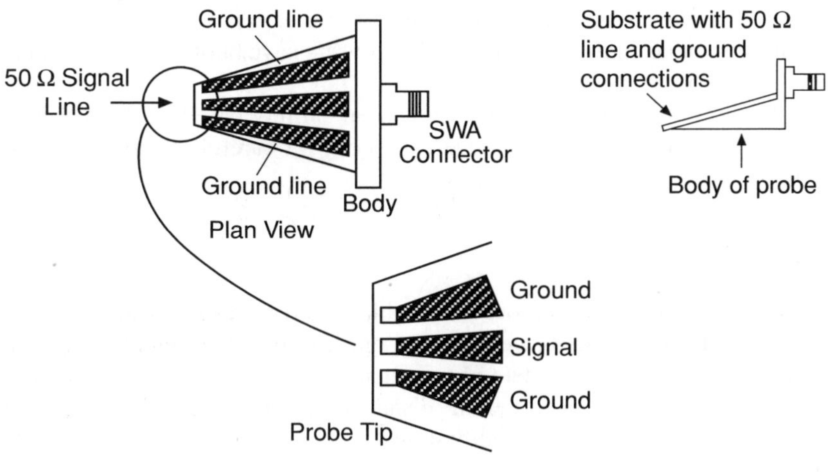

Figure 3.76 Co-planner microwave probe

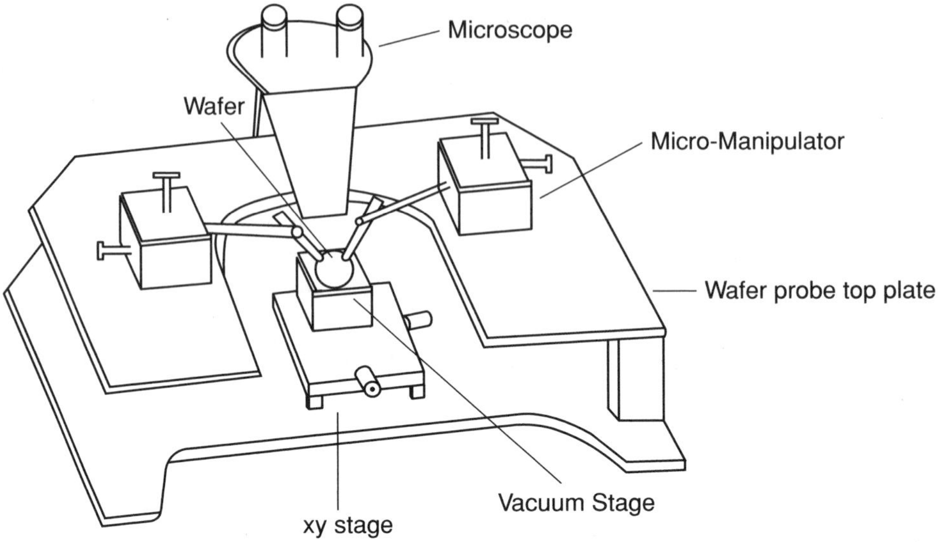

Figure 3.77 Wafer prober with micromanipulators and probes

in Figure 3.77. The wafer is held firmly on the stage by a vacuum pump and the stage is moved up or down (to make or break contact with the probes) with air pressure. The stage sits on an xy platform to enable different parts of the wafer to be probed without having to move the probes or the wafer. The micromanipulators are held firmly on the top plate of the wafer prober using a magnetic base. The positioning of the micromanipulator on the wafer prober is not critical since the fine adjustment is done with the micromanipulator controls and the xy stage of the wafer prober.

The calibration of the system is not as difficult as one might think. Many foundry houses place calibration cells on the same wafer that contain the devices. These calibration cells (open, through, 50 ohm load, short, etc.) employ the same co-planar design and dimensions as the devices. Manufacturers of the microwave probes also provide calibration cells on a wafer to aid the circuit designer. Some network analysers have built-in calibration procedures which recognise such calibration cells.

Although the discussions in this section have been entirely based on the requirements for measuring individual semiconductor devices, the need for careful attention to detail is also important in the characterisation of circuits.

3.10.4 Calibration Approaches

In previous sections the calibration procedure highlighted requires four standards for the error correction process. These standards are the short, open, load and through line (SOLT), which enable the 12 terms of the error model to be evaluated. While this is the most common form of calibrating system, it does have certain disadvantages that arise from the accuracy of the standards themselves. Clearly the calibration standards must be accurately modelled and consistent if the error correction process is to be accurate. In this respect the need for the open and short standards presents a problem. For example, the open standard for on-wafer measurements is usually realised by the probe tip being placed on an open circuit cell. For this to work well, however, the capacitance of the probe tip must be accurately known. This becomes more difficult to evaluate as the operating frequency increases. The same is true of the short standard whose inductance must be accurately known for on-wafer measurements. The 50 ohm load and through-line standards, on the other hand, are less difficult to model and are more consistent. The SOLT method is best used in situations where the measurements are carried out in a 50 ohm characteristic impedance environment.

The 12 terms of the error model can also be evaluated using the Through-Reflect-Line (TRL) standard. As the name implies, this requires a through line and one or more transmission lines of varying offset-lengths. This procedure avoids the problems with the open and short standards needed for the SOLT method. The TRL approach therefore provides better accuracy than the SOLT approach in on-wafer measurements where good quality line standards are available. A disadvantage of the TRL approach is that, because the bandwidths of the lines are limited, several lines may be required for broadband measurements. Some of these may be physically long if low frequencies are to be included.

A third possibility is to use the Line-Reflect-Match (LRM) technique. In this case the standards required are a transmission line for the through standard and broadband loads for defining the impedance references. This procedure avoids the disadvantages of TRL and SOLT since open, short or long line standards are not required and the standards that are needed can be accurately fabricated and modelled.

The choice of calibration procedure is determined by the application, the availability of standards and the availability of calibration models in the network analyser. Whichever procedure is used, however, it should be remembered that the quality of the s-parameter data for a device or circuit being characterised is determined by the accuracy with which the system is calibrated.

3.11 Summary

Transmission lines, network methods and impedance matching techniques are all fundamental tools for microwave circuit design. Transmission lines are principally used to connect other components. They are also, however, passive devices in their own right, being the raw material for couplers, power splitters and distributed filters. They can be implemented in a variety of technologies including twisted and parallel pair, waveguide, coaxial cable, finline, microstrip, slotline and other, more exotic, variations. Twisted pair, parallel pair, and coaxial cable can support electric and magnetic fields that are purely transverse to the direction of propagation and are therefore referred to as transverse electromagnetic (TEM) lines. Two-conductor lines that have non-uniform dielectric constant in the plane orthogonal to the line are quasi-TEM lines. TEM and quasi-TEM transmission lines are generally easy to analyse because they allow distributed circuit theory to be applied. The analysis of other non-TEM technologies is generally more difficult since it requires the application of field theory.

Microstrip is the most important transmission line technology in the context of circuit design. The microstrip design problem is typically that of establishing the line geometry (width and length) given the required operating frequency, characteristic impedance and parameters of the substrate on which the line is to be implemented (i.e. relative permittivity and thickness). This is usually achieved using semi-empirical design formula (or curves) often implemented within a CAD software package.

Dispersion in microstrip cannot be avoided completely but can be predicted using field theory or approximated using semi-empirical models. Other microstrip limitations such as surface waves and transverse resonance may also need to be considered in some designs.

A discontinuity occurs whenever the geometry of a microstrip changes. The most common types of discontinuity are the foreshortened open end, microstrip via, the right-angled bend and the T-junction. The open end results in fringing capacitance that is often modelled by an additional (hypothetical) section of line. The microstrip via is a plated through hole that 'short-circuits' the microstrip conductor to the ground plane. The discontinuity of a right-angled bend can be mitigated by appropriately mitring (chamfering) the outside corner of the bend and the optimum effective widths of the main and branch lines comprising a T-junction can be found using semi-empirical models.

Microstrip can be used to implement a range of low-Q filters. Designs are typically realised by transforming the lumped elements (capacitances and inductances) of a conventional filter into equivalent segments of parallel and series connected microstrips.

The electromagnetic coupling that occurs between closely spaced microstrip lines can be used to realise a variety of devices including filters and directional couplers. The design of these devices typically relies on the theory of even and odd modes – the former referring to the component of electric field that has the same polarity on each microstrip and the latter referring to the component of field that has opposite polarity.

A general two-port device or circuit can be described from a system's point of view by any of several sets of network parameters. These include impedance- or z-parameters, admittance- or y-parameters, hybrid- or h-parameters, ABCD parameters and scattering- or s-parameters. The latter are by far the most important at microwave frequencies although, in principle, all sets contain identical information and transformation between sets is possible using simple matrices. The advantage of s-parameters relates to the practical difficulty of engineering good open- and short-circuit loads (precisely) at the terminals of the device or circuit under test, such loads being required for all except s-parameters. Their interpretation as travelling wave reflection and transmission coefficients is also particularly appropriate and intuitive at microwave frequencies.

Impedance matching refers to the design of circuits and subsystems such that their output and input impedances over a required frequency band realise some overall performance criteria such as maximum power transfer, minimum return loss or minimum noise figure (see Chapter 4). The Smith chart provides a graphical representation of impedance and is commonly used as a design tool for impedance matching problems.

Network analysers are used to measure the s-parameters of a circuit or device automatically. Accurate measurements require prior calibration of the network analyser and associated cables, connectors and mounts.

References

[1] S.Y. Liao, *Microwave Circuit Analysis and Amplifier Design*, Prentice Hall Inc., Englewood Cliffs, NJ, 1987.

[2] F.A. Benson and T.M. Benson, *Fields Waves and Transmission Lines*, Chapman & Hall, London, 1991.

[3] C. Bowick, *RF Circuit Design*, H.W. Sams & Co., Indianapolis, 1982.

[4] G.D. Vendelin, A.M. Pavio and U.L. Rohde, *Microwave Circuit Design Using Linear and Nonlinear Techniques*, Wiley & Sons, New York, 1992.

[5] V.F. Fusco, *Microstrip Circuits*, Prentice-Hall, Englewood Cliffs, NJ, 1989.

[6] T.C. Edwards and H.B. Steer, *Foundations of Interconnect and Microstrip Design*, 3rd edition, John Wiley & Sons, Chichester, 2000.

4

Amplifier Design

N. J. McEwan and D. Dernikas

4.1 Introduction

Everybody knows that the job of an amplifier is in some way to increase power, and would have no hesitation in defining gain as output power over input power. To make these ideas precise enough to use in amplifier design work, we have to think more clearly and define all the power quantities carefully.

Section 4.2 will first give the basic definitions of gain, and derive expressions for gain, working in terms of s-parameters. There will then be a comment on the mathematical origin of the two types of expressions that give rise to graphical representations in terms of circles. In the final part of this section, the idea of gain circles will be briefly introduced.

We then digress in Section 4.3 to look at some basic ideas of stability – a major consideration for any amplifier design is to avoid instability or oscillation. After that, we return to discussing the appearance of gain circles, and some further concepts relating to power gain. Section 4.4 looks into techniques for developing broadband amplifiers, while Section 4.5 concentrates on what is typically the first element in any amplifier – the low noise amplifier.

The limitations and features of implementing amplifiers in practice are discussed in Section 4.6, as at microwave frequencies the capacitance, inductance and resistance of the packages holding the devices can have very significant effects. As we increase our operating frequency, alternative circuit layout techniques can be used in place of discrete inductors or capacitors, and some of these are discussed. In dealing with this we see that even relatively simple amplifier circuits are described by a large number of variables. The current method of handling this amount of data and achieve optimum designs is to use a CAD package, and some of the more popular packages are described in Section 4.7.

4.2 Amplifier Gain Definitions

Consider first the situation shown in Figure 4.1, where a linear two-port network is connected to a given load and a given generator. Remember that we are working in the frequency

Microwave Devices, Circuits and Subsystems for Communications Engineering Edited by I. A. Glover, S. R. Pennock and P. R. Shepherd

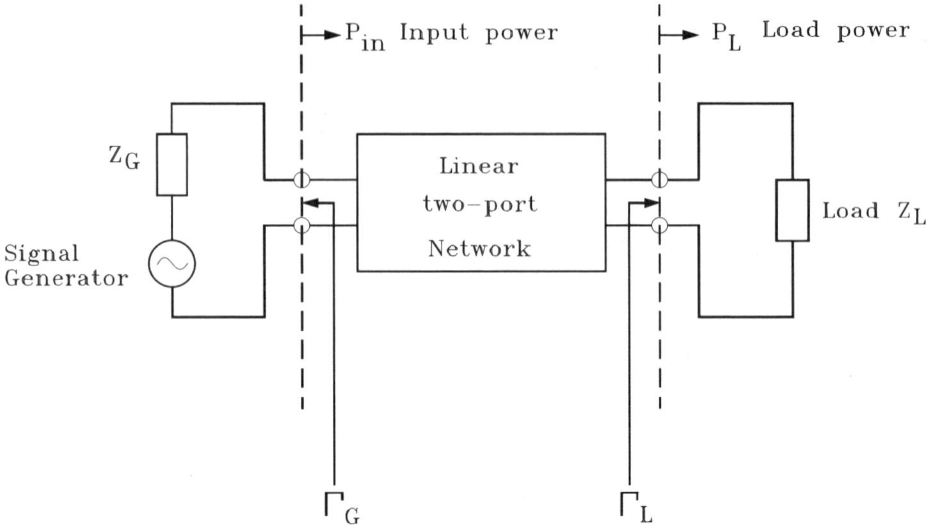

Figure 4.1 Layout of two-port with generator and load

domain. We are considering what is happening at a single frequency; *all* the quantities we are using are functions of frequency, but we do not spell this out.

It should be remembered that, while an impedance is an intrinsic property of a one-port device on its own, its reflection coefficient relates to its interaction with its environment, and is only defined when the reference impedance of that environment has been specified. The same comments apply, for a multi-port device, to its impedance matrix and its s-parameter set. All the reflection coefficients and s-parameters in Figure 4.1 are assumed to have been expressed relative to a specified reference impedance, in practice, probably 50 Ω.

The two-port could just be a single transistor, maybe a complete amplifier, or one of many other devices. In a typical amplifier problem we have an active element as a central block, and this is most commonly a single transistor, possibly with some feedback components. On each side of the device we have a pair of impedance matching networks designed to couple the device to a given generator and load. Usually it will be convenient to take the reference planes in Figure 4.1 as lying on each side of the active device. We can then choose the generator and load reflection coefficients, or equivalently their impedances, as we please, by varying the matching networks, but the s-parameters of the two port will be fixed.

We could, as an alternative, move the reference planes out to the final generator and load positions, and include the matching networks inside the central two-port. In this case the source and load Γs would be fixed and the s-parameters of the two-port would change if the matching networks were varied.

However, it is probably simplest to use the inner reference planes and visualise the effect of changing the load and generator Γs. Fortunately the matching networks are mostly chosen to be nominally lossless, meaning that in practice their losses will usually be negligible, at least as a first approximation. In this case the power gains defined at the inner reference planes will then automatically be the same as those defined at the outer ones.

Referring to Figure 4.1, at the input plane, we can define two power quantities:

1. The power *available* from the generator, $P_{A,G}$.
2. The *actual* input power P_{in} accepted by the two-port from the generator.

Note that P_{in} is the net power flow from left to right at the input reference plane, and it will be represented as the difference in powers carried by the incident (right travelling) wave and reflected (left travelling) waves at that plane.

It is very important to be clear what parameters of the system the various power quantities depend on. For example, note that $P_{A,G}$ is a property of the generator alone, while P_{in} depends on its interaction with the two-port. In the general case where the two-port is not unilateral, i.e. has some coupling from its output to its input, varying the *load* impedance on the output port can change the apparent impedance or reflection coefficient at the input port, and hence can affect P_{in}. So P_{in}, in the general non-unilateral case, depends on all four s-parameters of the two-port, on Γ_L, and on both $P_{A,G}$ and Γ_G.

Now looking at the output port, we can clearly define a similar pair of quantities:

1. The actual power P_L accepted by the load from the two-port.
2. The available power $P_{A,out}$ at the output port of the two-port.

The physical meaning of P_L is clear, but what does it depend on? The output port of the two-port becomes a well-defined generator if, and only if, its output impedance and available power are defined. These in turn are well defined if and only if generator properties Γ_G and $P_{A,G}$ are well defined. The load power P_L also obviously depends on how much power is being reflected back from the load, and hence on Γ_L.

In fact, P_L depends on *all* the parameters of the system – all four s-parameters of the two-port, Γ_G, $P_{A,G}$ and Γ_L. However, the quantity $P_{A,out}$ is, by definition, the quantity P_L when it is optimised with respect to Γ_L while keeping all the other parameters fixed. Thus $P_{A,out}$ is a function of *all* parameters except Γ_L. These points will become clearer when the flow graph analysis of the system is undertaken.

4.2.1 The Transducer Gain

It turns out that the most fundamental definition of power gain is the ratio of *actual* load power to *available* power from the generator. This is called the transducer gain:

$$\text{Transducer gain} \quad G_t = \frac{P_L}{P_{A,G}} \tag{4.1}$$

Why is this is the most fundamental definition? To do something practically useful, an amplifier must deliver power into a real, specified load, which might be some signal processing device such as a demodulator, another amplifier, an antenna, a transmission line for communicating signals to some distance point, etc. It must do this when fed from an actual specified signal source, and the measure of its practical usefulness is that the *actual* power we get into our load is greater than the maximum power we *could have obtained* from the specified source. The practical usefulness does not, however, depend on whether the amplifier is actually accepting all the power available from the source.

You should by now be convinced that the transducer gain depends on all the parameters of the system: all four s-parameters of the two-port, Γ_G, and Γ_L. We can write:

$$G_T = G_T(\Gamma_G, [S], \Gamma_L) \tag{4.2}$$

where $[S]$ denotes the two-port's s-parameter matrix, to emphasise this functional dependence. This again suggests that G_T is the most fundamental definition, as it takes all the relevant parameters of the system into account.

4.2.2 *The Available Power Gain*

This quantity, as its name suggests, is simply defined as the *available* power at the two-port output over the *available* power at the input

$$\text{Available power gain} \quad G_A = \frac{P_{A,out}}{P_{A,G}} \tag{4.3}$$

To interpret this, imagine that everything is fixed except the load, which we vary to make P_L as large as possible. G_A is then this load power divided by the available generator power. G_A is thus the same as G_T optimised with respect to Γ_L while keeping everything else fixed. Because it has been optimised with respect to Γ_L, it is no longer a function of it; thus, we can write:

$$G_A = G_A(\Gamma_G, [S]) \tag{4.4}$$

Why might we wish to use this alternative definition? There are three possible reasons:

1. It gives us a simpler expression to work with, by allowing us to 'shelve' one of the dependencies. It is easy to find an explicit expression for G_A and it tells us the transducer gain that will be achieved on the assumption that the output *is* matched for optimum power transfer, but without needing to worry about the actual value of Γ_L that is required to do it. The problem of actually obtaining the output match is thus put aside for the time being.
   When we use the expression for G_A to examine its variations with Γ_G, it is as though the output matching network is constantly being adjusted to maintain an output match by tracking the variations in the output impedance of the two-port caused by the variations of the generator impedance.
2. It can be used in cascading. If we cascade a pair of two-ports, labelled 1, 2, 3 . . . N with 1 at the input, the overall transducer gain can be written as $G_{A,1} \cdot G_{t,2}$. This is not quite as simple as it looks, because the value of the second gain term depends on the output reflection coefficient of the first stage – which in turn depends on the generator impedance at the input to the first stage. In system design, however, it is usual to perform a straightforward multiplication which will be approximately correct provided the cascaded ports can be taken as reasonably well matched to the reference impedance level.
3. As will be discussed elsewhere, G_A appears in several expressions relating to noise theory, especially the formula for the overall noise figure of cascaded networks, and it is the appropriate quantity to use when choosing Γ_G to optimise the trade-off between gain and noise figure.

4.2.3 The Operating Power Gain

This is the most obvious definition and it is simply the *actual* output power over the *actual* input power:

$$\text{Operating power gain} \quad G_P = \frac{P_L}{P_{in}} \tag{4.5}$$

The reasons for using this definition are just like the reasons (1), (2) given above for using the available gain definition, but the reason (3) relating to noise does not apply:

1. G_p allows us to evaluate G_T and its dependence on Γ_L on the assumption that an input match is always maintained, and without worrying about the value of Γ_G actually needed.
2. G_p can be used in other ways of writing formulae for cascaded power gain, which is left as a self-assessment exercise, see Self-assessment Problem 4.1.

Self-assessment Problem

4.1 For two cascaded two-ports:
   (a) Show that the overall available power gain is the product of the individual available power gains. Does the same hold for operating power gains?
   (b) Show that the overall transducer gain, which we earlier wrote as $G_{A,1} \cdot G_{t,2}$, could also be written as $G_{t,1} \cdot G_{P,2}$.

4.2.4 Is There a Fourth Definition?

You may have noticed that there is a fourth combination of powers we could use to try to define a power gain, namely, $P_{A,out}/P_{in}$. This is well defined if we imagine the two-port with a fixed generator, and the output load being adjusted to receive the maximum power from it. It turns out not to be a particularly useful definition.

We might, however, re-interpret this definition by forgetting about the generator and just adjusting the load to maximise the ratio of P_L/P_{in}. In that case we would obtain G_p optimised with respect to Γ_L, or G_T simultaneously optimised with respect to both Γ_G and Γ_L. Either way, the result would depend on $[S]$ only, i.e. it would be an intrinsic property of the two-port. This does prove to be an important quantity and it is discussed further after considering the issue of stability.

4.2.5 The Maximum Power Transfer Theorem

Before proceeding to derive expressions for the various power gains, we shall first do two preliminary derivations. The first will be to revisit the well-known maximum power transfer theorem, but examining it in terms of s-parameters.

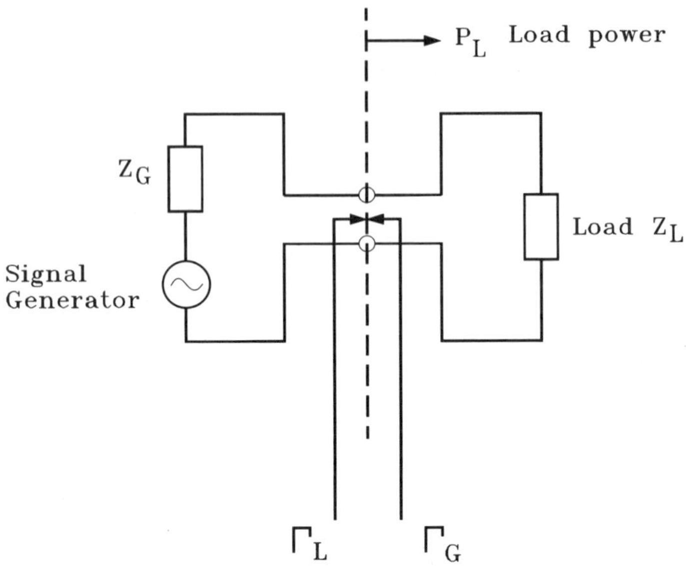

Figure 4.2 Generator connected to a load

Consider a generator and load connected as in Figure 4.2. As normally stated, the optimum or matched load which can extract the maximum power (or 'available' power) from the generator must be chosen to have:

$$Z_L = Z_G^* \tag{4.6}$$

Possibly you only learnt this theorem in the simple case where both the source and load impedances are real, i.e. purely resistive. If so, you can easily see what the complex statement is saying. Let us write $Z_G = R_G + jX_G$, $Z_L = R_L + jX_L$. Now suppose that R_L has been fixed. Then, if the *total* reactance in the circuit is non-zero, it can only reduce the current flowing and hence reduce the power delivered to the load. Hence, part of the optimisation must be to choose $X_L = -X_G$. This is saying that the load reactance will be chosen to cancel out the generator internal reactance, making the total reactance zero. Then, since we know that the final optimisation must have this value of load reactance, we can just fix it at that value, and the problem then reduces exactly to its simple form involving only resistances. We then know from the elementary derivation (which need not be repeated, as it only needs simple calculus to find the maximum) that the optimum load resistance is the same as the source resistance. It has thus been shown that the optimum load is $Z_L = R_G - jX_G = Z_G^*$.

The problem will now be described in s-parameter terms. While impedances in a system have absolute values, the s-parameters only acquire definite values in relation to some specified reference impedance level. Thus a reflection coefficient Γ, which is the single s-parameter of a one-port network, is related to the associated impedance by the equation:

$$\Gamma = \frac{Z - Z_0}{Z + Z_0} \tag{4.7}$$

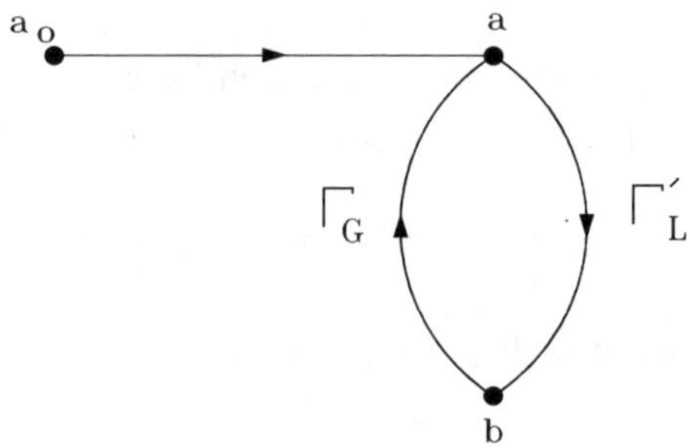

Figure 4.3 Flow graph for generator connected to load

where Z_0 is the reference impedance. The convention will be adopted here that the reference impedance Z_0 is always assumed to be real, and of course in practice usually 50 Ω. This convention makes it unnecessary to distinguish between power waves and travelling waves and, in any case, we know that transmission lines at high frequencies have a real Z_0 to a very good approximation.

Now imagine that the generator is connected, not to the actual load, but to an infinite transmission line whose characteristic impedance is Z_0. A parameter a_0 will be defined as the complex wave amplitude, using the generator terminals as the reference plane, that the generator would launch into that line. There is no reflection on the line because it is infinite, but the generator has a non-zero reflection coefficient Γ_G looking back into its terminals from the line, because its impedance in general is not equal to Z_0.

If the generator is now re-connected to the actual load, a flow graph can be drawn as in Figure 4.3. The two coefficients a, b are the coefficients of forward and reverse waves. There is no need to have any transmission line in the system at all. If it is helpful you imagine a very short section of line interposed between the load and generator terminals, and visualise a and b as specifying the amplitudes and phases of the forward and reverse waves on that line.

Mason's rule (Appendix II) applied to this flow graph gives:

$$a = \frac{a_0}{1 - \Gamma_G \Gamma_L}, \quad b = \Gamma_L a \tag{4.8}$$

The load power is now the forward wave power minus the reverse wave power:

$$\text{Load power} = |a|^2 - |b|^2 = |a|^2 \cdot (1 - |\Gamma_L|^2) \tag{4.9}$$

Therefore, using the expressions for a and b:

$$\text{Load power} = \frac{|a_0|^2 \cdot (1 - |\Gamma_L|^2)}{|1 - \Gamma_G \Gamma_L|^2} \tag{4.10}$$

Now we know that the optimum load is $Z_L = Z_G^*$, and if we insert this value into the $\Gamma - Z$ relationship, equation 4.7, we shall find that the corresponding optimum load reflection is also the conjugate of the generator reflection coefficient:

$$\text{Optimum (matched) } \Gamma_L = \Gamma_G^*$$

NB The assumption that Z_0 is real is essential to this last step. Then if this value of Γ_L is inserted into the expression for load power, we shall have an expression for the available load power:

$$\text{Available power} = \frac{|a_0|^2}{1 - |\Gamma_G|^2} \tag{4.11}$$

This result will be used later in finding expressions for power gain.

Self-assessment Problem

4.2 (a) Combine Equations (4.10) and (4.11) to obtain an expression, in terms of Γ_G, Γ_L for the fraction of the available power that is delivered to the load.

(b) If a 50 Ω source is connected to a 100 Ω load (both being resistive), use your equation, assuming 50 Ω reference impedance, to show that the load power is 0.51 dB less than the available power.

(c) Verify that the same result is obtained if the reference impedance is taken as 25 Ω.

(d) In Equation (4.10), suppose that $|\Gamma_L|$ is fixed but you are allowed to change its argument (phase angle). How would you choose its phase to obtain maximum power transfer? (Think about a phasor diagram representing the denominator of the equation.) Could you use this result to finish proving the maximum power theorem?

It is very important to be aware of the limitations under which the maximum power theorem was derived. It was assumed that everything stays absolutely linear no matter what load impedance is chosen, the generator's internal excitation stays fixed, and the load is being varied to extract the maximum possible output power for this fixed excitation.

This optimisation can be very different from the one which applies when optimising the load in a power amplifier. The object there is usually to maximise the saturated load power, the power at which non-linearity sets in, but the input excitation level can also be varied at will in achieving the best saturated power – provided the power gain does not become unacceptably low. Slight variations of this optimisation are to optimise the DC to RF conversion efficiency, or power added efficiency, both near saturation. In all these cases the two main conditions for deriving the simple theorem fail. Hence, in power amplifiers designed to work near saturation, the load impedance may be far from the conjugate of the output impedance of the active device.

4.2.6 Effect of Load on Input Impedance

Before proceeding to derive expressions for the various power gains, we shall do one final simpler derivation, which is connected with the gain expressions and is used later on in the theory of stability. If we have a two-port device or network, which is not unilateral, there is some coupling from the output port back to the input. The effect is to make the input

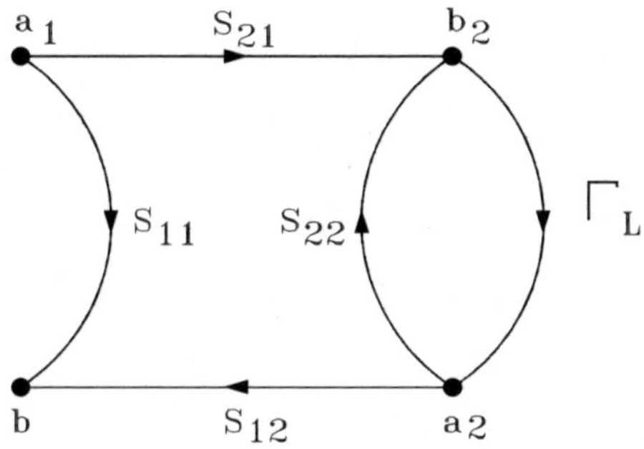

Figure 4.4 Flow graph for considering effect of load on input reflection coefficient

impedance dependent on the load impedance, and the output impedance dependent on the generator impedance. Single transistors always show this effect, which is crucial when considering stability.

It is a good exercise in flow graph analysis to derive the expression which describes the effect of a varying load impedance, connected at the output side of a two-port, on the apparent impedance seen looking into the input port. We work, however, in terms of reflection coefficients rather than impedances, as shown in Figure 4.4. When terminated by a specified load, the two-port acquires a definite value for the apparent reflection coefficient seen at its input port. (Taken together with its load, it becomes in effect a one-port.)

N.B. The notation Γ_{in} will be used throughout for the input reflection coefficient of a two-port, when made definite by specifying the load. Note, however, that some books use S'_{11} for the same quantity. Obviously the analogous notation Γ_{out} (or S'_{22}) will be used when considering how the output reflection of the two-port is modified by the generator impedance at its input.

In this simple graph we can cut out a step or two from the formal Mason's rule recipe. There is only one first-order loop in the system. There are clearly just two paths from a_1 to b_1.

If we just had the second path via S_{21}, Γ_L and S_{12}, Mason's rule would give the transfer function of this path as $S_{21}\Gamma_L S_{21}/(1 - S_{22}\Gamma_L)$. In fact, we have a second path via S_{11} but this has no associated loops and is obviously unaffected by the loop on the other path. Hence the total transfer function is just the sum of the two individual ones and we get:

$$\Gamma_{in} = S_{11} + \frac{S_{21}\Gamma_L S_{12}}{1 - S_{22}\Gamma_L} = \frac{S_{11} - \Delta\Gamma_L}{1 - S_{22}\Gamma_L} \tag{4.12}$$

The first expression here is the one obtained by summing the two transfer functions, and the last one is just a neater but exactly equivalent way of writing it – the symbol Δ is a standard notation for the determinant of the s-parameter matrix, which is $S_{11}S_{22} - S_{21}S_{12}$.

It should be obvious, by symmetry, that we could get the corresponding expression for Γ_{out} by just interchanging suffices '1' and '2' in all the expressions, and of course replacing Γ_L by Γ_G. (As far as the maths is concerned, it doesn't matter which way round the ports of the two-port are labelled!)

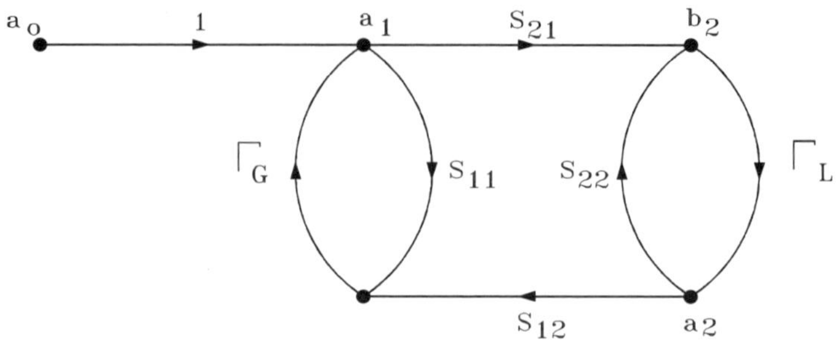

Figure 4.5 Flow graph for two-port with generator and load

4.2.7 The Expression for Transducer Gain

To derive the expression for transducer gain we now need a slightly more elaborate flow graph, Figure 4.5, which is like Figure 4.4 except that the generator has been added. This graph describes the physical set-up shown in Figure 4.1. The relationships between the generator parameters $P_{A,G}$, a_0 and Γ_G have already been discussed. At one frequency, just two parameters – Γ_G, and either $P_{A,G}$ or a_0 – are sufficient to characterise the generator, provided it remains linear. The main step in finding the transducer gain is to write down the transfer function b_2/a_0. We also need to know a_2, but this is just Γ_L times b_2.

There are three first-order loops in this graph, which are the two obvious end loops, and the outermost loop formed by the arrows S_{21}, Γ_L, S_{12}, and Γ_G. A second-order loop is defined as the product of any two non-touching first-order loops – in this case there is just one, the product of the two simple loops at each end. All of these loops touch the single path from a_0 to b_2 which is formed by the arrows 1 and S_{21}.

Since all the loops touch the path, Mason's rule says that the transfer function is the path product divided by (1 – sum of all first order loops + sum of all second order loops.)

This reads:

$$\frac{b_2}{a_0} = \frac{S_{21}}{1 - S_{11}\Gamma_G - S_{22}\Gamma_L - S_{21}\Gamma_L S_{12}\Gamma_G + S_{11}\Gamma_G S_{22}\Gamma_L} \tag{4.13}$$

On inspection, a slightly neater way of writing this can be seen:

$$\frac{b_2}{a_0} = \frac{S_{21}}{(1 - S_{11}\Gamma_G)(1 - S_{22}\Gamma_L) - S_{21}\Gamma_L S_{12}\Gamma_G} \tag{4.14}$$

Now the load power is:

$$P_L = |b_2|^2 - |a_2|^2 = |b_2|^2 \cdot (1 - |\Gamma_L|^2) \tag{4.15}$$

so that Equation (4.12) will now give:

$$\frac{P_L}{|a_0|^2} = \frac{|S_{21}|^2(1 - |\Gamma_L|^2)}{|(1 - S_{11}\Gamma_G)(1 - S_{22}\Gamma_L) - S_{21}\Gamma_L S_{12}\Gamma_G|^2} \tag{4.16}$$

Table 4.1 Transducer gain expressions

Three equivalent expressions:

$$G_T = \frac{(1-|\Gamma_G|^2)|S_{21}|^2(1-|\Gamma_L|^2)}{|(1-S_{11}\Gamma_G)(1-S_{22}\Gamma_L)-S_{21}\Gamma_L S_{12}\Gamma_G|^2}$$

$$G_T = \frac{(1-|\Gamma_G|^2)}{|1-\Gamma_{in}\Gamma_G|^2}\cdot|S_{21}|^2\cdot\frac{(1-|\Gamma_L|^2)}{|1-S_{22}\Gamma_L|^2}$$

$$G_T = \frac{(1-|\Gamma_G|^2)}{|1-S_{11}\Gamma_G|^2}\cdot|S_{21}|^2\cdot\frac{(1-|\Gamma_L|^2)}{|1-\Gamma_{out}\Gamma_L|^2}$$

Unilateral approximation:

$$G_{TU} = \frac{(1-|\Gamma_G|^2)}{|1-S_{11}\Gamma_G|^2}\times|S_{21}|^2\times\frac{(1-|\Gamma_L|^2)}{|1-S_{22}\Gamma_L|^2}$$

= Product of 3 separable 'gain' factors:

One dependent on input match	One intrinsic	One dependent on output match

Note: S'_{11}, S'_{22} as alternative notation for Γ_{in}, Γ_{our} in some texts.

Now refer back to Equation (4.11), and recall the definition of transducer gain, and it should be obvious that:

$$G_T = \frac{(1-|\Gamma_G|^2)|S_{21}|^2(1-|\Gamma_L|^2)}{|(1-S_{11}\Gamma_G)(1-S_{22}\Gamma_L)-S_{21}\Gamma_L S_{12}\Gamma_G|^2} \tag{4.17}$$

Now refer to Table 4.1 which contains the above important expression for transducer gain, and two further ways of writing it. These are exactly equivalent, as can be shown after some algebra, using Equation (4.12) or the corresponding expression for Γ_{out}.

Rather than doing the algebra, there are two ways of re-drawing the flow graph in Figure 4.5, which makes the equivalence of the three expressions obvious, see Figure 4.6.

In this version of the flow graph, we have removed the feedback path and represented instead its effect on the input reflection coefficient by substituting Γ_{in} for S_{11}. The four paths representing the two-port are no longer *generally* equivalent to [S], because Γ_{in} depends on

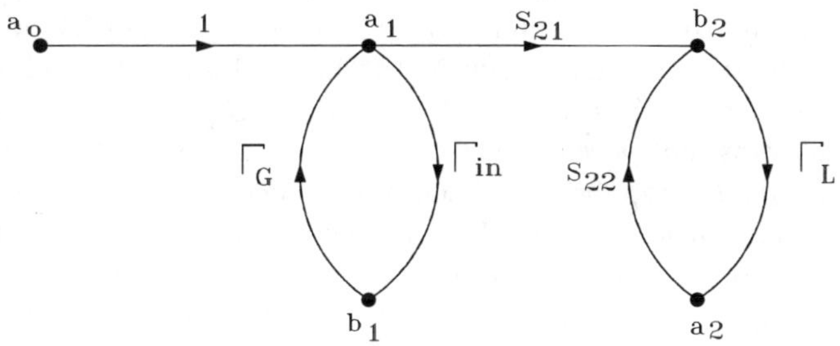

Figure 4.6 Modified flow graph

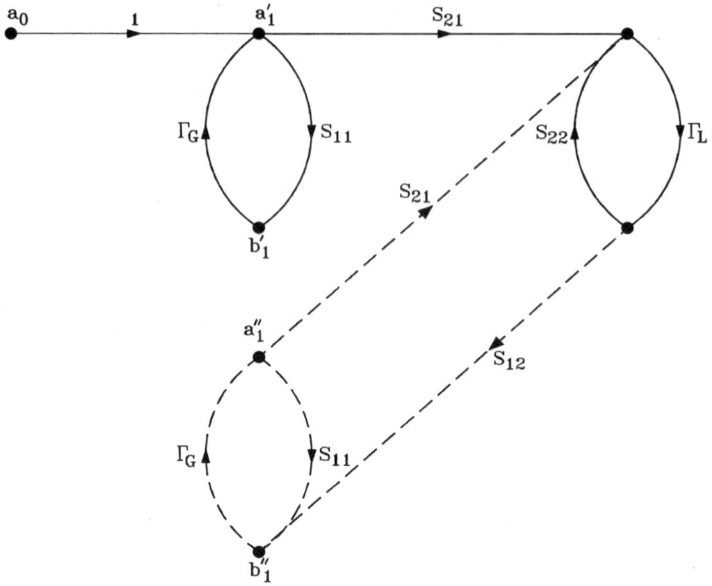

Figure 4.7 Second modified flow graph

Γ_L, but they are equivalent for the particular Γ_L in question. Looking into the input port, the real system and the one described by Figure 4.6 are indistinguishable. This implies that the input a_1, b_1 variables must be the same. However, if a_1 is the same, then b_2 and a_2 must also be $-a_1$ as the only input to the right-hand section of the diagram; the feedback path only leads away from that section and its removal cannot affect the variables on the output side as long as the input variable a_1 stays the same. It should by now be obvious that the second form of the expression for G_T could be derived immediately from the modified flow graph.

Another way of re-drawing the flow graph is shown in Figure 4.7. The feedback loop causing Γ_{out} to be affected by Γ_G has been separated out. Knowing that the system is linear, we could superpose the nodes a_1'', b_1'' on top of a_1', b_1' and we would have the original flow graph of Figure 4.5 again, with the relations: $a_1 = a_1' + a_1''$, $b_1 = b_1' + b_1''$. It would then be possible to reduce the dashed section of Figure 4.7 to a single path with a transfer function Γ_{out}, so that the graph could be drawn yet again as Figure 4.8. Although the a and b variables on the left-hand side are now no longer the same as in Figure 4.5, the output variables a_2 and b_2 are the same, and so are the output power and the available generator power. Thus, Figure 4.8 could be used in a simple derivation of the third form of the G_T expression.

The final expression in Table 4.1 is the form that the G_T expression would take if the device could be taken as unilateral, $S_{12} = 0$. The validity of this *unilateral approximation* depends both on the device and the source and load impedances, and often works well for transistors at high frequencies and/or when designing for gains considerably below the maximum that the transistor could give at the frequency in question.

In the unilateral approximation the G_T expression breaks up simply into three factors: the first dependent only on the input match, the coupling factor $|S_{21}|^2$ which is intrinsic to the two-port, and the third dependent on the output match. The first and third factors are

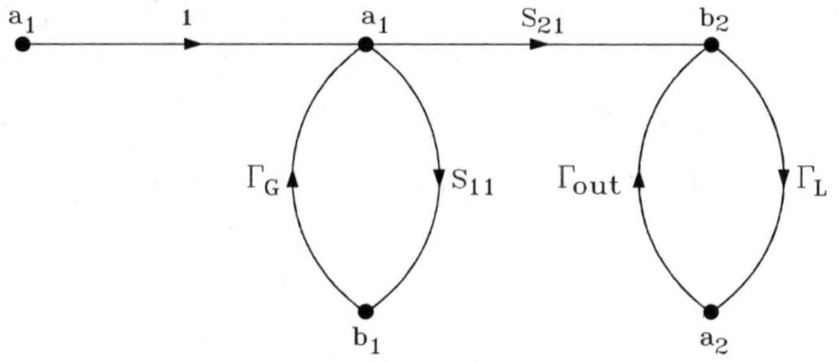

Figure 4.8 Final modification to flow graph

sometimes referred to as the input and output matching 'gains'. They can be larger than unity, allowing G_T to exceed $|S_{21}|^2$.

It is important to keep in mind the possibility of using (with caution) the unilateral approximation, as it simplifies the design problem by separating the input and output matching processes. Even if the approximation is not too good, it may help us to make a rough design which a CAD package, working with the accurate expressions, could be allowed to optimise.

4.2.8 The Origin of Circle Mappings

In RF and microwave engineering, important quantities frequently appear in the form of circle diagrams. This happens in two general ways, and we pause here for a short comment on the mathematics involved.

The first general type of circle diagram arises from a *bilinear mapping*:

$$\zeta = \frac{Az + B}{Cz + D} \tag{4.18}$$

In this equation A, B, C, D are complex constants, and z should be thought of as the variable. For any value of z in the complex plane, a complex value of ζ is generated. The equation can thus be pictured as mapping the z plane into the ζ plane.

The two main things to remember about the bilinear mapping are:

1. The bilinear transformation always maps circles in one plane into circles in the other. (The term circles includes straight lines as a special case i.e. circles of infinite radius.)
2. The area *inside* a given circle in the z plane may be mapped into either the *interior* of the corresponding ζ plane circle, or the *entire region exterior to it.*

Thus, if we draw a circle in the z plane which encloses the point $z = -C/D$, ζ will go off to infinity at this point, so we know that the interior of the z circle is mapped into the exterior of the ζ circle. Any z circle not enclosing this critical point would have to be mapped interior to interior, because ζ would have to stay finite at all interior points of the z circle. A familiar example is the $\Gamma \leftrightarrow Z$ relation (Equation 4.7), leading to the Smith chart in which straight lines of constant resistance or reactance in the impedance plane correspond to the circles in the reflection coefficient (Γ) plane.

Another important application is the stability circles to be discussed later. Many situations in RF engineering give rise to bilinear expressions; error correction in a network analyser is another example.

The second type of circle diagram arises in a different way. Let z be a complex number, and write x and y for its real and imaginary parts: $z = x + jy$. Now consider an expression of the form:

$$X = \frac{Ax^2 + Ay^2 + Bx + Cy + D}{Ex^2 + Ey^2 + Fx + Gy + H} \tag{4.19}$$

In this expression, all the constants are real, and so is the result X. The expression can be visualised as associating a real number with any point in the complex z plane.

What are the loci in the z plane along which X takes a constant value? Again, the answer is a circle. This is easy to prove, as is discussed in Appendix I, both for these and for the bilinear mappings.

The distinguishing feature of the above expression which gives rise to the circle feature is that the coefficients of x^2 and y^2 in the numerator are equal, i.e. both A, and likewise their coefficients are both E in the denominator. If x^2 and y^2 had unequal coefficients on either top or bottom, the loci of constant X would become ellipses.

4.2.9 Gain Circles

The gain circles are simply the loci, in either the Γ_G or Γ_L planes, of any one of the various power gain quantities defined above. It can easily be seen that the various expressions always have the form of the second type of expression discussed in the previous section, so the locus of constant gain is always a circle.

The following are the types of gain circle we might consider plotting:

1. Circles of constant G_T in the Γ_G plane, for a fixed value of Γ_L.
2. Circles of constant G_T in the Γ_L plane, for a fixed value of Γ_G.
3. Circles of constant G_A in the Γ_G plane.
4. Circles of constant G_P in the Γ_L plane.
5. Circles of constant value of input matching gain, in the Γ_G plane.
6. Circles of constant value of output matching gain, in the Γ_L plane.

The input and output matching gains are the factors discussed above in connection with the unilateral approximation. For a strictly unilateral device, i.e. $S_{12} = 0$, all the Γ_G plane circles mentioned will reduce to just one family, and likewise all the Γ_L circles.

The gain circles are readily plotted by CAD packages and are a useful design aid. It is useful to have a group of circles plotted at a selection of frequencies spanning the range we are trying to design for. The trajectory of Γ, as frequency varies, generated by a projected matching circuit, can be plotted against the background of the circles, and used to visualise how the gain will vary. It may be possible to guess a matching circuit topology which will generate a locus that tracks across the gain circles in a way that gives a flat gain or a desired rate of gain tapering. An example will be discussed later. The general appearance of gain circles will be discussed further after considering the question of stability.

Self-assessment Problem

4.3 (a) G_T will become equal to G_A when Γ_L is set equal to Γ^*_{out}. Use the third form of the expression for G_T to show that:

$$G_A = \frac{(1 - |\Gamma_G|^2)}{|1 - S_{11}\Gamma_G|^2} \cdot |S_{21}|^2 \cdot \frac{1}{(1 - |\Gamma_{out}|^2)}$$

(b) We showed earlier that:

$$\Gamma_{out} = \frac{S_{22} - \Delta\Gamma_G}{1 - S_{11}\Gamma_G} 1.$$

Use this to prove that:

$$G_A = (1 - |\Gamma_G|^2) \cdot |S_{21}|^2 \cdot \frac{1}{|1 - S_{11}\Gamma_G|^2 - |S_{22} - \Delta\Gamma_G|^2}$$

(c) If a and b are any two complex numbers, the modulus squared of their sum can be obtained as:

$$|a + b|^2 = (a + b) \cdot (a + b)^* = a \cdot a^* + a \cdot b^* + b \cdot a^* + b \cdot b^*$$
$$= a \cdot a^* + b \cdot b^* + a \cdot b^* + (a \cdot b^*)^* = |a|^2 + |b|^2 + 2\text{Re}(ab^*)$$

Try using this result to expand one of the numerator terms in the expression just obtained for G_A, for example, the term $|S_{22} - \Delta\Gamma_G|^2$. Writing $\Gamma_G = x + jy$, you should find that the only terms of quadratic order occur in $|\Delta|^2 \cdot |\Gamma_G|^2$ which is $|\Delta|^2 \cdot (x^2 + y^2)$. This, of course, is in the form required to generate circles as mentioned before.

You should be able to convince yourself that the same thing happens in the other terms in this expression, and indeed all the various expressions for gain quantities, and hence that the loci of constant gain are always circles.

4.3 Stability

Unintentionally creating an oscillator is the most notorious pitfall of amplifier design. The resulting circuit is usually quite useless for its intended purpose, and effort put into an elaborately optimised design, e.g. for low noise or flat gain, will have been wasted. An indifferently performing but stable amplifier would be more useful.

Stability is therefore the first and most vital issue that must be addressed in amplifier design. There are several reasons why unintended oscillation is an easy trap to fall into:

1. We are tempted to do our design work, e.g. of impedance matching networks, only for the intended operating frequency band of our circuit but oscillation might occur at any frequency, often one far removed from the design frequencies.
2. Virtually all single transistors are potentially unstable at some frequencies, which are often outside the design frequency band.

3. The source and load impedances are a critical factor in producing oscillation, but are never wholly predictable. Oscillation is most likely with highly reactive source or load impedances, and these are very likely to occur outside the operating band.

The general problem of the predictability of the impedance environment is very important. The external impedances are least predictable outside the design operating band, especially in the 'building block' approach to system design where cables of arbitrary length (producing unpredictable impedance transformation) may be used to interconnect the functional units. Outside their specified operating band, the external devices will often have nearly unity reflection coefficients, especially components such as antennas and filters.

Even within the operating band, the external impedances may be uncertain. We could, for example, be designing an amplifier to feed into a nominally 50 Ω load, but then connect it to another amplifier made by someone else which is designed to be *fed* from 50 Ω but whose input does not itself present a close approximation to 50 Ω – some mismatch may have been designed in to produce flatter gain or better noise figure.

When designing amplifiers we should therefore try to have some feel for what regions of the Smith chart our source and load Γ's could possibly lie in, and especially in the frequency ranges of potential instability. Even when designing for an ideal 50 Ω external environment, this must still be subject to some tolerance. The following sections will consider how oscillation can occur, and how it can be avoided.

4.3.1 Oscillation Conditions

A criterion is now needed to establish the conditions under which a circuit will oscillate, both for the purpose of avoiding oscillation in amplifiers, and to deliberately design oscillators when we need to.

When designing amplifiers it is mostly sufficient to consider a 'one-port' oscillation condition, and the following discussion will be limited to this. First, consider a one-port network (one pair of terminals) at which we know the impedance $Z(f) = R(f) + jX(f)$ as a function of frequency (Figure 4.9(a)). Under what conditions will the network oscillate if the terminals are short-circuited?

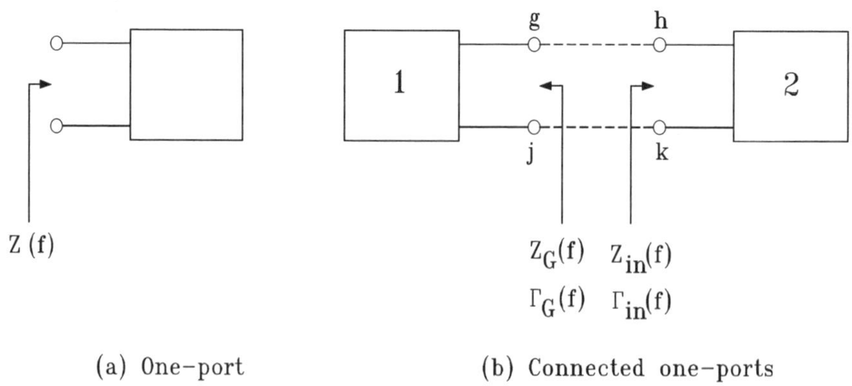

Figure 4.9 Tests for oscillation: definitions of quantities

When the network is completely general, it is quite difficult to state a rigorous condition for oscillation to occur. However, we can limit our discussion to simple networks composed of just one transistor, with common emitter or source, and associated passive matching networks. For these simple cases a criterion which will usually work is that the circuit will be oscillating with short-circuited terminals if the plot in the complex plane of $Z(f)$, as f runs from $-\infty$ to ∞, crosses the negative real axis at a finite frequency with the reactance increasing with frequency at the crossing point. In other words there is a finite frequency where $R < 0$, $X = 0$, and $dX/df > 0$. At this frequency we would also find conductance $G < 0$, susceptance $B = 0$ and $dB/df < 0$. (Using admittance $Y(f) = 1/Z(f) = G(f) + jB(f)$.)

The condition described, where the impedance plot crosses the negative real axis, is the condition that the circuit when first switched on will exhibit exponentially growing oscillations, and of course in a real circuit the oscillation amplitude will always be limited by the onset of non-linearities after a very short time. You might find it somewhat easier to think of the special case where the impedance plot passes exactly through the origin at some finite frequency – in this case, the circuit is in the critical condition where oscillations at that frequency will, if started, neither grow or decay in amplitude. Physically we are saying that the total circuit impedance at that frequency is zero, and hence that a current at that frequency can exist without any voltage to drive it.

Now consider the situation shown in Figure 4.9(b). Block 2 represents a transistor, or other active device. The terminals k, h could be the emitter and base, or source and gate, and there will be a passive load network connected at the output terminals (usually emitter and collector, or source and drain, i.e. using the common emitter or common source connection). The input impedance $Z_{in}(f)$ will be determined once the s-parameters of the device are known, and the load network has been specified. Block 1 with impedance $Z_G(f)$ will be a passive network, generally the generator with associated matching/filtering circuitry. (The discussion will, however, be the same if Block 1 is taken as the output circuit and terminals k, h are, for example, the source and drain of a device which has been equipped with a specified input network.)

What is the condition that the circuit will be oscillating if terminals g and h, and j and k, are connected together? Assuming that j, k are already connected, the total impedance between terminals g, h is $Z(f) = Z_G(f) + Z_{in}(f)$. Clearly, when g, h are shorted together, the circuit will be oscillating if the total impedance $Z(f)$ obeys the condition already stated above in connection with Figure 4.9(a).

The following deductions can now be made:

1. The generator circuit, Block 1, has been assumed to be passive, so it can only have positive resistance. Hence if Block 2 containing the active device only exhibits positive resistance, the plot of total impedance cannot enter the left-hand half of the impedance plane and certainly cannot cross the negative axis. Oscillation is therefore impossible.
2. If Block 2 has $R_{in} < 0$ at some frequencies, this does not necessarily imply that oscillation is present. Even if $R = R_G + R_{in}$ remains < 0 at some frequencies, it may be that the total circuit reactance prevents the Z plot from crossing the negative real axis in the manner required for oscillation.
3. If Block 2 has negative resistance at some frequency f, then there will always be some choice of Block 1 that can make the circuit oscillate at that frequency. At a single frequency, a passive circuit can be designed to have any value of X (positive or negative)

and any value of $R > O$. So R_G, X_G could be chosen to make $R_G + R_{in} = 0$, $X_G + X_{in} = 0$ at that frequency, at which the circuit would then just be starting to oscillate.

4. If a pure resistance R_d is placed between terminals g and h, then seen at terminal g there is effectively a new value of Z_{in} with R_d added to the original value. The effect is to move the impedance plane plot of Z_{in} bodily to the right by an amount R_d. If the original R_{in} becomes negative at some frequencies, but the negative values stay finite at all frequencies (including as $f \to \pm\infty$), then a sufficiently large value of R_d can make oscillation impossible for all possible choices of Block 1 (provided, of course, this remains passive.)
5. If a shunt conductance G_d were connected across terminals h and k, its value would be added to G_{in} and this would move the plot of $Y_{in} = 1/Z_{in}$ bodily to the right by the amount G_d. If any negative values of G_{in} remain finite at all frequencies including $\pm\infty$, a sufficiently low shunt resistance can prevent the admittance plot from entering the left half admittance plane, thus making oscillation impossible for any choice of Block 1.

Case (5) is the one which most often arises with typical transistors.

Self-assessment Problem

4.4 (a) Show that replacing Z by $-Z$ in the fundamental $\Gamma - Z$ relation is equivalent to replacing Γ by $1/\Gamma$. You can now see that $Z_G + Z_{in} = 0$ is equivalent to $\Gamma_G \cdot \Gamma_{in} = 1$.

(b) Prove that an oscillation condition at the input of a two-port network is equivalent to an oscillation condition at the output. This means that if $\Gamma_G \cdot \Gamma_{in} = 1$, then $\Gamma_L \cdot \Gamma_{out} = 1$, and conversely. (It is assumed that S_{12} and S_{21} are nonzero.) This can be proved in a few lines of algebra from Equation (4.12) and the analogous expression for Γ_{out}. The mathematical result is saying what is physically obvious, that an oscillating system is oscillating everywhere, except possibly where there is no coupling between two parts of the system.

Additional comments:

1. To really understand the principles of oscillation it is necessary to generalise the idea of frequency to include the idea of complex frequency. This is likely to seem an unfamiliar idea unless you are well used to network theory, but a rough intuitive idea of it will be enough for the time being. Consider the function of time e^{st}. If we make s a complex number $s = \sigma + j\omega$, then $e^{st} = e^{\sigma t} \cdot e^{j\omega t}$. The imaginary s axis ('real frequency') corresponds to the usual idea of frequency as the purely oscillatory function $e^{j\omega t}$. Values of s with $\sigma > 0$, i.e. with s in the right-hand side of the s-plane, correspond to exponentially growing oscillations, and those with $\sigma < 0$ to exponentially decaying ones. All quantities such as impedances, reflection coefficients, etc., that we normally think of as functions of 'ordinary' frequency can in fact be defined over the entire s-plane.

The oscillation condition is actually that the total impedance in the circuit is zero, and equivalently $\Gamma_G \cdot \Gamma_{in} = 1$, at a value of s in the right-hand side of the s-plane. A zero total impedance means that a current can exist with no driving voltage, and an exponentially growing current described by that value of s will occur. (Until of course the system starts to become non-linear!)

2. You may have come across Nyquist's criterion for oscillation in a control system with negative feedback. The criteria used here in terms of impedances or Γs are similar, so that for example the system will be oscillating if the plot of $\Gamma_G \cdot \Gamma_{in}$ for a real frequency increasing from $-\infty$ to ∞ encloses the real point +1 in a clockwise sense.

Summarising, the main point is that the possibility of oscillation in simple single-device circuits always requires the existence of negative resistance, in some frequency ranges, at either input and output terminals. Avoiding negative resistance will always make oscillation impossible. When negative resistance does exist, this does not necessarily mean the circuit is oscillating, but it is a safer strategy not to have practical designs working in this condition.

4.3.2 Production of Negative Resistance

Modern RF design procedures using CAD try to make very accurate predictions, including many small effects such as 'stray' components, line discontinuities, etc. Modelling of devices often uses elaborate equivalent circuits that aim to reproduce the measured parameters very accurately. Relying on the computer, and trying to work to high accuracy, we can sometimes forget how useful it is to find a simple physical understanding of an effect, and how a drastically simplified model can often reveal the essence of a problem. An example occurs in the production of negative resistance by transistors. We might know from the measured s-parameters that a transistor is capable of showing negative resistance, but this would not reveal the mechanism by which it occurs.

The most basic mechanism producing negative resistance is internal capacitive feedback, and for MESFETs and other FETs this can be explained very simply. Figure 4.10 shows a ruthlessly simplified model of a MESFET. Assuming that a source impedance Z_G is connected across terminals G and S, the problem is to find the output impedance seen between drain and source. The method is to apply a voltage V_{DS} and calculate the drain current I_D that results. Even a simple equivalent circuit would normally include a capacitance C_{DS} from the drain terminal to ground, but this is initially omitted because, at the first level of

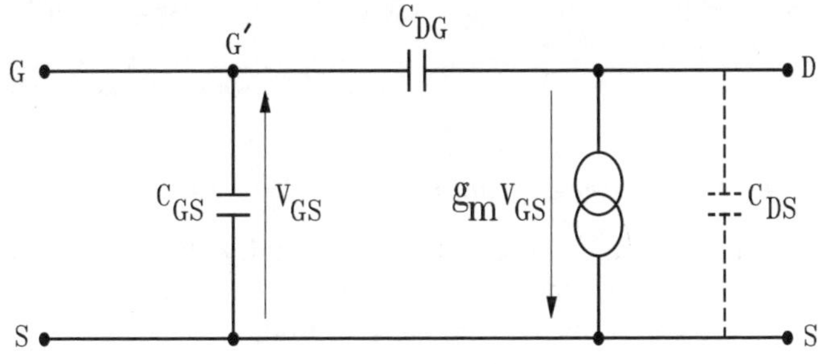

Figure 4.10 Highly simplified MESFET model

approximation, it would only add susceptance at the output and hence would not affect the question of the sign of the output resistance.

The applied voltage is coupled via the feedback capacitance to the parallel combination of C_{GS} and Z_G, and develops a voltage across them. This controls the flow of drain current. Now suppose that the source impedance is a pure inductor L. If the applied frequency is below the resonant frequency of the L/C_{GS} combination, this combination will appear inductive. Also if the frequency is kept low enough, we can be sure that the reactance of the feedback capacitance C_{DG} is greater than the inductive reactance of the L/C_{GS} combination. The impedance of the whole feedback path (C_{DS} in series with L/C_{GS}) is now capacitive, and the current in C_{DG} is 90° *leading* the applied voltage V_{DS}. However, the inductive reactance of the L/C_{GS} combination will develop a voltage across it that is 90° leading the current fed into it. Hence V_{GS} has a total lead of 180° with respect to V_{DS}, i.e is in antiphase with it. The drain current $G_m V_{GS}$ is thus in antiphase with V_{DS} and we have an apparent negative resistance at the output.

Self-assessment Problem

4.5 (a) In Figure 4.10, let Z_a be the impedance of the parallel combination of the generator and C_{GS}, Z_b the impedance of the feedback capacitor C_{DG}, and Y_G the generator admittance. Show that, in the case where $g_m Z_a \gg 1$, the output admittance Y_{out} is given by:

$$Y_{out} = \frac{g_m \cdot Z_a}{Z_a + Z_b}$$

Multiply throughout by $Y_a = 1/Z_a$ to show that this can be re-written as:

$$Y_{out} = g_m\left(\frac{1}{1 + Y_a Z_b}\right) = g_m\left(\frac{1}{1 + (j\omega C_{GS} + Y_G)/j\omega C_{DG}}\right)$$

(b) Remembering that g_m is real and > 0, show that Y_{out} is a negative resistance if Y_G is a sufficiently large inductive susceptance.

(c) A simple improvement of the transistor model is to include a series resistor $R_{GG'}$ between the gate capacitor at G' and the external gate terminal G. The same expression for Y_{out} can be used if $R_{GG'}$ is included as part of the generator impedance. Show that negative resistance can now only occur below a certain frequency.

4.3.3 Conditional and Unconditional Stability

A two-port network or device is said to be unconditionally stable, at a given frequency f_0, if the real part of its input impedance remains positive for any passive load connected to its output port. In other words, the device cannot produce a negative input resistance at frequency f_0 when any load with positive (or zero) resistance is connected at the output. Expressed in terms of Γs, in the notation of Section 4.2.6, the two-port is unconditionally stable if $|\Gamma_{in}| \leq 1$ whenever any load with $|\Gamma_L| \leq 1$ is connected at its output. If the two-port is not unconditionally stable, it is called conditionally stable or potentially unstable.

The condition could have been stated equivalently that no passive *source* could give rise to a negative *output* resistance. A device that is potentially unstable can definitely be made to oscillate at the real frequency f_0 by selecting any load Γ_L that makes $|\Gamma_{in}| > 1$. A source can then be chosen so that $\Gamma_G\Gamma_{in} = 1$ at f_0, i.e. the device is just starting to oscillate at that frequency. Now $\Gamma_G\Gamma_{in} = 1$ with $|\Gamma_{in}| > 1$ implies $|\Gamma_G| < 1$. Furthermore it was already seen in Self-assessment Problem 4.4 that an oscillation condition at the input is equivalent to one at the input, i.e. the condition $\Gamma_G\Gamma_{in} = 1$ implies also that $\Gamma_L\Gamma_{out} = 1$. Hence we know that the chosen Γ_G is making $|\Gamma_{out}| > 1$. Therefore a device that is potentially unstable using the original load-to-input definition could not be unconditionally stable by the source-to-output definition, and vice versa. The two possible definitions must be exactly equivalent.

If a device is unconditionally stable in a band of frequencies containing f_0, it cannot be made to oscillate at frequencies in that region. When the stable band contains the target operating band for our amplifier, this is the easiest design regime but it is still essential to think about what is happening at frequencies far out of the operating band; most single transistors are potentially unstable at some frequencies.

4.3.4 Stability Circles

The stability circles are an almost indispensable graphical aid in amplifier design. The output stability circle of a two-port, at a given frequency, is the locus in the Γ_L (load reflection coefficient) plane which gives rise to Γ_{in} (input reflection coefficient) values lying on the unit circle $|\Gamma_{in}| = 1$. (i.e. purely reactive input impedances.) We know that this locus is indeed a circle because of the general property of bilinear mappings that they map circles into circles. The stability circle is the image of the Γ_{in} unit circle under the bilinear relation, Equation (4.12), which expresses Γ_{in} as a function of Γ_L.

The stability circle therefore divides the Γ_L plane into the stable region, giving rise to positive resistance on the input side, and the potentially unstable region giving rise to negative resistance. If the two-port is unconditionally stable, the potentially unstable region must have no points lying within the unit circle. This means that either (a) the stability circle lies entirely outside the unit circle, without enclosing it, and the potentially unstable region is its interior; or (b) the stability circle encloses the unit circle, and the potentially unstable region is its exterior. These two cases are illustrated in Figure 4.11.

Another circle that may be considered is the locus of points produced in the Γ_{in} plane by points on the unit circle in the Γ_L plane. If the two-port is unconditionally stable, this circle must lie wholly within the unit Γ_{in} plane, and be mapped interior to interior. This is also shown in Figure 4.11, where different shadings have been used to indicate the regions which map into each other.

Obviously the input stability circles in the Γ_G plane are defined in an exactly analogous way. Stability circles can be plotted on demand by the various CAD packages. Usually it will be indicated whether the interior or exterior of the stability circle is the potentially unstable region. However, if needed, a simple check can be made as follows. The point $\Gamma_G = 0$ by definition makes Γ_{out} equal to S_{22}. Thus, the origin of the Γ_G plane must lie in the stable region if $|S_{22}| < 1$ and in the potentially unstable region if $|S_{22}| > 1$. Of course, the known value of $|S_{11}|$ can be used in the same way to identify the two regions in the Γ_L plane.

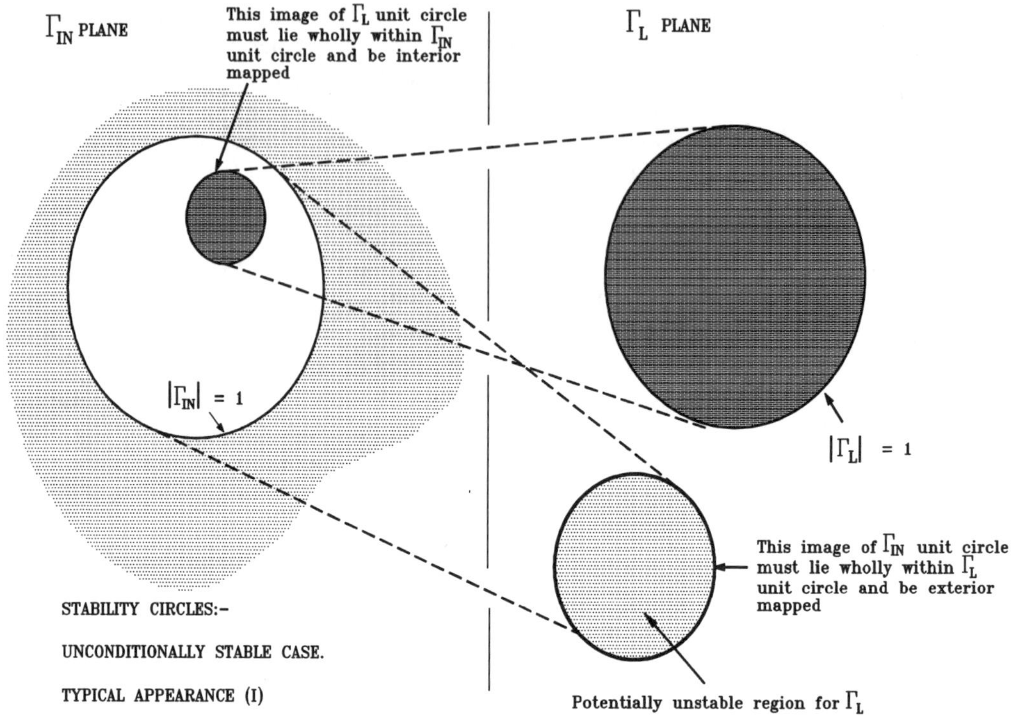

Figure 4.11 Typical appearance of stability circles in the unconditionally stable case

4.3.5 Numerical Tests for Stability

It is useful to have a quick numerical test for whether a transistor or other two-port is unconditionally stable at a given frequency, without having to plot out the stability circles. This is provided by the stability factor which is conventionally written 'k' or less commonly 'K', and defined as:

$$k = \frac{1 - |S_{11}|^2 - |S_{22}|^2 + |\Delta|^2}{2|S_{12}||S_{21}|} \tag{4.20}$$

The basic test is that $k > 1$ is a necessary, and usually sufficient, condition for unconditional stability. The definitions of k and a group of related parameters are probably not worth committing to memory, but one should be aware of them because they appear in several important expressions.

The mathematics shows that $k > 1$ is not by itself an absolute test of unconditional stability for arbitrary two-ports. In the most general case, it is a necessary but not sufficient condition, and one or two extra parameters must be checked. Table 4.2 shows three alternative sets of necessary conditions that can be applied. For example, to use test A, it is necessary and sufficient for unconditional stability to have $k > 1$ and $|B_1| > 0$. B_1 and C_1 are standard notations for two further parameters that are important in amplifier theory, and defined in the table for completeness. (Two further parameters B_2, C_2 are defined like B_1 and C_1 but with ports 1 and 2 interchanged.)

Table 4.2 Tests for unconditional stability

Definitions:

$$\text{stability factor } k = \frac{1 - |S_{11}|^2 - |S_{22}|^2 + |\Delta|^2}{2|S_{21}||S_{12}|}$$

$$B_1 = 1 + |S_{11}|^2 - |S_{22}|^2 - |\Delta|^2$$

$$C_1 = S_{11} - \Delta S_{22}^*$$

$$\Delta = S_{11}S_{22} - S_{12}S_{21}$$

Note: Δ, also written D in some texts, is the determinant of the S parameter matrix.

For single transistors, however, the conditions $|\Delta| < 1$ and $B_1 > 0$ are virtually always obeyed at all frequencies and it is only necessary to look at the value of k to check stability. Some transistors will be found to violate the auxiliary conditions (involving $|S_{21}||S_{12}|$) in test A at some frequencies, but normally only where k is in any case less than 1.

A major factor in determining k is the denominator. For typical transistors S_{21} usually falls with frequency and S_{12} rises, so that their product is greatest in a mid-range of frequencies. Thus transistors are usually unconditionally stable at very low and very high frequencies.

When our two-port in question is cascaded at its output with another two-port network, such as a matching or stabilising network (of which more presently), we shall use the notation Γ_L for the reflection coefficient seen by the original device and Γ_L' for the value presented to the second network by the final load. Thus Γ_L is a function of Γ_L' defined by the s-parameters of the intermediate network. The commonest design problem is of course to choose the second network to obtain the right Γ_L when $\Gamma_L' = 0$.

Let us define the *accessible* region in the Γ_L plane as all the values that Γ_L could take if Γ_L' is allowed to range over its entire unit circle. A very important property of any strictly lossless network, provided it has $S_{12} \neq 0$, is that the accessible region at its input is also the unit circle. In other words, for a truly lossless network, we can obtain any passive ($R \geq 0$) impedance looking into its input with a suitable choice of a passive load impedance. This makes it obvious that the unconditional stability (or otherwise) of device is unchanged if it is cascaded with lossless networks at its input and/or its output. In fact, it can be shown that the stability factor k is invariant under such cascading.

Any one of the three following sets of conditions are necessary and sufficient for unconditional stability.

A $k > 1$ and $|S_{12}||S_{21}| < 1 - |S_{22}|^2$ and $|S_{12}||S_{21}| < 1 - |S_{11}|^2$
(Note that the last two conditions also imply $|S_{11}| < 1$ and $|S_{22}| < 1$.)
B $k > 1$ and $B_1 > 0$ and $|S_{11}| < 1$, $|S_{22}| < 1$
C $k > 1$ and $|\Delta| < 1$ and $|S_{11}| < 1$, $|S_{22}| < 1$

4.3.6 Gain Circles and Further Gain Definitions

Now that stability has been defined, it is convenient to return to discussing the appearance of gain circles, but we first need to look at the concept of the simultaneous conjugate match (SCM).

Under the SCM condition, the two-port is conjugately matched to both its generator and its load, so that:

$$\Gamma_G = \Gamma_{in}^* \quad \text{and} \quad \Gamma_L = \Gamma_{out}^* \tag{4.21}$$

A two-port can be simultaneously conjugately matched if it is unconditionally stable at the frequency in question. If conditionally stable, oscillation will occur if an attempt to set up SCM is made.

For a two-port where the feedback S_{12} is not negligible, the input and output matching processes are not separable and two simultaneous equations must be solved:

$$\Gamma_G^* = \frac{S_{11} - \Delta\Gamma_L}{1 - S_{22}\Gamma_L} \tag{4.22}$$

$$\Gamma_L^* = \frac{S_{22} - \Delta\Gamma_G}{1 - S_{11}\Gamma_G} \tag{4.23}$$

After eliminating one variable, a quadratic equation results which can be solved explicitly. Following considerable re-arranging of terms, the SCM solution can be reduced to:

$$\Gamma_G = \frac{B_1 + -\sqrt{B_1^2 - 4|C_1^2|}}{2C_1} \tag{4.24}$$

$$\Gamma_L = \frac{B_2 + -\sqrt{B_2^2 - 4|C_2^2|}}{2C_2} \tag{4.25}$$

An unconditionally stable device has $B_1^2 > 4|C_1|^2$ and $B_2^2 > 4|C_2|^2$. Using the minus signs gives a solution with $|\Gamma_G| < 1$ and $|\Gamma_L| < 1$ which is of course the practically useful one. Using the plus signs a second solution with $|\Gamma_G| > 1$ and $|\Gamma_L| > 1$ is also obtained, and corresponds to using an active (i.e. negative resistance) generator and load.

Under SCM conditions, G_T has been optimised with respect to both Γ_G and Γ_L (recall Section 1.1.5), and then becomes equal to both G_A and G_p. This peak gain value is called the maximum available gain of the two-port at the given frequency, and written G_{ma}. By substituting the SCM values of Γ_G, Γ_L into the expression for transducer gain, after considerable re-arranging it can be shown that G_{ma} is given by:

$$G_{ma} = \frac{|S_{21}|}{|S_{12}|} \cdot (k - \sqrt{k^2 - 1}) \tag{4.26}$$

G_{ma} may be defined as the highest gain that the two-port device can possibly achieve at the frequency in question, subject to the proviso that no feedback path exists between the input and output networks. The gain could, in principle, be increased by introducing positive feedback. This was common in early radio receivers but is never done nowadays. G_{ma} can only be defined where the two-port is unconditionally stable at the given frequency.

It is now possible to describe the general appearance of gain circles. Taking first the unconditionally stable case, we consider the circles of constant G_A in the Γ_G plane. The expression derived in Self-assessment Problem 3 (b) shows that $G_A = 0$ ($-\infty$ in dB) on the unit circle. We therefore see a set of nested (obviously non-intersecting) circles such that G_A is zero on the unit circle. As the constant value of G_A is increased from zero, the circle shrinks and converges on the single point at which the maximum value G_{ma} occurs, and at which Γ_G is given by Equation (4.19). The centres of the circles all lie on the diameter of the circle which passes through this point. The general appearance of G_p circles in the Γ_L plane, also converging on the SCM point, is similar.

If the two-port is potentially unstable, the gain circles have a different appearance, whose general form is shown in Figure 4.12. This diagram shows the mappings: $\Gamma_L \leftrightarrow \Gamma_{in}$ in

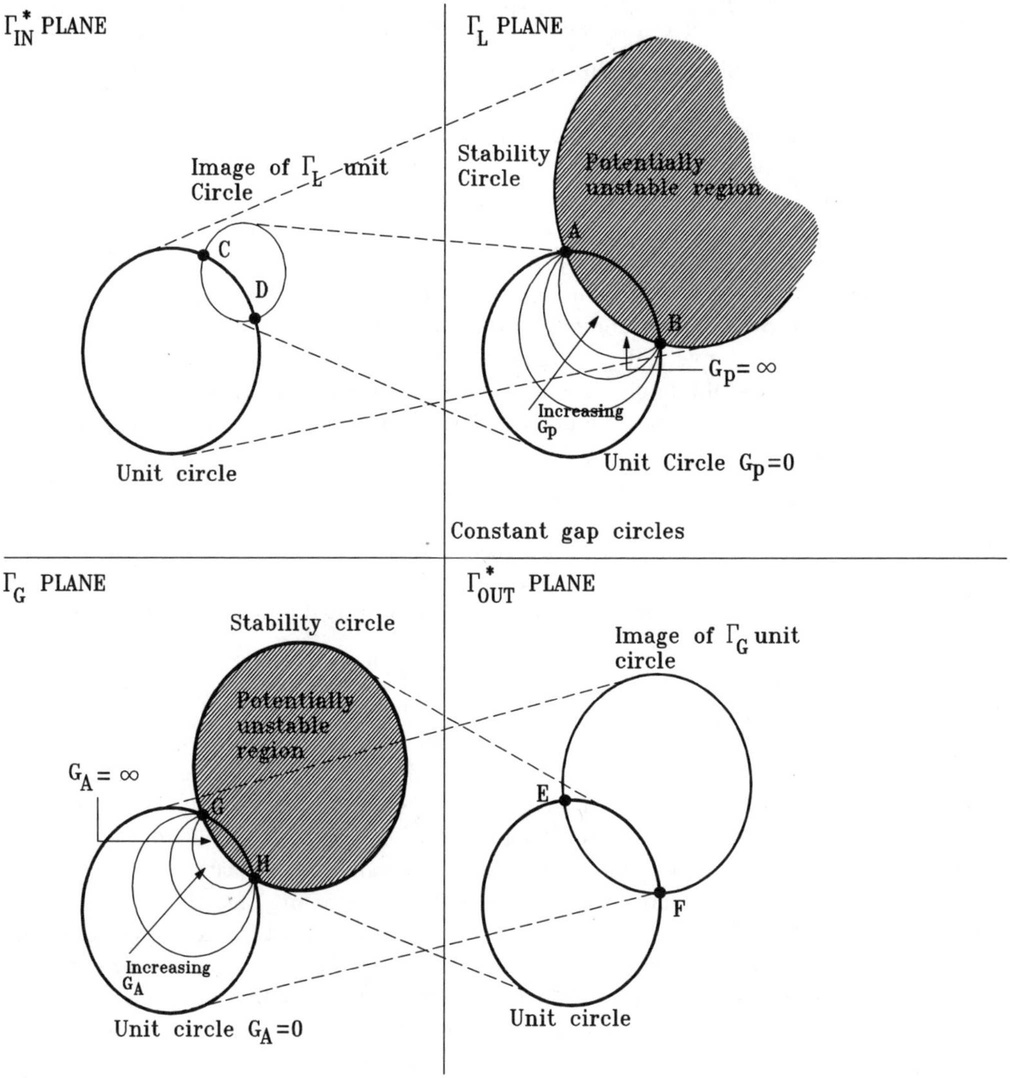

Figure 4.12 Typical appearance of gain and stability circles for potentially unstable two-port

the top half of the diagram, and $\Gamma_G \leftrightarrow \Gamma_{out}$ in the lower half. In fact, it is slightly neater to use the Γ^*_{in} and Γ^*_{out} planes, which are simply the Γ_{in} and Γ_{out} planes reflected about the horizontal axis.

The Γ_L plane unit circle is mapped into a certain circle (its 'image') in the Γ^*_{in} plane, and the unit circle in this plane maps to the stability circle in the Γ_L plane. Because the device is potentially unstable, we know that the stability circle and the other image circle intersect the unit circles, and the intersection points A, B map into the points C, D. A similar set of relationships exist in the lower half of the diagram.

Interestingly the intersection points A, B turn out to be the same as points E, F, and C, D the same as G, H. These points are 'pathological' points which obey both the oscillation condition and the SCM. They are the SCM solutions given by Equations (4.19, 4.20) in the potentially unstable case where $B_1^2 < 4|C_1|^2$ and $B_2^2 < 4|C_2|^2$. At these points we have, for example, $\Gamma_G = \Gamma^*_{in}$, but this also implies $\Gamma_G = 1/\Gamma_{in}$ or $\Gamma_G\Gamma_{in} = 1$ (the oscillation condition) since $|\Gamma_G| = 1$.

It is now easy to deduce the form the gain circles must take. In the lower left part of Figure 4.12, the section of the unit circle outside the potentially unstable region is easily shown from the G_T expressions to have $G_A \equiv 0$ ($-\infty$ dB), so is itself one of the gain circles. The non-zero G_A circles must not intersect either this, the stability circle, or each other, and it is clear that the only way they can be fitted into the space between the unit circle and the stability circle is if they all pass through the pathological points as shown. We see the gain circles moving towards the stability circle as G_A increases. G_A becomes infinite on the stability circle and in the potentially unstable region. At the pathological points, the circle of infinite gain intersects the circle of zero gain. Gain becomes undefined at these points but can take any value in their neighbourhood. The design points should obviously be kept well away from them.

Similar comments apply to the G_p circles in the top right of Figure 4.12. Although G_p becomes infinite on the stability circle, it actually has finite but *negative* values (not negative dBs, but negative numbers) within it. This means simply that a load in this region produces negative resistance at the input side, and there is consequently a net power flow back into the generator as well as out into the load. It is not recommended to operate in this condition but it is possible in principle.

This section concludes with two final definitions of special power gain quantities. It is normally possible to *resistively* load a potentially unstable two-port device, at its input and/or its output, to make the whole network become unconditionally stable. (This would only be impossible where the potentially unstable region enclosed the entire unit circle, which would not happen for normal transistors.) Imagine the original two-port, cascaded with any loading networks, as a single new two-port, and suppose the loading is adjusted to the point where this new two-port just becomes unconditionally stable. The maximum stable gain of the original two-port, written G_{ms}, is defined as being the value of G_{ma} for the entire network, and is a useful figure of merit. G_{ms} can be defined as the largest gain that the potentially unstable device (again disallowing feedback) can give under the requirement that the generator and load must be *simultaneously* matched at the outermost reference planes and this can of course only be achieved by the introduction of *loss* between the original two-port and at least one of the outer reference planes.

The resistive loading to the point where the network is just unconditionally stable makes $k = 1$ and the above expression for G_{ma} then shows that G_{ms} would be given by the ratio

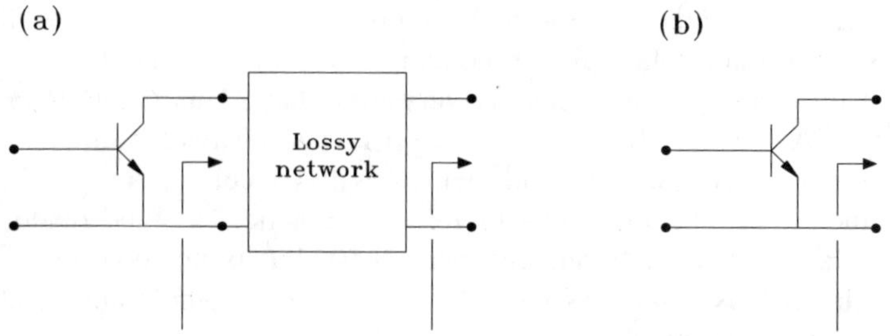

Figure 4.13 (a) Transistor with general lossy output loading; (b) a specific example

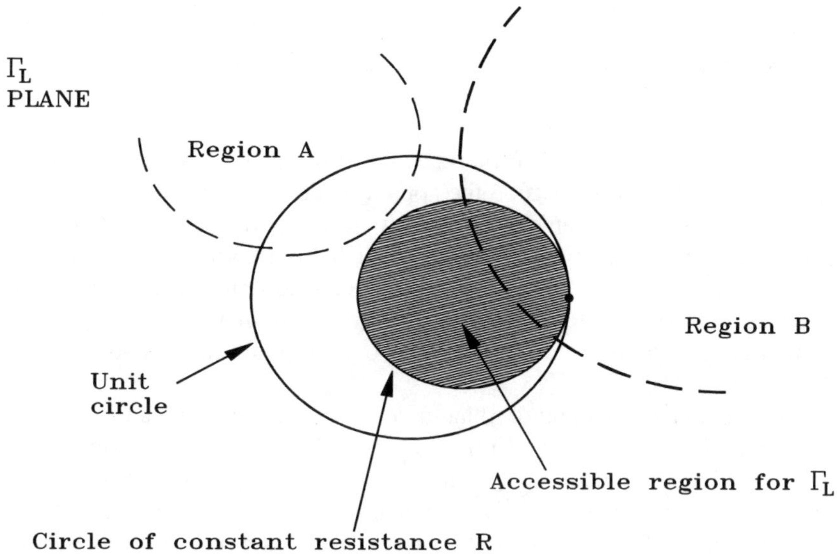

Figure 4.14 Accessible region for Γ_L with series resistive loading

S_{21}/S_{21} for the entire loaded network. In fact, this remains the same as S_{21}/S_{21} evaluated for the original two-port itself.

It is important to realise that not all loading topologies can produce the stability in any given case. Consider Figure 4.13(a) where the two-port (in this case shown as a transistor) is cascaded with a lossy network at its output. Γ_L is the reflection coefficient seen by the transistor output and Γ_L' is that presented by the external load at the outermost reference plane. If the lossy network is just a series resistance (R) as in Figure 4.13(b), connecting any passive load can only give a Γ_L corresponding to a larger value of resistance. The region accessible to Γ_G is therefore the interior of the circle of resistance R in the Smith chart, Figure 4.14. As G is increased from 0, this circle will shrink down towards the open circuit point $\Gamma_L = +1$. If the device's region of potential instability were B, no amount of series resistance would prevent the accessible region from intersecting the potentially unstable region. However, if the potentially unstable region were A, a sufficiently large R would make the regions non-intersecting, thus producing unconditional stability.

One further gain quantity is theoretically very important. This is the unilateralised gain, usually written U. At a single frequency, a two-port can always be unilateralised, i.e. provided with a *lossless* feedback network which will cancel its own internal feedback, making S_{12} of the whole network equal to zero. (A practical method of doing this called 'neutralisation' was used quite frequently in the past, but is not now used in low power transistor amplifiers.) U is defined as the value of G_{ma} of the whole unilateralised network, and can be shown to be given by:

$$U = \frac{1/2|S_{21}/S_{21} - 1|^2}{k|S_{21}/S_{12}| - \mathrm{Re}(S_{21}/S_{12})} \tag{4.27}$$

(Caution: U is not to be confused with G_{TU}, the last entry in Table 4.1, which is simply an estimate of G_T obtained by neglecting S_{12}, or with G_{TUmax}, the value of G_{TU} maximised by setting $\Gamma_G = S_{11}^*$, $\Gamma_L = S_{22}^*$. U would equal G_{TUmax} for a device that actually had $S_{12} = 0$. Note also that some books use symbol U for a related quantity called 'unilateral figure of merit', which gives an indication of how good the approximation of neglecting S_{12} is.)

U is not a very useful parameter in circuit design procedures, but is important in characterising devices. U turns out to be invariant if the two-port device is embedded in any lossless network whatsoever, provided two ports are still presented to the outside world. This embedding also includes change of the common terminal, so that a transistor in the usual common emitter connection has the same U if operated in common base configuration, etc.

The condition $U > 1$ is the condition that a device is 'active', i.e. capable in principle of giving power gain if surrounded by sufficiently ingenious networks. A device with $U < 1$ is incapable of giving any power gain no matter what circuitry surrounds it. For any device U must fall to zero at some sufficiently high frequency f_{max}, which is called the maximum frequency of oscillation of the device. For transistors U decreases continuously with frequency, though other more exotic amplifying devices might also have $U > 1$ only above a minimum frequency, or even in more than one separate frequency band. Notice that if we had a device unilateralised to make $S_{12} = 0$, we could also equip it with matching networks so that we could realise a device with $S_{11} = S_{22} = S_{12} = 0$, $|S_{21}|^2 = U$. If $U < 1$ it appears physically obvious that there is nothing we could do to make this network give any gain, but if $U > 1$ it is obvious that an oscillation condition could be created just by connecting a cable of suitable length, and impedance equal to the reference impedance Z_0, from its input to its output. Hence the name given to f_{max}, which is clearly the highest frequency at which the device could possibly be made to oscillate.

In conclusion, it would be useful to inspect Figure 4.15 showing the typical frequency variations of gains for a typical MESFET. Notice that the device becomes unconditionally stable above about 7 GHz, and the G_{ma} curve definable above this frequency is continuous with the G_{ms} curve below it. f_{max} is 80 GHz, and it is interesting to note from the U curve that significant gain could still be obtained at about 45 GHz where G_{ma} has fallen to unity (0 dB). This would be rather hard to do in practice, as it would require some kind of feedback network, and it would be easier to find a higher frequency transistor.

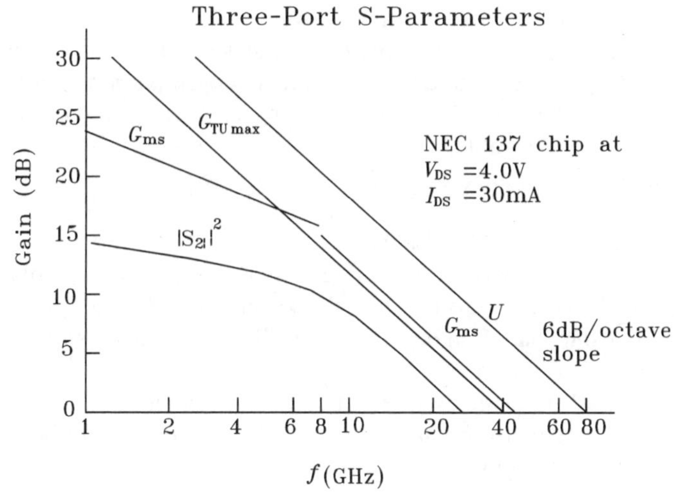

Figure 4.15 Dependence of gain on frequency for typical MESFET

Self-assessment Problem

4.6 In connection with resistive loading of a two-port to make it unconditionally stable, we considered the accessible regions for Γ_L under series resistive loading. What would the accessible region be if the loading network were:
(a) a shunt resistor of conductance G?
(b) a matched 10 dB attenuator, i.e. with $S_{11} = S_{22} = 0$, $|S_{21}| = |S_{12}| = 1/\sqrt{10}$?
Could either or both of these loading types produce unconditional stability if the potentially unstable region were B in Figure 4.14? How could you realise the 10 dB attenuator as a T-network of three ideal resistors?

4.3.7 Design Strategies

We are now in a position to summarise some actual design procedures for designing an amplifier to give useful transducer gain at a single target frequency f_0, leaving aside for the moment the issues of optimising its noise performance or obtaining substantial bandwidth. The design problem is as follows: given values Γ'_G, Γ'_L (most commonly zero) for the generator and load reflections that our amplifier will see at its interfaces with the outside world, we must design matching networks between these interfaces and the active device's ports so that these are transformed into the correct Γ_G, Γ_L as seen by the device.

One useful point to note is that for the circuit to be oscillating, *both* Γ_G and Γ_L must be in their potentially unstable regions, and the device must be showing negative resistance at both ports. It was shown earlier that an oscillation condition at the input is equivalent to an oscillation condition at the output. Thus, if (e.g.) Γ_L were in its stable region, the input resistance would be > 0 ($|\Gamma_L| < 1$), no oscillation would then be possible at the input, and hence none would be possible at the output – even if Γ_G were in its potentially unstable region producing a negative resistance on the output side. Thus, it is only necessary to ensure one of Γ_G and Γ_L is in its stable region to prevent oscillation. It can thus be seen that operation without oscillation, yet with negative resistance present on one side, is possible in principle, though not very desirable.

The following procedures can be applied:

1. If the device is unconditionally stable at f_0:
   (i) Select Γ_G, Γ_L for a simultaneous conjugate match at f_0; this not only maximises the transducer gain but also ensures that the generator and load will also see a match when looking into our designed amplifier. Nominally lossless matching networks can be used and G_T can be made equal to G_{ma}.
   (ii) Ensure that at all frequencies where the device is potentially unstable, normally in a band somewhere below f_0, at least one of Γ_G, Γ_L (preferably both) is outside the potentially unstable region, to prevent oscillation.
   (iii) If Γ'_G and Γ'_L are not sufficiently predictable in the potentially unstable bands to be sure of meeting condition (ii), consider introducing resistive loading networks. These can be designed so that resistive damping only appears in the potentially unstable band. For example, a shunt resistor might be connected from gate/base to ground, and placed in series with a parallel tuned circuit or quarter wave resonant line which will remove the damping in the neighbourhood of f_0. Alternatively, it might be placed in series with an inductor which would effectively remove its influence at all sufficiently high frequencies.
2. If the device is conditionally stable at f_0, then:
   (i) a possible strategy is to load it resistively, using an appropriate topology as discussed in the previous section, to the point where it becomes unconditionally stable, then proceed as in (1). This has the advantage that the outside world can see an impedance match looking into *both* of the outermost ports, i.e. at the reference planes which enclose the resistive loading and the matching circuits we have added. It also has the advantage that oscillation cannot be produced if the Γ'_G and Γ'_L at f_0 should happen to be a long way from their nominal values. The maximum transducer gain is limited to G_{ms} and there would be some penalty in the saturated power handling with output resistive loading, or noise figure with input loading.
   (ii) If alternatively lossless matching networks are retained, an impedance match will not be possible at both of the outermost ports, but can still be obtained at either the input or the output.
   (iii) Suppose that the choice is made to have a match at the input. The operating power gain (G_p) circles (Figure 4.12) can be plotted to find a value of Γ_L which is outside the potentially unstable region and not too near the unit circle. The resulting Γ_{in} can then be calculated from Equation (4.12), and the input matching circuit can be designed to make Γ_G conjugate matched to it. The indicated G_p value will then of course be the G_T that is actually achieved.
   (iii) In principle, the G_p selected can be arbitrarily large, by selecting Γ_L close enough to the stability circle. In practice an upper limit on the selected G_p can be imposed by the requirement that the Γ_G required to perform the input match also remains outside its potentially unstable region. It is also possible to set a limit by considering the

likely tolerance on the Γ_L that will actually be presented (because of the tolerance on Γ'_L relative to its nominal value, and manufacturing tolerances in the matching circuit that is actually made) to ensure that the actual Γ_L in practice can never move into its potentially unstable region. Finally, a limit on G_p might be imposed when the design is required to have substantial bandwidth – too large a value will make the required bandwidth impossible to obtain.

(iv) The same procedure can be used making the alternative choice that a match is to exist at the output, and then using the G_A circles and stability circle in the Γ_G plane.

In all cases the circuit will present mismatches outside its operating band at both of the interfaces with the outside world. In case (2),(i) the circuit must also present a mismatch to the generator within the operating band, or likewise to the load in case (2),(ii). These mismatches may cause unpredictable gain or gain variation with frequency, or possibly oscillation, when the circuit is interfaced to external circuits that are also imperfectly matched. The balanced amplifier technique (discussed later) is a useful way of avoiding these problems.

4.4 Broadband Amplifier Design

The broadband design problem arises when an amplifier or other radio frequency component is required to work over a band of frequencies that is a substantial fraction of the frequency at the centre of the band. (This is commonly expressed as a 'percentage bandwidth'.) A design made for a single frequency will usually give at least a few percentage of useful bandwidth, over which there is negligible gain variation, without making any further effort. In a broadband design the transducer gain must have a specified variation over a substantial percentage bandwidth. By far the most common case is that the variation should be minimised, i.e. the gain should ideally be 'flat' over the desired band.

There are four main techniques for achieving broadband operation:

1. negative feedback;
2. distributed amplifiers;
3. compensated matching.
4. special connections of two or more transistors as a unit.

The compensated matching technique simply consists of designing for a controlled degree of mismatch at the input and/or output which is designed to vary in such a way as to compensate for the inherent tendency of the active device's gain to vary (usually fall) with frequency.

Associated with these methods is another, the balanced amplifier. This is not strictly a broadbanding technique in its own right, but rather one which reduces the interactions of the amplifier with the outside environment, and hence makes the compensated matching approach easier to apply.

The distributed amplifier is an important method for achieving extremely high (multi-octave) bandwidths. The price of its high performance is that many active devices are used in the one amplifier. This specialised method is outside the scope of this discussion, which is restricted to methods that could be used to improve the bandwidth achievable with a single device. There are also arrangements where a small number of transistors may be connected together as a single unit.

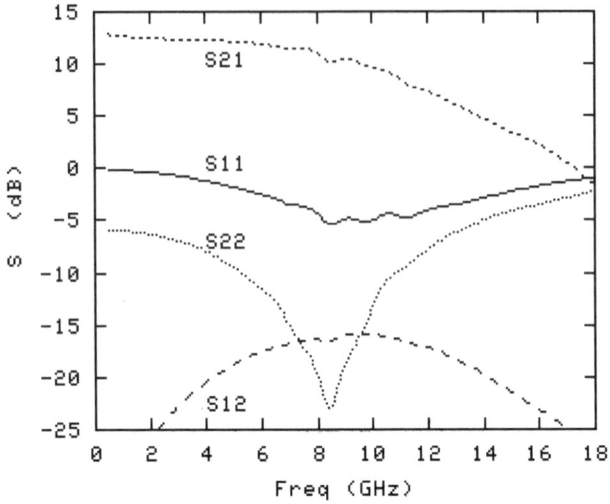

Figure 4.16 Scattering parameters of the ATF-36163 FET

4.4.1 Compensated Matching Example

Taking an example transistor, ATF-36163, its s-parameters are as shown in Figure 4.16. The S_{21} of the transistor gradually decreases with frequency while S_{11} is always greater than S_{22}, which naturally has a good match near 8 GHz. We can then add reactive matching networks to reduce the reflections and therefore increase the S_{21} of the complete circuit. The equations in Table 4.1 show that in the unilateral case the gains of the input and output networks have a maximum value (when conjugately matched) of $1/(1 - |S|^2)$. The greatest reflection, S_{11} in this case, will produce the greatest increase in gain when matched out.

Requiring the input network to produce a match at 11 GHz and the output network to match at 12 GHz the complete circuit response becomes that shown in Figure 4.17. Clearly

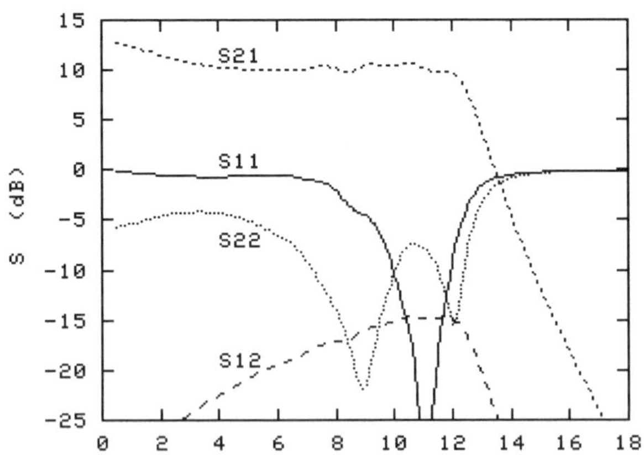

Figure 4.17 Scattering parameters of the ATF-36163 FET with matching circuit

the removal of the reflections in the 10 GHz to 12 GHz region has increased S_{21} in that region, producing a relatively flat profile up to 12 GHz, but then drops off quite sharply. Such an amplifier is well matched at the frequencies where we have compensated for the natural roll-off of the transistor, but is poorly matched at the lower frequencies.

4.4.2 Fano's Limits

Design of amplifiers and other components for broadband operation depends on being able to perform impedance matching over wide ranges of frequency, and this operation proves to be subject to fundamental limits. Consider the situation shown in Figure 4.18(a), where a *lossless*, passive matching network is to be used to match a given load to a purely resistive generator. Γ will denote the input reflection coefficient using the source resistance R_0 (usually 50 Ω) as the reference impedance.

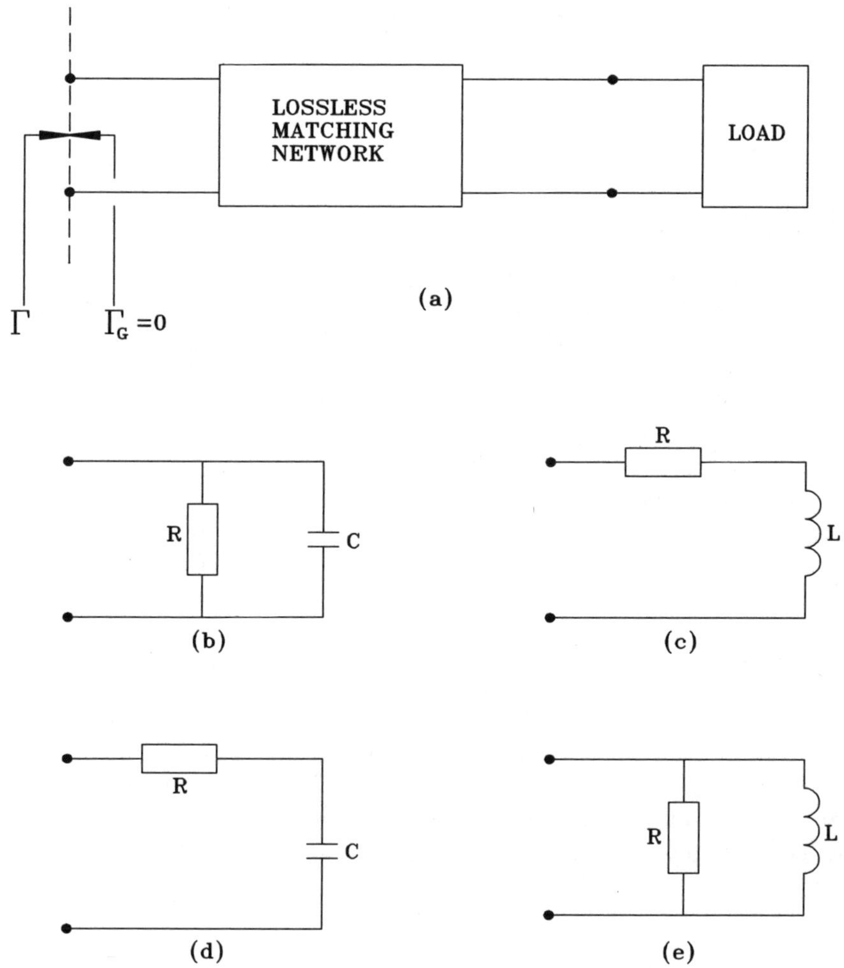

Figure 4.18 Topologies for defining Fano's limits

If the load is purely resistive, Γ can be made as small as desired over an arbitrarily large frequency range, even if the load resistance is not equal to R_0. (In principle, an ideal transformer would do this; other methods such as a transmission line with a gradually tapered or multi-stepped Z_0 are more practical at UHF and above.) If the load includes reactive components, it can be shown, as one would expect intuitively, that Γ cannot be made as small as we like over unlimited bandwidths.

For certain simple loads, shown in Figure 4.18 (b) to (e), the theoretical limits on matching have been found, and these are known as Fano's limits:

$$\int_0^\infty \ln\left(\frac{1}{|\Gamma|}\right) d\omega \leq \frac{\pi}{\tau} \quad \text{for load topologies (b), (c)} \tag{4.28}$$

$$\int_0^\infty \frac{1}{\omega^2} \ln\left(\frac{1}{|\Gamma|}\right) d\omega \leq \pi\tau \quad \text{for load topologies (d), (e)} \tag{4.29}$$

τ is the time constant of the load, given by RC or L/R as appropriate. To understand the meaning of one of these expressions, notice that the term $\ln(1/|\Gamma|)$ in the integrand is a measure of the 'goodness' of the matching scheme at the frequency in question. Smaller values of $|\Gamma|$ make the integrand larger, zero values make it infinite, and a complete mismatch with $|\Gamma| = 1$ makes it zero. The integrals are saying that the total 'goodness' integrated over frequency cannot exceed a certain limit set by the load. Thus if we attempt to improve the match in certain frequency ranges, we must expect it to worsen at others. Since the normal design problem is to produce a system which functions just over some specified frequency band, and the performance out of band is immaterial, it is clearly a good strategy to make (ideally) $|\Gamma| = 1$ outside the operating band, so we can have the maximum 'goodness' within it.

It is also immediately obvious from the expressions that no matching scheme can make $|\Gamma| = 0$ over any finite range of frequency, as this would automatically make the integral infinite, but it is possible to have $|\Gamma| = 0$ at a finite number of *isolated points* on the frequency axis, and in this case the integral will converge to a finite value even though the integrand is tending logarithmically to infinity at these frequencies.

The first two topologies (b) and (c) clearly have the property that they become more difficult to match over wide bandwidths as the time constant of the load network increases; in the limit where the capacitor or inductor tend to zero, they just become pure resistors again. On the other hand, circuits (d), (e) become easier for broadband matching as the time constant increases; for example, in (d), as $C \rightarrow \infty$ it looks like a short circuit and reduces the load to a resistor again. This is an easy way to remember whether the time constant appears on the top or bottom of the right-hand side of the equation: for example, for circuit (b), the integrated goodness must be smaller for a larger time constant, which therefore appears on the bottom of the right-hand side.

It is also easy to check whether the integral should include the $1/\omega^2$ factor by looking at the dimensions of the expressions. For example, in Equation (4.28), the logarithm term is a pure number, and the dimensions of the integral are those of $d\omega$, namely, a frequency or equivalently a reciprocal of time, and this is also the form of the right-hand side.

Self-assessment Problem

4.7 Show that the dimensions are also correct for the other form of the integral containing the $1/\omega^2$ term.

No attempt will be made here to prove the limits formally. However, the issue can be simplified a little by pointing out that, although there appear to be four cases of the limit, there is really only one. If one of the limits is taken as given, the others could all be deduced from it by duality arguments. This is explained in a Self-assessment Problem at the end of this section. The results can also be made more plausible by pointing out each integral is in fact exactly equal to the limit value if the matching network is omitted entirely, provided the resistor in the load is equal to the source resistance. (This of course makes the match perfect at either zero or infinite frequency.) This result is easy to prove by a straightforward integration of the expression for Γ for the no-matching case. Hence we can have no more 'integrated goodness' than is possible for no matching device at all; the best the matching can do is to re-distribute the goodness to frequencies where it is more useful to us.

It has already been mentioned that compensated matching is one of the most important techniques for broadband amplifier design. The tendency of transistor gain to fall with frequency is compensated by making the input and/or output mismatch worse at lower frequencies. When this technique is used, the need to intentionally create controlled mismatch eases the demands of the Fano limits.

The limits can be applied where an approximate model for the device exists, and they really need the approximation that the device is unilateral. Fortunately this usually holds reasonably well for high frequency problems. Figure 4.19 shows the very simplified model of a GaAs MESFET that could be used for a rough assessment of the gain bandwidth limitations.

In this simplified model we see that the input network is of type (d) in Figure 4.18 while the output network is of type (b).

4.4.3 Negative Feedback

Negative feedback is a familiar technique in low frequency amplifiers where it offers the benefits of:

1. gain and phase responses that have less frequency variation;
2. reduced non-linear distortion;
3. performance that is less sensitive to variations between individual samples of the active devices, because it is at least partially controlled by external passive components.

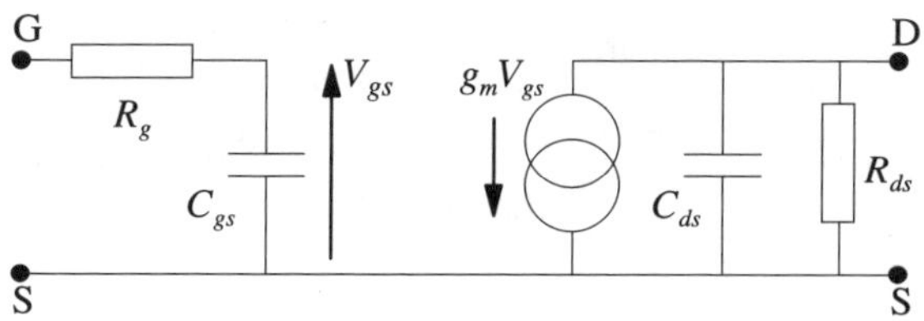

Figure 4.19 Simplified MESFET model for assessing the Fano limits

These principles have long been applied in high fidelity audio amplifiers, and they are developed to their fullest extent in the vast number of precision circuits based on operational amplifiers.

The same benefits can often be obtained in radio frequency amplifiers, and with the added advantage that it may be possible to produce a good input and output match to a standard impedance (usually 50 Ω) over wide ranges of frequency.

However, three basic problems reduce the applicability of negative feedback in RF amplifiers, especially at the higher microwave frequencies:

1. The amplifier noise figure is usually degraded.
2. The phase of a transistor's S_{21}, which is 180° at low frequency lags increasingly in phase as the frequency is raised. Transistors in the common emitter or common source configurations produce phase inversion at low frequency, i.e. the phase angle of S_{21} is 180°. This makes it easy to provide negative feedback by a simple shunt or series resistor. However, as the frequency is raised, the phase departs from 180° and can eventually reach 90° where the feedback starts to be positive (destabilising) rather than negative.
3. Negative feedback is most effective when it is made to produce a large reduction in the gain of a device whose intrinsic gain is high. This condition is not met where the operating frequency is so high that the transistor's G_{ma} is approaching unity.

This can only be a short introduction to the theory of negative feedback in RF amplifiers. The conclusion can be summarised that negative feedback is worth considering where some noise figure degradation can be tolerated, where the required bandwidth is large, and where the upper limit of the require operating frequency range is either below or comparable to the turn-over frequency at which S_{21} moves into the first quadrant.

4.4.4 Balanced Amplifiers

The 3 dB couplers are required to be quadrature couplers. An input at Port 1 produces outputs at 2 and 3 only, and the output at 2 is 90° behind that at 3, at the operating frequency. Of course, for 3 dB coupling these outputs are also equal in magnitude.

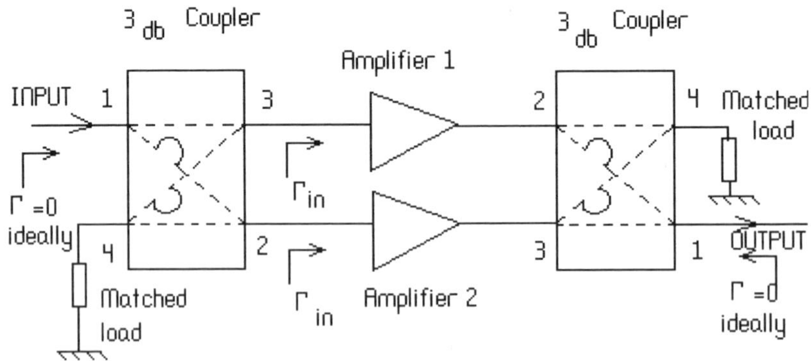

Figure 4.20 Basic balanced amplifier configuration

A lossless matched reciprocal coupler with this property must have the same property for an input at 4, which must divide equally between Ports 2, 3 only, with the output at 3 being 90° behind that at 2. Thus, in the diagram, the loop Ω on a coupling path represents an excess phase lag of 90°.

4.4.4.1 Principle of operation

Two nominally identical amplifiers are used then, if the coupler is ideal, $\Gamma = 0$ at Port 1 regardless of the Γ_{IN} of the amplifiers. With an input at 1, the wave incident on amplifier 2 is 90° behind that incident on 1. On retracing its path back to Port 1, it receives an additional excess lag of 90°. Therefore for equal reflections at the amplifiers, reflections at Port 1 are 180° out of phase and cancel (all reflected power appears at Port 4). On the output side, all the output appears on Port 4 and there is nominally a perfect match.

4.4.4.2 Comments

1. Typically couplers also have the property that couplings from 1 → 3 and 2 → 4 are of equal phase (arg S_{31} = arg S_{42}); in this case the coupler at either side could be flipped about a vertical line. With 3 on the left, 1 on the right, 2 on the right and 4 on the left, the arrangement would still work.
2. Basic coupled lines, properly matched, have the required isolation and quadrature property. The coupling ratio, however, is frequency dependent and it is also hard to realise 3 dB coupling in simple microstrip structures.
3. The Lange coupler is a most common realisation for balanced amplifiers. This is an improved microstrip coupler giving the required high coupling (3 dB) and quadrature properties over greater bandwidths – an octave or more. Realisation on a circuit board can, however, be problematic.
4. A possible alternative coupler is a Wilkinson divider with a + 90° excess path in one arm (Figure 4.21). This gives a narrower bandwidth, but is simple.

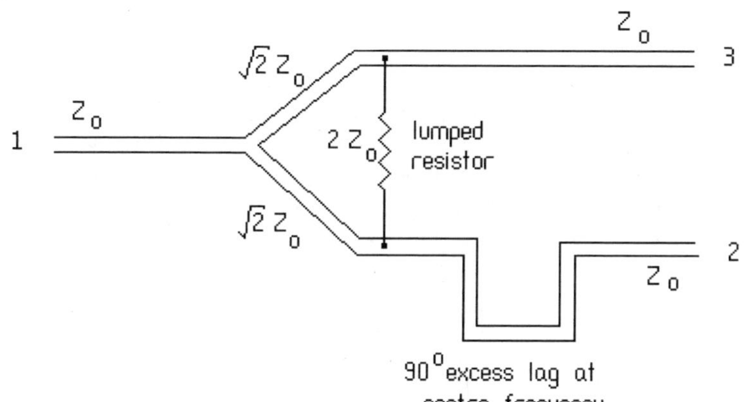

Figure 4.21 A Wilkinson divider with additional phase lag in one arm

4.4.4.3 Balanced amplifier advantages

1. The balanced configuration gets rid of reflections, and therefore interactions between stages. It is then possible to adopt a simple 'building block' approach to system design by cascading such elements together. Nevertheless the active devices in the complete amplifier can see the mismatches that are needed to realise gain flatness, stability, noise optimisation, or optimised saturated output.
2. 3 dB more output power at saturation.
3. Intermodulation products are about 9 dB down in the balanced configuration as compared with an unbalanced equivalent working at same total output power.
4. Greatly improved stability.
5. If one amplifier fails, system may still continue working, with about 6 dB less gain (provided it does not oscillate).

4.4.4.4 Balanced amplifier disadvantages

More board space and more power are required.

4.5 Low Noise Amplifier Design

4.5.1 Revision of Thermal Noise

This section begins with a brief review of the fundamentals of thermal noise generation. This should be familiar to readers who have already studied electronics to degree level, but may still be useful revision. Probably you will have been introduced to thermal noise and shot noise as fundamental processes producing noise in electronic circuits. We are concerned here with additive noise, i.e. the random fluctuations added to the signals we are trying to amplify or otherwise process, and which degrade their information content. Noise can be generated by several other processes, both within electronic circuits and externally. For example, a signal collected from an antenna may be accompanied by both thermal and non-thermal noise generated in the external environment, which includes both the earth and outer space.

It is common to specify the level of noise at any point in a system, regardless of its actual origin, in terms of an equivalent thermal generator. It is therefore necessary to know how much thermal power a device at a given temperature can generate.

Any passive device with a pair of terminals (i.e. a passive one-port) has a random noise voltage across its terminals arising from the thermal agitation of the charge carriers (mainly electrons) within it. If a load is connected to the device, some noise power can be collected from it. The fundamental property is that the spectral density of the available thermal power from the device, i.e. the power which could be collected at a given frequency from the device by a matched load, is given by:

$$S(f) = \frac{hf}{e^{hf/kT} - 1} \text{ Watts Hz}^{-1} \tag{4.30}$$

where $S(f)$ denotes the power spectral density, T is the *physical* temperature of the passive device, h is Planck's constant and B is Boltzmann's constant. The values of the constants are:

$h = 6.626 \times 10^{-34}$ Joule second $\quad k = 1.381 \times 10^{-23}$ Joule Kelvin^{-1} (both to 3 d.p.)

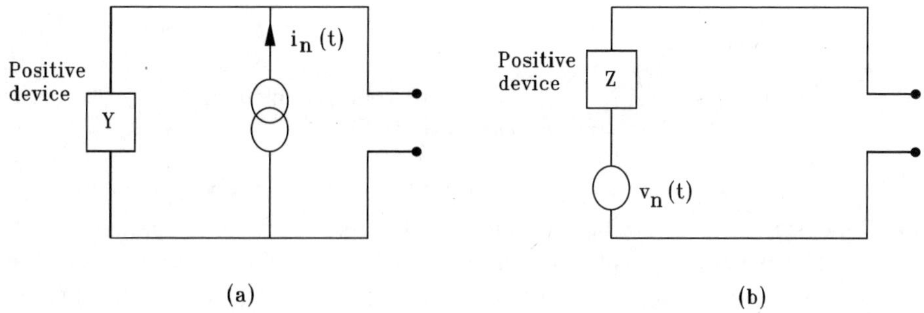

Figure 4.22 Equivalent circuits for thermal noise: (a) Norton; (b) Thévenin

Note carefully the units of $S(f)$, which is a *power per unit bandwidth*. A random signal has a continuous spectrum in which the average power collected in the neighbourhood of a chosen frequency is directly proportional to the bandwidth over which power is accepted, provided this bandwidth is kept small enough. This is the meaning of the term spectral density.

The full expression for $S(f)$ may look unfamiliar because it contains the quantum theory correction. The graph of $S(f)$ against frequency is virtually flat at low frequencies, then it falls off very rapidly at frequencies where the quantum energy hf becomes comparable with the characteristic thermal energy kT. In almost all radio and microwave engineering, hf is very much less than kT, and the expression for $S(f)$ can be simplified to a more familiar form, which is usually an extremely good approximation:

$$S(f) = kT \text{ Watts Hz}^{-1} \tag{4.31}$$

The approximate expression will be used here. The exact expression occasionally needs to be used for more exotic applications, such as remote sensing or radio astronomy at very high millimetric frequencies, and where very low temperatures are involved.

It is often useful to have Norton and Thévenin equivalent circuits for the thermal noise fluctuations, as shown in Figure 4.22.

In both cases the real passive device is replaced with a noise-free equivalent which has the same impedance Z (at the frequency f in question), this being written as $Z = R + jX$ or as $Y = 1/Z = G + jB$. In the Norton form (a) the noise is generated in a separate constant current source, with a random noise current $i_n(t)$ which is also the short-circuit terminal current. Likewise in (b), the random noise voltage $v_n(t)$ is the open-circuit terminal voltage.

$S_i(f)$ and $S_v(f)$ will denote the spectral densities (i.e. mean square values per unit bandwidth) of the current and voltage generators in these equivalent circuits, and are measured in Amps^2/Hz and Volts^2/Hz respectively. They are given by:

$$S_i(f) = 4kTG, \quad S_v(f) = 4kTR \tag{4.32}$$

(Note that, in connection with noise, the term 'passive' as applied above to a passive one-port, must be used in a strict sense to mean a device which has no external power supplies to it. It must have no inputs of energy other than heat absorbed from its surroundings. An example would be a resistor. The term passive is used elsewhere in a wider sense to mean incapable of giving power gain, so that for example a transistor becomes passive in this sense at frequencies above its maximum frequency of oscillation.)

Self-assessment Problem

4.8 Using the approximation $e^x \approx 1 + x$ for small x, derive the approximate expression $S(f) = kT$ from the exact one, Equation (4.30).

4.9 If $T = 3$ K, show that $hf = kT$ at $f = 63$ GHz. Comment: $T = 3$ K, the temperature of the cosmic microwave background radiation, is the lowest temperature we can ever see in the environment, and well below the temperatures usually encountered. At this very low temperature, the quantum correction becomes noticeable at a fairly high microwave frequency. 63 GHz is not too exotic a frequency – 94 GHz radar systems are now quite common, for example.

4.10 Explain why a factor of 4 appears in Equation (4.33). (This is not specifically a question about noise, but applies to any generator. The available power from a generator, whose rms open circuit voltage is V and internal resistance is R, is given by $V^2/4R$.)

4.5.2 Noise Temperature and Noise Figure

Two quantities commonly used to characterise the noise performance of two-port devices, particularly amplifiers, are the noise temperature and the noise figure of the device. These are simply related and contain equivalent information. The noise temperature of the two-port device, which will be written T_e (for equivalent temperature), is simply a convenient way of referring the noise generated in the two-port to an equivalent thermal noise generator at its input. It is defined as follows: let the two-port be connected to a specified generator, which itself may be producing signals and noise. At the output of the two-port, noise power is available of which some is generated by the two-port itself.

Now imagine that the two-port is replaced by a noise-free equivalent which is otherwise identical, and that the generator is replaced by a passive device of the same impedance. Then if the noise temperature of the real two port T_e is defined as the physical temperature, the generator impedance would have to produce the same noise power at the output as is actually produced by the internal noise sources of the two-port. This rather cumbersome verbal definition could just be summed up in the equations:

$$S_{out} = G_t k T_e \tag{4.33}$$

where S_{out} is the spectral density of noise delivered to a specified load, *due to the internal noise sources of the two-port only*, G_T is the two-port's transducer gain, and T_e is its noise temperature. We could remove dependence on the load by writing:

$$S_{A,out} = G_A k T_e \tag{4.34}$$

where $S_{A,out}$ is the available noise output spectral density, i.e. what could be extracted by a matched output load. If we include also the output noise due to the real generator, it would read:

$$S_{A,out}(total) = G_A k(T_e + T_G) \tag{4.35}$$

where the real generator has been an assigned an equivalent temperature T_G such that kT_G is the available noise spectral density from its output port – whether or not that noise is actually generated thermally.

It is very important to note that T_e of a two-port depends on the source impedance from which its input is fed, but does not depend in any way on the loading at its output.

The noise figure of the two-port will be written F and it quantifies the idea that it degrades the signal to noise ratio (SNR) of the signal we pass through it. The two-port gain is G_A, linking the input and output signal power. The most familiar definition of noise figure is usually stated:

$$F = \frac{SNR_{in}}{SNR_{out}} = \frac{P_{in}}{kT_G} \times \frac{G_A k(T_e + T_G)}{G_A P_{in}} = 1 + \frac{T_e}{T_G} \tag{4.36}$$

The problem with this definition is that it only makes sense if the absolute noise power level accompanying the input signal is specified. We could have the same S/N ratio with different absolute levels of the input noise from the generator and of course at higher input noise levels, the added noise in the following two-port would have less effect on the S/N ratio.

To make the definition meaningful, the input noise level must be specified, and unless otherwise stated it is usually taken as corresponding to a standard room temperature. This is written T_0 and is usually taken as 290 K. In other words, this amounts to assuming $T_G = T_0$ in Equation (4.35).

Although F seems to have the advantage that it gives a quick calculation of the degradation of S/N ratio, this simplicity is often lost because, more often than not in practical situations, the input noise level does not actually correspond to T_0. This is particularly true of antenna noise temperatures, which can be far above T_0 at low frequencies and well below it in the low noise region of the microwave spectrum between about 1 and 10 GHz. The noise temperature is in many ways a more elegant definition, but the use of F is very entrenched. (F is usually given in dB, but remember to convert back to an ordinary number when the calculations need it!)

A very important expression gives the noise temperature of a cascade of two-ports, in an obvious notation:

$$T_{e(total)} = T_{e1} + \frac{T_{e2}}{G_{A1}} + \frac{T_{e3}}{G_{A1}G_{A2}} + \ldots \tag{4.37}$$

The interesting feature of this expression is that it can be expressed purely in terms of available power gains. However, the impedance matching at the output of stage 1 is important, although it does not affect T_{e1} or G_{A1}, because $\Gamma_{out,1}$ affects the second term via its effect on T_{e2} (And, if significant, the third term via G_{A2}.) Typically in a well-designed low noise system the first term should be dominant, the second term small and the third term negligible.

Both high gain and low noise figure contribute to the 'goodness' of an amplifier. A quantity called the noise measure M of a two-port is occasionally useful as a single figure of merit that includes both factors. M is defined as the noise figure of an infinite cascade of identical copies of the two-port in question. It is useful when all the stages in a receiver that contribute significantly to F are based on the same active device, or at least have similar noise figure and gain, but this is often not the case.

Self-assessment Problem

4.11 For three two-ports in cascade, show that the available noise power spectral density at the final output is $kT_{e1} \cdot G_{A1} \cdot G_{A2} \cdot G_{A3} + kT_{e2} \cdot G_{A2} \cdot G_{A3} + kT_{e3} \cdot G_{A3}$. How would you deduce the expression for $T_{e,total}$ from this? How would you write the cascading formula in terms of noise figures?

4.12 Show that the noise measure M of a two-port is given in terms of its noise figure and gain *by* $M = (G_A F - 1)/(G_A - 1)$. (*Use* $1 + x + x^2 + x^3 + \ldots = 1/(1 - x)$, provided $|x| < 1$.)

4.5.3 Two-Port Noise as a Four Parameter System

To describe fully the noise properties of a two-port at a single frequency, four real parameters are needed. In the following sections the reason for this will be explained, and will be used to deduce a general form for the dependence of a device's noise temperature on the impedance of the generator that feeds it.

The fundamental point is that the internal noise processes in a two-port cause random noise waveforms to appear at *both* its ports. We can in fact extract noise power from both its input and its output. Furthermore, the noise at the two ports can arise in the same internal physical process which is coupled out in different ways to both the ports. Hence there is usually some degree of statistical correlation between the noise waveforms produced at the two ports.

If we had a random waveform $x(t)$, you will have some idea of characterising it by its frequency spectrum, and you should understand that this can be described precisely by its spectral density. However, if we had two (real) random waveforms $x_1(t)$, $x_2(t)$, what information would we need then? Obviously we would need to specify their individual spectral densities, but this would only be enough when they are completely uncorrelated statistically, as would happen if they arose from physically separate apparatus. Where some correlation exists, another quantity called the cross-spectral density must also be specified.

This is a complex quantity, so it consists of two real numbers which together with the individual spectral densities make up the four real parameters needed to specify the two noise generators at a single frequency.

To explain the idea of the correlation, imagine that the random waveforms $x_1(t)$, $x_2(t)$ are each passed into a filter whose passband is centred on the frequency f in question, and which has a small bandwidth δf which is very much less than f. At the output of each filter, we would see a waveform which, on a short time scale, would look like a sinusoid at frequency f. On a longer scale, the amplitude and phase of this sinusoid would be seen to be varying randomly. We could write:

$$x_1(t)_{filtered} = \sqrt{2}\mathrm{Re}(X_1(t) \cdot e^{j2\pi ft}), \qquad x_2(t)_{filtered} = \sqrt{2}\mathrm{Re}(X_2(t) \cdot e^{j2\pi ft}) \tag{4.38}$$

where $X_1(t)$, $X_2(t)$ are random complex numbers or phasors (expressed as RMS values) which vary on a longer time scale of $1/\delta f$. We would have:

$$\langle |X_1|^2 \rangle = \sigma_{11} \cdot \delta f, \qquad \langle |X_2|^2 \rangle = \sigma_{22} \cdot \delta f \tag{4.39}$$

where $\langle\ \rangle$ denotes an average over time, and σ_{11} and σ_{22} are just a convenient notation for the spectral densities of x_1, x_2 at frequency f. Note how this equation expresses the idea that mean power is proportional to bandwidth.

When there is some correlation between x_1, x_2, there would be some tendency for their amplitudes of X_1, X_2 to vary in sympathy, and, although their individual phases would be quite random, their *relative* phase would tend to favour some preferred value. This can be expressed by introducing the cross-spectral density σ_{12} defined by the relation:

$$\langle X_1 \cdot X_2^* \rangle = \sigma_{12} \cdot \delta f \tag{4.40}$$

The magnitude of σ_{12} is related to the degree of correlation between the variables, and its phase is the average of the relative phase.

In order to use these ideas to calculate noise temperatures, we need an expression for the spectral density of a filtered and weighted sum of the two variables. Working in frequency domain in the normal way, if X is a complex variable expressed as $AX_1 + BX_2$, where the complex numbers A, B are transfer functions of any kind, then the spectral density of X is given by:

$$\sigma = |A|^2\sigma_{11} + |B|^2\sigma_{22} + 2\mathrm{Re}(AB^*\sigma_{12}) \tag{4.41}$$

4.5.4 The Dependence on Source Impedance

Armed with the idea of a two-port device as containing two noise generators which will usually exhibit some degree of correlation, it is straightforward to find the general form of the dependence of noise temperature on the generator impedance.

The operation of the two-port's internal noise generators could be observed in at least three ways: (1) we could short-circuit its terminals, and look at the random currents flowing at each port; (2) we could leave the terminal's open circuit, and look at the random voltages appearing across the terminals; or (3) we could connect infinitely long transmission lines of specified reference impedances to each of the two ports, and watch the two-port launching random waves outwards onto these lines. These three approaches correspond to the Y, Z and S matrix approaches to characterising a two-port.

Rather surprisingly, the first method, based on random currents, will be used here. This is a definite departure from the usual approach of doing everything with s-parameters. It turns out to be rather simpler, and most importantly, it leads to the expression for the noise figure in its most standard form.

The concept of noise temperature is based on referring all the noise generated in the two-port to a single equivalent noise source in the generator. Using the model described Figure 4.23, this is done in two steps:

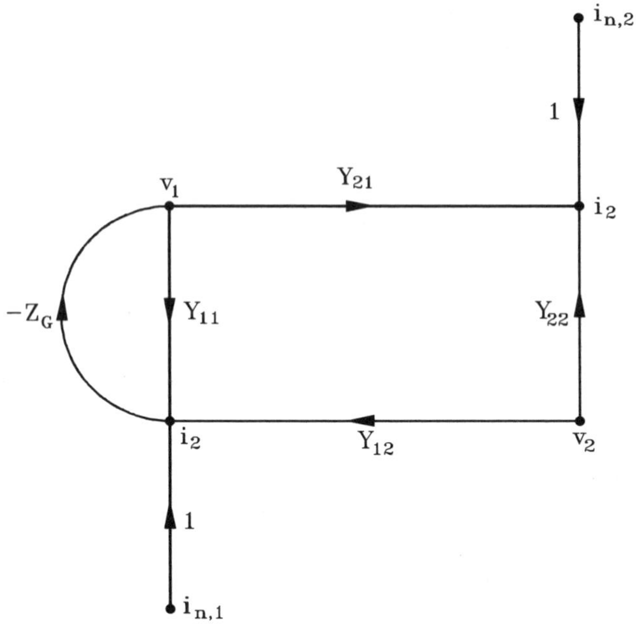

Figure 4.23 Noise model for two-port device

1. $i_{n,2}$ is replaced by an equivalent current generator at the input port.
2. The two-port is now imagined to be noise-free, and the total noise current at the input port is considered to arise instead from thermal noise in the generator. The noise temperature T_e of the two-port is the physical temperature of the generator impedance required to do this.

Of course in the real system, the generator might not just be a passive impedance, e.g. it could be another amplifier, but for the purpose of calculating T_e we imagine it to be replaced by a strictly passive component of the same impedance.

Referring to Figure 4.23, Mason's rule (see Appendix II) shows that the contribution $i_{n,1}$ makes to the current i_2 is given by:

$$i_{n,1} \times \frac{-Z_G Y_{21}}{1 + Z_G Y_{11}} \tag{4.42}$$

This expression can be made slightly neater by multiplying throughout by Y_G (= $1/Z_G$) and becomes:

$$i_{n,1} \times \frac{-Y_{21}}{Y_G + Y_{11}} \tag{4.43}$$

Hence to replace $i_{n,2}$ by an equivalent current generator at the input, we would have to multiply it by the reciprocal of this transfer function:

$$i'_{n,2} = -i_{n,2} \times \frac{Y_G + Y_{11}}{Y_{2,1}} \qquad (4.44)$$

$i'_{n,2}$ is the equivalent of $i_{n,2}$ referred to the input side, as shown in the modified flow graph of Figure 4.23.

The total equivalent current at the input side is now given by:

$$i'_n = i_{n,1} - i_{n,2}\left(\frac{Y_G + Y_{11}}{Y_{2,1}}\right) \qquad (4.45)$$

The expression (4.41) can now be used to write down the spectral density of this total equivalent noise current:

$$\sigma = \sigma_{11} + \sigma_{22}\left|\frac{Y_G + Y_{11}}{Y_{21}}\right|^2 - 2\mathrm{Re}\left[\sigma^{12} \cdot \left(\frac{Y_G + Y_{11}}{Y_{21}}\right)^*\right] \qquad (4.46)$$

The object now is just to reduce this to a standard form, treating $Y_G = G_G + jG_B$ as the independent variable and all the other quantities as given constants which describe the device. To save a good deal of clutter we shall just use symbols for most of the constants which arise in the working, without bothering to write out their exact expressions.

Inspecting Equation (4.46), the bracket contains no terms in Y_G higher than quadratic, and the quadratic term arises only from expanding out the modulus squared term and is $|Y_G|^2 \cdot \sigma_{22}/|Y_{21}|^2$. Also from expanding this term we get some terms of first order in G_G and jG_B, and some constants. The term in Re(. . .) gives two further terms of first order in G_G and jG_B. Grouping together all terms of the same order, the whole expression could obviously be reduced to:

$$T_e = \frac{1}{G_G}[AG_G^2 + AB_G^2 + HG_G + JB_G + K] \qquad (4.47)$$

where A, B, H, J, K are suitably chosen constants. Note that the squared terms have the same coefficient A.

Now let an arbitrary temperature T' be subtracted from both sides of Equation (4.47), and on the right-hand side place this term as $-T'G_G$ inside the square bracket. The equation now reads:

$$T_e - T' = \frac{1}{G_G}[AG_G^2 + AB_G^2\ H'G_G + JB_G + K] \qquad (4.48)$$

where the constant $H' = H - T'G_G$.

The method of 'completing the square' can now be applied to this equation, rewriting it as follows:

$$T_e - T' = \frac{1}{G_G}[A(G_G + H'/2A)^2 + A(B_G + J/2A)^2 + K - H'^2/4A - J^2/4A] \qquad (4.49)$$

Finally, since $K - H'^2/4A > 0$, the constant H' can be adjusted to make the total constant term $K - H'^2/4A - J'^2/4A$ equal to zero. The value of T' required to do this will be called $T_{e,min}$. The expression has now been reduced to:

$$T_e - T_{e,min} = \frac{1}{G_G}[A(G_G + H'/2A)^2 + A(B_G + J/2A)^2] \tag{4.50}$$

Finally, the terms $H'/2A$ and $J/2A$ will be rewritten as $-G_{G,opt}$, $-B_{G,opt}$ respectively, and we now can write:

$$T_e - T_{e,min} = \frac{1}{G_G}[A(G_G - G_{G,opt})^2 + A(B_G - B_{G,opt})^2] \tag{4.51}$$

It is now obvious that $T_{e,min}$ is indeed the minimum noise temperature and that $Y_{G,opt} = G_{G,opt} + jB_{G,opt}$ is the source admittance at which it occurs. We now have:

$$T_e = T_{e,min} + \frac{A}{kG_G}[(G_G - G_{G,opt})^2 + (B_G - B_{G,opt})^2] \tag{4.52}$$

Finally, re-written in terms of noise figures, this becomes:

$$F = F_{min} + \frac{R_n}{G_G}[(G_G - G_{G,opt})^2 + (B_G - B_{G,opt})^2] \tag{4.53}$$

where R_n is a new constant. This expression is the one which is useful in practice. Notice that, as we would expect, it still contains four real parameters, but it has been re-arranged into a more convenient form. The four parameters: F_{min}, R_n, and the real and imaginary parts of $Y_{G,opt}$.

Note that the parameter R_n, when multiplied by a quantity which has the dimensions of admittance, gives F which is a pure number. R_n therefore has the dimensions of a resistance, which is why it was so written. It is called the noise resistance of the two-port. Like the other noise parameters it is a function of frequency.

4.5.5 Noise Figure Circles

The meaning of the noise figure circles is now obvious. Equation (4.53) for F conforms to the second class of expression described in Section 4.2.8, and in fact is a special case where the denominator contains no quadratic terms. The loci of constant F are therefore circles in the Y_G plane, and hence also in the Γ_G plane, since Y_G and Γ_G are bilinearly related.

The noise figure circles always have the same general appearance. The expression for F becomes infinite when $G_G = 0$, i.e. for purely reactive generator impedances, which means all points on the unit circle in the Γ_G plane. Thus the circles of constant finite noise figure cannot cross the unit circle and they must all be nested within it, and shrink down as F falls onto the optimum point $\Gamma_{G,opt}$ at which $F = F_{min}$. Their centres all lie on the diameter which passes through $\Gamma_{G,opt}$.

Noise circles can be drawn for a given device by CAD packages when the four defining parameters have been supplied. In some cases they are given in device manufacturers' data sheets. If really needed, expressions for the centres and radii, though cumbersome, are straightforward to find by the method described in Appendix I.

4.5.6 Minimum Noise Design

A typical receiving system will consist of a cascade of amplifiers, and possibly other components such as filters, mixers, etc. We shall consider the problem of designing the first stage to minimise the overall noise performance, which can be expressed as:

$$F_{(total)} = F_1 + \frac{(F_2 - 1)}{G_{A1}} \tag{4.54}$$

In this case F_2 is the total noise figure of all stages except the first, treated as a single unit. Assuming that F_2 is known at least approximately, the result depends only on F_1 and G_{A1}, and these depend only on the Γ_G seen by the input of the first stage.

For nearly all transistors and other amplifying devices, the value of Γ_G that minimises F is not the same as the value that maximises G_A. This can be seen in Figure 4.24, which gives specimen noise and gain circles for actual transistors in both unconditionally stable and potentially unstable regimes. (The figure also illustrates comments made earlier about the general appearance of gain circles in the two cases.) The optimisation problem is to obtain the best trade-off between F and G_A for the first stage.

Suppose that we choose a trial value of G_A for our optimisation. Then the selected Γ_G should be one that minimises the value of F for that value of G_A. If we imagine sliding our Γ_G round the chosen G_A circle, it will reach a minimum value of F at a point where that circle is tangent to one of the F circles. Whatever G_A is finally chosen, the Γ_G should ideally obey this tangency condition. Thus, as the trial G_A is varied, we can generate a corresponding set of Γ_G points obeying the condition, and giving a definite best value of F for each G_A. The trial pairs of values of G_A and F can then be inserted in the cascading formula to find the unique Γ_G which optimises the overall noise figure.

To perform this optimisation precisely, the noise figure of the second and subsequent stages needs to be specified. Usually in practice, the gain of the first stage can be made reasonably high; then the contribution of the second stage will be relatively small, provided its noise performance is not drastically worse, and the contribution of third and subsequent stages will be almost negligible. In this condition the optimisation is not very critical, and it may not be necessary to know F_2 very accurately. It will also be found that modest departures from the tangency condition are not too serious, and that only a few values of G_A will need to be tried when working by hand. (If no estimate of F_2 is available in advance, it may be necessary to assume that all early stages can be made the same and to optimise the noise measure, as above, of the first stage.)

The optimisation could in any case be done precisely by the optimiser of a CAD package, but it is a good idea to find a rough solution by inspection first. In principle, every stage in a receiver chain should be optimised by the same method of trading off gain against noise figure, but the return from this effort becomes insignificant when the accumulated gain of all previous stages is large. The first stage should be optimised carefully, and some care taken over the second, but when both these stages have substantial gain, the contribution from later stages is likely to be insignificant.

Generally a receiver will include a mixer for conversion to a (usually lower) frequency. A resistive mixer (i.e. one based on non-linear resistive characteristics like normal diode mixers) will have conversion loss ($G_A < 1$) and will therefore make noise from later stages more, rather than less, significant. Care should then be taken that sufficient gain is provided before the mixer stage. (A well-designed resistive mixer has noise figure equal to its conversion loss.) Another important point here is that, with normal mixers, input frequencies both above and below the local oscillator frequency can produce the same IF (intermediate frequency). Only one of these frequency bands normally contains a wanted signal, but noise in both bands can contribute to the IF noise. To avoid a near-doubling of the system noise level, a filter to pass only the wanted one of the input frequency bands must be placed at some point before the mixer. If the IF is quite high, the performance required of this filter need not be very exacting, as it only needs to have substantial rejection at a frequency at a frequency twice the IF distant from its passband. Precise channel shaping will be left to IF filters.

A few special receivers use a 'mixer front end', meaning that the signal is down-converted to a low frequency before any amplification. This tends to be done either where an inexpensive receiver is required and noise is not critical, or at millimetric frequencies where amplifiers are either extremely expensive or not available at all. In the latter case the mixer loss must be carefully minimised, and the noise figure of the first IF amplifier is crucial.

In a narrowband design, the input matching network can be designed to give the optimum Γ_G, subject only to manufacturing tolerances and any error in specifying the external source impedance that feeds into this network. In a broadband design, the optimum $\Gamma_G(f)$ for the device will be some function of frequency and it may not be a realistic target to design an input matching circuit which tracks it precisely.

4.6 Practical Circuit Considerations

4.6.1 High Frequencies Components

It is a common misconception to assume, at high frequencies that all components perform ideally. In practice, really there is no such thing as an ideal inductor, ideal resistor or ideal capacitor. All have associated with them either stray inductances, stray capacitances or stray resistances, meaning that by altering the operation frequency the behaviour of a component can change quite dramatically. Capacitors may look actually inductive, while inductors may look like capacitors and resistors may tend to be a little of both.

4.6.1.1 Resistors

The resistor is probably the most commonly used component in electrical engineering. Everybody is familiar with a resistor. They are used everywhere in circuits, as transistor bias networks, pads and signal combiner, etc. However, very rarely a thought is given to how a resistor actually behaves once we depart from DC. In some cases, although the resistor is employed to perform a DC function, it can severely affect the RF performance. The equivalent circuit of a resistor at radio frequency is shown in Figure 4.24.

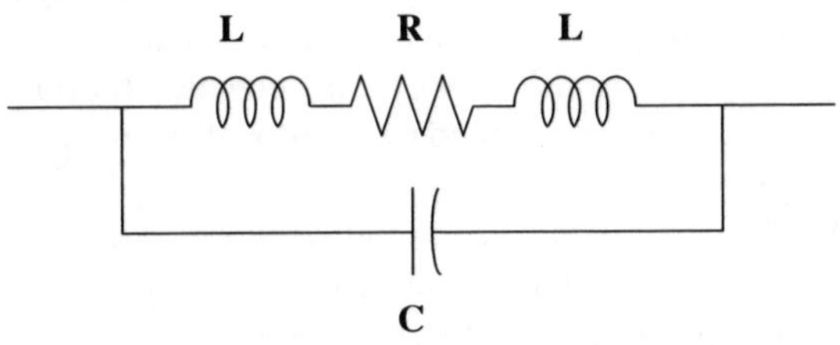

Figure 4.24 Resistor equivalent circuit

R: The resistor value.
L: The lead inductance.
C: The parasitic capacitance. It is the combination of all the internal parasitic capacitances and can vary depending on the resistor structure.

Bearing this model in mind, it easy to appreciate that the expected impedance performance is the one shown in Figure 4.25.

As the frequency increases, the effect of the parasitic inductance will also increase, thus raising the total impedance of the component, until f_r where the inductance resonates with the shunt capacitance. Any further increase in the frequency will decrease the total impedance since the capacitance will progressively be more significant.

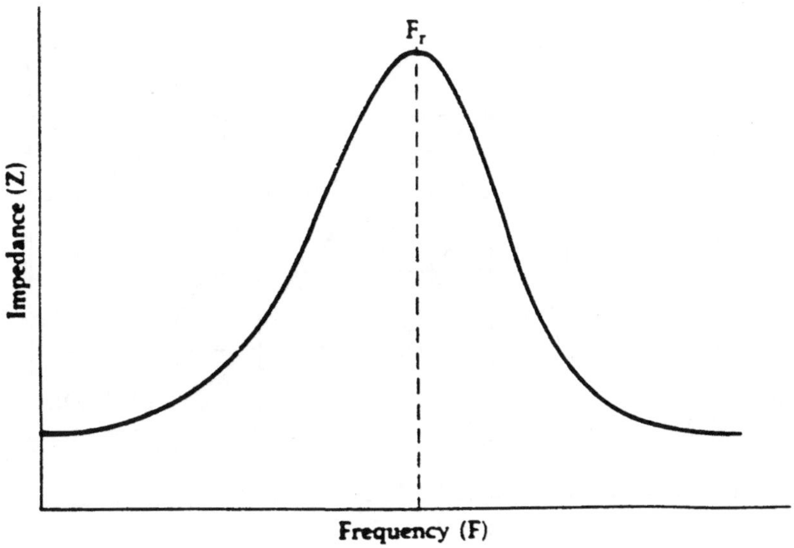

Figure 4.25 Frequency characteristic of a practical resistor

There are several types of resistors to choose from but the most important ones are:

1. *Carbon composition resistor.* They consist of densely packed carbon granules. When current flows through the resistor they are charged, thus forming a very small capacitor between each pair of carbon granules. This parasitic reactance influences the behaviour of the component giving a notoriously poor high frequency performance. A carbon composition resistor can experience an impedance drop of more than 70% of the nominal value even at 100 MHz.
2. *Wirewound resistors.* They are basically a coil of wire. These resistor types have poor high frequency performance too. They exhibit widely varying impedances over various frequencies. This is particularly true of the low resistance values in the frequency range of 10 MHz to 200 MHz. The inductor L shown in the equivalent circuit is much larger for a wirewound resistor than for a carbon-composition resistor. Its value can be calculated in the same manner as for a single layer air core inductor.
3. *Metal film resistors.* Metal film resistors exhibit the best characteristics over frequency. The equivalent circuit remains the same, but the actual values of the parasitic elements in the equivalent circuit decrease.

Figure 4.26 presents the impedance characteristic of the metal film resistor. It can be seen that with resistor values up to 1 K they can operate within the same limit up to 600 MHz. The impedance of the metal film resistor tends to decrease especially for high values above 10 MHz. This is due to the shunt capacitance in the equivalent circuit. For very low resistor values under 50 Ohms the lead inductance and skin effect may become noticeable. The lead inductance produces a resonance peak and the skin effect decreases the slope of the curve as it falls off with frequency.

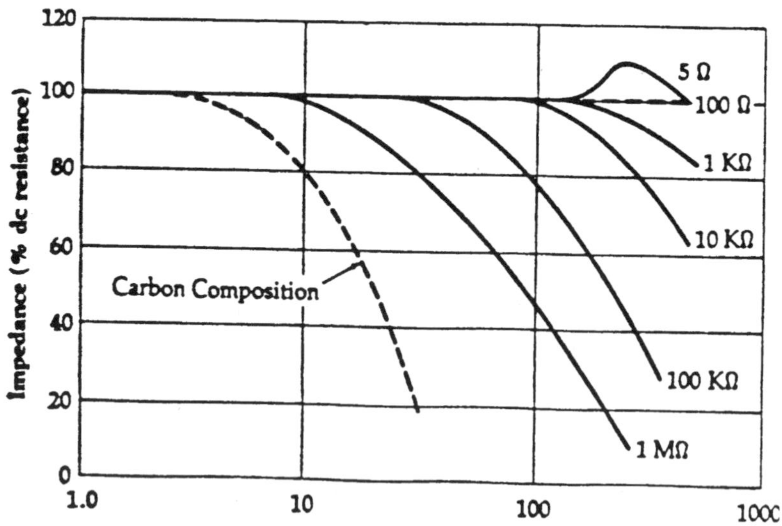

Figure 4.26 Frequency characteristics of metal film and carbon composition resistors

For high frequency applications thin film chip resistors have been developed. They exhibit very little parasitic reactance (approximately 1.5 nH) and with care can be used up to 2 GHz. Typically they are produced on alumina or beryllia substrates.

An alternative at high frequencies is to build up circuits by depositing resistors, capacitors, inductors and interconnectors in the form of metal films on insulating substrates (MIC and MMIC). Two available technologies exist associated with the thickness of the metal film:

1. *Thick films* are usually made by silk screening brushing, dipping or spraying. This provides a cost-effective method that has been known to operate up to at least 20 GHz.
2. *Thin films* are made by vacuum, sputtering or chemical deposition. Although more expensive and complicated it is the only alternative for critical designs and very high frequencies (can operate above 100 GHz).

Self-assessment Problem

4.13 Using the resistor equivalent circuit shown in Figure 4.24 determine the RF impedance at 200 MHz and 400 MHz for a 10 KΩ Resistor when:

(a) A 0.5 W metal film resistor having leads of length 21.5 mm and diameter 0.8 mm and stray capacitance of 0.2 pF is used.
(b) A 0.5 W carbon film resistor with leads of 28 mm length and 0.7 mm diameter and having stray capacitance 0.8 pF is used.

Comment on their comparative performance at both frequencies.

4.6.1.2 Capacitors

Capacitors are used extensively in RF applications such as matching, bypassing, interstage coupling, DC blocks and in resonant circuits and filters. Although in theory an ideal capacitor can perform equally well in real life, the major task of the designer is to use the right capacitive structure as well as the right value. The reason for the discrepancy between the theoretical performance and practice is basically the appearance of additional parasitics that alter the overall performance. These parasitics have to be taken into account in order to avoid any surprises in the final design.

The usage of a capacitor is primarily dependent upon the characteristic of its dielectric. The dielectric characteristics also determine the voltage levels and the temperature extremes at which the device may be used. Capacitors that are to be used at high frequency must have low dielectric losses, since as the frequency increases more energy will be dissipated in the dielectric. This causes a reduction of the overall efficiency or in the extreme even overheating failure at high and medium power applications.

The equivalent circuit is presented in Figure 4.27.

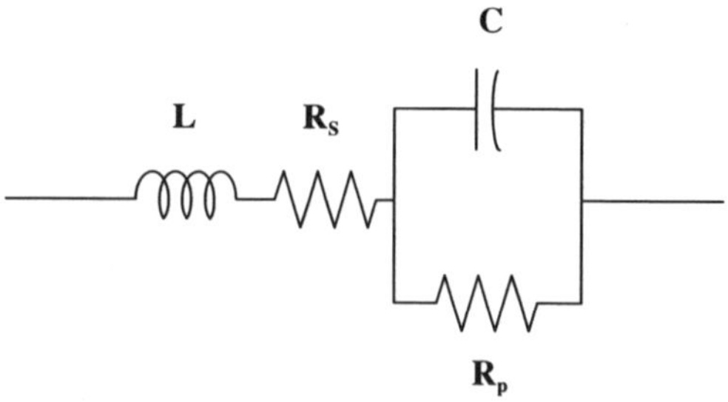

Figure 4.27 Capacitor equivalent circuit

C: Capacitance.
L: Inductance of leads and plates.
R_S: Heat dissipation loss resistance. It models the power transferred to heat due to the imperfections in the dielectric.
R_P: Insulation resistance. It models the power loss due to the leakage through the dielectric.

The effect of the imperfections in capacitor can be seen in Figure 4.28.

Here the impedance characteristic of an ideal capacitor is plotted against that of a practical one. As the frequency of operation increases, the lead inductance becomes more and more important. At f_r the inductance resonates in series with the capacitance. Above that frequency, the capacitor behaves as an inductor. Generally larger-value capacitors tend to exhibit more internal inductance than smaller-value capacitors, due to the larger size of the plates and leads.

The implications of the internal parasitics can be further clarified if we consider the following. Assume that a capacitor is used in a bypass application, then using $X_c = \dfrac{1}{\omega C}$ we

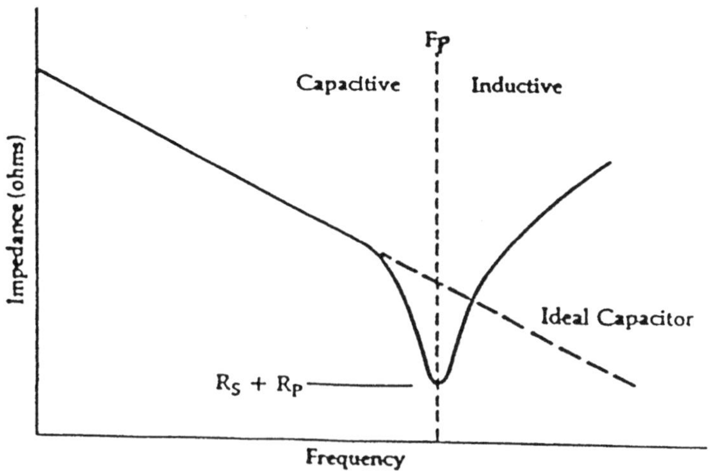

Figure 4.28 Impedance characteristic of a practical capacitor

can expect that the larger capacitor will result in smaller impedance value and thus better performance. At high frequencies though the opposite may be true since a larger capacitor has in general, lower self-resonant frequency. This is something that should be taken into account when designing an RF circuit at frequency above 100 MHz.

Moreover, since at frequency above f_r the capacitor looks inductive, it can be used as one. When the Q of the capacitor is high and the resistive losses are small, a capacitor operated in this mode can represent a low-cost high-Q inductor. An additional advantage is the built-in block by virtue of its capacitance.

The quality factor Q of a capacitor is given by:

$$Q = \frac{X_c}{\mathrm{ESR}} \tag{4.55}$$

where ESR is the effective series resistance effectively being the resistance presented by the capacitor at the specific frequency for which the Q is calculated.

Practical capacitors are usually characterised in terms of the following parameters apart from the nominal:

- *Capacitance tolerance*. The maximum change on the nominal capacitive value.
- *Temperature range*. The range of temperatures in which the capacitor can safely operate.
- *Power factor (PF)*. This is a measure of the imperfections of the capacitor dielectric. It is given by:

$$\mathrm{PF} = \cos\phi \Leftrightarrow \mathrm{PF} = \cos(\angle V - \angle I) \tag{4.56}$$

- *Temperature coefficient*. Relates the expected variation in the nominal value for a given temperature change. Typically is given in ppm/°C.
- *Dissipation factor (tan δ)*. It is the ratio of the effective series resistance at a given frequency over the reactance of the capacitor at the same frequency:

$$\tan\delta = \frac{\mathrm{ESR}}{X_C} \tag{4.57}$$

4.6.1.3 Capacitor types

There are several different dielectric materials used in the fabrication of capacitors, such as paper, plastic, ceramic, mica. polystyrene, polycarbonate, teflon, oil, glass, porcelain. Each material has its advantages and disadvantages. The decision of the chosen structure is based on the application requirements and, of course, the cost.

A table is included that gives a number of capacitor types along with their maximum usable frequency as well as other important considerations. It is because of the behaviour shown earlier that sometimes in applications such as wideband decoupling a number of capacitors of varying value, plus even varying types, have to be used.

4.6.1.3.1 Ceramic capacitor

Ceramic dielectric capacitors vary widely in both dielectric constant (ε_r = 5 to 10,000) and temperature characteristics. A rule of thumb is, 'The higher the dielectric constant ε_r, the worse the temperature characteristic.'

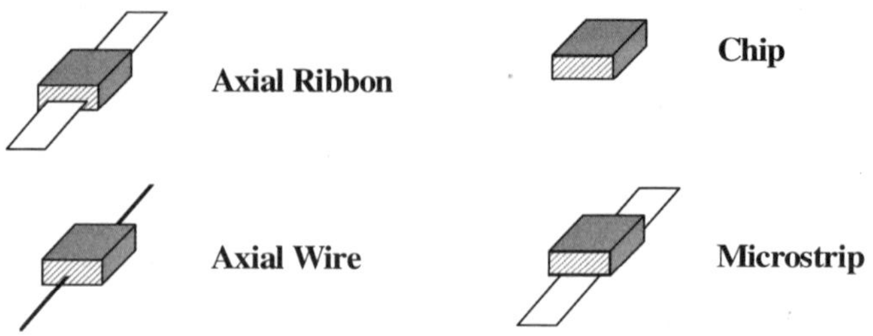

Figure 4.29 Radio frequency capacitor types

Low ε_r ceramic capacitors tend to have linear temperature characteristics. These capacitors are generally manufactured using both magnesium titanite which has a positive temperature coefficient and calcium titanite which has a negative temperature coefficient. By combining the two materials in varying proportions a range of controlled temperature coefficient can be generated. The capacitors are called temperature-compensating capacitors or NPO (Negative Positive Zero) ceramics.

They can have temperature coefficients that range anywhere from +150 to −4700 ppm°/C. Because of their excellent temperature stability, NPO ceramics are well suited for oscillator, resonant circuits or filter applications.

There are ceramic capacitors available on the market which are specifically designed for RF applications. Such capacitors are typically high-Q devices with flat ribbon leads or no leads at all, see Figure 4.29.

The lead material is usually solid silver or silver-plated and contain very low losses. At VHF these capacitors exhibit very low lead inductance due to the flat ribbon leads. At higher frequency where lead cannot be tolerated, chip capacitors are used (not lead capacitors). The common chip capacitor with care can be used up to 2 GHz, though their high internal inductance of the order of 1.5 nH means care should be taken to avoid the frequency of self resonance. This point can sometimes make a normally working circuit behave strangely such as producing a very peaky response. Higher frequency chip capacitor with inductances of 0.3 nH are readily available from companies such as American Technic Ceramics or Dielectric Lab. Inc.

4.6.1.3.2 Mica capacitors

Mica capacitors typically have a dielectric constant of about 6, which indicates that for a particular capacitance value, mica capacitors are typically large. The low ε_r though produces an extremely good temperature characteristic. Thus mica capacitors are used extensively in resonant circuits and in filters, when board space is of no concern.

To further increase the stability, silvered mica capacitors are used. In silvered mica capacitors the ordinary plates of foil are replaced with silver plates. These are applied by a process called vacuum evaporation which is a very exacting process. This produces even better stability with very tight and reproducible tolerances of typically +20 ppm/°C over a range of −60°C to + 89°C, see Figure 4.30.

4.6.1.3.3 Metallised film capacitors

This is a broad category of capacitors encompassing most of the other capacitors not listed previously. This includes capacitors that use teflon, polystyrene, polycarbonate, etc. As the name implies, they consist of a thin film of dielectric contained between two thin metalic films.

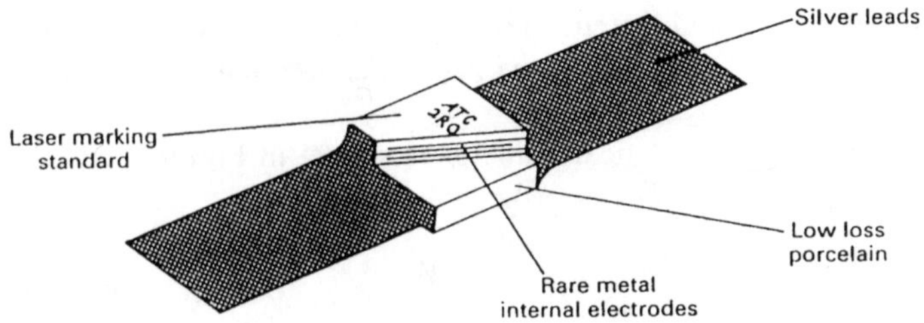

Figure 4.30 Low loss interdigitated capacitor

Most of the polycarbonate, polystyrene and teflon styles are available in very tight (±2%) capacitance tolerances over their entire temperature range. Polystyrene, however, typically cannot be used above +85°C as it is very temperature-sensitive above this point. (It melts beyond that point.) Most of the capacitors in this category are typically larger than the ceramic type equivalent value and are used when space is not a constraint.

Self-assessment Problem

4.14 What is the quality factor of a capacitor with a dissipation factor.
(a) $\tan\delta = 2\ 10^{-3}$;
(b) $\tan\delta = 1\ 10^{-4}$?
Comment on the connection between the quality factor and the dissipation factor.

4.6.1.4 Inductors

4.6.1.4.1 Straight wire inductors

From basic electromagnetic theory we know that even a straight piece of wire will exhibit some self-inductance at radio frequencies. The value of the inductance associated with a straight wire is given by:

$$L = 0.002\,\ell\,2.3\ \log\left(\frac{4\ell}{d} - 0.75\right) \tag{4.58}$$

L: is the inductance in μH.
ℓ: is the length of the wire in cm.
d: is the diameter of the wire in cm.

Inductance becomes more and more important as the frequency increases. At high frequencies every conductor in a circuit exhibits inductive performance, thus changing the behaviour of the design.

4.6.1.4.2 Practical inductors

An inductor is usually a wire of some form wound or coiled in such a manner as to increase the magnetic flux linkage between the turns of the coil. The increased flux through each turn results in a self-inductance which can reach values well beyond that of the plain wire.

Inductors are widespread in radio frequency applications in resonant circuits, filters, phase-shifters and delay networks, matching networks and as frequency chokes used to control the flow of radio frequency energy.

The equivalent circuit of a practical inductor is shown in Figure 4.31.

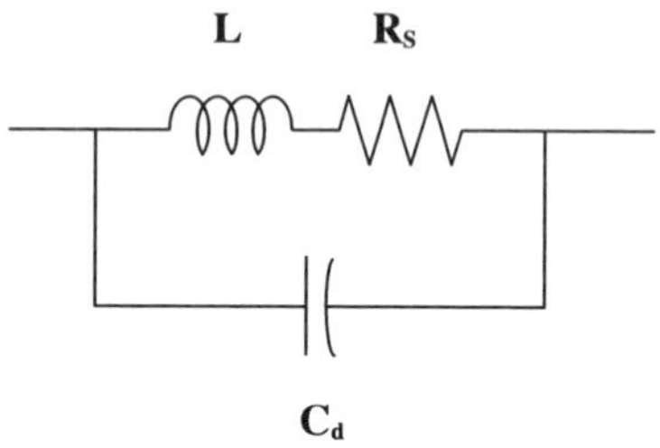

Figure 4.31 Inductor equivalent circuit

R_s: The resistance of the wire.
L: The inductance of the coil.
C_d: The aggregate of the individual parasitic capacitances of the coil.

The physical interpretation is more evident if we consider Figure 4.32. As mentioned earlier, in order to achieve higher inductance values the proximity of the turns should increase. That essentially will result in two conductors (here the conducting wound) being separated by a dielectric material which is a capacitor, assuming there is a voltage drop between the conductors. Since in real world we always experience wire resistance, (and at RF this is even more apparent due to the skin effect), a voltage drop will exist between the windings and C_d will appear as marked in Figure 4.32. The aggregate of all these small capacitances is called the distributed capacitance.

These parasitics and especially the resistance make the inductor probably the component that experiences the most drastic changes over frequency.

Based on the equivalent circuit the performance of a practical inductor can be evaluated with frequency (Figure 4.33). At low frequencies the inductor's reactance follows the ideal

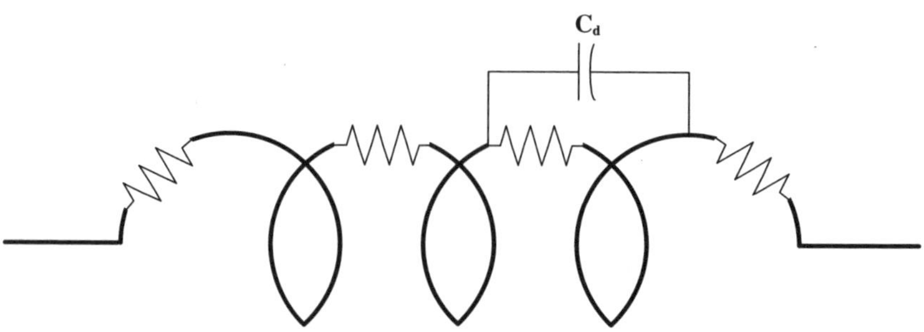

Figure 4.32 Illustration of parasitics on a practical inductor

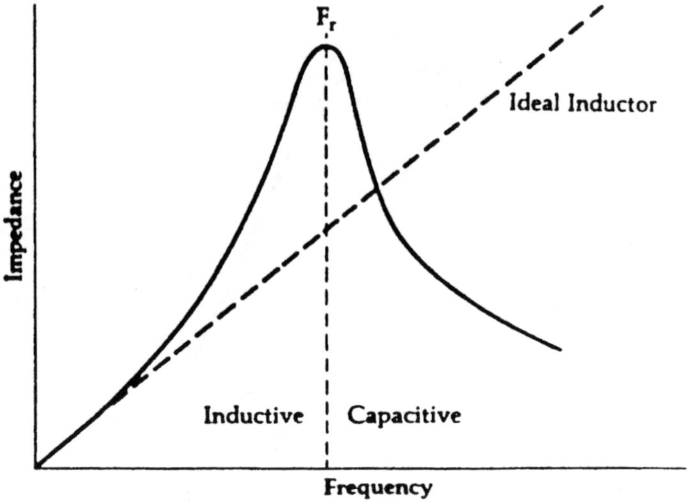

Figure 4.33 Impedance characteristic of inductor

value. This area is basically where we would want the inductor to operate. As the frequency increases, the reactance begins to deviate from the ideal curve and starts increasing at a much faster rate. At f_r the inductor's impedance reaches a peak when the inductance resonates with the parallel capacitance C_d. At still higher frequencies the inductor begins to look capacitive.

The ratio of the inductor's reactance to its series resistance is used to measure the quality of the inductor. The larger the ratio, the better is the inductor. This quality factor is the Q:

$$Q = \frac{X}{R_s} \tag{4.59}$$

The effect of the resistance therefore is two-fold:

1. The peak at the resonance remains finite.
2. The resonance peak broadens as the resistance increases.

From Equation (4.59) it is evident that the variation of the parasitic elements will result in a change of the quality factor of the inductor (Figure 4.34).

At low frequency the Q of an inductor increases almost linearly due to the linear increase of the reactance X_L with frequency. As the frequency increases the skin effect becomes more and more significant and as the resistance increases the slope of the Q gradually decreases. The flat portion of the curve occurs when the series resistance and the coil reactance change with the same rate. Above that point, the shunt capacitance and skin effect of the windings combine to decrease the Q until the resonant frequency of the inductor is reached where it becomes zero.

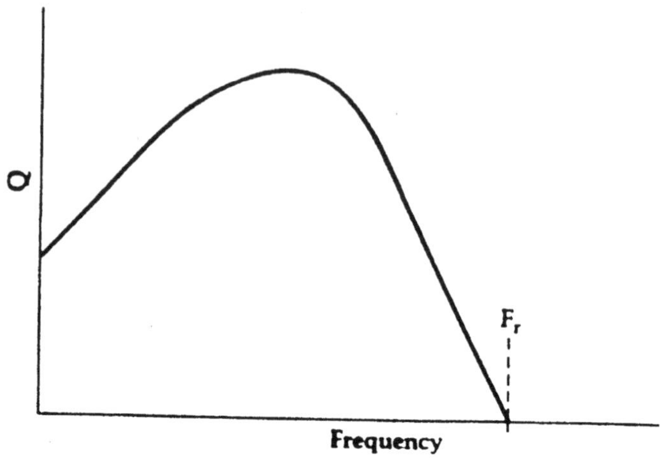

Figure 4.34 The Q variation of an inductor with frequency

Figure 4.35 Type of high frequency lumped inductors

The aim of the design of an inductor in most applications is to achieve high Q, compact size and low capacitance (high self-resonant frequency). In order to design a high Q inductor, the following rules should be observed:

1. Decrease the AC and DC resistance by increasing the diameter of the wire.
2. Increase the separation of the windings, thus reducing the interwinding capacitance.
3. Increase the permeability of the flux linkage path. This is mostly done by winding the inductor round a magnetic core material such as iron or ferrite. These types of core, however, are used primarily for applications below 800 MHz.

At high frequencies, especially in microstrip applications, the inductive values needed are usually very small (less than 10 nH). It is therefore easier to employ a straight piece of wire or length on a narrow track, since it becomes impractical to design coil inductors at this value. In order to make the resulting inductors more compact, designs like the one shown in Figure 4.35 are employed.

4.6.1.4.3 Single Layer Air Core Inductor Design

In the simplest form a practical inductor consists of a single layer of turns and uses air as the dielectric, as is shown in Figure 4.36.

A practical formula generally used to design single-layer air core inductors is:

$$L = \frac{0.394r^2N^2}{9r + 10\ell} \qquad (4.60)$$

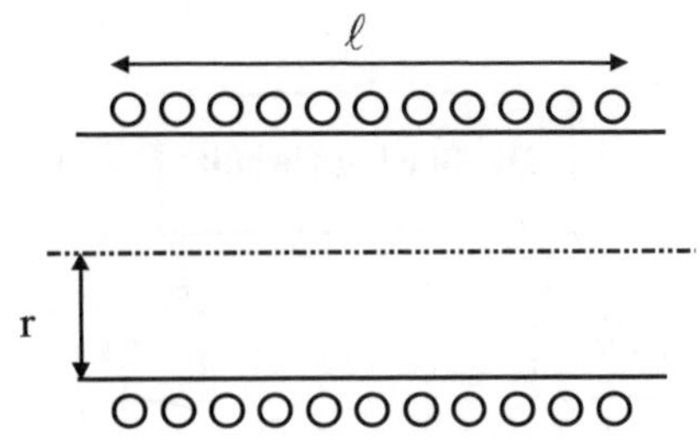

Figure 4.36 A practical inductor

r: Is the coil radius in cm.
ℓ: The coil length in cm.
N: The number of turns.
L: The inductance in μH.

This formula is accurate within 1% when the coil length is greater than 0.67r. The optimum Q for such an inductor is achieved when the length of the coil is equal to its diameter (2r). However, this is not always practical and in many cases the length has to be much larger than the diameter. In such cases a compromise can be reached when the following solution are examined:

1. Use a smaller diameter wire, while keeping the length the same, in order to decrease the interwinding capacitance. However, that will decrease the Q of the inductor since it will increase the resistance of the wire.
2. Extend the length of coil while retaining the same wire just enough to leave some air gap between the windings. Nevertheless, the effect again will be that Q will decrease somewhat (since L will decrease) but the capacitance will decrease even more.

Self-assessment Problem

4.15 You are required to design a 10 mm air core having 500 nH inductance. You can select any one of the following wires (all made from the same material):
(a) a wire with 1 mm diameter;
(b) a wire with 0.5 mm diameter;
(c) a wire with 0.1 mm diameter.
Justify your answer.

4.6.2 Small Signal Amplifier Design

A microwave transistor amplifier can be represented in general by the block diagram of Figure 4.37.

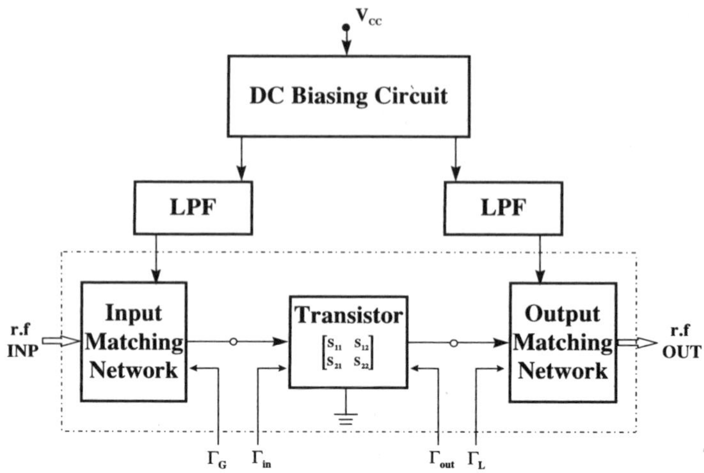

Figure 4.37 Block diagram of amplifier at radio frequencies

The main blocks of this configuration are:

- *Transistor.*
- *Biasing circuit.* The DC biasing circuit determines the quiescent point of the amplifier and *does not* contain any microwave components.
- *Low pass filter.* The LPF is used in order to decouple the microwave circuit from the DC.
- Input matching network.
- *Output matching network.*

The realisation of the input and output networks determines the performance of the amplifier. These networks can be called optimising rather than matching networks since the values of Γ_G and Γ_L are chosen in order to determine the performance for low-noise or moderate to high gain. It is important to realise that the I/O matching networks are there not in order to minimise the input/output reflection coefficient but rather in order to define the behaviour of the transistor.

4.6.2.1 Low-noise amplifier design using CAD software

The design process of a low noise amplifier is basically the construction of the various sub-blocks that make up the amplifier. The steps are tabulated in the following:

1. Select a transistor using as guidelines the design specification. The specifications will include parameters like the noise figure, the gain, the power output, the input and output reflection coefficients and the price.
2. Calculate the maximum stable gain or available gain and the stability factor within the band of interest.
3. For unconditionally stable transistors input matching will be determined by selecting a suitable compromise between the gain and the noise figure. The point that is usually chosen lies on the tangent point of a constant noise figure and a gain circle.

4. When the transistor is conditionally stable, plot the regions of instability, by drawing the input and output stability circles. The matching network is again chosen to compromise between the gain and the noise figure but additionally the unstable areas must be avoided.
5. Once the input matching network is selected, the output reflection coefficient is obtained (usually by means of a CAD software) and the output is matched to that value. If $k < 1$, then the output stability circle should be drawn to establish that the output reflection coefficient does not lie in an unstable region.
6. Replace the ideal elements with real-life components. It is up to the designer to include those parasitic phenomena that make the performance deviate from the ideal case. This step is critical in order to enable a close approximation of the actual performance. The design is then optimised using a software tool to determine the optimum configuration.
7. Add the DC biasing and check the stability for the whole frequency range. If the design is unstable at low frequency, add stability resistors. Fine tune the DC biasing circuit if necessary. If the performance is degraded, fine tune the input/output matching networks as well.

4.6.2.2 Example

Design a LNA to operate between 1.55 GHz–1.80 GHz using the Avantek AT41485 transistor with the following specifications:

Noise Figure	< 2 dB
Gain	> 12.5 dB

The first step in the actual design is to establish the stability of the amplifier. The input and output stability circles are therefore drawn so that the behaviour of the amplifier can be determined. The result is shown in Figure 4.38.

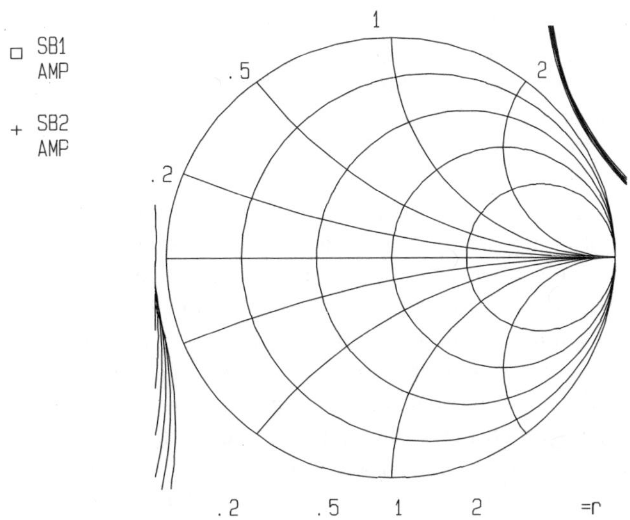

Figure 4.38 Input and output stability circles of the low noise amplifier in the operational bandwidth

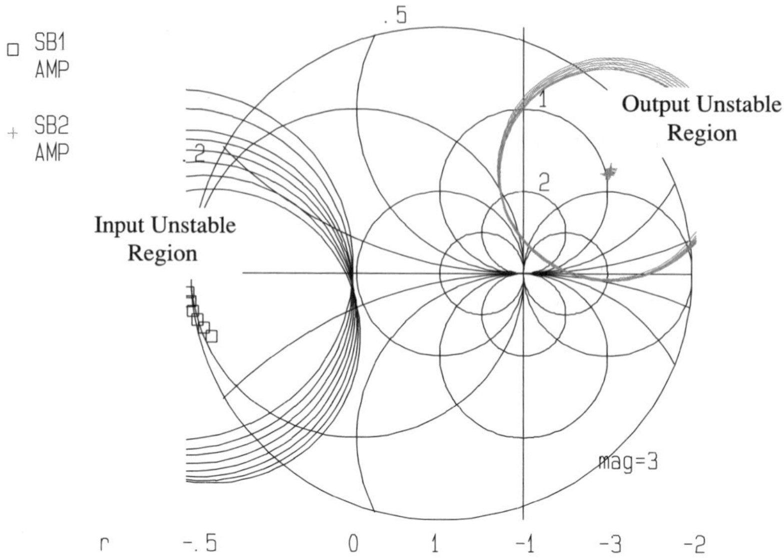

Figure 4.39 Input output unstable areas on compressed Smith chart

This stability check should be performed for a larger frequency range since any existing harmonics can render the amplifier unstable. However, the introduction of a biasing network in the input and output introduces a resistive load that stabilises the amplifier at low frequency. Only an initial check is performed at this stage, since a more vigorous analysis is due at later stages.

Once the unstable areas are determined, as shown in Figure 4.39, the operation point is chosen so that these are avoided. The operation point is selected as a compromise between two antagonistic parameters noise figure (NF) and the available gain (G_A). If we determine the optimum noise figure (NF_o) and the maximum available gain G_{max}, then it is easy to realise that these are two separate points on the Γ_G plane. A compromise, therefore, is essential in order to achieve the optimum performance which is application-oriented. In Figure 4.40 the noise figure circles and the gain circles are presented.

The adjoining point of any two circles results in the optimum performance for both parameters. Which set of circles is used depends on the designer and the application. Attention should be drawn though to the fact that a margin should be allowed for manufacturing degradation. Thus, a choice at this point of a gain of 12.5 dB (in our example) or close by will certainly result in a circuit that does not satisfy this condition. Allowing for this, therefore, the selected values are presented in Figure 4.41.

Because the plane on which the calculation is performed is the Γ_G plane, any reflection coefficient calculated by the CAD software will be a source reflection coefficient. If, therefore, the normal matching procedure is to be followed, the conjugate of Γ_G should be used.

Once the required source reflection coefficient Γ_G is determined, the output reflection coefficient Γ_{out} can be calculated from this and the s-parameters of the transistor, using:

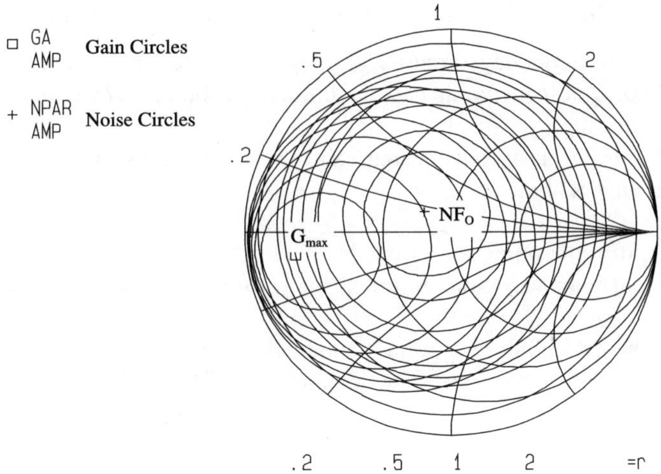

Figure 4.40 Constant noise figure and gain circles at 1.675 GHz

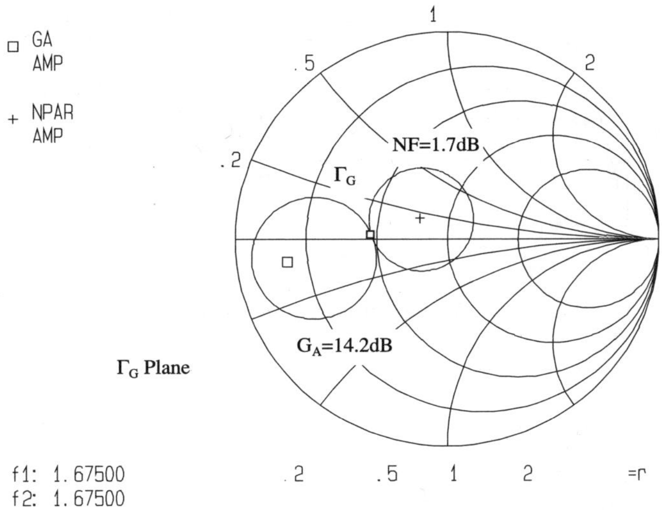

Figure 4.41 Selected noise figure and gain for the low noise amplifier design

$$\Gamma_{out} = s_{22} + \frac{s_{12} \cdot s_{21} \cdot \Gamma_G}{1 - s_{11} \cdot \Gamma_G} \tag{4.61}$$

or the appropriate CAD command (which performs the same calculation). When these two reflection coefficients are established, the input and output matching networks are designed.

4.6.3 Design of DC Biasing Circuit for Microwave Bipolar Transistors

The biasing for high frequency circuits, although essential, has often been overlooked. The selection of the quiescent point heavily influences the characteristics of the amplifier Essentially high gain, high power, high efficiency and low-noise are directly influenced by the biasing. Furthermore, hasty bias design can render the amplifier unstable.

The aim of a good DC bias design is:

- to select the appropriate quiescent point;
- to keep it constant over variations in transistor parameters and temperature.

The selection of the DC quiescent point for an HF transistor depends very much on the application. Thus, the most common selections are presented in Figure 4.42.

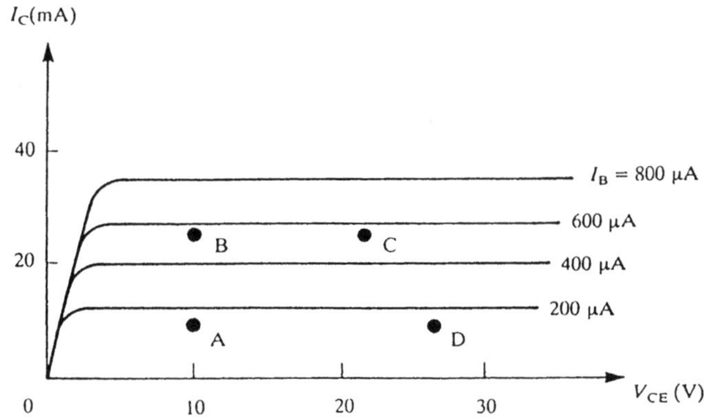

Figure 4.42 Quiescent points for bipolar transistors

A → Low noise and low power.
B → Low noise and high gain.
C → High power in class A.
D → High power in class AB or B.

Depending on the components used, biasing circuits are divided in two categories, passive biasing circuit (purely resistive) and active biasing circuits. Resistor bias networks can be used with good results over moderate temperature changes. However, active bias network are necessary when large temperature changes are encountered.

4.6.3.1 Passive biasing circuits

The two most common configurations are presented in Figure 4.43:

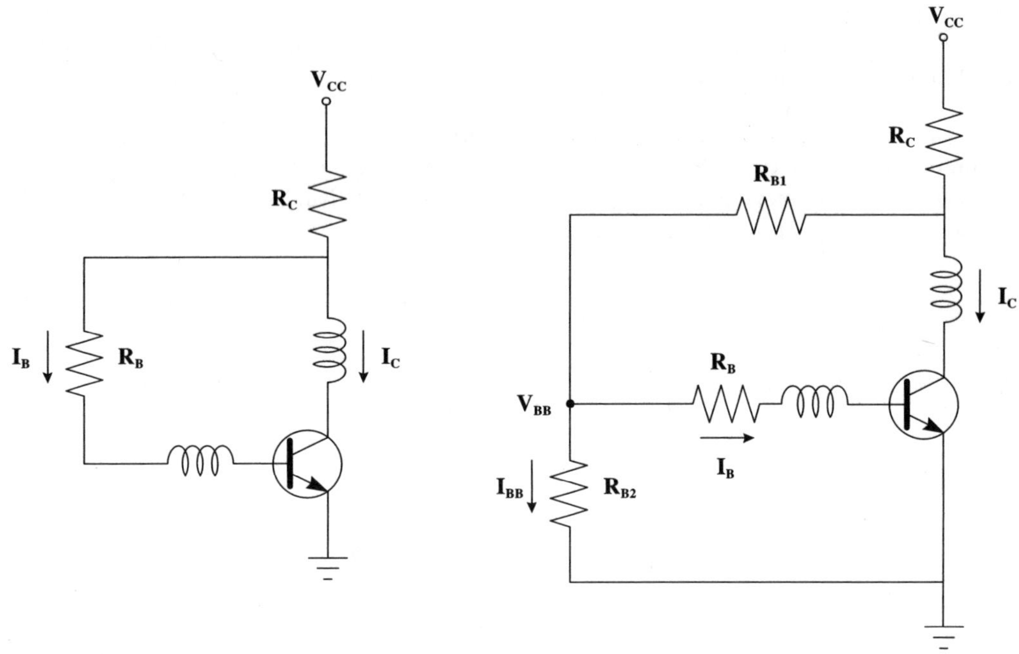

Figure 4.43 Passive biasing circuits for bipolar transistors

These two configurations are often used at microwave frequencies. The voltage feedback with constant base current circuit produces smaller resistive values which makes it more compatible with thin or thick film technology. In order to improve stability a bypass resistor is added at the emitter of the transistor. This method, however, is useful only at the low range of microwave frequencies. A more detailed analysis of the two circuits is given in the following paragraphs.

4.6.3.1.1 Voltage feedback biasing circuit

The starting point is as always the selection of the quiescent point. Once this is chosen, the typical value for the forward current gain (β), and the base-emitter junction voltage (V_{BE}) are determined from the transistor data sheets. The following set of equations is then used to calculate the values of the resistors:

$$I_B = \frac{I_C}{\beta} \tag{4.62}$$

$$R_B = \frac{V_{CE} - V_{BE}}{I_B} \tag{4.63}$$

$$R_C = \frac{V_{CC} - V_{CE}}{I_C + I_B} \tag{4.64}$$

The last parameter needed is the value of the voltage source V_{CC}. This parameter is selected by the designer so that adequate power is provided to the design. A value larger than V_{CE} should be selected but not much larger since this reduces the efficiency of the amplifier.

4.6.3.1.2 Voltage feedback with constant base current

Again the analysis starts with the selection of the quiescent point from the specifications for the scattering parameters and the noise figure. The values of β and V_{BE} are then determined.

The current at the base of the transistor is:

$$I_B = \frac{I_C}{\beta} \tag{4.65}$$

Also using Kirchhoff's Laws we have:

$$R_B = \frac{V_{BB} - V_{BE}}{I_B} \tag{4.66}$$

$$R_{B_1} = \frac{V_{CE} - V_{BB}}{I_B + I_{BB}} \tag{4.67}$$

$$R_{B_2} = \frac{V_{BB}}{I_{BB}} \tag{4.68}$$

$$R_C = \frac{V_{CC} - V_{CE}}{I_C + I_B + I_{BB}} \tag{4.69}$$

For the correct operation of the biasing network it is essential that $I_B \ll I_{BB}$. This enables the biasing circuit to retain a stable quiescent point. In order to certify this a relation is chosen between the current in these branches. A satisfactory value is:

$$I_{BB} \approx 10 \cdot I_B \tag{4.70}$$

Even with this assumption it is evident that the number of unknowns still exceeds the number of equations. Thus, an appropriate value is selected for one parameter. Usually the choice is between R_B or V_{BB}. It is recommended that R_B should be a few KΩhms in order to establish that I_B is quite small. Usually a practically available value is chosen, e.g. 22 KΩhms.

Thus, after determining V_{BB} from

$$V_{BB} = V_{BE} + I_B R_B \tag{4.71}$$

we can substitute in Equations (4.61) to (4.63) to determine the other resistors.

4.6.3.2 Active biasing circuits

When large temperature variations are expected and close control of the quiescent point is required, usually an active biasing circuit is employed. The active biasing circuit is actually a feedback loop that secures the collector current of the microwave transistor and adjusts the

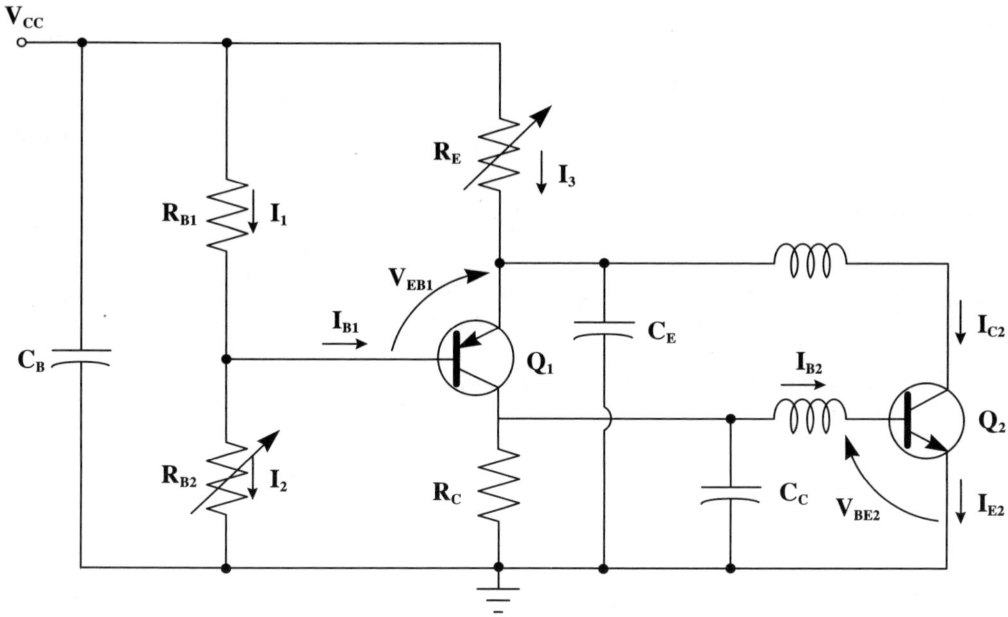

Figure 4.44 Active biasing circuit for bipolar transistors

base current to hold the collector current constant. For improved temperature performance the two devices should be situated as close as possible, in order to experience the same temperature variations.

A commonly used biasing network is presented in Figure 4.44. A P-N-P BJT (Q_1) is used to stabilise the operating point of the microwave transistor (Q_2) over any variations of the quiescent point. The actual value of the quiescent point can be regulated by R_{b_2} and R_E, R_{B_2} controls the voltage V_{CE} and R_E controls the collector current I_C.

The operation of the network is qualitatively explained as follows. If I_{C_2} tends to increase, the current I_3 increases and V_{EB_1} decreases. The reduction of V_{EB_1} decreases I_{E_1}, which in turn decreases I_{C_1} and I_{B_2}. Thus, it follows that the tendency of I_{C_2} to increase is combated from the biasing network, therefore producing a closer bias stability.

The mathematical analysis for the circuit of Figure 4.44 is presented next:

The base current I_{B_2} is:

$$I_{B_2} = \frac{I_{C_2}}{\beta_2} \tag{4.72}$$

Assuming $I_{C_1} \gg I_{B_2}$ we can give a value to I_{C_1} such that the inequality is enforced.

$$I_{C_1} \approx 10 \cdot I_{B_2} \tag{4.73}$$

Therefore:

$$R_C = \frac{V_{BE}}{I_{C_1}} \tag{4.74}$$

Also for Q_1:

$$I_{B_1} = \frac{I_{C_1}}{\beta_1} \tag{4.75}$$

$$I_{E_1} = \frac{(\beta_1 + 1)}{\beta_1} I_{C_1} \tag{4.76}$$

At the emitter of Q_1:

$$I_3 = I_{C_2} + I_{E_1} \tag{4.77}$$

and therefore:

$$R_E = \frac{V_{CC} - V_{CE_2}}{I_3} \tag{4.78}$$

As earlier we should satisfy $I_1 \gg I_{B_1}$ and therefore:

$$I_1 \approx 10 \cdot I_{B_1} \tag{4.79}$$

And also:

$$R_{B_1} = \frac{V_{EB_1} + I_3 R_E}{I_1} \tag{4.80}$$

Finally:

$$\left.\begin{array}{r} R_{B_2} = \dfrac{V_{CC} - I_1 R_{B_1}}{I_2} \\ \text{But} \quad I_1 \gg I_{B_1} \Rightarrow I_1 \approx I_2 \end{array}\right\} \Rightarrow R_{B_2} = \frac{V_{CC} - I_1 R_{B_1}}{I_1} \tag{4.81}$$

Note: The assumption that the base current is much smaller than the vertical branch is a general one and has to be satisfied to extract the optimum performance. The choice of the base current being an order of a magnitude smaller is just an approximation that usually yields good results. It is up to the designer to decide whether it is adequate for the particular application or a more stringent one is required.

From the standpoint of temperature compensation, passive biasing circuits are inexpensive to build and can provide satisfactory results. However, when the DC current gain variation is large, automatic compensation can be achieved only by an active biasing circuit. Furthermore, the active biasing circuit can provide better operating point stability, especially for low noise or high power amplifier. The major drawbacks to the active biasing circuit though are added complexity and cost.

Self-assessment Problem

4.16 For a bipolar transistor having $V_{BE} = 0.7$ and β varying between 30 (min) and 300 (max) and typical value 150, design:

(a) A voltage feedback biasing circuit.

(b) A voltage feedback with constant base current biasing circuit.

(c) An active biasing circuit.

for a quiescent point of $V_{CE} = 5\text{V}$ and $I_C = 6$ mA. The voltage supply is $V_{CC} = 20$ V.

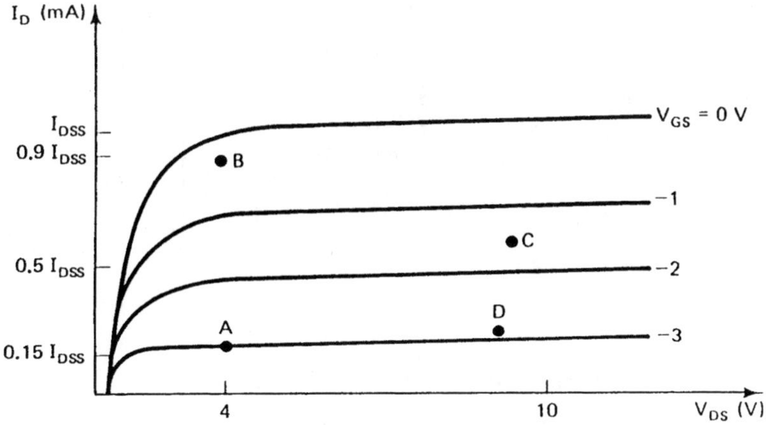

Figure 4.45 Quiescent points for GaAs FET

4.6.4 Design of Biasing Circuits for GaAs FET Transistors

The general aspects for the design of the biasing circuit are valid for both GaAs and silicon transistors. There are differences in the actual layouts since the two transistor types have different characteristics. The selection of the quiescent point in a GaAs FET is again application dependent. Figure 4.45 shows typical GaAs FET characteristics and biasing points.

(A) → *Low power low noise operation*
It operates at a relative low drain source voltage (V_{ds}) and current (I_{ds}), with I_{ds} typically equal to 0.15 I_{dss}. The type of operation is usual in Class A mode (low noise).

(B) → *High gain low noise*
The drain voltage remains low but the drain current is increased to 0.90 I_{dss} for high power gain.

(C) → *High power high efficiency. Class A operation*
To achieve higher power output the drain voltage V_{ds} must also be increased. In order to maintain linear operation in Class A the drain current must be decreased. The recommended values are $V_{ds} \approx 8 - 10$ V, $I_{ds} \approx 0.5\ I_{dss}$.

(D) → *High power high efficiency. Class AB or C operation*
For higher efficiency or to operate in Class AB or B the drain to source current is decreased further.

The most typical passive designs for GaAs FET amplifiers are presented next.

4.6.4.1 Passive biasing circuits

1. Bipolar power supply

This DC biasing circuit requires two power supplies. When the source is connected directly to the ground terminal the source inductance can be made relatively small. By doing so low noise high gain, high power and high efficiency can be achieved at higher frequency.

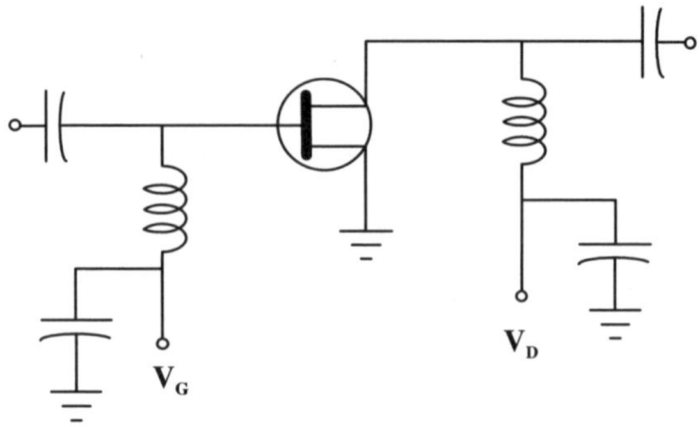

Figure 4.46 Bipolar power supply circuit

2. One sign power supply
To remove the requirement for two sources with opposite signs, the following two configurations are used (Figure 4.47):

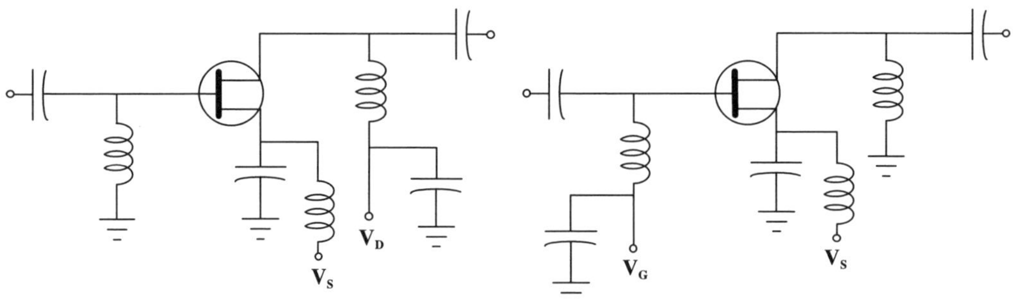

(a) Positive Supply Circuit (b) Negative Supply Circuit

Figure 4.47 Single power supply biasing circuits

These types of biasing circuits need very good microwave bypass capacitors at the source. This may cause a problem at higher frequency because any small series source impedance could cause high noise and possible oscillations.

Note: The previously mentioned configurations require caution during the start-up phase in order to prevent the burnout of the transistor. The proper start-up procedure is to establish first that a negative voltage is applied at the gate before the drain is connected to the supply. To avoid these difficulties delay circuit is included to certify that the V_{gs} is applied before V_{ds} or a Unipolar power source is employed (Type 3).

3. Unipolar power supply
These two DC biasing circuits require only one power source. They employ a source resistor to provide the turn-on turn-off transient protection. However, the efficiency and noise figure will be degraded, since the source resistor will dissipate some power and generate noise. Additionally low frequency oscillations can appear due to the source bypass capacitor.

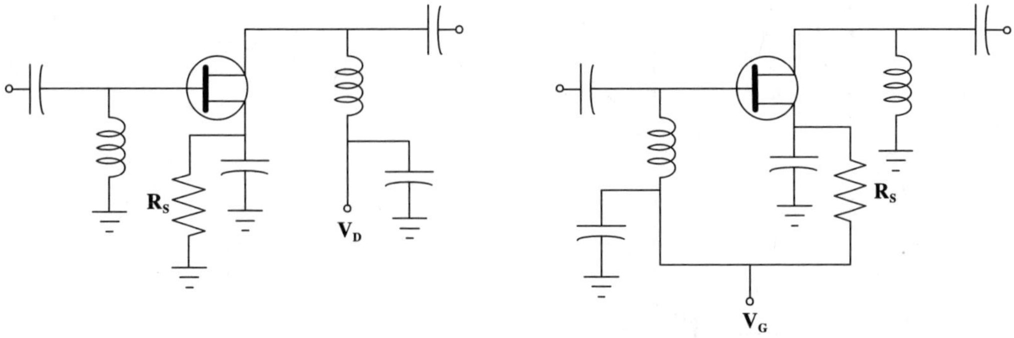

$V_S = I_{ds} R_S < +V_{DS}$

(a) Unipolar Positive Supply Circuit

$V_S = -I_{ds}\, R_S < -V_{gs}$

(b) Unipolar Negative Supply Circuit

Figure 4.48 Unipolar power supply biasing circuits

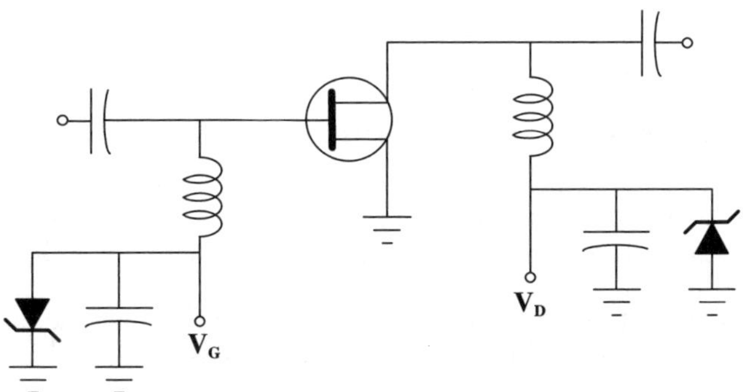

Figure 4.49 GaAs biasing circuit with Zener diodes

The value of R_S is adjusted to provide the right V_S for a proper quiescent point and transient protection.

For further protection it is common to shunt the decoupling capacitors with zener diodes. The diodes provide additional protection against transients, reversing biasing and overvoltage. The overall configuration is shown in Figure 4.49.

4.6.4.2 Active biasing circuits

Besides the aforementioned passive DC biasing circuits, when large parameter variations are expected it is usual to employ active circuits like the one in Figure 4.50.

The operation of the active network is similar to the silicon transistor case. The quiescent point is again controlled by R_{B_2} and R_E. R_{B_2} is adjusted to provide the required V_{ds} and R_E controls the drain current.

4.6.5 Introduction of the Biasing Circuit

A very important aspect of the implementation of the biasing circuit is introducing the biasing (DC power) to the radio frequency network. The application of the bias should be

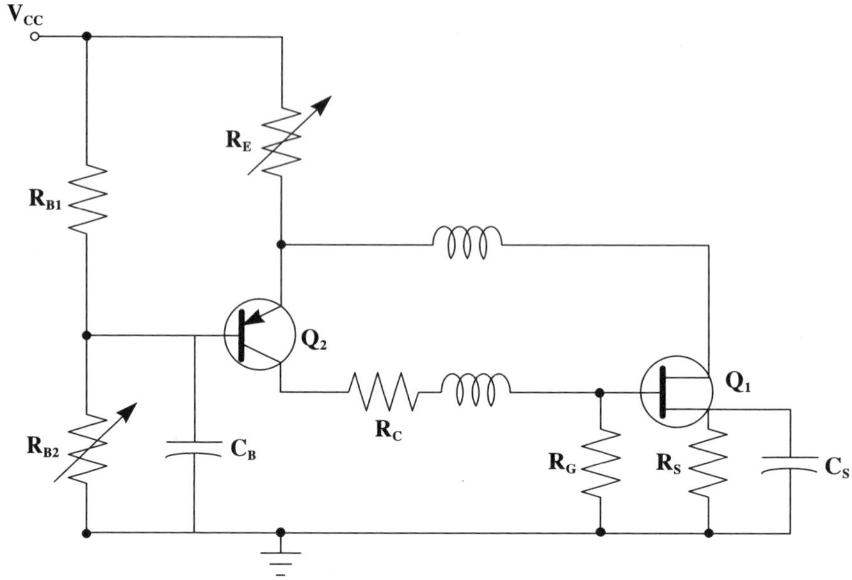

Figure 4.50 Active biasing circuit for GaAs transistors

performed in such a way that the performance of the amplifier is not affected over the intended operating band. There are two objectives in introducing the DC.

1. Minimise the amount of radio frequency power that is lost down the bias lines since this will obviously reduce the amplifier's performance (noise gain).
2. The impedance presented to the main input and output radio frequency lines, by the bias lines, should not affect the matching arrangements at the operating frequency.

The coupling of the DC and RF parts of the amplifier is achieved by the use of components like inductors and capacitors. The inductors usually known as radio frequency chokes (RFC) effectively operate as blocks for the high frequency signals, thus controlling the flow of RF power to the chosen path. Similarly the capacitors are used as DC blocks thus directing the DC power to the required nodes. With appropriate use of these components it is possible to design circuits that achieve very high efficiencies by maximising the DC to RF conversion power.

The use of ideal components for decoupling is illustrated in the following designs.

1. Common base

The requirement for a common base for the RF transistor makes the ground node different for the RF and the DC circuit. The solution is to employ inductors and capacitors to separate the two circuits as shown next (Figure 4.51).

- The capacitance C_b is used to block the DC grounding, but permitting the AC ground.
- The RFC_E is employed in order to ground the DC signal without presenting a ground for the AC signal.
- RFC_C, RFC_B are used to prevent the RF power from flowing in the biasing circuit.

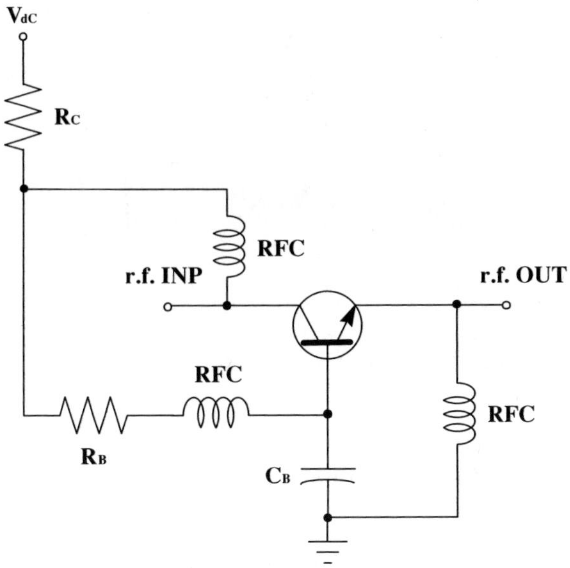

Figure 4.51 Common base configuration

2. Common collector
As in the previous case the necessary circuit is given in Figure 4.52.

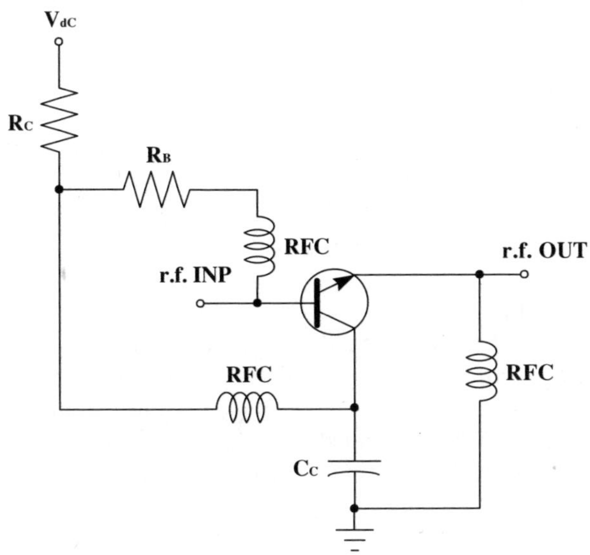

Figure 4.52 Common collector configuration

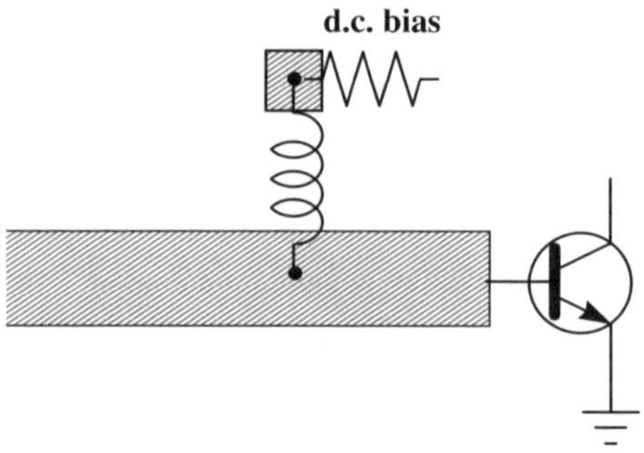

Figure 4.53 Introduction of biasing using practical inductors

4.6.5.1 Implementation of the RFC in the bias network

At the frequency range that most RF amplifiers operate, we cannot assume that the components are ideal. Therefore the ideal inductor employed earlier to provide the isolation between the DC and AC circuits is the obvious choice. Other techniques that can perform adequately even at high frequencies are presented next.

1. Inductor
In the simplest case when the operational frequency is sufficiently low (less the 600 MHz) then commercially available inductors can be used, i.e. SC30, SC10. The inductor is simply connected to the appropriate lead as is shown in Figure 4.53.

As the frequency increases the construction of simple inductors to operate well enough becomes progressively more difficult. For higher frequencies therefore it is common to use other techniques to provide the isolation between.

2. High impedance line
At high frequencies a length of a low impedance line is exhibiting inductive performance. Using this phenomenon we can employ a thin line as an inductor (Figure 4.54). The use of the high impedance line will allow the introduction of DC bias without affecting radio frequency performance significantly. Manufacturing limitation on the width of the line make this method difficult to implement if the frequency increases further.

3. Quarter wavelength shorted stub
From line transmission theory it is clear that a quarter wavelength line is basically a normalised impedance inverter, that is the normalised impedance looking in the input is the inverse of the output normalised impedance. If therefore a short-circuited quarter wavelength line is used, as shown in Figure 4.55, the impedance loading the RF circuit at point A will be infinite which means that there will be no RF power leakage to the biasing network.

To prevent the short circuit appearing at the biasing network a capacitor is used in series with the ground. The short circuit/capacitor combination can be implemented as a low impedance line (Figure 4.57a) acting as a parallel plate capacitor to the ground or a lumped

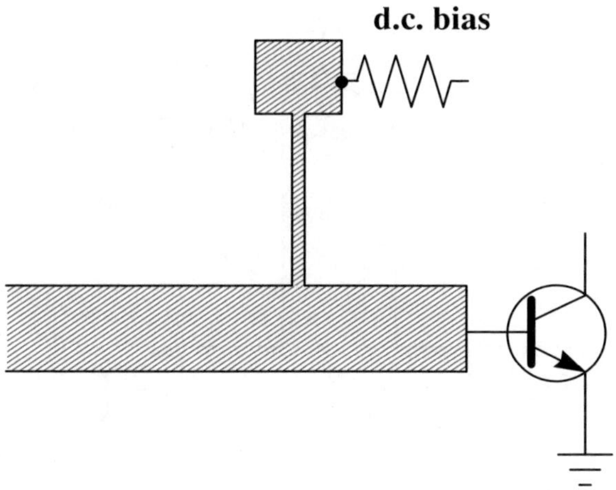

Figure 4.54 Introduction of biasing using a thin transmission line

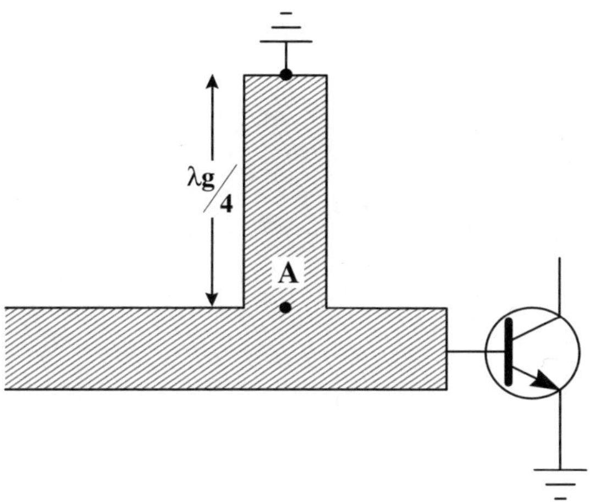

Figure 4.55 Quarter wavelength line configuration

bypass capacitor (Figure 4.56). If more wideband performance is required, then a radial stub can be used (Figure 4.57b). The gradual change of the width of the termination gives better performance over a larger frequency range.

When a distributed element is used for the construction of the short circuit, the dimensions of the patch should be selected in order to prevent the appearance of standing waves. The width of the short circuit therefore should not exceed half a wavelength ($\lambda_g/2$), to prevent the patch from becoming a microstrip antenna.

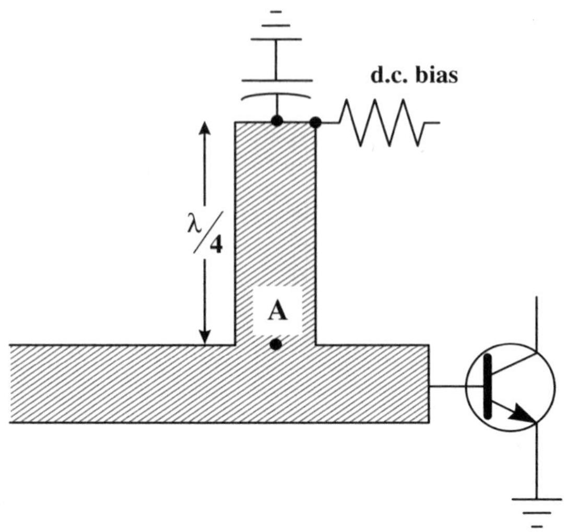

Figure 4.56 Introduction of biasing using short circuited $\lambda/4$ line and lumped capacitor

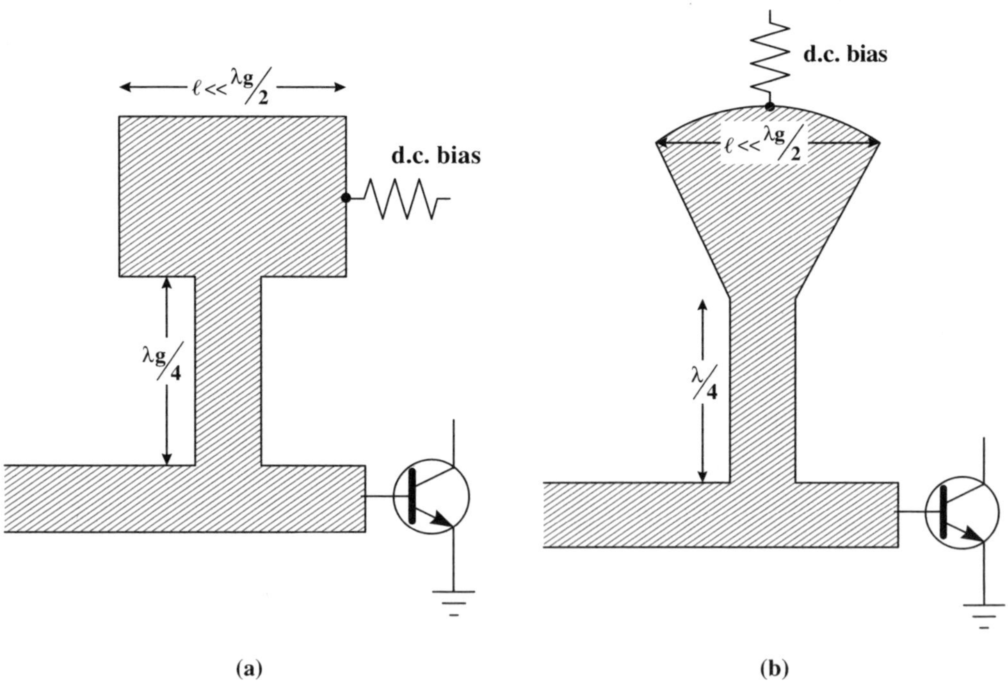

Figure 4.57 Introduction of biasing using short circuited $\lambda/4$ line and distributed capacitor

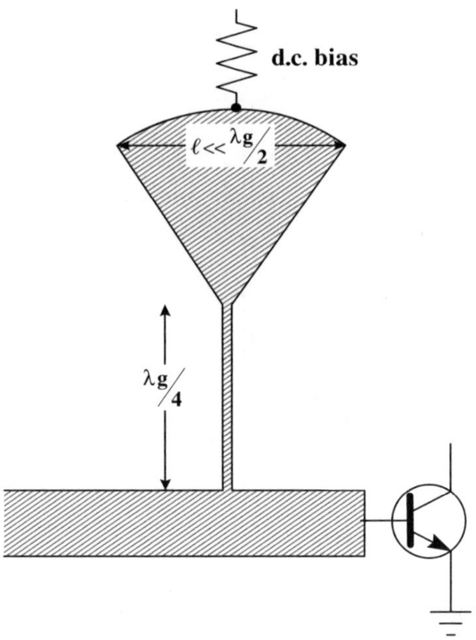

Figure 4.58 Introduction of biasing using thin $\lambda/4$ line and distributed capacitor as short circuit

An improvement of the high impedance line and $\lambda/4$ shorted stub can be achieved when $\lambda/4$ high impedance line is used (Figure 4.58). This is the most common technique in microwave design since it combines the advantages of both methods. The impedance of the line does not have to be very small (around 0.5 mm) which is well within most manufacturing procedures' capabilities. When short circuit stubs are used for input/output matching, the biasing can be introduced directly to them, instead of introducing additional stubs. This method should be employed with more care since any design errors in the short circuit would seriously affect the matching and subsequently the amplifier performance.

4. Filter network
For designs where the isolation is critical, low pass filters can be employed. This naturally means that the cost of the design will be much higher. The implementation of the filter is dependent on the operational frequency and bandwidth of the design. The general layout of the filter is presented in Figure 4.59.

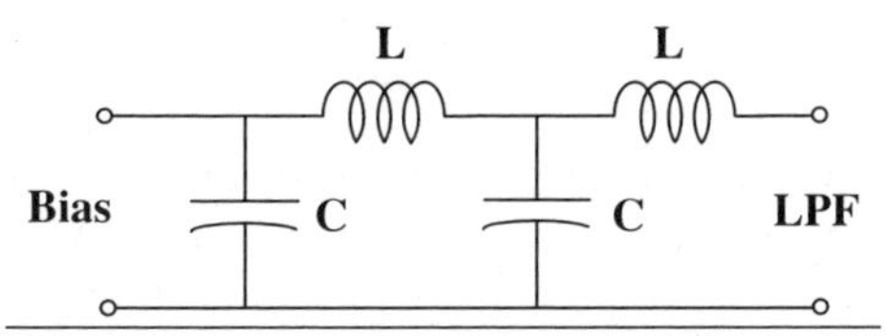

Figure 4.59 LPF network

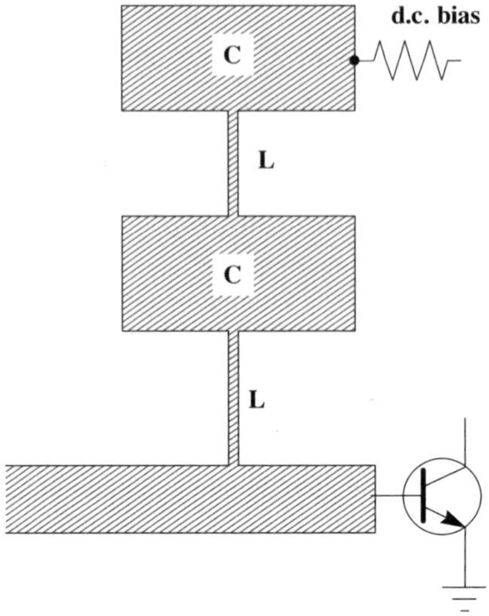

Figure 4.60 LPF using distributed elements

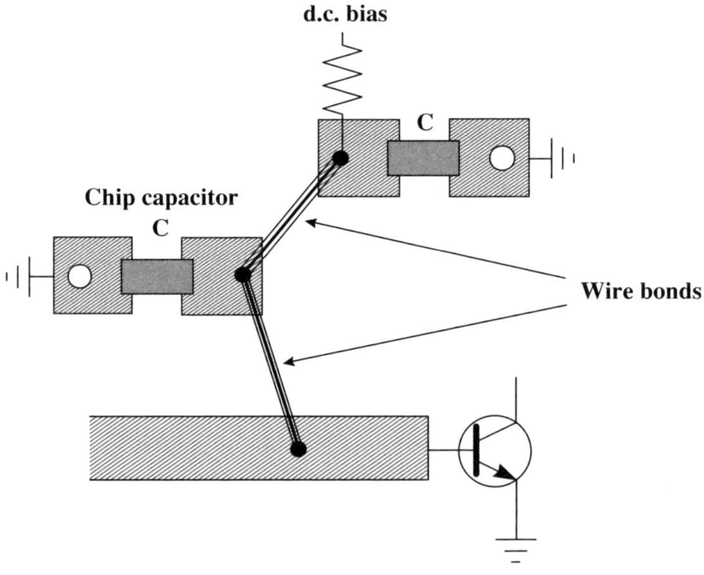

Figure 4.61 Semi-lumped LPF configuration

This can be implemented using either distributed or lumped components. In Figure 4.60 the design of the LPF is implemented using microstrip lines. Although easier to include in a microstrip board, the use of distributed elements can be limiting in their operating bandwidth and can cause spurious responses at higher frequencies. For multi-octave circuits it is necessary to use the semi-lumped filter arrangement shown in Figure 4.61.

Self-assessment Problem

4.17 What network would you use to introduce the biasing circuit when a narrowband amplifie's design is when:
(a) the centre frequency is 4 GHz;
(b) the centre frequency is 500 GHz?
Justify your choice.

4.6.5.2 Low frequency stability

Because the microwave transistors are potentially unstable at low frequency it is extremely important to ensure low frequency stability. Otherwise the appearance of any harmonic at low frequency can render the amplifier unstable.

The resistive loading from the biasing circuit is enough to extend the stable region to a few hundred MHz. In order to further ensure the stability of the amplifier it is common to include two additional load resistors so that they appear at low frequency as input, output loads.

The two available ways of configuring these resistors are presented in Figure 4.62.

Practical values for these resistors vary from 20–50 Ohms. In extreme cases even higher values can be included.

Note: In the design of Figure 4.57(b) the stability resistors are affecting the biasing circuit since they are connected in series to the base and collector resistors. It is therefore important that their values are included in the calculations. In the other configuration the stability resistors are not part of the biasing since they are grounded through a capacitor, which removes them from the DC equivalent circuit.

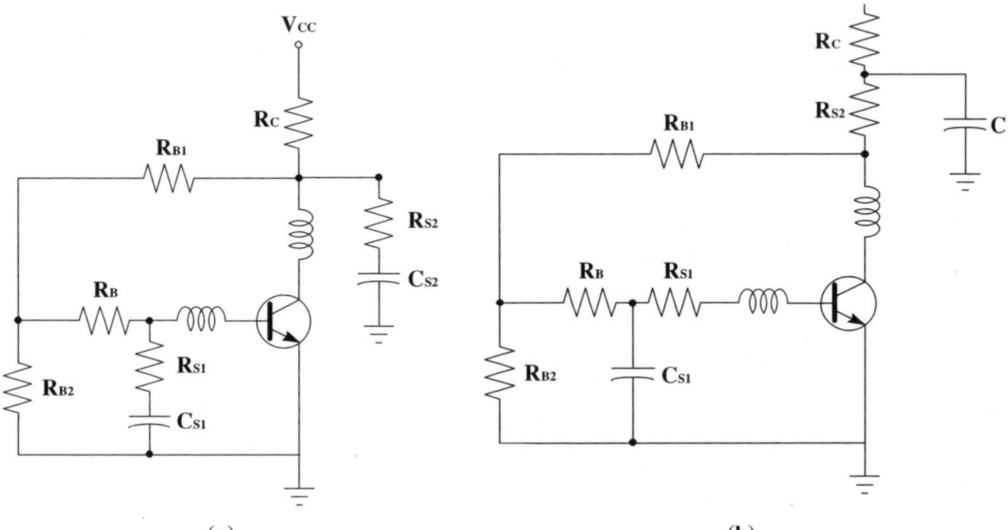

Figure 4.62 Configuration of stability resistors

4.6.5.3 Source grounding techniques

A major problem in the manufacturing of an RF amplifier is to find an effective method of grounding the emitter (or source in the case of GaAs transistors) without the introduction of significant parasitic inductance. From the earlier discussion it should be appreciated that any emitter inductance will result in series feedback tending to reduce the gain degrade the noise figure and potentially destabilise the amplifier.

Several possibilities for grounding the emitter are presented next. The common characteristic in all of them is the effort to minimise the length and/or maximise the diameter of the emitter lead.

1. Topside ground plane, tapered in to meet the source leads

Given the very small dimensions of the device and the possibility of coupling between the main radio frequency lines and topside ground plane, relatively long thin sections of line would be required to meet the source leads (Figure 4.63). A significant inductance would therefore result in making this a viable method only in a lower GHz region.

2. Plated through holes

In this method holes are drilled through to the ground plate, with the walls of the holes being metallised (Figure 4.64). This is an attractive option for use up to 10 GHz. However, it requires a more complicated manufacturing process for producing plated through holes, therefore being significantly more expensive.

3. Raised ground plane

This method introduces the least inductance. The device simply sits on the ground plane with the emitter (source) soldered to it (Figure 4.65). However, given the close proximity of the bias lines to the device, coupling between the ground plane and the bias lines would inevitably affect the low-pass performance of the biasing network. It is possible to reduce the amount of ground plane that is raised to lessen this problem. Additionally it necessitates specific preparation of the PCB which results in increased cost.

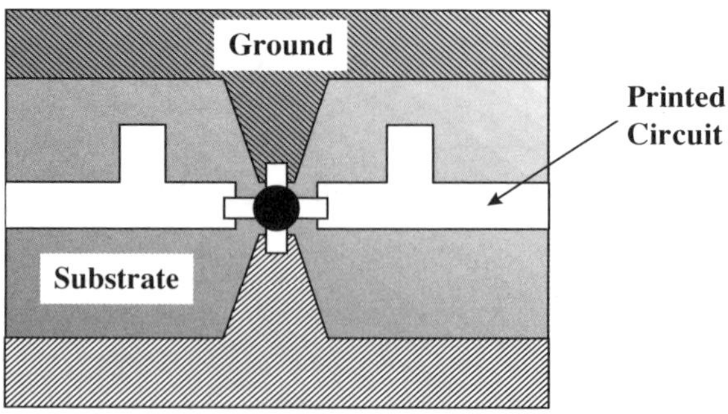

Figure 4.63 Topside ground plane

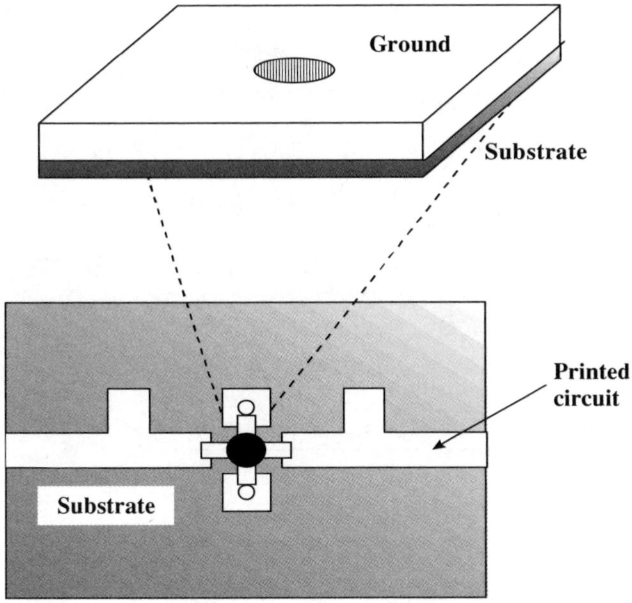

Figure 4.64 Plated through hole

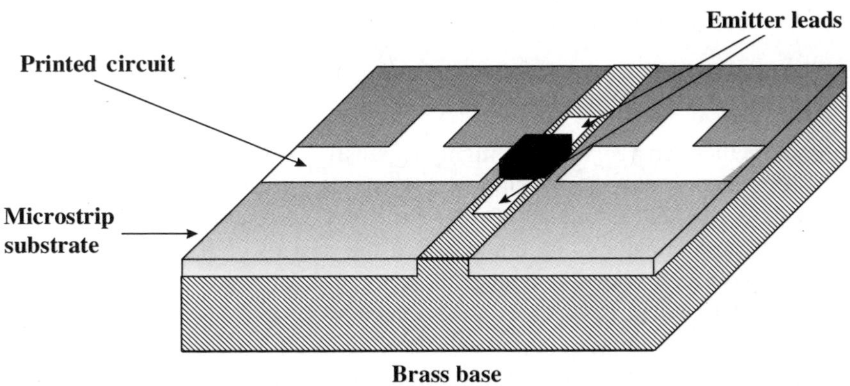

Figure 4.65 Raised ground plane

4. Metallic stud through holes
Metallic stud through holes are drilled in the board directly under the source leads. To achieve effective electrical grounding two bias studs are flush fitted into the holes in the main brass base, separated by the width of the device package. Holes, the same distance apart, are drilled through the substrate, directly below the position of the emitter (source) leads and as close to the package of the device as possible. The substrate is then firmly fixed to the main base using highly conductive silver loaded epoxy. The use of screws to hold

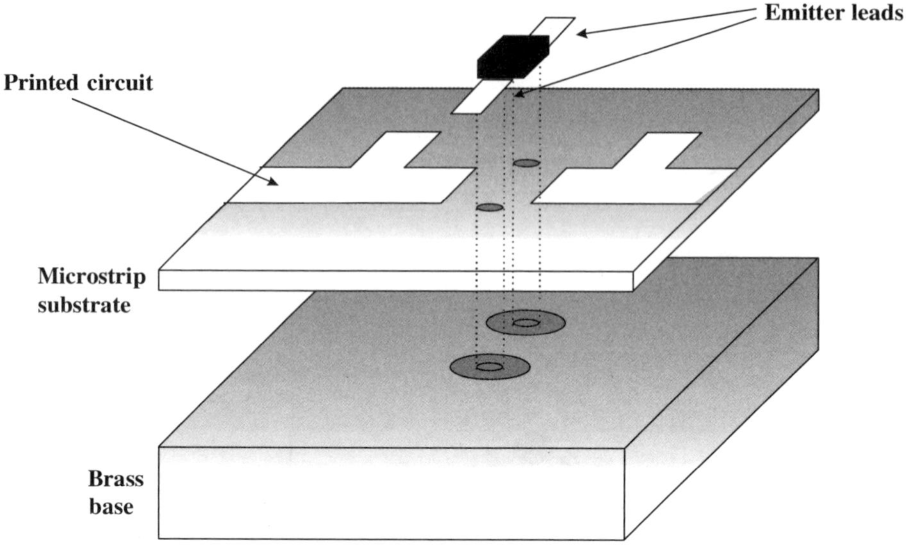

Figure 4.66 Metallic stud through holes

down the substrate should be avoided, since the resulting localised pressure has the tendency to bend the substrate. This can lead to air gaps between the substrate ground plane and the brass base and can give spurious transmission effects.

4.7 Computer Aided Design (CAD)

It took little more than ten years for the highly expensive software that was once only available on mainframe computers to develop in a relatively inexpensive tool that all engineers can have on their personal computer. In the early days Computer Aided Design was the prerogative of large companies that could afford the cost of both software and hardware. Thanks to the rapid pace of development in the computer hardware and software technology we are now at a stage when CAD tools are available even as freeware.

To understand the how the CAD approach revolutionised RF design a look into the early days of RF design is needed. Figure 4.67 illustrates the classical procedure for the design of microwave circuits. The first step is to identify the specification for the required application, using them to derive an initial circuit configuration. Available design data and previous experience are really the only tools in the first steps of the design. The circuit parameters are then determined, employing a series of analysis and synthesis techniques. A laboratory prototype is then constructed and its performance is measured and compared with the desired specifications. If the specifications are not met – which is the most probable outcome – the circuit is modified. Adjustment, tuning, trimming mechanisms are used in order to modify the obtained performance. Measurements are continuously carried out until the required specifications are met. The final configuration is then fabricated.

The above procedure had been in use for numerous years before the advent of computer technology. However, it was evident that as the technology of RF circuits was advancing,

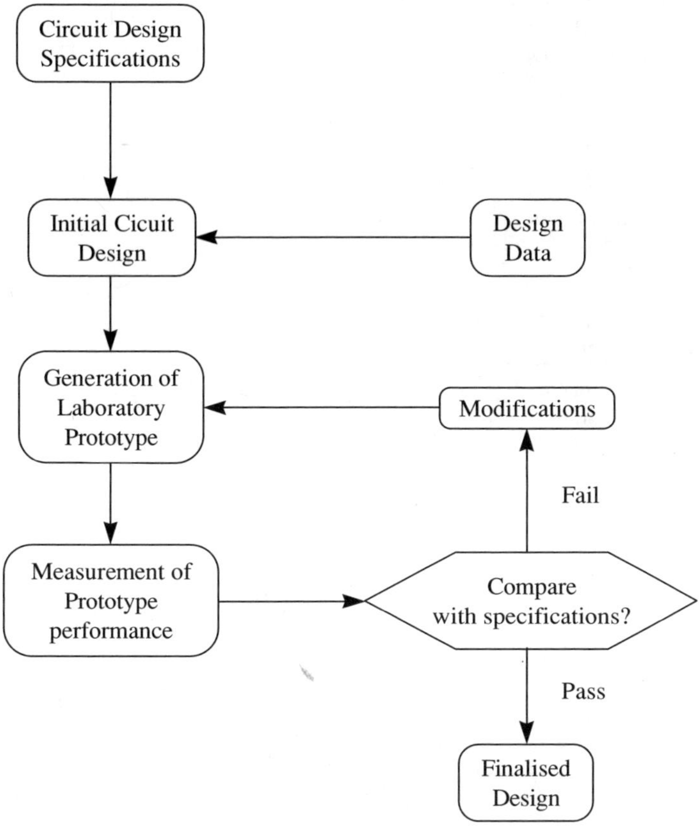

Figure 4.67 Classical RF circuit design approach

the application of this iterative and quite empirical method was becoming increasingly more difficult and costly, due to the following considerations:

- The increasing complexity of the modern RF circuits makes the analysis and the unaided design of the initial circuit if not impossible more time-consuming.
- The required specifications are usually more stringent, thus making the effect of tolerances more and more important.
- The huge variety of components that are currently in existence makes the choice of the appropriate device a very tedious task.
- The use of MIC and MMIC technology for the fabrication of most RF circuits makes any modifications to the prototype circuit very difficult, thus making prohibitive the cost of the iterative process.

4.7.1 The RF CAD Approach

To provide a solution to these problems computerised tools have been developed that enabled the prediction of the performance of the initial prototype without actually manufacturing it.

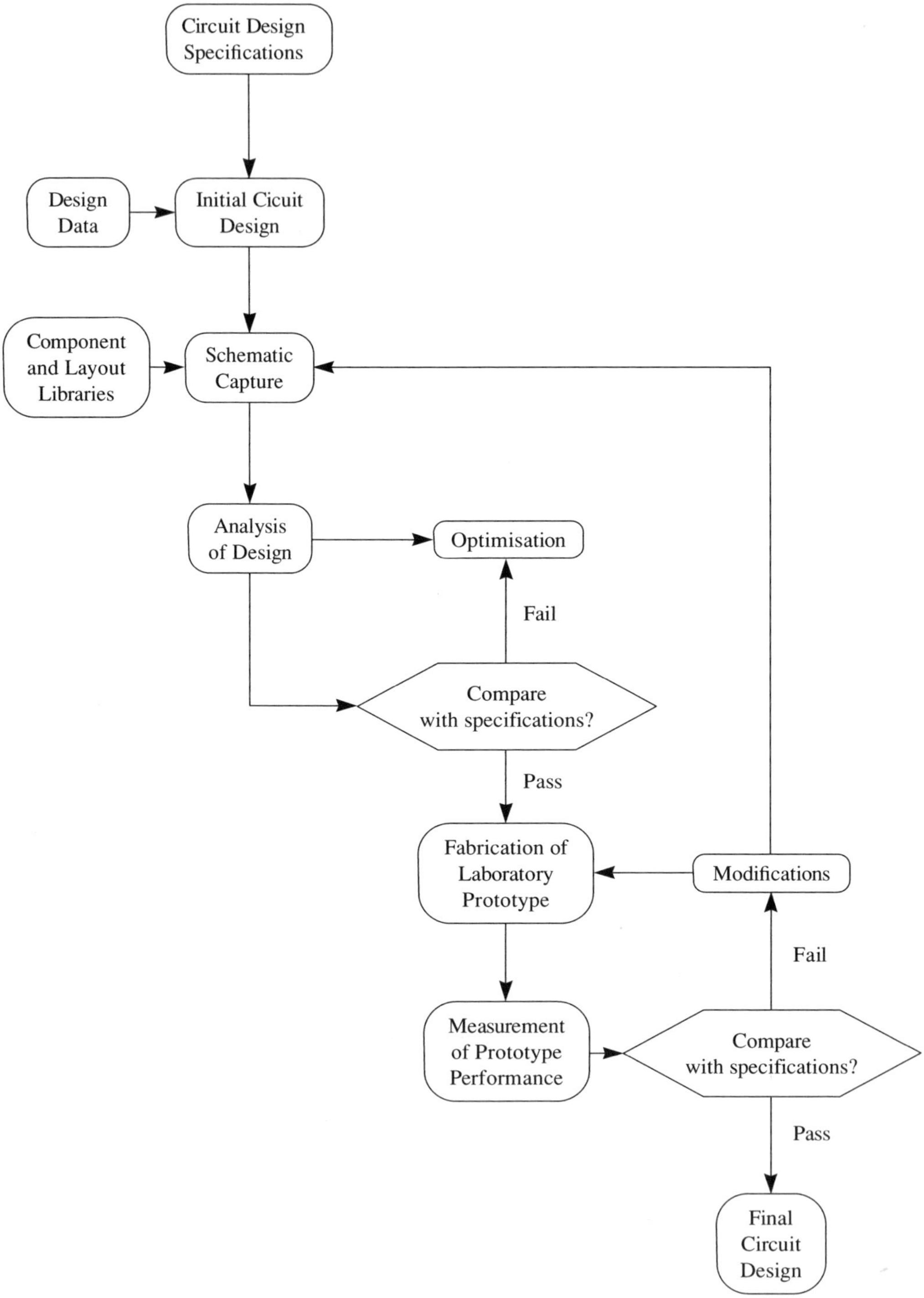

Figure 4.68 CAD RF circuit design procedure

In essence, by Computer Aided design it is implied that a number of the previous steps are now performed with the use of the computer thus making the whole process more reliable, more accurate, cheaper, less time-consuming and more comprehensive.

The amended design process that incorporates the computer as a tool is depicted in Figure 4.68. Effectively the process consists of two nested loops. In the inner one the computer is employed to evaluate the performance of the initial circuit. Numerical models for the various components are recalled from the libraries of subroutines, when needed in the analysis, now totally performed by the computer software. The estimated performance is then compared with the desired specifications and if these are not met, the circuit parameters are altered in a systematic manner. Several strategies exist that direct the manner in which the change occurs towards the optimum performance. The sequence of circuit analysis comparison with the desired performance and parameter modification is performed iteratively until the specifications are met or the best attainable performance (within the given constraints) is reached. The circuit is then fabricated and experimental measurements are obtained. Additional modifications may still be necessary if the modelling and/or analysis has not been sufficiently accurate. However, these if needed, are usually small, thus minimising the number of experimental iterations.

The process of CAD design as outlined comprises of three major segments:

- modelling
- analysis
- optimisation.

In order to appreciate the benefits and limitations of the CAD design approach, let's examine each of these segments in more detail.

4.7.2 Modelling

Modelling basically involves the characterisation of various active and passive components to the extent of providing a numerical model that can be analysed by the computer. This implies that a model exists in a pre-designed database that contains information about a variety of components that can be employed in an RF circuit. The capabilities of any software tool can be limited or enhanced depending on the extent of that library. Commercial packages nowadays contain data on any number of elements as shown in Table 4.3.

The list in Table 4.3 is by no means complete, it is really just an indication of available elements, each of which can contain hundreds of components. The extent to which one can take this list of RF elements evidences the difficulty of modelling. All these structures need to be characterised fully in terms of impedance, phase velocity losses, etc. Additionally, it is vital to model parasitic reactances as well, in order to obtain an accurate evaluation of the actual performance.

Difficulties in modelling have hindered the use of CAD techniques at RF frequencies. There is the need to identify simplified equivalent circuits and closed form expressions that possess sufficient accuracy for circuit design. Typical examples of this are microstrip discontinuities such as the T-junction shown in Figure 4.69. Quasi-analytical models published in the literature have been included in RF CAD tools to simulate this effect on microstrip circuits. So whenever a discontinuity effect is to be simulated, the equivalent representation

Table 4.3 Models found in commercial packages

Element	Features
Semiconductor devices	BJT MOSFET MESFET HEMT Point contact Schottky barrier detectors Varactor PIN diodes
Passive elements	Electron and avalanche devices Lumped elements Coaxial cables Microstrip lines Finlines Striplines Waveguides Coplanar waveguide lines Slot lines Lumped components YIG and DRO resonators Planar circuit elements
General RF elements	Air core inductors Chip capacitors Chip resistors Piezoelectric crystals Transmission line transformers Resistive pads Balun transformers Toroidal ferrite-core inductors
Fabrication specific elements	MMIC elements Surface mount technology elements
Signal sources	AC and DC voltage and current sources AC power sources Periodic waveform sources (sawtooth square wave, etc.) Noise sources

is introduced in the schematic whereupon the simulator will substitute for the analysis the equivalent electrical circuit.

For example, when a vertical connection between two transmission lines (very common in stub matching networks) is needed, the T-junction has to be introduced in the design in order to allow for the discrepancies on the Microstrip field. Figure 4.70 shows a schematic diagram of the configuration. In reality, whenever the representation shown in Figure 4.69a

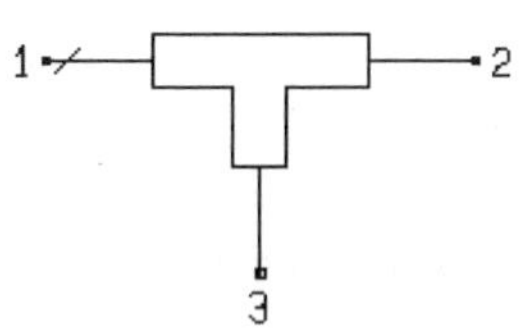

(a) Schematic representation

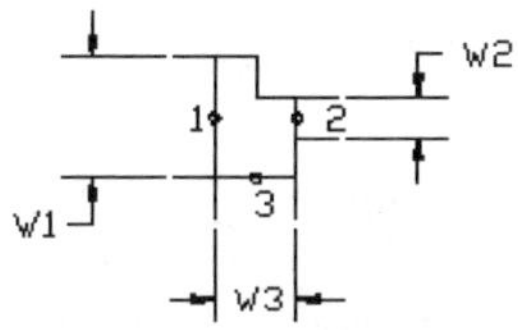

(b) Layout representation

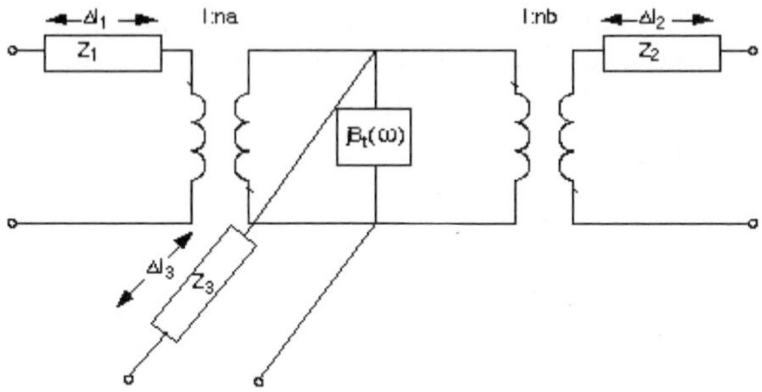

(c) Equivalent electric model

$$W_1 + W_3 \leq 0.5\lambda$$
$$W_2 + W_3 \leq 0.5\lambda$$
$$0.1 \cdot H \leq W_1 \leq 10 \cdot H$$
$$0.1 \cdot H \leq W_2 \leq 10 \cdot H$$
$$0.1 \cdot H \leq W_3 \leq 10 \cdot H$$
$$\varepsilon_r \leq 128$$

(d) Limitations of equivalent model

Figure 4.69 Microstrip T-junction in Libra Series IV

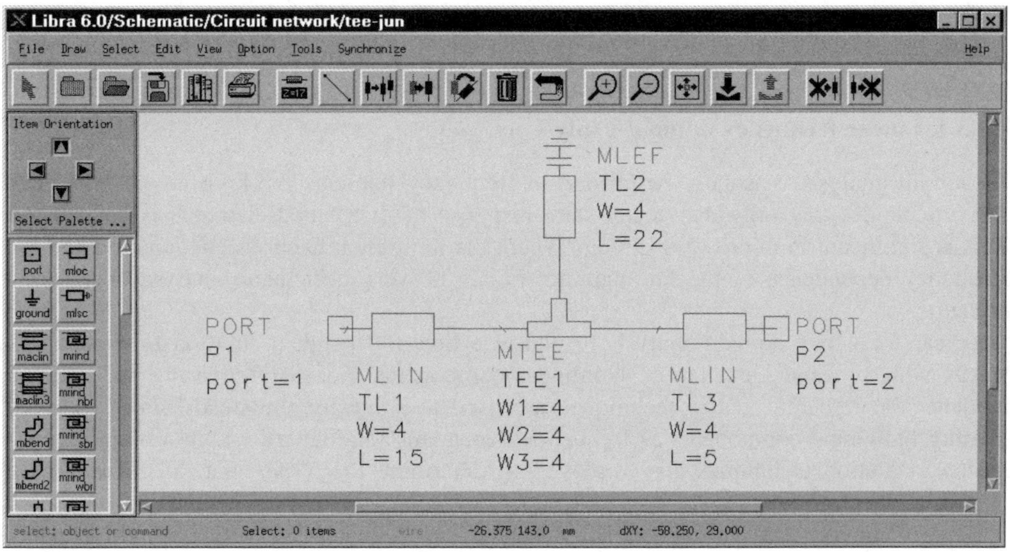

Figure 4.70 Schematic caption of single stub matching network

is encountered in a design, it is effectively substituted with the equivalent circuit shown in Figure 4.69c (the calculation of the parameters is performed using a set of semi-empirical closed expressions inherent in the model).

Modelling is a vital part of evaluation process. In reality, the error in the performance evaluation of a circuit is at least as large as the modelling error. It is essential therefore to understand the models employed as well as the way that these are introduced. If in Figure 4.69 the T junction is not inserted, then the actual performance will differ from the expected. Similarly if the limits of the model (e.g. if $w_1 < 0.1H$) are exceeded, the actual results can vary significantly.

4.7.3 Analysis

Analysis provides the response of a specified circuit configuration to a given set of inputs. This is probably the most developed and widely accepted aspect of CAD. Microwave circuit analysis usually involves evaluation of s-parameters of the overall circuit in terms of the given s-parameters of the constituent components.

Modern RF CAD tools contain several different strategies to allow the designer to analyse different circuits fast and accurately. The need for different techniques arises from the variety of issues involved in analysing complex RF circuits. It is self-evident that different types of problems require different approaches. That applies not only in terms of the required outcome (e.g. a different approach is needed when the input reflection coefficient and the third-order intercept point of a power amplifier are evaluated), but in terms of the actual circuit design (a low noise amplifier and an oscillator require different analysis techniques). The basic amplifier design can be carried out using only linear analysis based on the s-parameter matrix or small signal equivalent model. However, in applications where small signal assumption is not satisfied, such as mixers, oscillators or even power amplifiers, then a nonlinear simulation is needed.

The most common linear and nonlinear analysis techniques are summarised next.

4.7.3.1 Linear frequency domain analysis

RF circuit analysis is usually performed in frequency domain. Working in the frequency domain implies that only the steady state response of distributed elements is considered. This is significant in microwave circuits where the multiple reflections encountered and the frequency dependence of the line parameters can be very difficult to analyse in the time domain.

Linear frequency domain analysis results in a time-independent solution for linear networks with sinusoidal excitation. Nonlinear components are analysed using small-signal models. Y-parameters matrix techniques are used to solve for the steady state response, treating individual components as frequency dependent admittances within a nodal matrix. Matrix reduction techniques are employed to determine the Y-parameters of the overall circuit.

Linear analysis can perform noise measurements considering temperature-dependent thermal noise from lossy passive elements as well as temperature-dependent and bias-dependent thermal noise for nonlinear devices.

4.7.3.2 Non-linear time domain transient analysis

Non-linear components like transistors are contained in the equivalent circuits, amplitude-dependent. This necessitates the use of time domain analysis in order to obtain their true response. Spice-type transient analyses are commonly used in such cases. This involves the solution of a set of integro-differential equations that express the time dependence of the currents and voltages of the analysed circuit.

The need to consider losses, dispersion for distributed elements, and parasitics make the use of time domain simulators inefficient especially at high frequencies. The difficulty arises from the fact that in the time domain all of the RF elements are represented by simplified frequency independent models.

4.7.3.3 Non-linear convolution analysis

Convolution analysis is a time domain analysis strategy that circumvents the inefficiencies by modelling the frequency dependent elements in the frequency domain. The frequency domain representation is transformed to the time domain effectively giving the impulse response of the elements. This is subsequently convolved with input signal to provide the output. To reduce the computation time, elements with exact equivalent circuits are modelled in the time domain.

4.7.3.4 Harmonic balance analysis

The harmonic balance is an iterative process treating a nonlinear circuit as a two-part network comprising of a linear sub-network and a nonlinear element (Figure 4.71). The voltages and currents flowing from the interface into the linear part of the circuit are determined by using frequency domain linear analysis. Currents and voltages into the nonlinear part are calculated in the time-domain. Fourier analysis is used to transform the time domain to the frequency domain. Kirchhoff's Current Law states the currents should sum to zero at the interface. Any error that is encountered is reduced by successive iterations. When the method converges, the currents and voltages approximate the steady-state solution

The large number of calculations necessitated by each iteration combined with the requirement for several iterations before convergence imply that a fast processor and a lot of memory is needed. The strain on the computer resources increases as the number of harmonics increases. For example, as much as 300 Mbytes of RAM are needed to handle three independent frequencies with 12 harmonics.

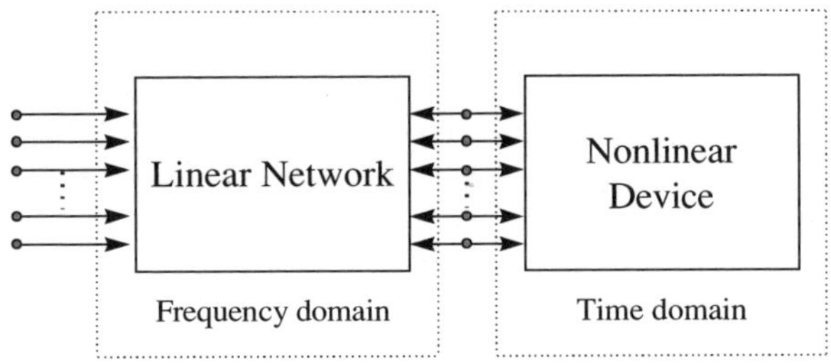

Figure 4.71 Representation of nonlinear circuit for harmonic balance analysis

4.7.3.5 Electromagnetic analysis

A different form of analysis often used at RF and microwave frequencies is electromagnetic analysis. Electromagnetic simulators basically solve Maxwell's equations for the analysed circuit in two and three dimensions. The electromagnetic simulators should be able to cope with the general metal-dielectric structure problem. To simplify the general case the transmission lines (metal structure) can be assumed to be infinitely thin. This assumption reduces the required computation time by an order of magnitude.

In the former case, a three-dimensional type analysis is used to determine the field patterns surrounding a metal structure embedded in various layers of dielectric materials. In the latter case only a two-dimensional analysis is needed to calculate the modal characteristics and electromagnetic fields for a cross-section of transmission lines embedded between layers of dielectric materials. The calculated characteristics can be impedances, voltages, currents, powers, propagation velocities and effective dielectric constants.

4.7.3.6 Planar electromagnetic simulation

This is usually used to solve for the s-parameters of arbitrary microstrip structures. By limiting the problem in two dimensions the necessary processing time is dramatically reduced. Some of the most recent tools can handle multi-layered structures with vias and varying dielectric thickness. Nevertheless the structures are still assumed to be planar and so the term $2^1/_2$-D is used (two-dimensional currents but 3-dimensional fields). The speed of planar electromagnetic simulators makes them a very attractive solution when non-standard structures are used.

4.7.3.6.1 3-D electromagnetic simulation

These simulators can analyse arbitrary-shaped metal structures positioned over multi-layered dielectric structures, in an enclosed metal shielded environment. They use finite element techniques to divide the problem space into a point mesh and calculate the field intensity at the vertices. Typically they are employed when analysing microstrip to stripline or microstrip to waveguide transitions.

All analyses techniques are employed to calculate the effect of variations in the circuit parameters on the overall performance. The results are useful for two purposes: optimisation of the circuit and statistical analysis.

4.7.4 Optimisation

Optimisation is the process of iterative modifications of a set of circuit parameters in order to achieve a specified performance, i.e. to meet a set of given specifications. The idea of optimisation is very simple and arises directly from the trial and error tuning procedures of the classical approach. Basically you alter some values of the circuit parameters and evaluate the overall performance if it is better you keep the values, otherwise you try a different set. The advantage of using a computer-aided approach is that the optimisation is performed

Figure 4.72 Typical partitioning of line transformer for planar electromagnetic analysis

in a systematic manner, thus covering a much larger set of input values than is possible with any other approach. Although one can envisage a scenario where the design is totally performed by trial and error with a very powerful computer this is far from the truth even with today's most advanced machines. In reality, the huge number of input parameters make the starting condition and the choice of variables to optimise critical both in terms of actual performance and design time.

Two critical issues arise when the optimisation of a circuit is attempted:

1. The search method. This identifies the systematic manner in which the optimisable parameters are altered.
2. The error function formulation. To determine whether the 'optimum' performance is achieved, not only a set of goals is required, but also a formula with which the distance from the objective is continuously assessed.

When an optimiser executes the chosen search method a new set of parameter values is calculated. The new values are used to recalculate the error function. The smaller the obtained value, the more closely the goals are met.

The following paragraphs contain a discussion of the most common search methods and error function formulas encountered in most RF CAD tools.

4.7.4.1 Optimisation search methods

4.7.4.1.1 Random search

The random optimisers arrive at new parameter values by using a random-number generator to select a number within the allowed range for each parameter. It is basically a trial and error process. Starting from an initial set of values with a known error function, a new set of parameters is obtained using the random numer generator. The error function for the new set is calculated and compared to the previous. If the result is better, then the new set is used as the initial point for the iteration, otherwise the new values are discarded and the process is repeated with the same initial point.

4.7.4.1.2 Gradient search

The gradient search involves the determination of the gradient of the network's error function. The advantage of this optimiser is that it progresses much faster to a point where the error function is minimised, although it is possible that a local minimum is reached rather than the optimum performance.

A cycle of the gradient optimiser starts with the calculation of the gradient of the error function for the initial parameter set. This points to a direction that reduces the value of the error function. Subsequently the initial set is moved in the direction that was determined, until the error function reaches a minimum. At that point the parameter set becomes the initial point and the gradient is recalculated.

A single iteration for the gradient search includes several error function evaluations, thus increasing the required processing time compared to the random. On the other hand, the gradient optimiser results in a more stable design (small changes of the actual design parameters do not significantly alter the overall performance).

4.7.4.1.3 Quasi-Newton search

The quasi-Newton optimisers employ second-order derivatives of the error function as well as the gradient in order to determine the direction of the search. The calculated second-order derivatives supplement the gradient pinpoint more accurately in the direction of the search. The optimisation terminates when the gradient is zero (a minimum is reached) or when the estimate variable change becomes to small (less than a predefined value, e.g. 14.68). Compared to both the gradient and random searches, each iteration of the quasi-Newton approach requires several more operations.

4.7.4.1.4 Direct search

This is a form of the random optimiser employed when some parameters are allowed to take only discrete values, for example the resistance of a resistor. This type of search is more of a trial of permissible combinations of values. So when a set of values is found that results in a reduction of the error function, it is stored until a better one is determined. Obviously the optimum performance is determined when all possible combinations are evaluated. The processing time, however, for such a venture is prohibitive unless a very simple circuit is investigated.

4.7.4.1.5 Genetic algorithm search

This is a form of direct search employed when the number of discrete parameters is so large that a direct search of all combinations is impossible. In a sense it is a development of the previous strategy but instead of blind trial of all combinations, a more intelligent guess is attempted, by combining some subsets of values rated best in the previous trials.

4.7.4.2 Error function formulation

4.7.4.2.1 Least squares

The least squares error function is calculated by evaluating the error of each specified goal at the operation range (usually frequency or power) for every point individually and then squaring the magnitudes of those errors. The error function is then the average of the individual errors over the whole range.

A mathematical expression for the error function is shown next:

Summation over all frequency subgroups

Summation over all optimisation groups

Summation over all frequencies in subgroup that contains f

Summation of all responses in subgroup

$$U = \sum_{opt\_groups}\left(\sum_{sub\_groups}\left(\frac{\sum_f(\sum_i w_i \cdot |R_j(f) - g_j|^2)}{N_f}\right)\right) \tag{4.82}$$

where:

U: is the least square error function
R_j: is the evaluated response at frequency f
g_j: is the goal to be achieved for the response R_j
w_i: is the weighting factor associated with the response $R_j(f)$
N_f: is the number of frequencies of the sub-group to which f belongs.

It is important to point out the significance of the weighting factor. It is evident that any change in the weightings applied in the various responses would completely alter the value of the error function. This is significant since it allows the designer to increase the significance of one response rather than the other. Furthermore the frequency can be substituted with other swept variables like power or an optimisation for more than one swept valuables can be attempted. Nevertheless care should be given not to increase excessively the complexity of the error function as this would result in significant increase of the required processing time.

4.7.4.2.2 Minimax

Minimax optimisation calculates the difference between the desired response over the entire measurement parameter range of the optimisation. It is the task of the optimiser then to minimise the difference for the point that has the maximum deviation between the evaluated and the desired response. In effect, minimax is the minimisation of the maximum of a set of functions denoted as errors.

The error function is now mathematically represented by the Equation (4.83):

$$U = \max_i(R_i - g_i) \tag{4.83}$$

A minimax optimiser always tries to minimise the worst case, giving a solution that has the goal specifications met in an equal ripple manner. This implies not only that the specifications should be met, but the error should be smoothed out for all the evaluated responses. In that sense the minimax solution would be the one that best fits the goals.

4.7.4.2.3 Least Pth

The least Pth error function formulation is similar in make-up to the least squares method discussed earlier. The difference arises in that the magnitudes of the individual errors are no longer squared but raised to Pth power, with $p = 2,4,8$ or 16. This effectively emphasises the errors that have large values.

As p increases, the least Pth error function approaches the minimax function. The least Pth formulation is an approximation of the minimax solution. Minimax error function contains infinite gradient changes (discontinuities) when the error contributions due to different goals intersect in the parameter space. By approximating the minimax with a least Pth, these discontinuities are smoothed out.

The mathematical formula that describes the least Pth function is given by:

$$U = \begin{cases} \left(\sum_i (R_i - g_i)^p\right)^{1/p} & \text{if } \max_i(R_i - g_i) > 0 \\ 0 & \text{if } \max_i(R_i - g_i) = 0 \\ \left(\sum_i -(R_i - g_i)^{-p}\right)^{-1/p} & \text{if } \max_i(R_i - g_i) < 0 \end{cases} \tag{4.84}$$

The least Pth method permits the error function to become negative. That implies that even if a solution is better than the goal (negative error), the solution will be improved further so that the error would be smooth for all the responses.

Note: It should be evident that the optimisation is not a trivial job, it requires long computation times especially when the design is complex. Some basic rules that simplify the task are presented next:

- Choose the optimisable variables sensibly. The complexity increases geometrically as the number of parameters to examine increases.
- Reduce the parameter space by limiting the input variables. The optimiser is a mathematical tool and it cannot appreciate the physical significance of the variables.
- Select optimisation goals carefully. Sometimes a non-critical response has to be degraded to allow an improvement in a more significant area.
- Be aware of circuits with marginal stability.
- Use more than one optimisation technique. For example, gradient optimisation can be combined with random to ensure that the achieved minimum is a global one.

4.7.5 Further Features of RF CAD Tools

4.7.5.1 Schematic capture of circuits

Schematic capture is one of the major advances in CAD brought about by the new fast graphic computers. The first CAD tools required a network list (like computer language listing) of all the elements comprising the circuit where the interconnections where defined by a series of nodes. The difficulty of introducing circuits with a few hundred nodes is self-evident. A typical view of a schematic editor is shown in Figure 4.73.

The problem is overcome with the schematic capture. Here the circuit is entered graphically using icons that represent circuit elements available in the CAD libraries. The additional benefit is that the designer can introduce the parameters required in a pop-up window invoked when the new element is positioned. The schematic capture results in a design that is easy to visualise since it retains the conventional circuit diagram format.

To avoid cluttering the display the concept of hierarchical design was introduced. In this the design is divided in sub-networks (in a manner similar to block diagrams, each of which is introduced independently and associated with a unique name and schematic symbol. The sub-networks can be interconnected together with other elements to create a more complicated design. The concept is similar to bench working where several instruments and test devices, each comprising of several electronic circuits are interconnected to create a more complex circuit.

4.7.5.2 Layout-based design

To further simplify the fabrication process modern RF CAD tools allow the designer to convert schematic diagrams into layout representations of the circuit. This capability allows for quick generation of the artwork required for the manufacturing of RF designs. Figure 4.74 contains the layout representation of the schematic shown in Figure 4.73. In essence it is a form of schematic representations since it can allow almost the same features. For example,

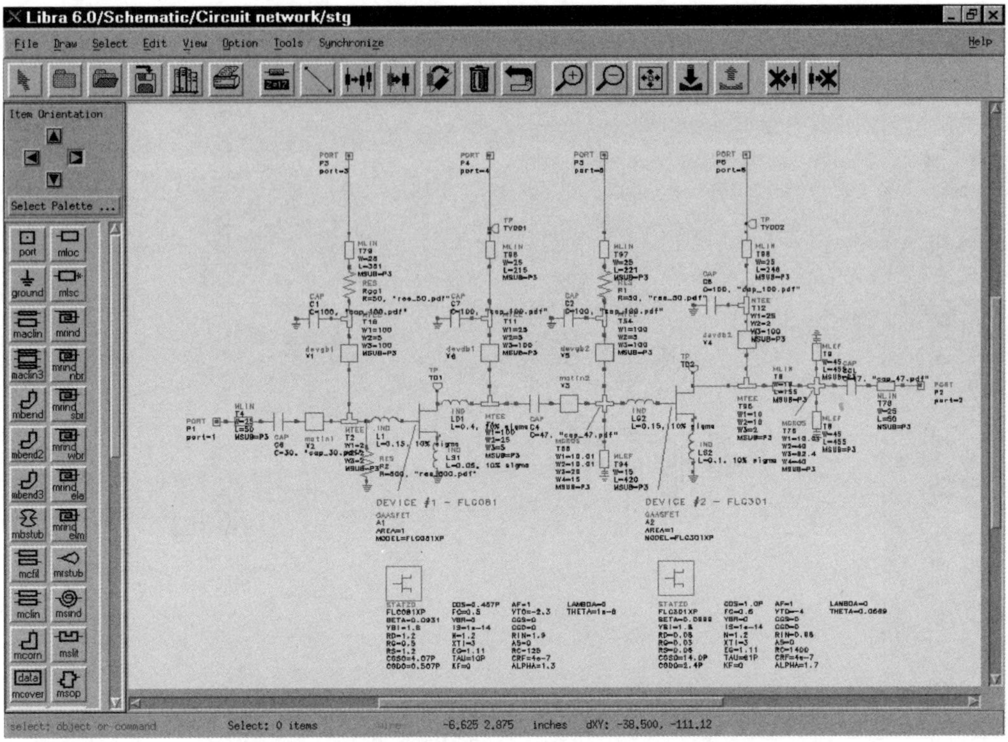

Figure 4.73 Schematic representation of a power amplifier

a circuit can be designed, simulated, and optimised completely in the layout form. This is achieved by associating each component available in the library with three representations:

- a schematic representation;
- an artwork representation;
- an electrical model.

In that sense the correspondence is direct, when the simulator encounters one, it can substitute it with the other. This allows direct conversion from the electrical design (schematic representation) to an executable program (for analysis purposes) or an artwork (layout representation). The direct transfer of data from an easy-to-read schematic representation to an artwork not only reduces the actual design time, but minimises the probability of an error creeping in at later stage.

4.7.5.3 Statistical design of RF circuits

Statistical design is the process of:

- accounting for the random variations of the parameters of a design;
- measuring the effect on the circuit performance of such variations;
- modifying the design to minimise these effects.

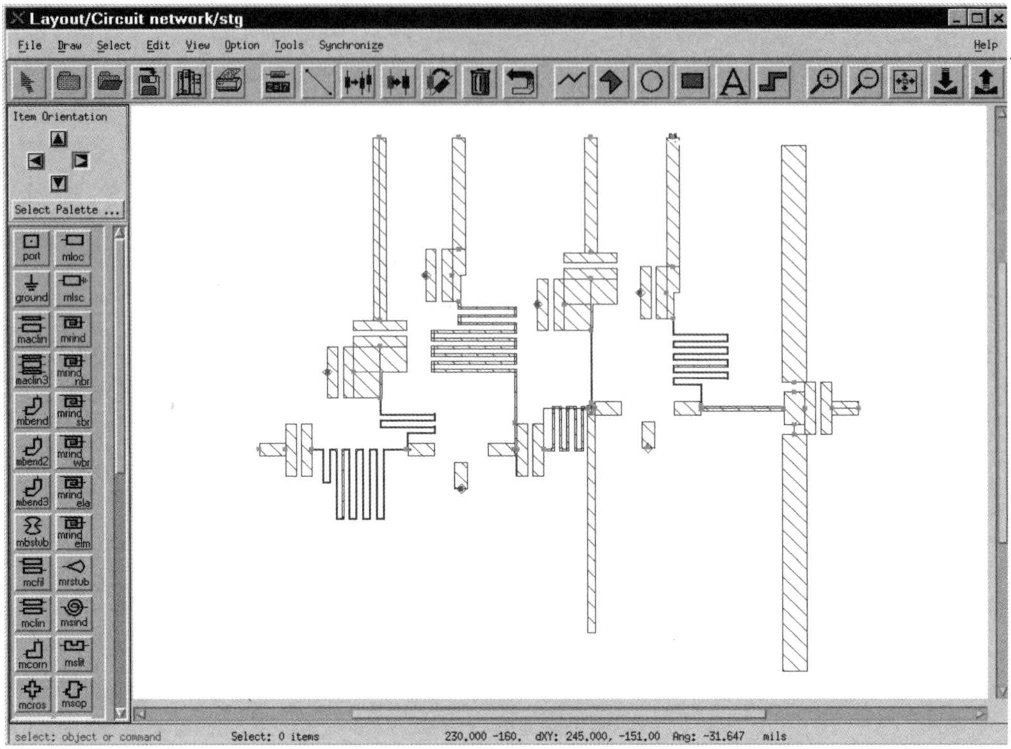

Figure 4.74 Layout representation of a power amplifier

In a practical application where RF circuits are mass-produced, it is important to identify the effect of the statistical variation of the nominal values of the individual components. In essence it is necessary to identify what are the acceptable deviations of the nominal value (tolerance) so that the resulting performance will lie within the specified range. The process of varying a set of input parameters using specified probability distributions to determine what percentage of resulting designs fall within the specifications is called yield analysis. The ratio of the number of circuits that pass the design specification to those produced is called the yield of the design.

To determine the yield of a circuit, one needs to identify which are the limits within which the performance of the actual design is considered acceptable. Figure 4.75 depicts a typical frequency response diagram which depicts the limits of this region (shaded area). To simplify the process let's assume that only two input parameters P_1 and P_2 can affect the outcome (the tolerances of these parameters, $\Delta P_1 \Delta P_2$, are displayed in Figure 4.76). By analysing the circuit for all the combinations of P_1 and P_2 we can determine a region that contains all the sets of P_1 and P_2 that result in acceptable performance, called the element constraint region (depicted in Figure 4.76). To obtain high yield the input parameters have to remain within the limits of this area. The yield of the circuit is the superimposed area.

To maximise the yield it is necessary to maximise the overlap area in the tolerance rectangle and the element constraint region. This can be done by shifting the tolerance square in the parameter space until the two are centred (Figure 4.77). This process is called

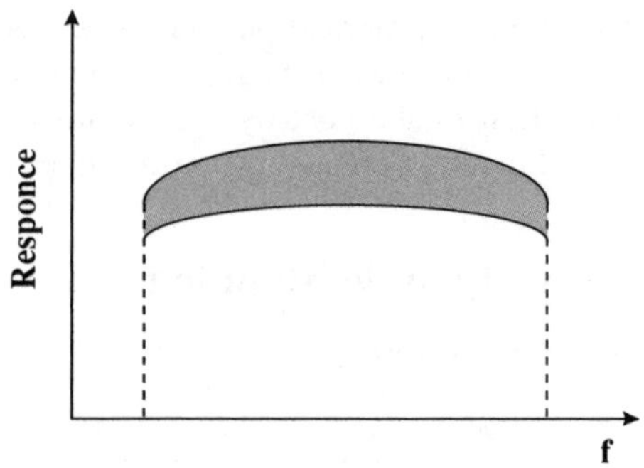

Figure 4.75 Range of acceptable performance for the designed circuit

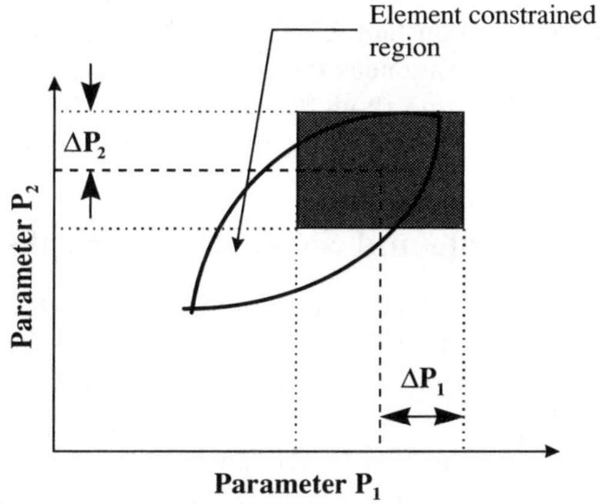

Figure 4.76 Determination of the yield for a given tolerance window

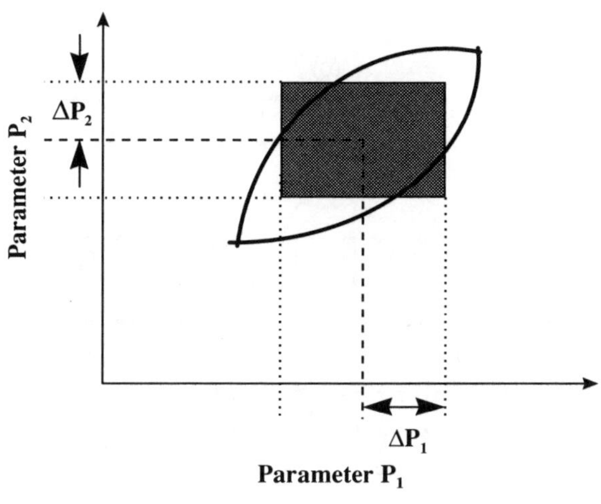

Figure 4.77 Design centring: maximisation of the circuit yield by centring the tolerance window and the element constrained region

yield optimisation or design centering. If fabrication yield is needed, it can be seen that more stringent limits must be set on the tolerances of the input parameters. It is the designer's task to determine whether the reduction of the cost by the minimisation of the failed circuits justifies the additional cost of obtaining components with higher tolerances.

Appendix I: Second Type of Circle Mapping

Suppose we have an expression of the form:

$$X = \frac{Ax^2 + Ay^2 + Bx + Cy + D}{Ex^2 + Ey^2 + Fx + Gy + H}$$

where A to F are a set of *real* constants. If we write a complex number z as $z = x + jy$, then this function generates a real number X associated with any point in the complex z plane. It is easy to show that the loci of constant X in the z plane are *circles*. The distinguishing feature of the expression which guarantees the circle property is that the coefficients of x^2 and y^2 in the numerator are the same (both A) and likewise in the denominator (both E). Without this condition, the loci of constant X would become ellipses.

Appendix II: Masons's Rule and Signal Flow Graphs

Signal flow graph techniques provide a graphical representation of the relationships between circuit parameters to which systematic manipulations can be applied to determine the performance of the network. The general solution to a signal flow graph can be obtained using Mason's rule.

Example networks are shown in Figure AII.1, where the networks consist of directed *branches* between *nodes*. Each branch has a *branch transmittance* that describes the relationships between the signals at the *node signals* seen at each end of the branch. A node signal is equal to the algebraic sum of all signals entering that node, and that signal is applied to each of its outgoing branches.

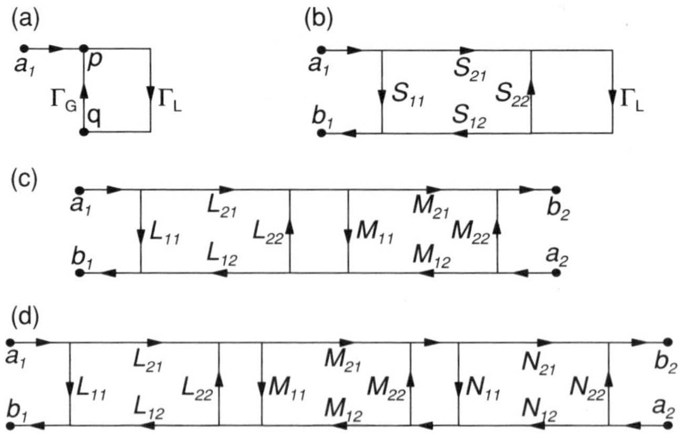

Figure AII.1 Example cascade circuits for analysis by Mason's Rule

AII.1 Mason's Non-Touching Loop Rule

Masons rule is a general method of determining the transmittance of an overall circuit from its flow graph. In these diagrams we have *sources* where the node has only outgoing branches, such as a_1 and a_2, and *sinks* where the node has only incoming branches, such as b_1 and b_2. *Paths* are formed by a continuous succession of branches that follow the direction of the arrows, in which any node is encountered only once.

A closed path that loops back to its starting node is a *loop*, and these are further characterised as:

- First order loop: a closed path looping back to a node without crossing the same node twice.
- Second order loop: the product of any two non-touching first order loops.
- Third order loop: the product of any three non-touching first order loops.
- Fourth order loop, etc . . .

The path transmittance, P, is the product of the branch transmittances which make up the path, while the loop transmittance, L, is the product of the branch transmittances which make up the loop. The graph transmittance, K, is the ratio of the signal appearing at the sink node to the signal applied from the source node (i.e. the overall desired circuit transmittance). Using the above definitions, Mason's rule defines K as:

$$K = \frac{1}{\Delta}\sum_{i=1}^{n} P_i\Delta_i \tag{A1}$$

where :

n = the number of forward paths from source to sink
Δ = 1 – (Sum of all first order loops) + (sum of all second order loops) – (sum of all third order loops) + (sum of all fourth order loops) – (. . .
P_i = transmittance of the ith forward path
Δ_i = the value of Δ for that portion of the graph not touching the ith forward path

AII.1.1 One Port

In this case, Figure AII.1(a), there is a single loop $\Gamma_G\Gamma_L$ and a unit transmittance path to the node signal p, giving

$$p = a_1\frac{1}{1-\Gamma_G\Gamma_L} \tag{A2}$$

AII.1.2 Two-Port – Single Element

Considering this circuit, Figure AII.1(b), we can identify the paths and loops:

Paths:	a_1 to b_1	$P_1 = S_{11}$, $P_2 = S_{21}\Gamma_L S_{12}$
Loops:	First Order	$L_1 = \Gamma_L S_{22}$
	Second Order	None

In this case $\Delta = 1 - S_{22}\Gamma_L$. The loop L_1 does not touch path P_1 so $\Delta_1 = 1 - S_{22}\Gamma_L$, but L_1 does touch path P_2 and $\Delta_2 = 1$. The overall transmittance from a_1 to b_1, or input reflection coefficient Γ_i, is then:

$$K = \Gamma_i = \frac{P_1\Delta_1 + P_2\Delta_2}{\Delta} = \frac{S_{11}(1 - S_{22}\Gamma_L) + S_{21}S_{12}\Gamma_L}{(1 - S_{22}\Gamma_L)} = S_{11} + \frac{S_{21}S_{12}\Gamma_L}{1 - S_{22}\Gamma_L} \qquad \text{(A3)}$$

AII.1.3 Two-Port – two Element Cascade

Considering the circuit, Figure AII.1(c), we identify the paths and loops:

Paths:	a_1 to b_1	$P_{A1} = L_{11}$, $P_{A2} = L_{21}M_{11}L_{12}$
	a_1 to b_2	$P_{B1} = L_{21}M_{21}$
	a_2 to b_1	$P_{C1} = L_{12}M_{12}$
	a_2 to b_2	$P_{D1} = M_{22}$, $P_{D2} = M_{12}L_{22}M_{21}$
Loops:	First Order	$L_1 = L_{22}M_{11}$
	Second Order	None

For all paths we have $\Delta = 1 - L_{22}M_{11}$.

The loop L_{A1} does not touch path P_{A1} so $\Delta = 1 - L_{22}M_{11}$, but L_{A1} does touch path P_{A2} and $\Delta_2 = 1$. The overall transmittance from a_1 to b_1, or as we are considering $a_2 = 0$ in these equations the overall S_{11} of the circuit, is then:

$$S'_{11} = \frac{P_{A1}\Delta_{A1} + P_{A2}\Delta_{A2}}{\Delta} = \frac{L_{11}(1 - L_{22}M_{11}) + L_{21}M_{11}L_{12}}{(1 - L_{22}M_{11})} = L_{11} + \frac{L_{21}L_{12}M_{11}}{1 - L_{22}M_{11}} \qquad \text{(A4)}$$

The forward transmission through the circuit from a_1 to b_2 has the path, P_{B1}, which touches L_1 so $\Delta_{B1} = 1$. The overall S_{21} of the circuit is:

$$S'_{21} = \frac{P_{B1}\Delta_{B1}}{\Delta} = \frac{L_{21}M_{21}}{(1 - L_{22}M_{11})} \qquad \text{(A5)}$$

Likewise, for the reverse transmission from a_2 to b_1 we have the path P_{C1}, and

$$S'_{12} = \frac{P_{C1}\Delta_{C1}}{\Delta} = \frac{L_{12}M_{12}}{(1 - L_{22}M_{11})} \qquad \text{(A6)}$$

The last path is similar to the first and gives the S_{22} of the circuit. The loop L_{D2} does not touch path P_{D1} so $\Delta_{D1} = 1 - L_{22}M_{11}$, but L_{D1} does touch path P_{D2} and $\Delta_{D2} = 1$. Hence:

$$S'_{22} = \frac{P_{D1}\Delta_{D1} + P_{D2}\Delta_{D2}}{\Delta} = \frac{L_{22}(1 - L_{22}M_{11}) + L_{12}M_{22}L_{21}}{(1 - L_{22}M_{11})} = L_{22} + \frac{L_{12}L_{21}M_{22}}{1 - L_{22}M_{11}} \qquad \text{(A7)}$$

AII.1.4 Two-Port – three element cascade

From the circuit in Figure AII.1(d) we identify the loops and paths:

Paths:	a_1 to b_1	$P_{A1} = L_{11}$, $P_{A2} = L_{21}M_{11}L_{12}$, $P_{A3} = L_{21}M_{21}N_{11}M_{12}L_{12}$
	a_1 to b_2	$P_{B1} = L_{21}M_{21}N_{21}$
	a_2 to b_1	$P_{C1} = L_{12}M_{12}N_{12}$
	a_2 to b_2	$P_{D1} = N_{22}$, $P_{D2} = N_{12}M_{22}N_{21}$, $P_{D3} = N_{12}M_{12}L_{22}M_{21}N_{21}$

Loops: First Order $L_1 = L_{22}M_{11}$, $L_2 = M_{22}N_{11}$, $L_3 = L_{22}M_{21}N_{11}M_{12}$
Second Order $L_{21} = L_{22}M_{11}M_{22}N_{11}$
Third Order None

from which we can determine the overall S-parameters of the cascade circuit.

For all paths we have: $\Delta = 1 - L_1 - L_2 - L_3 + L_{21}$.

For the paths from a_1 to b_1 we can identify:

P_{A1} is not touched by any loop, so $\Delta_{A1} = \Delta$.
P_{A2} is touched by L_1 and L_3, so $\Delta_{A2} = 1 - L_2$.
P_{A3} is not touched by all loop, so $\Delta_{A3} = 1$.

Hence the reflection parameter for the cascade is:

$$S''_{11} = \frac{P_{A1}\Delta_{A1} + P_{A2}\Delta_{A2} + P_{A3}\Delta_{A3}}{\Delta} = \frac{P_{A1}\Delta + P_{A2}(1 - L_2) + P_{A3}}{\Delta}$$

$$= P_{A1} + \frac{P_{A2}(1 - L_2) + P_{A3}}{\Delta} = P_{A1} + \frac{P_{A2}(1 - L_2) + P_{A3}}{1 - L_1 - L_2 - L_3 + L_{21}}$$

$$= L_{11} + \frac{L_{21}M_{11}L_{12}(1 - M_{22}N_{11}) + L_{21}M_{21}N_{11}M_{12}L_{12}}{1 - L_{22}M_{11} - M_{22}N_{11} - L_{22}M_{21}N_{11}M_{12} + L_{22}M_{11}M_{22}N_{11}}$$

$$= L_{11} + \frac{L_{21}L_{12}(M_{11}(1 - M_{22}N_{11}) + M_{21}N_{11}M_{12})}{1 - M_{22}N_{11} - L_{22}(M_{11} + M_{21}N_{11}M_{12} - M_{11}M_{22}N_{11})}$$

$$= L_{11} + \frac{L_{21}L_{12}(M_{11}(1 - M_{22}N_{11}) + M_{21}N_{11}M_{12})}{1 - M_{22}N_{11} - L_{22}(M_{11}(1 - M_{22}N_{11}) + M_{21}N_{11}M_{12})}$$

$$= L_{11} + \frac{L_{21}L_{12}\left(M_{11} + \dfrac{M_{21}N_{11}M_{12}}{1 - M_{22}N_{11}}\right)}{1 - L_{22}\left(M_{11} + \dfrac{M_{21}N_{11}M_{12}}{1 - M_{22}N_{11}}\right)} \tag{A8}$$

In the case of the second set of paths from a_1 to b_2 the path P_{B1} is touched by all loops, so $\Delta_{B1} = 1$. The transmission parameter for the cascade is therefore:

$$S''_{21} = \frac{P_{B1}\Delta_{B1}}{\Delta} = \frac{L_{21}M_{21}N_{21}}{1 - L_{22}M_{11} - M_{22}N_{11} - L_{22}M_{21}N_{11}M_{12} + L_{22}M_{11}M_{22}N_{11}}$$

$$= \frac{L_{21}M_{21}N_{21}}{1 - M_{22}N_{11} - L_{22}(M_{11} + M_{21}N_{11}M_{12} - M_{11}M_{22}N_{11})}$$

$$= \frac{L_{21}M_{21}N_{21}}{1 - M_{22}N_{11} - L_{22}(M_{11}(1 - M_{22}N_{11}) + M_{21}N_{11}M_{12})}$$

$$= \frac{\dfrac{L_{21}M_{21}N_{21}}{1 - M_{22}N_{11}}}{1 - L_{22}\left(M_{11} + \dfrac{M_{21}N_{11}M_{12}}{1 - M_{22}N_{11}}\right)} \tag{A9}$$

References

[1] H.W. Bode, *Network Analysis and Feedback Amplifier Design*, Van Nostrand, New York, 1945.
[2] R.M. Fano, 'Theoretical limitations on the broad-band matching of arbitrary impedances', *Journal of the Franklin Institute*, Vol. 249, pp. 57–83, January 1950, and pp. 139–154, February 1950.
[3] I.D. Robertson (ed.), *MMIC Design*, IEE, London, 1995.
[4] CDS Series IV, *Simulating and Testing*, Hewlett Packard, New Jersey, 1995.
[5] CDS Series IV, *Momentum User's Guide*, Hewlett Packard, New Jersey, 1995.
[6] G.D. Vendelin, A.M. Pavio, U.L. Rohde, *Microwave Circuit Design Using Linear and Nonlinear Techniques*, Wiley & Sons, New York, 1992.
[7] K.C. Gupta, R. Garg, R. Chadha, *Computer Aided Design of Microwave Circuits*, Artech House, 1991.
[8] T.C. Edwards, *Foundations for Microstrip Circuit Design*, Wiley & Sons, New York, 1992.
[9] S.R. Pennock, P.R. Shepherd, *Microwave Engineering for Wireless Applications*, Macmillan, Basingstoke, 1998.
[10] G. Kaplan, 'Special guide to software systems, packages and applications', *IEEE Spectrum*, November 1990, pp. 47–101.
[11] C. Bowick, *RF Circuit Design*, H.W. Sams & Co., London, 1982.
[12] D.M. Pozar, *Microwave Engineering*, Reading, MA, Addison-Wesley, 1990.
[13] S.Y. Liao, *Microwave Circuit Analysis and Amplifier Design*, Prentice Hall Inc., Englewood Cliffs, NJ, 1987.
[14] F.A. Benson, T.M. Benson, *Fields Waves and Transmission Lines*, Chapman & Hall, London, 1991.
[15] G.D. Vendelin, A.M. Pavio, U.L. Rohde, *Microwave Circuit Design Using Linear and Nonlinear Techniques*, Wiley & Sons, New York, 1992.

5

Mixers: Theory and Design

L. de la Fuente and A. Tazon

5.1 Introduction

Crystal detectors and mixers are the key circuits in receiver systems. At the start of the twentieth century detectors were very unreliable, they were built using a semiconductor crystal and a thin wire contact that had to be periodically adjusted to guarantee good behaviour. A significant advance in receiver sensitivity was the development of triodes since they made it possible to amplify before and after the detector. However, the real advance was obtained by the invention of the superregenerative receiver by Major Edwin Armstrong. Armstrong was also the first to use the vacuum tube as a frequency converter (mixer) to change the frequency from the received signal (RF) to an intermediate frequency (IF), which could be amplified, selected with low noise and detected. Armstrong is considered to be the inventor of the mixer. This kind of receiver (superheterodyne receiver) has been, up to now, the greatest advance in receiver architecture and it is used in practically all modern receivers.

During the Second World War the development of the microwave mixers took place due to radar development. At the beginning of the 1940s, single-diode mixers had very poor noise figures but by the 1950s, it was possible to obtain noise figures around 8 dB. Nowadays, mixers have this behaviour at frequencies greater than 100 GHz. Mixer theory has been developed and it is possible to reach noise figures of less than 4 dB up to 50 GHz.

Figure 5.1 shows a simplified scheme of a double-mix superheterodyne receiver. The signal received (RF) by the antenna is filtered, amplified by a low noise amplifier and mixed with the frequency of the first local oscillator to obtain the first intermediate frequency (1st IF). This IF is also amplified and filtered and mixed with the frequency of a synthesised oscillator (second local oscillator) to obtain the second intermediate frequency (2nd IF). This 2nd IF (base band) is filtered, amplified and detected to obtain the information.

5.2 General Properties

As we saw in the previous section, mixers are basic circuits in the emitter and receiver systems. A mixer is basically a multiplier, and if we suppose two sinusoidal signals at the multiplier input, we obtain the product of these signals at the output, as shown in Figure 5.2.

Microwave Devices, Circuits and Subsystems for Communications Engineering Edited by I. A. Glover, S. R. Pennock and P. R. Shepherd

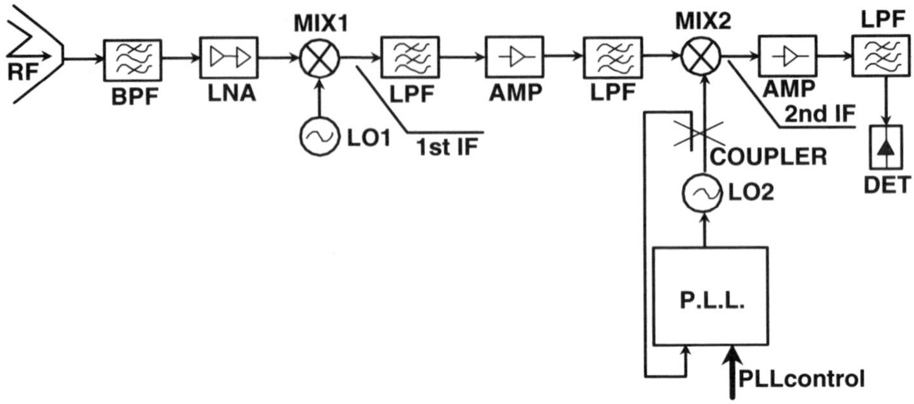

Figure 5.1 Double-mix synthesized receiver

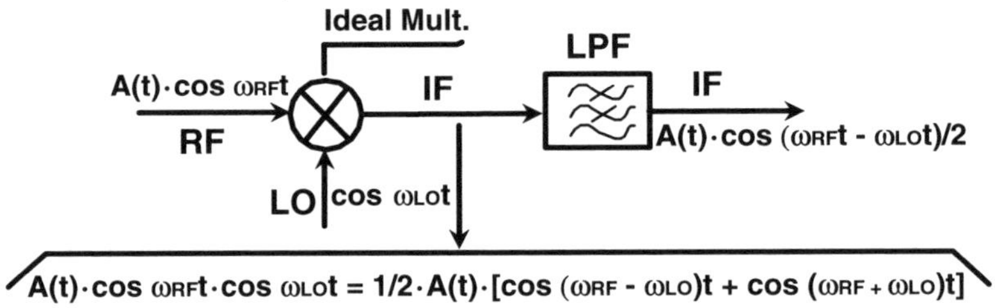

Figure 5.2 Signal components in an ideal mixer

Figure 5.2 shows an ideal mixer. The carrier is the RF signal and it is mixed with the LO signal and the information is transmitted by $A(t)$. The response of the multiplier is two sinusoidal IF components (mixing products). Normally, in communication systems, only one component is desired while the other is rejected by using appropriate filters.

However, electronic devices such as ideal multipliers do not exist, which provokes problems such as generation of superior harmonics and spurious intermodulation signals due to the non-linearity of the mixer. These signals, called spurious responses, must be eliminated later by filtering.

The mixers, even if they are ideal mixers, have a second frequency ($2f_{LO} - f_{RF}$) that can give an IF response. This kind of spurious response is called image frequency. For instance, if we suppose a typical VSAT reception frequency f_{RF} = 13.5 GHz and a frequency of the local oscillator f_{LO} = 12.3 GHz, we obtain an intermediate frequency f_{IF} = 13.5 − 12.3 = 1.2 GHz (typical value of the first IF). However, the non-linearity of the mixer gives the third band intermodulation product $2f_{LO} - f_{RF}$ = 24.6 − 13.5 = 11.1 GHz. The mixer can give, f_{LO} − 11.1 = 1 GHz, a spurious signal whose frequency is the same as IF. It is possible to develop hybrid circuits or combinations of mixers capable of rejecting the image response.

5.3 Devices for Mixers

In theory, any non-linear electronic device can be used as a mixer, however, there are only a few devices that have the practical requirements of mixers in communication systems. A device used as a mixer must satisfy the following characteristics:

- strong non-linearity;
- reliability and low dispersion of the devices;
- low noise;
- low distortion;
- good frequency response.

5.3.1 The Schottky-Barrier Diode

The Schottky-Barrier diode is most commonly used in mixer circuits and it is basically a metal-semiconductor junction. As well as its good behaviour as a mixer, the success of the Schottky-Barrier lies in that it can reach very high frequency (up to 1000 GHz) and it is relatively cheap. In the past, PN-junction diodes, mainly point-contact diodes, were used in mixers. Point-contact diodes are formed by pressing a contact wire onto a piece of semiconductor. This produces a primitive Schottky-Barrier or metal-to-semiconductor junction. However, the building process of modern Schottky-Barrier diodes (photolithography over an epitaxial substrate) gives more reliability than the point-contact diodes. The good characteristics of Gallium Arsenide (GaAs) Schottky-Barrier diodes have led to an important development of millimetre-wave mixers as GaAs has greater electron mobility and saturation velocity than silicon semiconductors. Figure 5.3 shows a Schottky-Barrier diode (a) and its equivalent circuit (b).

5.3.1.1 Non-linear equivalent circuit

The typical non-linear equation of the current source $i(v)$ is:

$$i(v) = I_{ss} \cdot \left[e^{\frac{q \cdot v}{n \cdot K \cdot T}} - 1 \right] \tag{5.1}$$

where I_{ss} is the saturation current, q is the electron charge, K is the Boltzmann's constant, T is the absolute temperature and n is a parameter called ideality factor whose value is between 1.0 and 1.25.

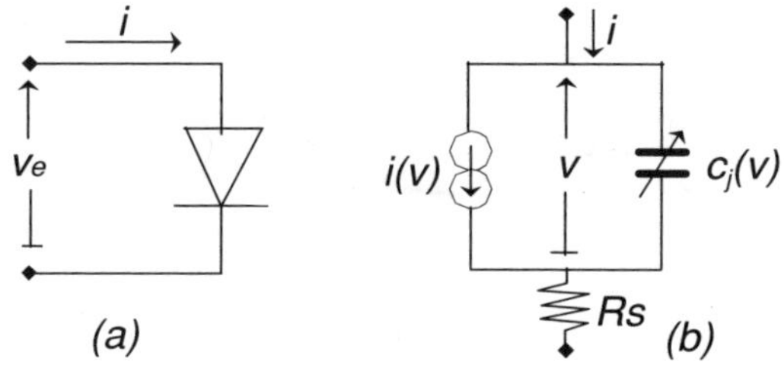

Figure 5.3 Schottky-Barrier diode equivalent circuit

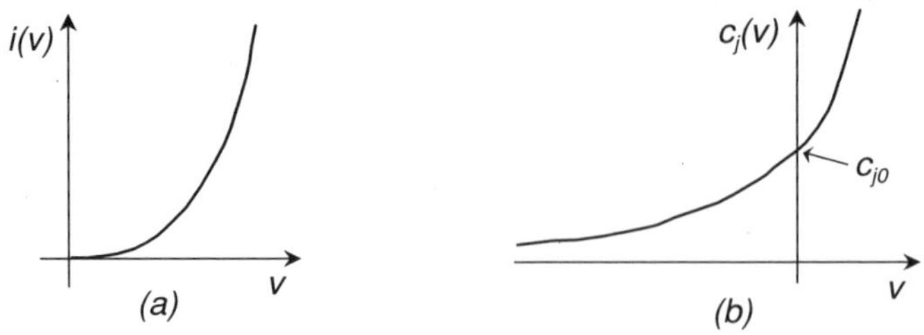

Figure 5.4 (a) Schottky diode current source characteristic
(b) Schottky diode junction capacitance characteristic

The non-linear junction capacitance can be expressed by the equation:

$$C_j = \frac{C_{j0}}{\left(1 - \frac{v}{\phi_{bi}}\right)^{\gamma}} \tag{5.2}$$

where, C_{j0} is the junction capacitance at zero bias voltage, ϕ_{bi} is the built-in voltage and its value typically varies between 0.7 and 0.9 volts, and γ is a parameter that varies between 0.5 and 0.33.

The series resistor of Figure 5.3 R_s is the loss resistance and its value is constant, around a few ohms. Figure 5.4(a) shows the typical curve of the current source $i(v)$ and Figure 5.4(b) shows the characteristic of the junction capacitance $C_j(v)$.

5.3.1.2 Linear equivalent circuit at an operating point

It is possible to deduce the small signal equivalent circuit of the current source (Figure 5.4a) if we suppose that v is a very small sinusoidal voltage around a DC voltage V_0 (operating point) as we can see in Figure 5.5, where the input signal around the operating point (V_0, I_0), is:

$$v = V_0 + V_1 \cdot \cos \omega_0 \cdot t \tag{5.3}$$

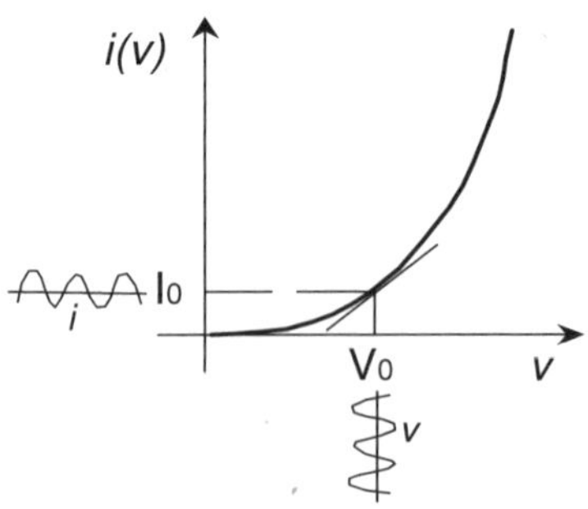

Figure 5.5 Linearization of the Schottky diode $i(v)$ characteristic

The approximated response of the current source $i(v)$ can be obtained by the first term of the Taylor series:

$$i(v) = i(V_0 + V_1 \cdot \cos\omega_0 \cdot t) =$$
$$= i(V_0) + \left.\frac{\partial i}{\partial v}\right|_{V_0} \cdot v + \ldots =$$
$$= I_0 + g(V_0) \cdot v = I_0 + i$$

where:

$$i = g(V_0) \cdot v \tag{5.4}$$

Equation (5.4) is a linear law (Ohm's Law) and the transconductance, taking into account Equation (5.1), can be written as:

$$g(v) = \frac{\partial i}{\partial v} = \frac{q}{n \cdot K \cdot T} \cdot Iss \cdot e^{\frac{q \cdot v}{n \cdot K \cdot T}} = \frac{q}{n \cdot K \cdot T} \cdot [i(v) + Iss] \tag{5.5}$$

The linear representation of the current source $i(v)$ and its linear spectral response can be seen in Figure 5.6.

In the case of the capacitance the problem is different because the non-linear characteristic is $Q(v)$ and the current response is given by Equation (5.6).

$$i(t) = \frac{dQ}{dt} = \frac{\partial Q}{\partial v} \cdot \frac{dv}{dt} = C_j(v) \cdot \frac{dv}{dt} \tag{5.6}$$

Where $C_j(v)$ is given by Equation (5.2).

When sinusoidal excitation of Equation (5.3) is very small, the approximate response of $Q(v)$ can be obtained using the first term of the Taylor series (Equation 5.7):

$$Q(v) = Q[V_0 + V_1 \cdot \cos(\omega_0 t)] = Q(V_0) + C_j(V_0) \cdot V_1 \cdot \cos(\omega_0 t) \tag{5.7}$$

and the approximate linear response is given by Equation (5.8):

$$i(t) = \frac{dQ}{dt} = \frac{d}{dt}[Q(V_0) + C_j(V_0) \cdot V_1 \cdot \cos(\omega_0 t)] = C_j(V_0) \cdot \omega_0 \cdot V_1 \cdot \cos\left(\omega_0 t + \frac{\pi}{2}\right) \tag{5.8}$$

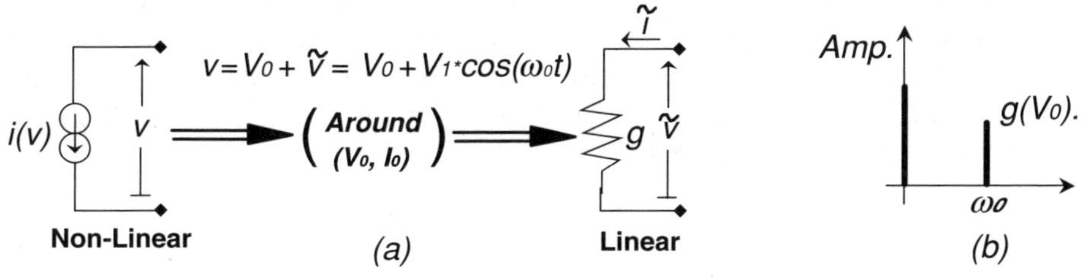

Figure 5.6 (a) Source current equivalent circuit (b) Spectral representation

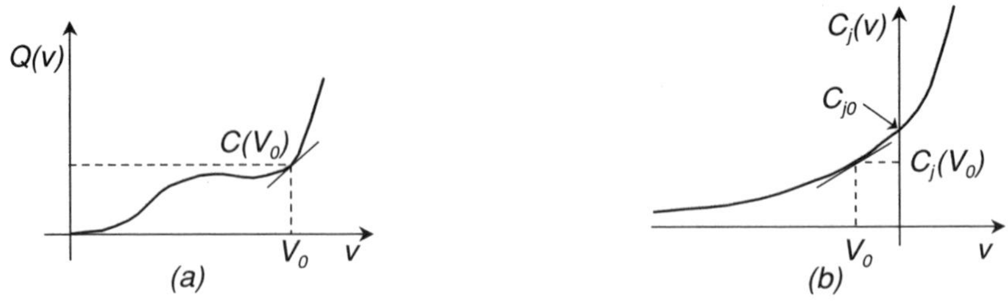

Figure 5.7 (a) Non-linear charge characteristic (b) Linearization of the Schottky capacitance at V_0 DC voltage

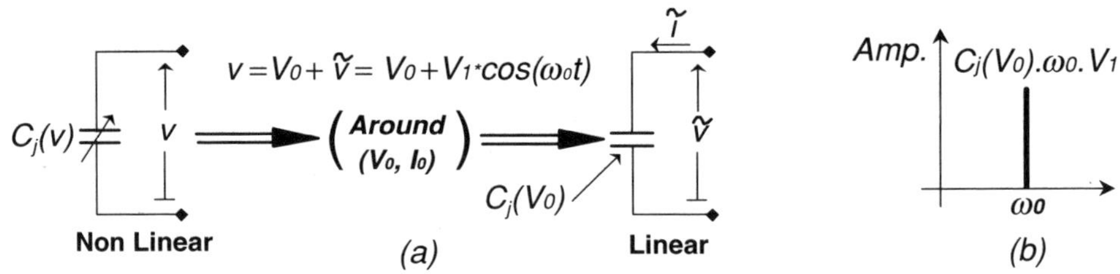

Figure 5.8 (a) Schottky capacitance equivalent circuit (b) Spectral representation

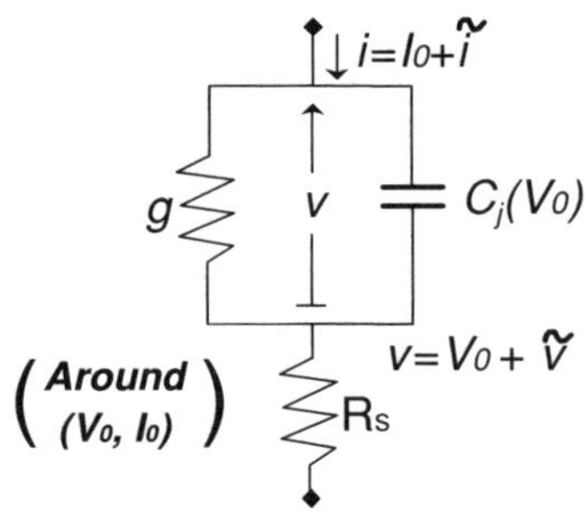

Figure 5.9 Linear equivalent circuit of Schottky diode

As we can see in Equation (5.8), there is no DC term and the amplitude of the ω_0 response is: $C_j(V_0) \cdot \omega_0 \cdot V_1$. Figure 5.7 shows the linearisation of any non-linear charge characteristic and the Schottky junction capacitance. The linearisation of the equivalent circuit of the Schottky capacitance and the spectral representation can be seen in Figure 5.8 where $C_j(V_0)$ is given by Equation (5.9). Therefore, the total linear circuit of a Schottky diode around V_0 operation point is given by Figure 5.9.

$$C_j(V_0) = \frac{C_{j0}}{\left(1 - \dfrac{V_0}{\phi}\right)^{\gamma}} \tag{5.9}$$

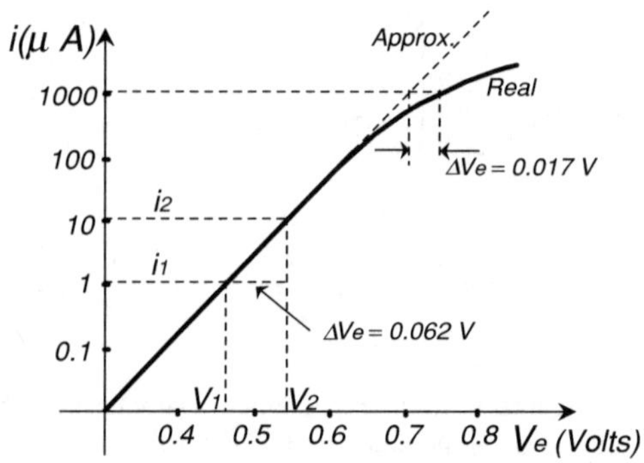

Figure 5.10 Experimental characteristic of a Schottky diode obtained by a curve tracer

5.3.1.3 Experimental characterisation of Schottky diodes

Taking into account the non-linear equivalent circuit of Figure 5.3, we can write the DC diode equation as:

$$v_e = v + i(v) \cdot R_s \tag{5.10}$$

and a semi-logarithmic representation of the diode characteristic obtained by a curve tracer is given in Figure 5.10. In the zone where i is very small but greater than I_{ss} and taking into account Equation (5.1) we can write:

$$\log i \approx \log I_{ss} + \frac{q \cdot v}{n \cdot K \cdot T} \cdot \log(e) \tag{5.11}$$

Considering Equation (5.11) and taking into account the points i_1 and $i_2 = 10 \cdot i_1$ of Figure 5.10, we can write:

$$\log(i_2) - \log(i_1) = 1 = \frac{q \cdot \Delta v_e}{n \cdot K \cdot T} \cdot \log(e) \tag{5.12}$$

and therefore:

$$n = \frac{q \cdot \Delta v_e \cdot \log(e)}{K \cdot T} \cong \frac{\Delta v_e}{0.05783} \tag{5.13}$$

because $\dfrac{K \cdot T}{q} \approx 0.025$ Volts at ambient temperature. In this case, the ideality is $n = 1.07$.

When $i(v) \cdot R_s$ is very small, Equation (5.10) represents the straight part of the diode characteristic of Figure 5.10 and, at any point of this part, we can obtain:

$$I_{ss} = I(v) \cdot e^{-\frac{q \cdot v}{n \cdot K \cdot T}} \tag{5.14}$$

In this case, $I_{ss} = 1.25\ 10^{-13}$ A.

Taking into account the deviation between the real and approximate diode characteristic of Figure 5.10, we can calculate R_s as:

$$v_e = v + R_s \cdot i(v) \tag{5.15}$$

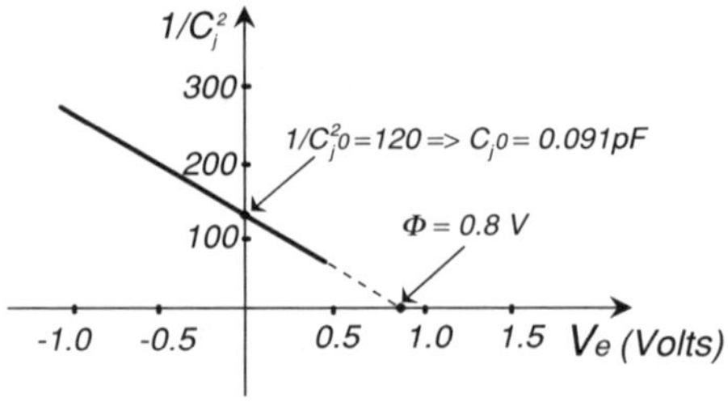

Figure 5.11 Schottky diode capacitance characteristic obtained by an impedance meter at 1 MHz

and

$$R_s = \frac{v_e - v}{i(v)} \tag{5.16}$$

Where $\Delta V_e = v_e - v = 0.017$ V (Figure 5.10) and therefore, $R_s = 17\ \Omega$.

On the other hand, taking into account the non-linear capacitance of the diode equivalent circuit of Figure 5.3, it is possible to obtain, by an impedance bridge meter, the characteristic of Figure 5.11.

From Equation (5.2) and Figure 5.11 we can obtain the Schottky junction capacitance as:

$$\frac{1}{C_j^2} = \frac{1}{C_{j0}^2} \cdot \left[1 - \frac{v}{\phi_{bi}}\right] \tag{5.17}$$

Where $\phi_{bi} = 0.8$ Volts and $C_{j0} = 0.091$ pF, in Figure 5.11.

An important merit figure is the cut-off frequency. If we suppose, in Figure 5.3, a voltage around zero, we can write the diode impedance as:

$$Z_d = R_S - \frac{j}{\omega \cdot C_{j0}} \tag{5.18}$$

the phase relationship is:

$$I = \frac{V}{\left|R_S - \dfrac{j}{\omega \cdot C_{j0}}\right|} \Rightarrow I_{\max} = \frac{V}{R_S} \tag{5.19}$$

As the cut-off frequency is defined at the 3 dB point we can write:

$$\frac{I}{\sqrt{2}} = \frac{V}{\left|R_S - \dfrac{j}{\omega_C \cdot C_{j0}}\right|} = \left\{R_S = \frac{1}{\omega_C \cdot C_{j0}}\right\} = \frac{I_{\max}}{|1 - j|} \Rightarrow f_C = \frac{1}{2 \cdot \pi \cdot R_S \cdot C_{j0}} \tag{5.20}$$

Another important merit figure is the quality factor. Taking into account that the circuit of Figure 5.3 around $v = 0$ volts is a $R_S C_{j0}$ series circuit, we can deduce the quality factor as:

$$Q = \frac{1}{2 \cdot \pi \cdot f_0 \cdot R_S \cdot C_{j0}} = \frac{f_C}{f_0} \tag{5.21}$$

5.3.2 Bipolar Transistors

Bipolar transistors (BJTs) have been used as mixers up to a few GHz but the development of heterojunction bipolar transistors (HBTs) at higher frequencies has increased the interest of these devices in the design of RF and microwave mixers. The bipolar transistors are also known as *homojunction bipolar transistors* but as the initials coincide with the HBT devices, they are called BJTs. BJTs are made in Silicon (Si) technology because it is difficult to control the *p*-type particles in Gallium Arsenide (GaAs) technology. Furthermore, this technology would need a weakly doped *p*-type to obtain high gain and this means a very high base resistance. All these problems lead to GaAs BJTs having worse behaviour than Si devices.

The great importance of the HBT devices in recent years is due to the high transconductance and output resistance values reached by these devices, along with the high power capability and the high breakdown voltages. Moreover, HBTs admit very low base resistance values with wide emitter terminals and these properties allow these transistors to reach very high operation frequencies. Thus, the HBTs are very useful devices for millimetre wave applications with high power levels.

In conclusion, we can say that the Si bipolar transistors (BJT) are used up to a few GHz, they use low cost technology and have high performance. GaAs and SiGe HBTs are used at high frequency. However, the HBT and BJT equivalent circuits and their I/V and Q/V characteristics are similar. The small differences between the two devices are:

- High doping densities of HBTs require less base width than BJTs and the high injection effects practically do not exist.
- Low current region in HBTs is greater than the same region in BJTs, this effect means that β increases monotonously with the collector current.
- The base current is less.

In most applications, the non-linearity base-emitter diode function is employed for mixing applications. Taking into account the generic non-linear model of Chapter 2, under forward bias conditions ($V_{be} \geq 0$ and $V_{bc} < 0$), the device behaviour can be approximated by the simplified circuit of Figure 5.12. This equivalent circuit along with the Gummel-Poon forward bias model permits the deduction of the currents through the collector and base terminals. These currents are given by Equation (5.22a, 5.22b):

$$I_{cc} = I_{sf}\left(\exp\left(\frac{V_{be}}{n_f KT}\right) - 1\right) \tag{5.22a}$$

$$I_{be} = I_{se}\left(\exp\left(\frac{V_{be}}{n_e KT}\right) - 1\right) \tag{5.22b}$$

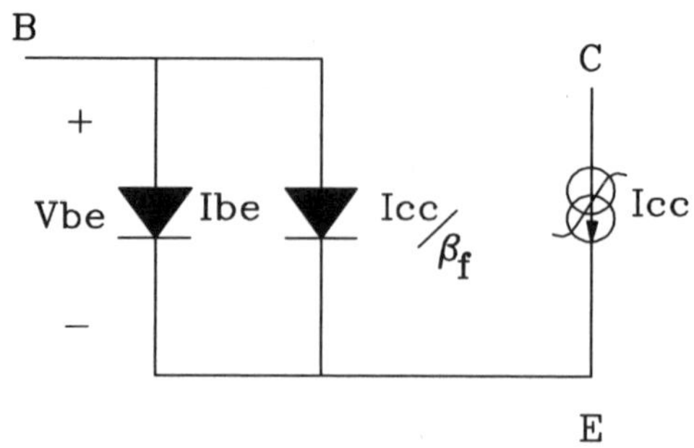

Figure 5.12 DC simplified equivalent circuit under forward bias condition of a HBT

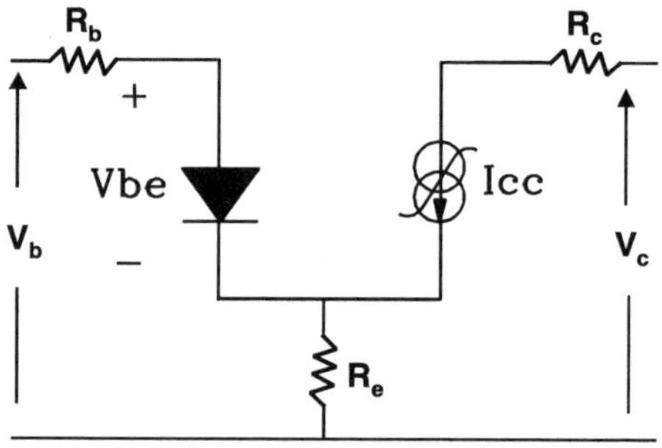

Figure 5.13 One diode simplified circuit under forward bias conditions

where the current source I_{cc} (eq. 5.22a) is the collector current while the base current is represented by two diodes whose current sources are: I_{be} (eq. 5.22b) and I_{cc}/β_f. β_f is the *forward DC gain* of the device.

The equivalent circuit of Figure 5.12 can be simplified by the equivalent circuit of Figure 5.13. In this case, the diode function can be represented by Equation (5.23):

$$I_{beT} = I_{seT}\left(\exp\left(\frac{V_{be}}{n_T KT}\right) - 1\right) \tag{5.23}$$

where:

I_{beT} is the total base-emitter current source.
I_{seT} is the equivalent saturation current.
N_T is the equivalent ideality factor.

The non-linearity base-emitter equivalent diode function is mainly responsible for the mixing function.

At low RF frequencies, the inductive and capacitive effects can be ignored but the access resistances of the HBTs and BJTs (Figure 2.58 of Chapter 2), R_b, R_c and R_e, play an important role in the behaviour of these devices because the experimental I/V and the base-emitter diode function characteristics of the transistor are measured as a function of the external voltages V_c and V_b but the voltages of the Equations (5.22) and (5.23) are functions of the internal (intrinsic) voltages.

These access resistances can be obtained from considerations of the geometrical and material properties or experimental measurements. The relationship between internal and external voltages is given by equations (2.148) and (2.149) of Chapter 2.

At high frequencies, access inductances and linear and non-linear capacitances (Figure 2.58 of Chapter 2) must be taken into account. Traditional extracting methods from scattering parameter measurements at different bias points can be employed using the small signal equivalent circuit of Figure 2.62 (Chapter 2).

5.3.3 Field-Effect Transistors

The *Field-Effect Transistors* (FET) currently play an important role in the RF and microwave circuit development, including mixer circuits for communication system applications. Furthermore, the monolithic technology has increased the importance of these devices.

Basically there are three kinds of field effect devices whose differences of behaviour is the used technology:

1. *Metal Oxide Semiconductor FET* (MOSFET): This is developed in silicon (Si) technology and its main characteristic is its power capabilities, mainly for amplification applications. These transistors find applications in mobile communications where their frequency bands can reach several GHz. These devices are not especially used in mixer design.
2. *MESFET*: Silicon (Si) and Gallium Arsenide (GaAs) technologies are normally used in these type of transistors. Si technology can be used up to several GHz while GaAs technology can reach up to the millimetre bands. MESFET devices are employed in the development of different circuits for communications systems including mixing function, mainly up to Ku band.
3. *HEMT (High Electron Mobility Transistor)*: The most recently developed field effect transistor is the HEMT. These devices are heterostructures in GaAs technology and the main difference with respect to the traditional FET is the transconductance compression. It has very good features of low noise and high gain at high frequencies (Ku, Ka and higher frequencies). HEMT devices are very usual in mixer applications mainly at Ka band and above.

If we analyse the three types of field effect transistors, we can deduce that MESFET and HEMT are the most interesting devices for mixing applications although, at high frequencies, HBTs (up to Ka band in GaAs technology and up to millimetre frequencies in SiGe technology) are a very important alternative. Also, dual-gate FETs are widely used as mixers because these devices have certain advantages with respect to conventional FET mixers:

- local Oscillator (LO) and RF signals can be applied in separated gates with an intrinsic isolation of 20 dB without filters or balanced structures;
- they have very linear transconductance and present low distortion.

Although there are differences between MESFET and HEMT devices, the equivalent circuit topology is basically the same for all types of FETs and the differences are apparent in the values in the equivalent circuit parameters. The most important non-linearity of the FET device for mixing is the channel current source I_{ds} (Figure 2.52 in Chapter 2). Another important non-linearity is the gate-to-source capacitance C_{ds} but its main effect is the limitation of useful bandwidth. The rest of the resistive and reactive parasitic elements of the FET equivalent circuit are of secondary interest in mixing applications.

The behaviour of the current source I_{ds} when we introduce two RF signals is similar to a near ideal multiplier as we saw in Section 5.2 (Figure 5.2). The Curtice I_{ds} non-linear current equation is given by:

$$I_{ds}(V_{gi}, V_{di}) = \beta(V_{gi} - V_{TO})^2(1 + \lambda V_{di}) \tanh(\alpha V_{di}) \tag{5.24}$$

where two signals (LO and RF) have been applied at the saturation region at any DC operating point, the RF behaviour of Equation (5.26) can be approximated by:

$$I_{ds} = \beta \cdot (v_{gi} - V_{TO})^2 \tag{5.25}$$

In this case:

$$v_{gi} = v_{RF} + v_{LO}$$

where:

$$v_{RF} = A \cdot \cos(\omega_{RF} \cdot t) \quad \text{RF signal}$$

$$v_{LO} = B \cdot \cos(\omega_{OL} \cdot t) \quad \text{LO signal}$$

and the current source of Equation (5.27) is given by:

$$\begin{aligned} I_{ds} = \beta \cdot [A' \cos(\omega_{RF} \cdot t) + B' \cdot \cos(\omega_{OL} \cdot t) - V_{TO}]^2 = a_0 + a_1 \cdot \cos(\omega_{RF} \cdot t) + \\ + a_2 \cdot \cos(\omega_{OL} \cdot t) + a_3 \cdot \cos(2 \cdot \omega_{RF} \cdot t) + a_4 \cdot \cos(2 \cdot \omega_{OL} \cdot t) + \\ + a_5 \cdot \cos(\omega_{RF} \cdot t) \cdot \cos(\omega_{OL} \cdot t) \end{aligned} \tag{5.26}$$

The last term of Equation (5.26) contains the product function that represents the sum or difference frequencies (mixing function). The rest of the terms can be eliminated by filtering.

5.4 Non-Linear Analysis

As we saw in Figure 5.2, a mixer is a multiplier and the non-linear devices used for this operation were studied in Section 5.3. The common mixing behaviour is basically frequency conversion in a time-varying parameter of a non-linear device. This parameter appears by applying a strong waveform to a non-linear device. In this part we will assume a general non-linear characteristic to calculate the intermodulation products when we apply two sinusoidal voltages. One of them, the LO, is stronger than the other one, the RF signal.

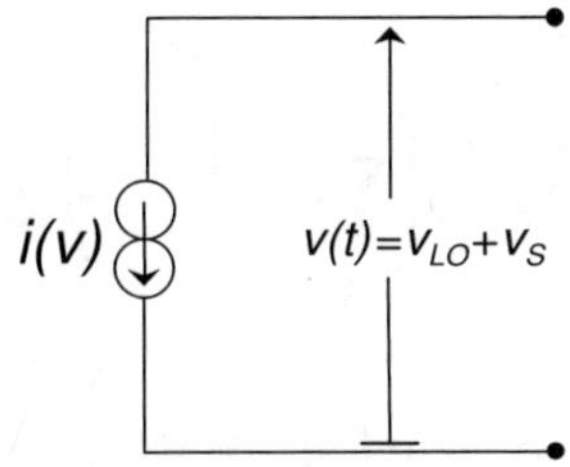

Figure 5.14 Non-linear characteristic

5.4.1 Intermodulation Products

The characteristic $i(v)$ of the non-linear device in Figure 5.14 can be described by:

$$i(v) = a_0 + a_1 \cdot v + a_2 \cdot v^2 + \ldots = \sum_{k=0}^{\infty} a_k \cdot v^k \tag{5.27}$$

The input signal at the circuit of Figure 5.14 is given by $v(t) = v_{LO}(t) + v_S(t)$, where the local oscillator signal (LO) is:

$$v_{LO}(t) = V_P \cdot \cos(\omega_P \cdot t + \phi_P) \tag{5.28}$$

and the RF signal:

$$v_S(t) = V_S \cdot \cos(\omega_S \cdot t + \phi_S) \tag{5.29}$$

To simplify, we suppose that $\phi_P = \phi_S = 0$. This does not mean a loss in generality of the study. In this case, the input signal is given by:

$$v(t) = v_{LO}(t) + v_S(t) = V_P \cdot \cos(\omega_P \cdot t) + V_S \cdot (\omega_S \cdot t) \tag{5.30}$$

If we suppose that $V_p \ll V_S$ and developing in Taylor series around v_{LO} the characteristic $i(v)$ can be written as:

$$i(v) = i(v_{LO} + v_S) = i(v_{LO}) + \frac{1}{1!} \cdot \left(\frac{di}{dv}\right)_{v_S=0} \cdot v_S + \frac{1}{2!} \cdot \left(\frac{d^2i}{dv^2}\right)_{v_S=0} \cdot v_S^2 + \ldots =$$

$$= i(v_{LO}) + V_S \cdot \cos(\omega_S \cdot t) \cdot g^{(1)}(v_{LO}) + \frac{V_S^2}{2} \cdot \cos^2(\omega_S \cdot t) \cdot g^{(2)}(v_{LO}) + \ldots \tag{5.31}$$

where:

$$g^{(0)}(v) = i(v) \quad g^{(1)}(v) = \frac{di}{dv} \quad g^{(2)}(v) = \frac{d^2i}{dv^2} \ldots g^{(n)}(v) = \frac{d^ni}{dv^n} \tag{5.32}$$

so:

$$i(v) = \sum_{n=0}^{\infty} \frac{V_S^n}{n!} \cdot \cos^n(\omega_S \cdot t) \cdot g^{(n)}(v_{LO}) \tag{5.33}$$

where:

$$g^{(n)}(v_{LO}) = \left.\frac{d^n i}{dv^n}\right|_{v_{LO}} = \frac{d^n}{dv^n}\left[\sum_{k=0}^{\infty} a_k v^k\right]_{v=v_{LO}} = \sum_{k=n}^{\infty} \frac{k!}{(k-n)!} \cdot a_k \cdot v_{LO}^{k-n} \qquad (5.34)$$

$$[v_{LO} = V_P \cdot \cos(\omega_P \cdot t)]$$

Applying the variable change $p = k - n$ in Expression (5.34), we can obtain the conductance $g^{(n)}$ as:

$$g^{(n)}(v_{LO}) = \left.\sum_{p=0}^{\infty} \frac{(p+n)!}{p!} \cdot a_{p+n} \cdot v_{LO}^{p}\right|_{v_{LO}=V_P\cdot\cos(\omega_P \cdot t)} =$$

$$= \sum_{p=0}^{\infty} \frac{(p+n)!}{p!} \cdot a_{p+n} \cdot v_P^p \cdot \cos^p(\omega_P \cdot t) \qquad (5.35)$$

From Expression (5.35) we can express the output characteristic $i(v)$ of Figure 5.14 by the equation:

$$i(t) = \sum_{n=0}^{\infty}\left[\frac{V_S^n}{n!} \cdot \cos^n(\omega_S \cdot t) \cdot \sum_{p=0}^{\infty} \frac{(p+n)!}{p!} \cdot a_{p+n} \cdot V_P^p \cdot \cos^p(\omega_P \cdot t)\right] =$$

$$= \sum_{n=0}^{\infty}\sum_{p=0}^{\infty}\left[\frac{(p+n)!}{p! \cdot n!} \cdot V_S^n \cdot V_P^p \cdot a_{p+n} \cdot \cos^n(\omega_S \cdot t) \cdot \cos^p(\omega_P \cdot t)\right] \qquad (5.36)$$

Taking into account the mathematical development of the $\cos^k(x)$ given in Appendix I, it is possible to write the current expression as:

$$i(t) = \sum_{n=0}^{\infty}\sum_{p=0}^{\infty} \frac{(p+n)!}{p! \cdot n!} \cdot V_S^n \cdot V_P^p \cdot a_{p+n} \cdot \left[\frac{1}{2^{n-1}} \cdot \left(\sum_{c=0}^{C} \frac{n!}{(n-c)! \cdot c!} \cdot \cos(n-2c) \cdot \omega_S \cdot t + b_n\right)\right] \cdot$$

$$\cdot \left[\frac{1}{2^{p-1}} \cdot \left(\sum_{d=0}^{D} \frac{p!}{(p-d)! \cdot d!} \cdot \cos(p-2d) \cdot \omega_p \cdot t + b_p\right)\right] \qquad (5.37)$$

where:

$$C = \begin{cases} \dfrac{n-2}{2} & \text{for } n \text{ even} \quad \text{and} \quad n \neq 0 \\ \dfrac{n-1}{2} & \text{for } n \text{ odd} \end{cases} \qquad \text{and} \quad b_n = \begin{cases} \dfrac{1}{2} \cdot \dfrac{n!}{(n/2!)^2} & \text{for } n \text{ even} \quad \text{and} \quad n \neq 0 \\ 0 & \text{for } n \text{ odd} \end{cases}$$

$$D = \begin{cases} \dfrac{p-2}{2} & \text{for } p \text{ even} \quad \text{and} \quad p \neq 0 \\ \dfrac{p-1}{2} & \text{for } p \text{ odd} \end{cases} \qquad \text{and} \quad b_p = \begin{cases} \dfrac{1}{2} \cdot \dfrac{p!}{(p/2!)^2} & \text{for } p \text{ even} \quad \text{and} \quad p \neq 0 \\ 0 & \text{for } p \text{ odd} \end{cases}$$

Operating on Equation (5.37), we can write:

$$i(t) = \sum_{n=0}^{\infty}\sum_{p=0}^{\infty}\frac{(p+n)!}{p!\cdot n!}\cdot V_S^n\cdot V_P^p\cdot\frac{a_{n+p}}{2^{n+p-2}}\cdot$$

$$\cdot\left\{\sum_{c=0}^{C}\sum_{d=0}^{D}\frac{n!}{(n-c)!\cdot c!}\cdot\frac{p!}{(p-d)!\cdot d!}\cdot\frac{1}{2}\cdot\cos[(\alpha\cdot\omega_S\pm\beta\cdot\omega_P)\cdot t]+\right. \tag{5.38}$$

$$\left. b_n\cdot\sum_{d=0}^{D}\frac{p!}{(p-d)!\cdot d!}\cdot\cos(\beta\cdot\omega_P\cdot t)+b_p\cdot\sum_{c=0}^{C}\frac{n!}{(n-c)!\cdot c!}\cdot\cos(\alpha\cdot\omega_S\cdot t)+b_n\cdot b_p\right\}$$

where $\alpha = n - 2c$ and $\beta = p - 2d$ and

$$\cos(\alpha\cdot\omega_S\cdot t)\cdot\cos(\beta\cdot\omega_P\cdot t) = \frac{1}{2}\cdot[\cos(\alpha\cdot\omega_S+\beta\cdot\omega_P)\cdot t+\cos(\alpha\cdot\omega_S-\beta\cdot\omega_P)\cdot t] =$$

$$= \frac{1}{2}\cdot\cos(\alpha\cdot\omega_S\pm\beta\cdot\omega_P)\cdot t$$

Observing Equation (5.38) we can obtain the following conclusions:

(a) α and β are equal to or greater than zero.
(b) $\alpha \neq 0$ and $\beta \neq 0$ at the first addend.
(c) $\alpha = 0$ and $\beta \neq 0$ at the second addend.
(d) $\alpha \neq 0$ and $\beta = 0$ at the third addend.
(e) The fourth addend $b_n \cdot b_p$ is the term where $\alpha = 0$ and $\beta = 0$.

Taking into account the expressions of b_n and b_p, given by Appendix I, the Equation (5.38) can be written as:

$$i(t) = \sum_{n=0}^{\infty}\sum_{p=0}^{\infty}\sum_{c=0}^{C}\sum_{d=0}^{D}\frac{(p+n)!}{(n-c)!\cdot c!\cdot(p-d)!\cdot d!}\cdot V_S^n\cdot V_P^p\cdot\frac{a_{n+p}}{2^{n+p-1}}\cdot\cos[(\alpha\cdot\omega_S\pm\beta\cdot\omega_P)\cdot t]+$$

$$+\sum_{n=0(n\ even)}^{\infty}\sum_{p=0}^{\infty}\sum_{d=0}^{D}\frac{(p+n)!}{(n/2!)^2(p-d)!\cdot d!}\cdot V_S^n\cdot V_P^p\cdot\frac{a_{n+p}}{2^{n+p-1}}\cdot\cos(\beta\cdot\omega_P\cdot t)+$$

$$+\sum_{n=0}^{\infty}\sum_{p=0(p\ even)}^{\infty}\sum_{c=0}^{C}\frac{(p+n)!}{(p/2!)^2(n-c)!\cdot c!}\cdot V_S^n\cdot V_P^p\cdot\frac{a_{n+p}}{2^{n+p-1}}\cdot\cos(\alpha\cdot\omega_S\cdot t)+$$

$$+\sum_{n=0(n\ even)}^{\infty}\sum_{p=0(p\ even)}^{\infty}\frac{(p+n)!}{(p/2!)^2\cdot(n/2!)^2}\cdot V_S^n\cdot V_P^p\cdot\frac{a_{n+p}}{2^{n+p}} \tag{5.39}$$

where the change $\sum_{k=0}^{\infty}\alpha_k$ $(k = 0, 2, 4, \ldots) = \sum_{k=0}^{\infty}\alpha_{2k}$ $(k = 0, 1, 2, \ldots)$ has been made in the last three terms of (5.39).

Next we will perform a variable change in the different terms of expression (5.40):

$$\begin{aligned} \alpha = n - 2c = x \Rightarrow n = 2c + x \\ \beta = p - 2d = y \Rightarrow p = 2d + y \end{aligned} \tag{5.40}$$

First term: $x \neq 0$ and $y \neq 0$

$$I_{xy} = \sum_{c=0}^{\infty}\sum_{d=0}^{\infty} \frac{(2c + 2d + x + y)!}{(x + c)! \cdot c! \cdot (d + y)! \cdot d!} \cdot \frac{V_S^{2c+x} \cdot V_P^{2d+y}}{2^{2c+2d+x+y-1}} \cdot a_{2c+2d+x+y} \cdot \cos(x \cdot \omega_S \pm y \cdot \omega_P) \cdot t \tag{5.41}$$

Second term: $x = 0$ and $y \neq 0$

$$I_{0y} = \sum_{n=0}^{\infty}\sum_{d=0}^{\infty} \frac{(2n + 2d + y)!}{(n!)^2 \cdot (d + y)! \cdot d!} \cdot \frac{V_S^{2n} \cdot V_P^{2d+y}}{2^{2n+2d+y-1}} \cdot a_{2n+2d+y} \cdot \cos(y \cdot \omega_P \cdot t) \tag{5.42}$$

The first summation is a complete sweep of the variable n and therefore, we can do $n \equiv c$ without losing the generalization capabilities of Equation (5.42). In this case $I_{0y} \equiv I_{xy}|_{x=0}$ obtained from the first term (Eq. (5.41)).

Third term: $x \neq 0$ and $y = 0$. This term is obtained like the second one. In this case $I_{x0} \equiv I_{xy}|_{y=0}$ obtained from the first term (Eq. (5.41)).

To calculate the last (DC) term we perform $n = c$ and $p = d$, so:

$$I_{00} = \sum_{c=0}^{\infty}\sum_{p=0}^{\infty} \frac{(2c + 2d)!}{(c!)^2 \cdot (d!)^2} \cdot \frac{V_S^{2c} \cdot V_P^{2d}}{2^{2c+2d}} \cdot a_{2c+2d} \tag{5.43}$$

We can observe that $I_{00} = \dfrac{1}{2} \cdot I_{xy}\Big|_{\substack{x=0 \\ y=0}}$.

The frequency components are given by

$$(n - 2c) \cdot \omega_S \pm (p - 2d) \cdot \omega_P \Rightarrow x \cdot \omega_S \pm y \cdot \omega_P$$

where

$$\begin{cases} x = 0, 1, 2 \ldots \\ y = 0, 1, 2 \ldots \end{cases} \text{ and } (x, y) \text{ cannot be zero simultaneously.}$$

The harmonic content and intermodulation products are:

$$\begin{matrix} \omega_S & \omega_P & \omega_S \pm \omega_P & 2\omega_S \pm \omega_P & \ldots \\ 2\omega_S & 2\omega_P & \omega_S \pm 2\omega_P & 2\omega_S \pm 2\omega_P & \ldots \\ 3\omega_S & 3\omega_P & \omega_S \pm 3\omega_P & 2\omega_S \pm 3\omega_P & \ldots \\ \vdots & \vdots & \vdots & \vdots & \end{matrix}$$

5.4.2 Application to the Schottky-Barrier Diode

In the diode of Figure 5.3, we will only consider the current source $i(v)$ which is given by:

$$i(v) = I_{SS} \cdot \left[e^{\alpha(v+V_0)} - 1\right] \tag{5.44}$$

where α is a diode parameter, v is time-varying voltage, V_0 is the DC voltage and I_{SS} is the saturation current. Taking into account the basic non-linear characteristic given in Equation (5.27), we will deduce the relationship between Equations (5.44) and (5.27). Developing Equation (5.44) in a Taylor series around V_0, we have:

$$i(v) = i(V_0) + \left.\frac{\partial i}{\partial v}\right|_{v=0} \cdot v + \frac{1}{2!} \cdot \left.\frac{\partial^2 i}{\partial v^2}\right|_{v=0} \cdot v^2 + \ldots \tag{5.45}$$

Using Equation (5.44), Equation (5.45) can be written as:

$$i(v) = I_{DC} + I_{SS} \cdot e^{\alpha \cdot V_0} \cdot \frac{\alpha}{1!} \cdot v + I_{SS} \cdot e^{\alpha \cdot V_0} \cdot \frac{\alpha^2}{2!} \cdot v^2 + \ldots \tag{5.46}$$

and comparing Equations (5.27) and (5.46), we can express the coefficients as:

$$\begin{cases} a_k = I_{SS} \cdot e^{\alpha \cdot V_0} \cdot \dfrac{\alpha^k}{k!} & \textit{for} \quad k \neq 0 \\ a_k = a_0 = I_{DC} = I_{SS} \cdot [e^{\alpha \cdot V_0} - 1] & \textit{for} \quad k = 0 \end{cases} \tag{5.47}$$

If we perform the variable change $a_k = a_{2c+2d+x+y}$ where $k = 2c + 2d + x + y$, we can write Equation (5.41) as:

$$I_{xy} = I_{SS} \cdot e^{\alpha \cdot V_0} \sum_{c=0}^{\infty} \sum_{d=0}^{\infty} \frac{(2c + 2d + x + y)!}{(x + c)! \cdot c! \cdot (d + y)! \cdot d!} \cdot \frac{\alpha^{(2c+2d+x+y)}}{(2c + 2d + x + y)!} \cdot \frac{V_S^{2c+x} \cdot V_P^{2d+y}}{2^{2c+2d+x+y-1}} \cdot \cos(x \cdot \omega_S \pm y \cdot \omega_P) \cdot t \tag{5.48}$$

Taking into account that the expressions:

$$I_x(\alpha \cdot V_S) \equiv \sum_{c=0}^{\infty} \frac{(\alpha \cdot V_S/2)^{2c+x}}{(c + x)! \cdot c!} \qquad I_y(\alpha \cdot V_P) \equiv \sum_{d=0}^{\infty} \frac{(\alpha \cdot V_P/2)^{2d+y}}{(d + y)! \cdot d!}$$

are the first class modified Bessel functions (Appendix II), Equation (5.48) can be expressed by:

$$I_{xy} = 2 \cdot I_{SS} \cdot e^{\alpha \cdot V_0} \cdot I_x(\alpha \cdot V_S) \cdot I(\alpha \cdot V_P) \cdot \cos(x \cdot \omega_S \pm y \cdot \omega_P) \cdot t \tag{5.49}$$

5.4.3 Intermodulation Power

Figure 5.15 shows a simplified scheme of a single diode mixer. The power P_{xy} of the intermodulation product (xy) can be expressed by:

$$P_{xy} = R_i \cdot \overline{I_{xy}^2} \tag{5.50}$$

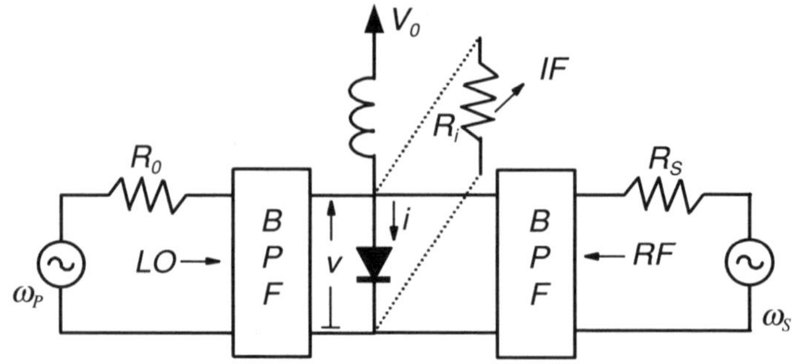

Figure 5.15 Simplified mixer scheme

where $\overline{I_{xy}^2}$ is the root mean square value given by Equation (5.51).

$$\overline{I_{xy}^2} = \frac{1}{2} \cdot [2 \cdot I_{SS} \cdot e^{\alpha \cdot V_0} \cdot I_x(\alpha \cdot V_S) \cdot I_y(\alpha \cdot V_P)]^2 \tag{5.51}$$

Taking into account Equations (5.50) and (5.51), the intermodulation power can be written as:

$$P_{xy} = 2 \cdot R_i \cdot (I_{SS} \cdot e^{\alpha \cdot V_0})^2 \cdot I_x^2(\alpha \cdot V_S) \cdot I_y^2(\alpha \cdot V_P) \tag{5.52}$$

The normal situation of the mixers is that the RF power is very small, in this case $\alpha \cdot V_S \ll 1$ and $I_x(\alpha \cdot V_S)$ can be approximated by the first term of the Bessel function (Appendix II), in this case:

$$I_x(\alpha \cdot V_S) \approx \frac{\alpha^x \cdot V_S^x}{2^x \cdot x!} \tag{5.53}$$

Therefore, Equation (5.52) can be written as:

$$P_{xy} = 2 \cdot R_i \cdot (I_{SS} \cdot e^{\alpha \cdot V_0})^2 \cdot \frac{\alpha^{2 \cdot x} \cdot V_S^{2 \cdot x}}{2^{2 \cdot x} \cdot (x!)^2} \cdot I_y^2(\alpha \cdot V_P) \tag{5.54}$$

where x and y are the RF and LO harmonics respectively and the harmonic frequency is $x \cdot \omega_S \pm y \cdot \omega_P$.

Equation (5.54) is only valid given sinusoidal voltage at the diode and given that P_{xy} decreases when the order of x, y grows independently of V_S and V_P. On the other hand, the mixer characteristics are a function of the LO amplitude.

As an example, supposing a sinusoidal voltage at the diode for a given LO value, the harmonic power from Equation (5.54) can be written as:

$$P_{xy} = \left[\frac{\alpha \cdot I_{SS} \cdot R_i \cdot e^{\alpha \cdot V_0} \cdot I_y(\alpha \cdot V_P)}{x!}\right]^2 \cdot \alpha^{2 \cdot x - 2} \cdot \left(\frac{R_i}{2}\right)^{x-1} \cdot P_1^x \tag{5.55}$$

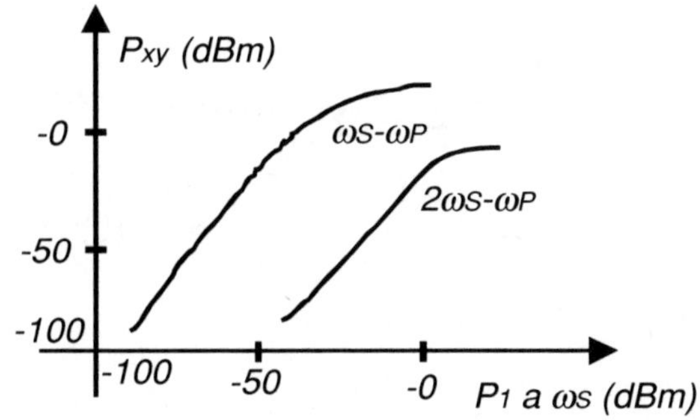

Figure 5.16 Example input power versus output power for two intermodulation products of a mixer

where P_1 is the input power of the RF frequency given by:

$$P_1^x = \left(\frac{1}{2} \cdot \frac{V_S^2}{R_i}\right)^x \tag{5.56}$$

Where V_S is the amplitude of the RF signal. Figure 5.16 shows the harmonic power behaviour of Equation (5.55).

5.4.4 Linear Approximation

Taking into account the characteristic of Expression (5.27) and $v_{LO}(t)$ and $v_S(t)$ given by (5.28) and (5.29), the input signal in the circuit of Figure 5.14 is given by Equation (5.30). If we suppose that $V_p \ll V_S$ (Figure 5.17), the first order of the Taylor series around v_{LO} of the characteristic $i(v)$, for a DC voltage V_0 (operating point), can be written as

$$i(v) = i(v_{LO} + v_S) \approx i(v_{LO}) + \frac{1}{1!} \cdot \left(\frac{di}{dv}\right)_{v_S=0} \cdot v_S = i(v_{LO}) + V_S \cdot \cos(\omega_S \cdot t) \cdot g^{(1)}(v_{LO}) \tag{5.57}$$

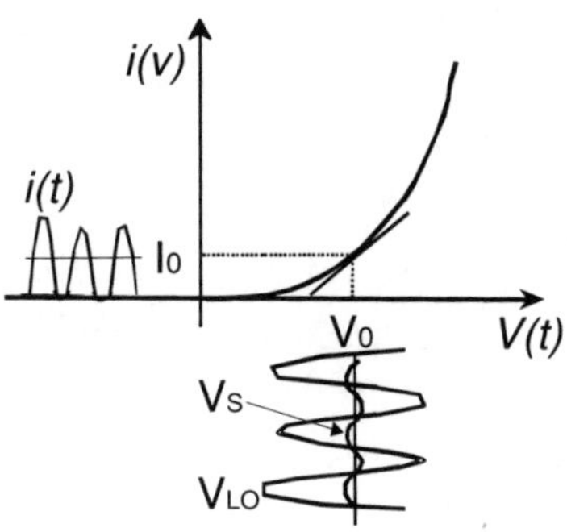

Figure 5.17 Linear approximation

where

$$i(v_{LO}) = \sum_{k=0}^{\infty} V_P^k \cdot a_k \cdot \cos^k(\omega_P \cdot t) \tag{5.58a}$$

$$g^{(1)}(v_{LO}) = \sum_{k=1}^{\infty} \frac{k!}{(k-1)!} \cdot a_k \cdot V_P^{k-1} \cdot \cos^{k-1}(\omega_P \cdot t) \tag{5.58b}$$

Putting Equations (5.58) into (5.57), we can write:

$$i(v) = \sum_{k=0}^{\infty} \alpha_k \cdot V_P^k \cdot \cos^k(\omega_P \cdot t) + V_S \cdot \cos(\omega_S \cdot t) \cdot \left[\sum_{k=1}^{\infty} \frac{k!}{(k-1)!} \cdot a_k \cdot V_P^{k-1} \cdot \cos^{k-1}(\omega_P \cdot t) \right] \tag{5.59}$$

It is important to observe that the DC contribution to the current I_0 is not written in Expression (5.59). I_0 is the value of the current at the operating point V_0. Developing Equation (5.59), we can obtain:

$$i(t) = a_0 + a_1 \cdot V_P \cdot \cos(\omega_P \cdot t) + a_1 \cdot V_S \cdot \cos(\omega_S \cdot t) + 2 \cdot a_2 \cdot V_S \cdot V_P \cdot \cos(\omega_S \cdot t) \cdot \cos(\omega_P \cdot t) +$$
$$+ \sum_{n=2}^{\infty} a_n \cdot V_P^n \cdot \cos^n(\omega_P \cdot t) + V_S \cdot \cos(\omega_S \cdot t) \cdot \sum_{n=2}^{\infty} \frac{(n+1)!}{n!} \cdot a_{n+1} \cdot V_P^n \cdot \cos^n(\omega_P \cdot t) \tag{5.60}$$

Looking at Appendix I, the penultimate term of Equation (5.60) can be expressed as follows:

$$\sum_{n=2}^{\infty} a_n \cdot V_P^n \cdot \cos^n(\omega_P \cdot t) = \sum_{n=2}^{\infty} a_n \cdot V_P^n \cdot \left[\frac{1}{2^{n-1}} \cdot \sum_{c=0}^{C} \frac{n!}{(n-c)! \cdot c!} \cdot \cos[(n - 2 \cdot c) \cdot \omega_P \cdot t \right] +$$
$$+ \frac{1}{2^n} \cdot \frac{n!}{\frac{n}{2}!} \Bigg|_{n\ even} \quad where \quad C = \begin{cases} \dfrac{n-2}{2} & for\ n\ even \\ \dfrac{n-1}{2} & for\ n\ odd \end{cases} \tag{5.61}$$

and the last term of Equation (5.60) will be:

$$\sum_{n=2}^{\infty} \frac{(n+1)!}{n!} \cdot a_{n+1} \cdot V_P^n \cdot \cos^n(\omega_P \cdot t) =$$
$$= \sum_{n=2}^{\infty} \sum_{c=0}^{C} a_{n+1} \cdot V_P^n \cdot \frac{(n+1)!}{n!} \cdot \frac{n!}{(n-c)! \cdot c!} \cdot \frac{1}{2^{n-1}} \cdot \cos[(n - 2 \cdot c) \cdot \omega_P \cdot t] +$$
$$+ \sum_{n=2\ (n\ even)}^{\infty} a_{n+1} \cdot V_P^n \cdot \frac{(n+1)!}{n!} \cdot \frac{1}{2^n} \cdot \frac{n!}{(n/2)!^2} \sum_{c=0}^{C} \dots \cos[(n - 2 \cdot c) \cdot \omega_P \cdot t]$$
$$where \quad C = \begin{cases} \dfrac{n-2}{2} & for\ n\ even \\ \dfrac{n-1}{2} & for\ n\ odd \end{cases} \tag{5.62}$$

Considering Equations (5.61) and (5.62), Equation (5.60) can be written as:

$$i(t) = a_0 + a_1 \cdot V_P \cdot \cos(\omega_P \cdot t) + a_1 \cdot V_S \cdot \cos(\omega_S \cdot t) + 2 \cdot a_2 \cdot V_S \cdot V_P \cdot \cos(\omega_S \cdot t) \cdot \cos(\omega_P \cdot t) +$$

$$+ \sum_{\substack{n=2 \\ n\ even}}^{\infty} \frac{n!}{(n/2)!} \cdot \frac{1}{2^n} a_n \cdot V_P^n + \left[\sum_{\substack{n=2 \\ n\ even}}^{\infty} \frac{(n+1)!}{(n/2)^2!} \cdot \frac{1}{2^n} a_{n+1} \cdot V_P^n \right] \cdot V_S \cdot \cos(\omega_S \cdot t) + \tag{5.63}$$

$$+ \sum_{n=2}^{\infty} \sum_{c=0}^{C} a_n \cdot V_P^n \cdot \frac{n!}{(n-c)! \cdot c!} \cdot \frac{1}{2^{n-1}} \cdot \cos[(n - 2 \cdot c) \cdot \omega_P \cdot t] + \quad \leftarrow \text{Harmonics of } \omega_P$$

$$+ \sum_{n=2}^{\infty} \sum_{c=0}^{C} a_{n+1} \cdot V_P^n \cdot V_S \cdot \frac{(n+1)!}{(n-c)! \cdot c!} \cdot \frac{1}{2^n} \cdot \cos[(n - 2 \cdot c) \cdot \omega_P \cdot t \pm \omega_S \cdot t] \quad \leftarrow \text{Intermodulation}$$

Harmonics of the local oscillator $[(n - 2c)\omega_P]$:

$n = 2$	$c = 0$	$2 \cdot \omega_P$		
$n = 3$	$c = 0, 1$	$3 \cdot \omega_P$	ω_P	
$n = 4$	$c = 0, 1$	$4 \cdot \omega_P$	$2 \cdot \omega_P$	
$n = 5$	$c = 0, 1, 2$	$5 \cdot \omega_P$	$3 \cdot \omega_P$	ω_P
$n = 6$	$c = 0, 1, 2$	$6 \cdot \omega_P$	$4 \cdot \omega_P$	$2 \cdot \omega_P$

Intermodulation products $[(n - 2c)\omega_P \pm \omega_S]$:

$n = 2$	$c = 0$	$\omega_S \pm 2 \cdot \omega_P$		
$n = 3$	$c = 0, 1$	$\omega_S \pm 3 \cdot \omega_P$	$\omega_S \pm \omega_P$	
$n = 4$	$c = 0, 1$	$\omega_S \pm 4 \cdot \omega_P$	$\omega_S \pm 2 \cdot \omega_P$	
$n = 5$	$c = 0, 1, 2$	$\omega_S \pm 5 \cdot \omega_P$	$\omega_S \pm 3 \cdot \omega_P$	$\omega_S \pm \omega_P$
$n = 6$	$c = 0, 1, 2$	$\omega_S \pm 6 \cdot \omega_P$	$\omega_S \pm 4 \cdot \omega_P$	$\omega_S \pm 2 \cdot \omega_P$

where $\omega_{FI} = \omega_S - \omega_P$. The spectral lines of the current $i(t)$ are given in Figure 5.18. The distance between the harmonics of ω_P and the intermodulation products is ω_{FI}. These intermodulation products are symmetric and occur at frequencies given by:

$$\omega_n = \omega_{FI} + n \cdot \omega_P \quad \text{where} \quad n = \ldots -3, -2, -1, 0, 1, 2, 3, \ldots$$

5.5 Diode Mixer Theory

We will start from the non-linear equivalent circuit of Figure 5.3 and from the dynamic model of Figure 5.9. The non-linear equations are given by Equations (5.1) and (5.2). Taking into account Equation (5.1), the associated incremental conductance $g(v)$ at each v point can be expressed by:

$$g(v) = \left. \frac{di}{dv} \right|_v = I_{SS} \cdot \frac{q \cdot v}{n \cdot k \cdot t} \cdot e^{\frac{q \cdot v}{n \cdot k \cdot t}} \approx \frac{q \cdot v}{n \cdot k \cdot t} \cdot i(v) \tag{5.64}$$

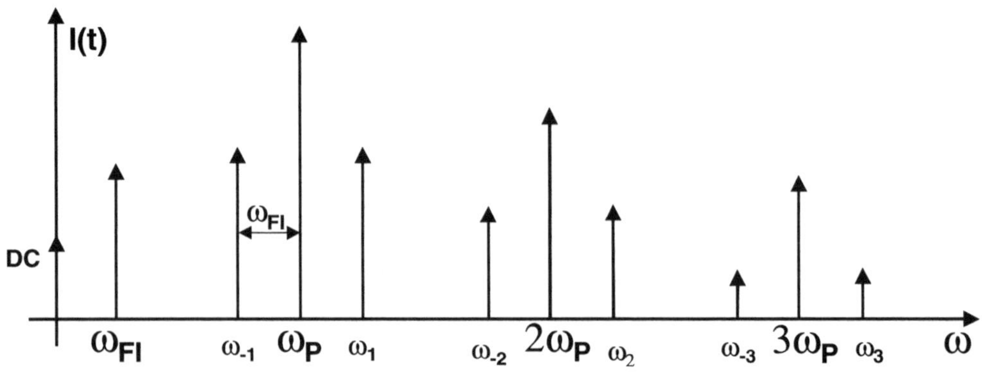

Figure 5.18 Harmonic representation of the $i(t)$ nonlinear current

Also, we have an incremental charge value, associated with the non-linear capacitor, in each v point. The alternating current through the capacitor can be written as:

$$i_c(t) = \frac{dQ}{dt} = \left.\frac{dQ}{dv}\right|_v \cdot \frac{dv}{dt} = C(v) \cdot \frac{dv}{dt} \tag{5.65}$$

We suppose quasi-linear excitation (RF amplitude ≪ LO amplitude). In this case, the diode can be considered to be non-linear with respect to LO signal and quasi-linear with respect to RF signal. Under LO excitation, the non-linear voltage at the diode terminals of Figure 5.3 is given by:

$$v_d = v(t) + R_S \cdot I_{SS} \cdot \left[e^{\frac{q \cdot v}{n \cdot k \cdot T}} - 1\right] + R_S \cdot C(v) \cdot \frac{dv}{dt} \quad \text{(Large signal)} \tag{5.66}$$

and the non-linear current:

$$i(t) = I_{SS} \cdot \left[e^{\frac{q \cdot v}{n \cdot k \cdot T}} - 1\right] + C(v) \cdot \frac{dv}{dt} \quad \text{(Large signal)} \tag{5.67}$$

Therefore, the way to solve the single-diode mixer problem is:

1. To solve the non-linear circuit for LO excitation and to obtain all the harmonics of ω_P (Figure 5.19).
2. To do linear superposition of the RF signal (Figure 5.19).

5.5.1 Linear Analysis: Conversion Matrices

We will apply two signals at the diode terminals: the RF signal, whose frequency is ω_S, and the LO signal, whose frequency is ω_P. We will suppose that the amplitude V_S of the ω_S signal is much less than the amplitude V_P of ω_P. That implies that V_S is a small perturbation around V_P. In this case, $C(v)$ and $g(v)$ from Equations (5.5) and (5.6) and Figure 5.3, vary harmonically under LO excitation.

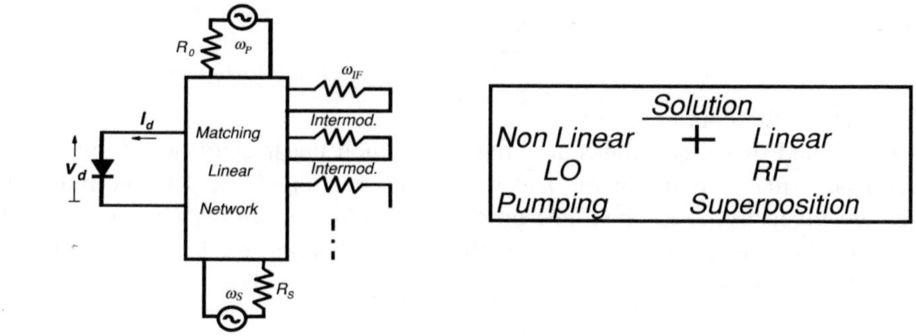

Figure 5.19 Single-diode mixer solution

If we know

$$i_d(t) = I_{do} + \sum_n \mathrm{Re}(I_{dn} \cdot e^{j \cdot n \cdot \omega_P \cdot t})$$
$$v_d(t) = V_{do} + \sum_n \mathrm{Re}(V_{dn} \cdot e^{j \cdot n \cdot \omega_P \cdot t}) \qquad (5.68)$$

we can find $v(t)$ (Figure 5.3) and in this case we know the Fourier series of:

$$C(v) \Leftrightarrow C(t)$$
$$g(v) \Leftrightarrow g(t)$$

5.5.1.1 Conversion matrix of a non-linear resistance/conductance

The excitation $i(t)$ is a harmonic temporal function (LO) therefore the response $v = f(i)$ is a harmonic temporal function (Figure 5.20). In this case, $r(i) = \dfrac{dv}{dt} = r(t)$ is also a harmonic temporal function that depends on LO and it is possible to develop it as a Fourier series:

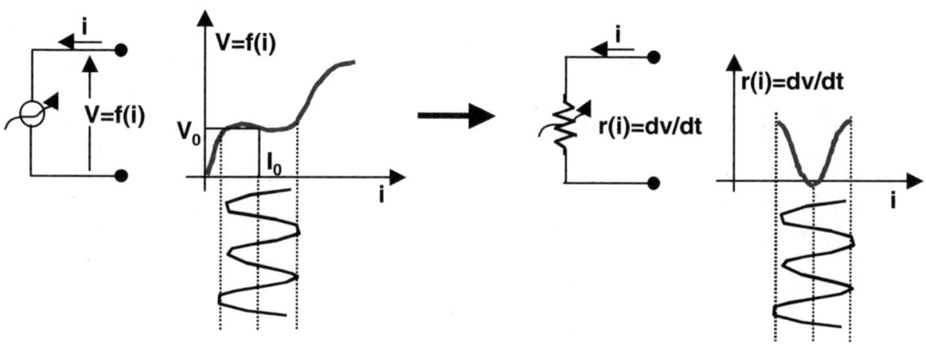

Figure 5.20 Non-linear resistance under LO excitation

$$r(t) = \sum_{-\infty}^{\infty} \overline{R_n} \cdot e^{j \cdot n \cdot \omega_P \cdot t} \tag{5.69}$$

where $\overline{R_n} = \overline{R_n^*}$ are the Fourier coefficients.

On the other hand, when two tones are applied in a non-linear element, intermodulation frequencies appear (linear approximation), as we saw in Equation (5.63). The intermodulation frequencies are given by $n \cdot \omega_S \pm m \cdot \omega_P$ or $\omega_n = \omega_{FI} + n \cdot \omega_P$ with $\omega_{FI} = \omega_S - \omega_P$. The generic form of the voltages and currents of the intermodulation products is given by the sum of all of their components:

$$\begin{aligned} \tilde{v}(t) &= \text{Re} \sum_{n=-\infty}^{\infty} \overline{V_n} \cdot e^{j \cdot (\omega_{FI} + n \cdot \omega_P) \cdot t} \\ \tilde{i}(t) &= \text{Re} \sum_{n=-\infty}^{\infty} \overline{I_n} \cdot e^{j \cdot (\omega_{FI} + n \cdot \omega_P) \cdot t} \end{aligned} \tag{5.70}$$

where $\tilde{v}(t)$ and $\tilde{i}(t)$ are very small signals. In this case, we can express Ohm's Law as:

$$\tilde{v}(t) = r(t) \cdot \tilde{i}(t) \tag{5.71}$$

and developing Equation (5.71) by Equation (5.70):

$$\begin{aligned} \sum_{l=-\infty}^{l=\infty} \overline{V_l} \cdot e^{j \cdot (\omega_{FI} + l \cdot \omega_P) \cdot t} &= \sum_{m=-\infty}^{m=\infty} \overline{R_m} \cdot e^{j \cdot m \cdot \omega_P \cdot t} \cdot \sum_{n=-\infty}^{n=\infty} \overline{I_n} \cdot e^{j \cdot (\omega_{FI} + n \cdot \omega_P) \cdot t} = \\ &= \sum_{m=-\infty}^{m=\infty} \sum_{n=-\infty}^{n=\infty} \overline{R_m} \cdot \overline{I_n} \cdot e^{j \cdot (\omega_{FI} + (n+m) \cdot \omega_P) \cdot t} \end{aligned} \tag{5.72}$$

Equating both parts of Equation (5.72), we obtain the conversion matrix of the non-linear resistor $r(t)$:

$$\begin{bmatrix} \bar{I}_{-N} \\ \bar{V}_{-N+1} \\ \vdots \\ \bar{V}_0 \\ \vdots \\ \bar{V}_{N-1} \\ \bar{V}_N \end{bmatrix} = \begin{bmatrix} \bar{R}_0 & \bar{R}_{-1} & \cdots & \bar{R}_{-N} & \cdots & \bar{R}_{-2N+1} & \bar{R}_{-2N} \\ \bar{R}_1 & \bar{R}_0 & \cdots & \bar{R}_{-N+1} & \cdots & \bar{R}_{-2N+2} & \bar{R}_{-2N+1} \\ \vdots & & & \vdots & & & \vdots \\ \bar{R}_N & \bar{R}_{N-1} & \cdots & \bar{R}_0 & \cdots & \bar{R}_{-N+1} & \bar{R}_N \\ \vdots & & & \vdots & & & \vdots \\ \bar{R}_{2N-1} & \bar{R}_{2N-2} & \cdots & \bar{R}_{N-1} & \cdots & \bar{R}_0 & \bar{R}_1 \\ \bar{R}_{2N} & \bar{R}_{2N-1} & \cdots & \bar{R}_N & \cdots & \bar{R}_1 & \bar{R}_0 \end{bmatrix} \cdot \begin{bmatrix} \bar{I}_{-N} \\ \bar{I}_{-N+1} \\ \vdots \\ \bar{I}_0 \\ \vdots \\ \bar{I}_{N-1} \\ \bar{I}_N \end{bmatrix} \tag{5.73}$$

The matricial Equation (5.73) can be written as:

$$[\bar{V}] = [\bar{R}] \cdot [\bar{I}] \tag{5.74}$$

where $[\bar{R}]$ is called the conversion matrix.

The intermodulation product response (voltage matrix) is a function of the excitation and the known conversion matrix. The conversion matrix $[\bar{R}]$ is a function of the circuit and the LO.

We have an analogous situation for calculating a non-linear conductance as in the diode case of Figure 5.3:

$$g(t) = \sum_{-\infty}^{\infty} \overline{G_n} \cdot e^{j \cdot n \cdot \omega_P \cdot t} \tag{5.75}$$

and the matricial equation is:

$$[\bar{I}] = [\bar{G}] \cdot [\bar{V}] \tag{5.76}$$

5.5.1.2 Conversion matrix of a non-linear capacitance

The excitation $v(t)$ is a harmonic temporal function (LO) and the response $Q(t) = f[(v(t)]$ is also a harmonic temporal function (Figure 5.21). The capacitance is given by:

$$C(v) = \frac{dQ}{dv} \tag{5.77}$$

and it will also be a harmonic temporal function. Developing this as a Fourier series, we have:

$$C(t) = \sum_{n=-\infty}^{\infty} \overline{C_n} \cdot e^{j \cdot n \cdot \omega_P \cdot t} \tag{5.78}$$

where $\bar{C}_n$ are the Fourier coefficients.

When we introduce a small signal ω_S along with the LO excitation ω_P, intermodulation frequencies appear as in the resistance case where $\omega_n = \omega_{FI} + n \cdot \omega_P$ with $\omega_{FI} = \omega_S - \omega_P$. Now we can apply the linear approximation:

$$\tilde{\imath}(t) = C(t) \cdot \frac{d\tilde{v}}{dt} \tag{5.79}$$

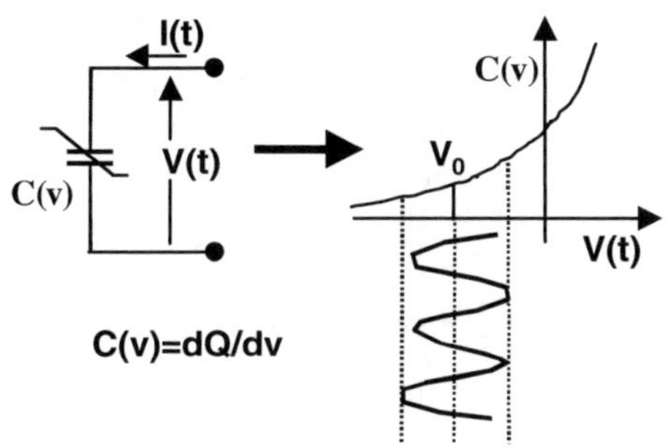

Figure 5.21 Non-linear capacitance under LO excitation

The generic form of the voltages and currents of the intermodulation products is given by Equation (5.70) and therefore, we can write:

$$\sum_{m=-\infty}^{m=\infty}\overline{I_m}\cdot e^{j\cdot(\omega_{FI}+m\cdot\omega_P)\cdot t}=\sum_{k=-\infty}^{k=\infty}\overline{C_k}\cdot e^{jk\cdot\omega_P\cdot t}\cdot\sum_{n=-\infty}^{n=\infty}j\cdot(\omega_{FI}+n\cdot\omega_P)\cdot\overline{V_n}\cdot e^{j\cdot(\omega_{FI}+n\cdot\omega_P)\cdot t} \quad (5.80)$$

Equating both parts of Equation (5.80), we obtain the matricial Equation (5.81):

$$\begin{bmatrix}\bar{I}_{-N}\\ \bar{I}_{-N+1}\\ \vdots\\ \bar{I}_0\\ \vdots\\ \bar{I}_{N-1}\\ \bar{I}_N\end{bmatrix}=\begin{bmatrix}\bar{C}_0\cdot\omega_{-N} & \bar{C}_{-1}\cdot\omega_{-N+1} & \cdots & \bar{C}_{-2N}\cdot\omega_N\\ \bar{C}_1\cdot\omega_{-N} & \bar{C}_0\cdot\omega_{-N+1} & \cdots & \bar{C}_{-2N+1}\cdot\omega_N\\ \vdots & \vdots & & \vdots\\ \bar{C}_N\cdot\omega_{-N} & \bar{C}_{N-1}\cdot\omega_{-N+1} & \cdots & \bar{C}_{-N}\cdot\omega_N\\ \vdots & \vdots & & \vdots\\ \bar{C}_{2N}\cdot\omega_{-N} & \bar{C}_{2N-1}\cdot\omega_{-N+1} & \cdots & \bar{C}_0\cdot\omega_N\end{bmatrix}\cdot\begin{bmatrix}\bar{V}_{-N}\\ \bar{V}_{-N+1}\\ \vdots\\ \bar{V}_0\\ \vdots\\ \bar{V}_{N-1}\\ \bar{V}_N\end{bmatrix} \quad (5.81)$$

Each current response sub-index is obtained from the capacitor sub-index plus the voltage sub-index because the sub-index of the frequency is the same as that of the voltage.

$$I_k=C_j\cdot\omega_n\cdot V_n \quad j+n=k \quad (5.82)$$

The matricial Equation (5.81) can be written as:

$$[\bar{I}]=j\cdot[\bar{C}]\cdot[\bar{\Omega}]\cdot[\bar{V}] \quad (5.83)$$

where the matricial current response is a function of the RF excitation and matrix $[\bar{C}]$ is a function of the circuit and the LO signal. The product $[\bar{C}]\cdot[\bar{\Omega}]$ is called the conversion matrix of the a non-linear capacitance. The matrices $[\bar{C}]$ and $[\bar{\Omega}]$ are given by:

$$[C]=\begin{bmatrix}\bar{C}_0 & \bar{C}_{-1} & \cdots & \bar{C}_{-2N}\\ \bar{C}_1 & \bar{C}_0 & \cdots & \bar{C}_{-2N+1}\\ \vdots & \vdots & & \vdots\\ \bar{C}_N & \bar{C}_{N-1} & \cdots & \bar{C}_{-N}\\ \vdots & \vdots & & \vdots\\ \bar{C}_{2N} & \bar{C}_{2N-1} & \cdots & \bar{C}_0\end{bmatrix} \quad [\Omega]=\begin{bmatrix}\omega_{-N} & 0 & 0 & \cdots & 0\\ 0 & \omega_{-N+1} & 0 & \cdots & 0\\ 0 & 0 & \omega_{-N+2} & \cdots & 0\\ \vdots & \vdots & \vdots & & \vdots\\ 0 & 0 & 0 & \cdots & \omega_N\end{bmatrix}$$

5.5.1.3 Conversion matrix of a linear resistance

In the case of the linear resistance of the diode equivalent circuit of the Figure 5.3, the matricial equation voltage/current can be written as:

$$[\overline{V_S}]=[\overline{R_S}]\cdot[\overline{I_d}] \quad (5.84)$$

where $[\overline{V_S}]$ is the matrix of the difference of potential at R_S terminals, $[\overline{I_d}]$ is the matrix of the total current trough the diode and the $[\overline{R_S}]$ matrix is given by:

$$[\overline{R_S}] = \begin{bmatrix} R_S & 0 & 0 & \cdots & 0 \\ 0 & R_S & 0 & \cdots & 0 \\ 0 & 0 & R_S & \cdots & 0 \\ \vdots & \vdots & \vdots & & \vdots \\ 0 & 0 & 0 & \cdots & R_S \end{bmatrix} \tag{5.85}$$

The same criterion we will apply for any linear impedance.

5.5.1.4 Conversion matrix of the complete diode

Taking into account the Schottky diode model of Figure 5.3, the voltage/current matricial relationship is:

$$[\overline{V_d}] = [\overline{Z_d}] \cdot [\overline{I_d}] \tag{5.86}$$

where $[\overline{V_d}]$ and $[\overline{I_d}]$ are the matrices of the total voltages and currents at the external diode terminals and $[\overline{Z_d}]$ is the complete impedance conversion matrix given by Equation (5.87) and it is only valid for the intermodulation product frequencies.

$$[\overline{Z_d}] = [\overline{R_S}] + ([\bar{G}] + j \cdot [\bar{C}] \cdot [\bar{\Omega}])^{-1} \tag{5.87}$$

5.5.1.5 Conversion matrix of a mixer circuit

The complete pumped diode behaves as a frequencial multiport network as we can see in Figure 5.22.

Z_{en} is the impedance that the diode sees at each frequency ω_n. $[\overline{Z_d}]$ is an exclusive function of the diode parameters and the power at the frequency of the local oscillator, for a given LO. Therefore, our interest is the relationship between RF and IF ports. For this purpose, we open each port as indicated in Figure 5.23.

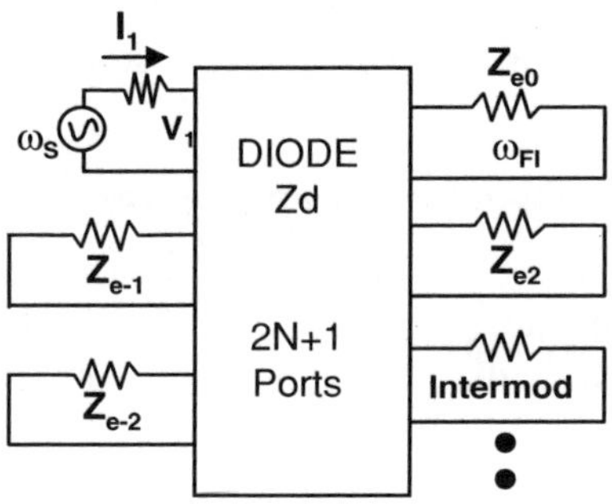

Figure 5.22 Diode mixer

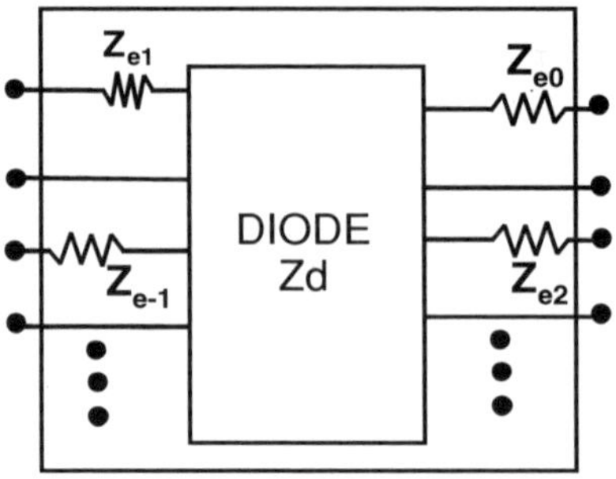

Figure 5.23 $[\overline{Z_d}]^a$ matrix

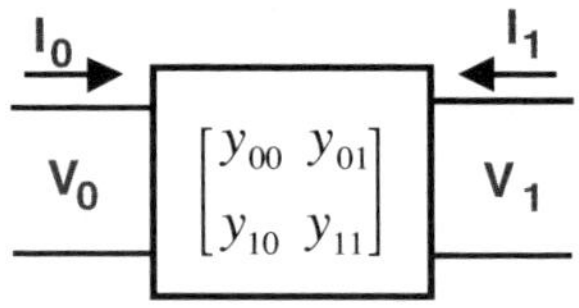

Figure 5.24 Relationship between IF (0) and RF (1) ports

The process to obtain the relationship between RF and IF is the following:

1. Sum the diagonal linear matrix $[\overline{Z_{en}}]$ with the conversion matrix $[\overline{Z_d}]$:

$$[\overline{Z_d}]^a = [\overline{Z_{en}}] + [\overline{Z_d}] \tag{5.88}$$

2. Calculate the inverse matrix of $[\overline{Z_d}]^a$: $[\overline{Y}] = ([\overline{Z_d}]^a)^{-1}$.
3. Null all unwanted intermodulation frequencies. In this case we have:

$$\begin{bmatrix} I_0 \\ I_1 \end{bmatrix} = \begin{bmatrix} y_{00} & y_{01} \\ y_{10} & y_{11} \end{bmatrix} \cdot \begin{bmatrix} V_0 \\ V_1 \end{bmatrix} \tag{5.89}$$

where 0 corresponds to the IF port and 1 corresponds to the RF port (Figure 5.24).

5.5.1.6 Conversion gain and input/output impedances

From the circuit of Figure 5.24, we can obtain the circuit of Figure 5.25 introducing the RF generator and drawing in explicit form the RF and IF impedances.

Taking into account Figure 5.25, we can define the transference power gain, or gain conversion, as:

$$G_C = \frac{P_{IF}}{P_{RF}} \tag{5.90}$$

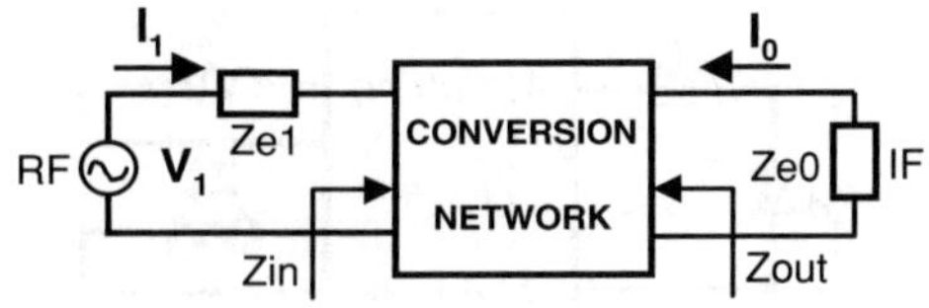

Figure 5.25 Mixer circuit

where P_{IF} is the dissipated power at the IF port and P_{RF} is the available power at the RF generator.

From Equation (5.89) when $V_0 = 0$, we can express the matrix parameters as:

$$y_{01} = \left.\frac{I_0}{V_1}\right|_{V_0=0} \quad y_{11} = \left.\frac{I_1}{V_1}\right|_{V_0=0} \quad \Leftrightarrow \quad \begin{cases} I_0 = y_{01} \cdot V_1 \\ I_1 = y_{11} \cdot V_1 \end{cases} \tag{5.91}$$

In this case, the dissipated power P_{IF} will be:

$$P_{IF} = \frac{1}{2} \cdot I_0^2 \cdot \mathrm{Re}(Z_{e0}) = \frac{1}{2} \cdot |y_{01}|^2 \cdot V_1^2 \cdot \mathrm{Re}(Z_{e0}) \tag{5.92}$$

and the available power P_{RF}:

$$P_{RF} = \frac{1}{8} \cdot \frac{V_1^2}{\mathrm{Re}(Z_{e1})} \tag{5.93}$$

Taking into account Equations (5.90), (5.91) and (5.92), the conversion gain, supposing the circuit is coupled without losses, is:

$$G_C = 4 \cdot |y_{01}|^2 \cdot \mathrm{Re}(Z_{e1}) \cdot \mathrm{Re}(Z_{e0}) \tag{5.94}$$

From Figures 5.24 and 5.25, we can deduce the expressions of the input and output impedance:

$$Zin = \frac{1}{y_{11}} - Z_{e1} \quad Zout = \frac{1}{y_{00}} - Z_{e0} \tag{5.95}$$

If we generalise the gain conversion and the impedances at any intermodulation products, we obtain the following expressions:

$$G_{Cmn} = 4 \cdot |y_{mn}|^2 \cdot \mathrm{Re}(Z_{en}) \cdot \mathrm{Re}(Z_{em}) \tag{5.96}$$

$$Zin = \frac{1}{y_{nn}} - Z_{en} \quad Zout = \frac{1}{y_{mm}} - Z_{em} \tag{5.97}$$

5.5.2 Large Signal Analysis: Harmonic Balance Simulation

Under large signal excitation (LO signal) the diode behaviour follows the non-linear Equations (5.67) and (5.68) and the objective of the large signal analysis is to calculate the Fourier components $V(n \cdot \omega_P)$ of the internal voltage $v(t)$. When the components $V(n \cdot \omega_P)$

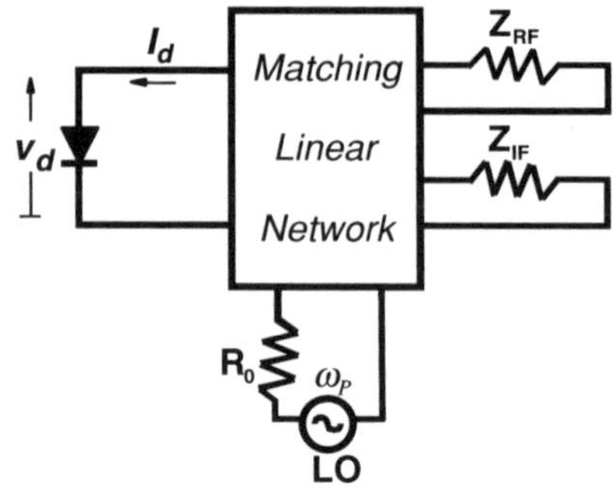

Figure 5.26 Mixer circuit under LO excitation

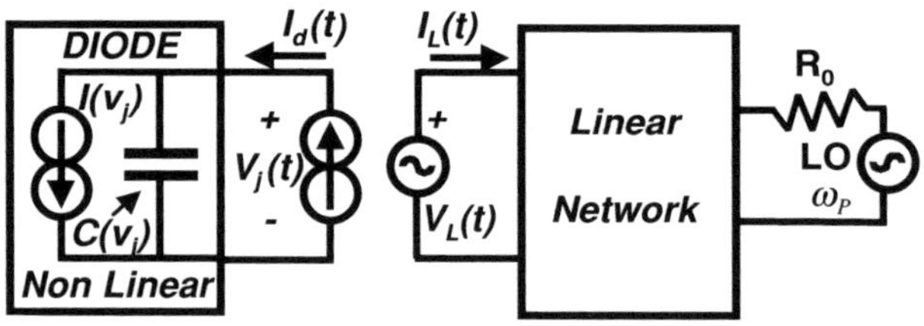

Figure 5.27 Linear and non-linear mixer networks under LO excitation

are known, $v(t)$ is known. Taking into account Equations (5.65) and (5.66), we can know the Fourier components of Equations (5.76) and (5.78) and therefore the complete conversion matrix (5.87).

The solution of the multi-harmonics $i_d(t)$ and $v_d(t)$ given in Equation (5.69), through the non-linear device (diode) under large signal excitation can be obtained by harmonic balance analysis method. This analysis can be implemented in several ways but all of them can be explained under the same idea:

The mixer circuit under large signal excitation (Figure 5.26) can be divided in two parts: the linear and non-linear circuits as we can observe in Figure 5.27. $I(v_j)$ and $C(v_j)$ are the junction current source and junction capacitance respectively, which depend on the junction voltage v_j. $I_d(t)$ is the external non-linear current of the diode and $V_L(t)$ and $I_L(t)$ are the voltage and current of the linear circuit.

The basic idea of the harmonic balance method is quite simple. For a given local oscillator and an initial condition of the linear network, we can calculate the linear voltage and current $V_L(t)$ and $I_L(t)$. The voltage $V_L(t)$ must be equal to the junction voltage $V_j(t)$, with this voltage we can calculate the non-linear current $I_d(t)$ and this current is compared with $I_L(t)$. We will reach the stationary solution when the currents have the same amplitude and a phase difference of 180 degrees.

Normally, the linear network is calculated in the frequency domain and the non-linear one is calculated in the time domain. A good Fourier transforming algorithm is necessary to change from the temporal to the frequency domains and the frequency to time domain. The number of harmonics that we take into account should be chosen carefully, if the number of harmonics is low, the solution will not be correct, but if the number is high, the calculation time can be very high.

Finding the stationary solution is an optimisation problem. It is necessary to define an appropriate error function and to search for the zeros of this function by an optimisation algorithm. The advantage of optimisation is that a great number of optimisation subroutines are already available in computer mathematics libraries.

One of the commonly used optimisation techniques is the Newton's method or Newton-Raphson method. It is an iterative method for finding the zeros of the error function and it needs to know the gradient of the error function. It is an efficient method when the derivatives are easily evaluated and we can make a good initial estimation of the solution.

There are other optimisation methods (relaxation and reflection algorithms, for instance) but it is not the objective of this chapter present a study of them. In any case, all the modern non-linear and large signal simulators have implemented these methods.

5.6 FET Mixers

Any device used in a mixer must have a strong non-linearity, low noise, low distortion and an adequate frequency response. Traditionally, the Schottky diode has been the most used non-linear device for mixers. The field-effect-transistors (FET) have become very popular in the past few years as mixer devices. The main reason is due to the great advantage that GaAs' microwave monolithic integrated circuit can produce. FET transistors are more suitable than diodes in this technology, since non-planar structures as phase shift circuits are required in the balanced mixers. Nevertheless, the most important characteristic of the FET mixers is that they can exhibit conversion gain well into the millimetre-wave region.

FET mixers can be designed with good noise performance as well as conversion gain, and lower LO power is required, if compared to that of diode mixers. Since FETs are available as dual-gate devices, the LO and RF can be applied to separate ports, improving the isolation between these ports. Balanced FET mixers are also possible, and they have the same LO noise rejection and spurious properties as balanced diode mixers.

5.6.1 Single-Ended FET Mixers

5.6.1.1 Simplified analysis of a single-gate FET mixer

The square-law characteristic of a FET can be used in a frequency conversion. Figure 5.28 shows a simplified scheme of a FET mixer where both RF and LO signals are injected through the gate of the transistor.

The drain current can be expressed as:

$$I_d = I_{dss}\left(1 - \frac{V_{gs}}{V_p}\right)^2 \tag{5.98}$$

The transconductance is defined as: $g_m = \dfrac{\partial I_d}{\partial V_{gs}}$ and substituting I_d into the last expression:

$$g_m = \frac{-2I_{dss}}{V_p}\left(1 - \frac{V_{gs}}{V_p}\right) \tag{5.99}$$

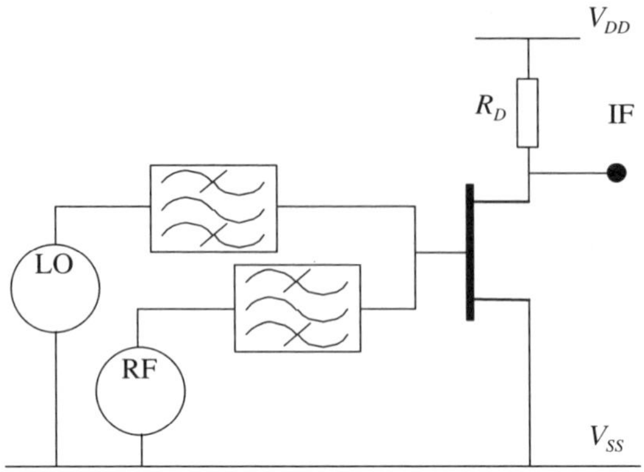

Figure 5.28 FET mixer with RF and LO both applied at the gate

However, $V_{gs} = V_g + V_{LO} \cos \omega_{LO} t$ ($V_{LO} \gg V_{RF}$) where V_{LO} is the local oscillator voltage amplitude, V_{RF} is the radiofrequency voltage amplitude and V_g is the gate-source bias voltage. The transconductance is now:

$$g_m = \left(\frac{-2I_{dss}}{V_p}\right)\left(1 - \frac{V_g + V_{LO}\cos\omega_{LO}t}{V_p}\right) = \left(\frac{-2I_{dss}}{V_p}\right)\left(1 - \frac{V_g}{V_p}\right) + \left(\frac{-2I_{dss}}{V_p^2}\right)V_{LO}\cos\omega_{LO}t \quad (5.100)$$

The small-signal drain current is:

$$i_d(t) = g_m(t)V_{RF}(t) = \left(\frac{-2I_{dss}}{V_p}\right)\left(1 - \frac{V_g}{V_p}\right)V_{RF}\cos\omega_{RF}t + \left(\frac{-2I_{dss}V_{LO}V_{RF}}{V_p^2}\right)\cos\omega_{LO}t\cos_{RF}t \quad (5.101)$$

The large-signal LO modulates [1] the transconductance (g_m) of the device and when a small-signal RF is applied simultaneously, the small-signal drain current is proportional to their product (V_{LO} V_{RF}).

As the time-varying transconductance is the main contributor to mixing, these mixers are called transconductance mixers, and the mixing products attributable to parametric 'pumping' of the gate-source capacitance, gate-drain capacitance and drain-source resistance can be considered negligible. In order to get the maximum conversion gain is important to maximise the range of the MESFET's transconductance variation and, in particular, the magnitude of the fundamental frequency component of the transconductance. To maximise the magnitude of the fundamental component of g_m, the device must be biased close to the pinch-off value, V_p, and must remain in the saturation region throughout the LO cycle. The best way to ensure this can be achieved by short-circuiting the drain terminal at the LO frequency and all LO harmonics.

The input matching circuit must match the RF source to the MESFET's gate, and short-circuit at the IF frequency to avoid the amplification of any input noise at this frequency. A good output matching circuit is important since an inadequate output network can cause instablility. As well, this network must be a good IF filter in order to get good isolation between the ports. In many cases it is possible for the IF circuit to provide both impedance transformation and filtering functions via a single structure, which always minimises circuit loss.

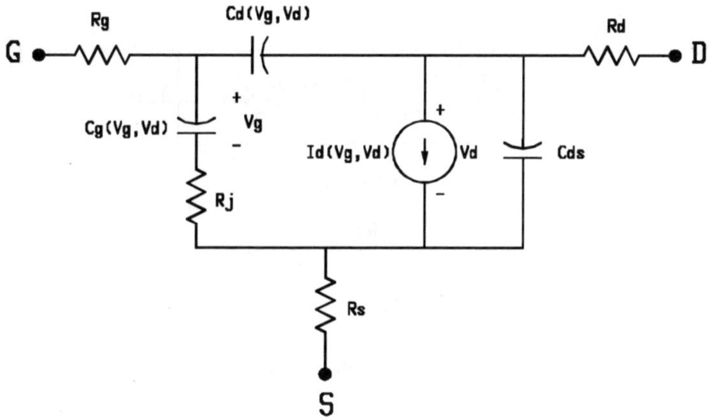

Figure 5.29 Large-signal non-linear MESFET equivalent circuit

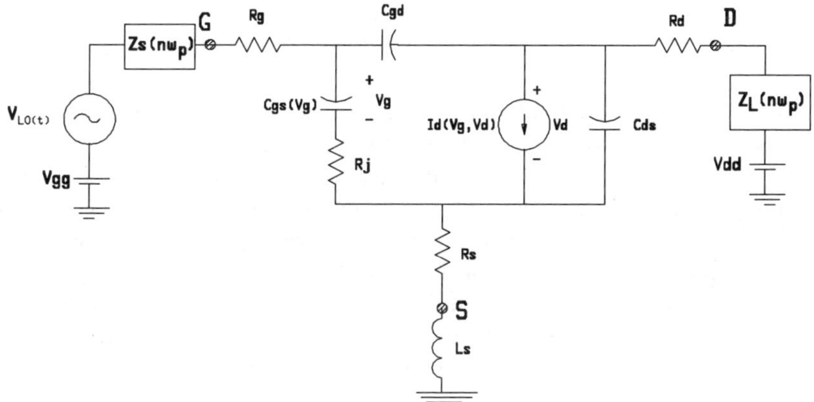

Figure 5.30 Large-signal equivalent circuit of the FET mixer

5.6.1.2 Large-signal and small-signal analysis of single-gate FET mixers

The FET mixer can be analysed via the large-signal procedure [2] similar to that described in Section 5.3. The large-signal non-linear MESFET equivalent circuit is shown in Figure 5.29. If the transistor is biased in its saturation region through the LO cycle, the non-linearities of I_d and C_{gs} can be simplified a lot, and C_{gd} can be considered as a linear element. In this case, the large signal equivalent circuit of the FET mixer is shown in Figure 5.30, where $Z_S(n\omega_p)$ and $Z_L(n\omega_p)$ are the embedding impedances of the source and load respectively, and ω_p is the LO frequency.

The most used method to perform large-signal analysis of FET mixers is the harmonic-balance method. This method calculates only the steady-state solution for the circuit. The non-linear circuit is divided into linear and non-linear subcircuits. The linear subcircuit can be treated as a multiport and described by its y-parameters, s-parameters, or some other multi-port matrix. The non-linear elements are modelled by their global I/V or Q/V characteristics, and must be analysed in the time domain.

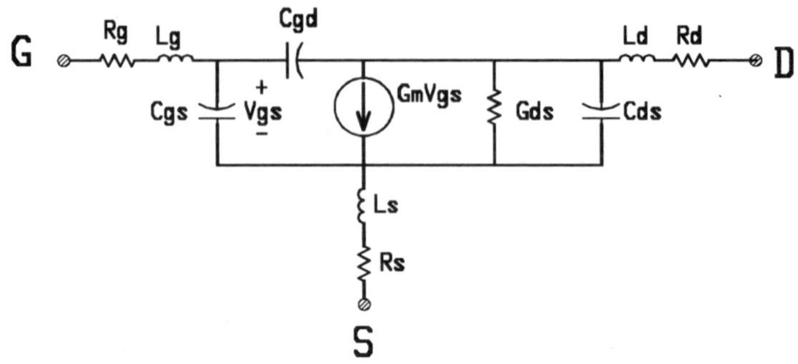

Figure 5.31 Small-signal equivalent circuit of the FET

The idea of harmonic balance is to find a set of port voltage waveforms (or, alternatively, the harmonic voltage components) that gives the same currents in both the linear-network equations and the non-linear-network equations. When that is satisfied, we have the solution.

The aim of a small-signal analysis is to calculate the conversion gain and input and output impedances of the mixer, using a linear small-signal FET equivalent circuit, as shown in Figure 5.31.

The mixing takes place in the transistor when the small-signal elements are varied periodically by a large LO signal, which is applied between the gate and source terminals. For a GaAs MESFET, the major dependence with the gate bias is produced by the transconductance, g_m. The mixing products produced by the C_{gs} capacitance and the R_i resistance are considered negligible. For the gate-pumped mixers the drain resistance, R_{ds}, variation is small, so the time-averaged value is used. As the main contributor to the frequency conversion is produced by the variation of the tranconductance, these mixers are called 'transconductance mixers'.

When a large LO signal is applied between the gate and source terminals, the transconductance becomes in a time-varying function $g_m(t)$ with a period equal to that of the LO. If ω_o is the LO frequency:

$$g_m(t) = \sum_{k=-\infty}^{\infty} g_k e^{jk\omega_o t} \tag{5.102}$$

where:

$$g_k = \frac{1}{2\Pi} \int_0^{2\Pi} g_m(t) e^{-jk\omega_o t} d(\omega_o t) \tag{5.103}$$

are the Fourier coefficients of the transconductance. Of these coefficients g_1 is the most important, which corresponds to the fundamental component of g_m in the frequency domain. This coefficient is a function of the LO signal amplitude, of the gate bias, and of the shape of the curve g_m/V_{gs}. The value of g_1 is not greater than g_{mo}/A, where g_{mo} is the maximum value of g_m. For the ideal case, when g_1 is a step function of the gate voltage:

$$\frac{g_1}{g_m} = \frac{1}{\Pi} \tag{5.104}$$

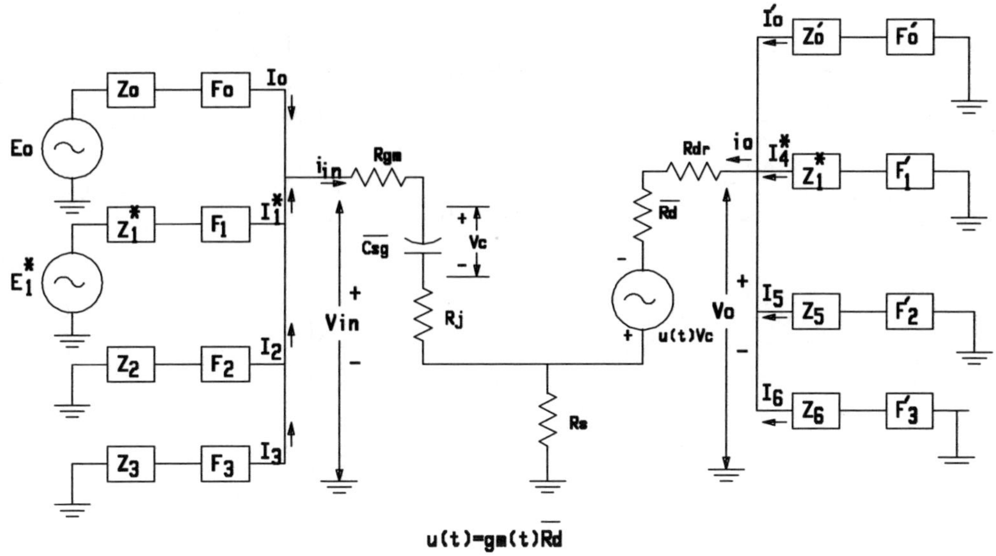

Figure 5.32 Small signal equivalent circuit of a FET mixer

Figure 5.32 shows the equivalent circuit of a FET mixer [3], where ω_1 corresponds to the RF frequency and ω_o to the LO signal. V_1, V_2, V_3 and I_1, I_2, I_3 are the complex voltage and current amplitudes of the signal, image and intermediate frequency in the gate circuit, and V_4, V_5, V_6 and I_4, I_5, I_6 the corresponding voltage and current amplitudes of the drain circuit. E_1 represents the voltage source for the RF signal, with internal impedance Z_1, and the rest of components are terminated in complex impedances.

The circuit shown in Figure 5.32 can be analysed with the loop equations for each frequency component. In matrix notation these equations are written as:

$$[E] = [V] + [Z_t][I] = [Z_m][I] + [Z_t][I] \tag{5.105}$$

Where:

$$[E] = \begin{bmatrix} E_1^* \\ 0 \\ 0 \\ V \\ 0 \\ 0 \end{bmatrix} \qquad [V] = \begin{bmatrix} V_1^* \\ V_2 \\ V_3 \\ V_4^* \\ V_5 \\ V_6 \end{bmatrix} \qquad [E] = \begin{bmatrix} I_1^* \\ I_2 \\ I_3 \\ I_4^* \\ I_5 \\ I_6 \end{bmatrix}$$

and $[Z_m]$ and $[Z_t]$ are, respectively, the matrices representing the proper mixer and its terminations. They are given by:

$$[Z_m] = \begin{bmatrix} Z_{11}^* & 0 & 0 & Z_{14}^* & 0 & 0 \\ 0 & Z_{22} & 0 & 0 & Z_{25} & 0 \\ 0 & 0 & Z_{33} & 0 & 0 & Z_{36} \\ & 0 & Z_{43} & Z_{44}^* & 0 & 0 \\ 0 & Z_{52} & Z_{53} & 0 & Z_{55} & 0 \\ Z_{61}^* & Z_{52} & Z_{63} & 0 & 0 & Z_{66} \end{bmatrix}$$

$$[Z_t] = \begin{bmatrix} Z_1^* & 0 & 0 & 0 & 0 & 0 \\ 0 & Z_2 & 0 & 0 & 0 & 0 \\ 0 & 0 & Z_3 & 0 & 0 & 0 \\ 0 & 0 & 0 & Z_4^* & 0 & 0 \\ 0 & 0 & 0 & 0 & Z_5 & 0 \\ 0 & 0 & 0 & 0 & 0 & Z_6 \end{bmatrix}$$

The available conversion gain between the RF input (port 1) and the IF output (port 6) is:

$$G_{av} = \frac{|I_6|^2 \mathrm{Re}(Z_6)}{|E_1|^2 / 4\mathrm{Re}(Z_1)} \tag{5.106}$$

When the IF frequency is small compared to the input signal frequency, many simplifications can be made. The conversion gain is maximum when the source and load are conjugately matched to the FET. In this case the conversion gain is given by:

$$G_{av,max} = \frac{g_1^2}{4\omega_1^2 \bar{C}^2} \frac{\overline{R_d}}{R_{in}} \tag{5.107}$$

where $\bar{R}_d$ is the time-averaged value of R_{ds}, $\bar{C}$ is the time-averaged value of C_{gs}, and $R_{in} = R_{gm} + R_i + R_s$.

5.6.1.3 Other topologies

The gate-pumped FET mixer is the most used topology and the best known. Nevertheless, in many applications it is not possible, nor convenient, to apply the LO signal through the gate. Drain-pumped mixers are frequently used when LO and RF signals are separate in frequency and it is not possible to design a combiner with sufficient performance. Then, the LO signal is injected through the drain terminal and the IF signal is extracted at the same port with the aid of a diplexer.

Figure 5.33 shows the small-signal equivalent circuit for a drain mixer [4]. When a large signal is applied between two FET terminals, the small signal elements are varied periodically. The main non-linearities are the transconductance g_m and the channel resistance R_{ds}. For

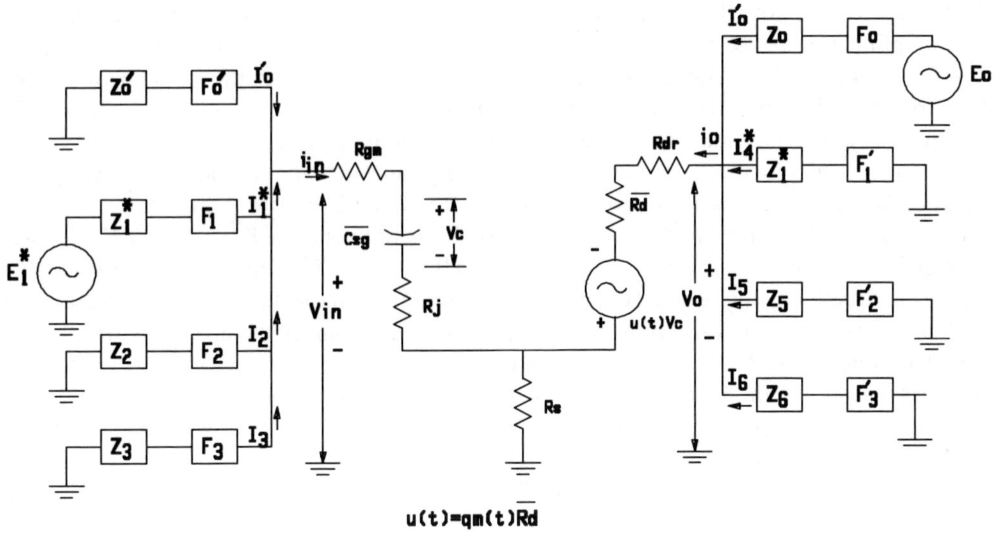

Figure 5.33 Small-signal equivalent circuit of a drain mixer

a gate-pumped design and under saturation drain bias, the g_m non-linearity is more significant than that of the R_{ds}. Therefore a time-averaged value is taken for R_{ds}. But for a drain-pumped mixer, the transconductance and the channel resistance are modulated by the LO signal and become in a time-varying functions with a period equal to that of the LO. Therefore, the amplification factor $\mu = g_m R_{ds}$ becomes in a time-varying function as well.

When a small-signal of frequency ω_1, is applied to the gate-source terminals, both non-linearities $\mu(t)$ and $R_{ds}(t)$ are contributing to the mixing. Only the intermediate frequency $\omega_3 = \omega_{LO} - \omega_1$ and the image frequency $\omega_2 = 2\omega_{LO} - \omega_1$ are considered as generated product by the mixing. The rest of the mixing products are considered to be eliminated by the filters F_k, $k = 1, \ldots 6$, which are supposed as ideal band-pass filters. g_m and R_{ds} are the only non-linearities of the circuit. R_{gm}, R_s, R_{dr}, C_{gs} and C_{gd} are considered constants in the analysis.

If the non-linearities are developed in Fourier series, with μ_n and R_{dn} as coefficients of both series, the conversion matrix can be obtained (similar to the development in Section 5.6.1.2). When C_{gd} is considered zero, the conversion gain is:

$$G_M = 4R_G R_L \left| \frac{I_6}{E_1} \right|^2 \tag{5.108}$$

where R_G is the real part of the signal source impedance and R_L is the real part of the IF load impedance. When Z_3, Z_4 and Z_5 have a high value and the input and output circuits are conjugately matched, the conversion gain is:

$$G_M = \frac{|\mu_1|^2}{4\omega_1^2 C_{gs}^2 (R_{gm} + R_s)(R_{dr} + R_s + R_{do})} \tag{5.109}$$

where R_{do} is the DC component of R_{ds}.

In this last expression, the non-linearity of R_{ds} only appears in the fundamental component of the amplification factor μ, which differs from the expression of the conversion gain for a gate-pumped mixer. For gate LO pumping the gain is a function of the fundamental component of g_m, and a time-averaged value is taken for R_{ds}.

A third way to inject the LO signal is using the source terminal. This topology is less common than the other two ones, but it is used when LO and RF are very distant in frequency and can be filtered properly. Normally, RF signal is injected through the gate and the IF signal is extracted from the drain. Because C_{gs} is not very small, bad LO/RF isolation will result if filters are not added in the RF and LO ports. For this reason, LO and RF bands cannot be close in frequency. For a source-pumped mixer, the transconductance and the channel resistance become time-varying functions, so both of them must be taken into account in order to design the mixer properly.

Recently several authors have studied resistive FET mixers. These have conversion losses and noise figures comparable to those of the diode mixers, but can achieve much better IM and spurious signal performance.

Mixers are usually the most non-linear devices of a receiver front end. For this reason the intermodulation (IM) performance is often limited by the mixer, and furthermore, this device is the only stage that generates spurious signal responses. For some applications, where broadband behaviour is required, these characteristics may be more limiting than noise. On the other hand, low intermodulation levels are required in some digital systems (like OFDM modulation systems), which has led to an improvement in the distortion level required of the mixers.

Traditionally, mixers have been constructed with a large LO signal and a small RF signal applied to a non-linear device. The large LO voltage changes the union impedance between a very small value and nearly an open circuit. This time-varying resistance is responsible for the mixing, and these mixers are called 'resistive mixers'. The non-linearity of the union presents a time-varying resistance, since the slope of the I/V curve is changed when pumped by the LO signal. If a linear time-varying resistance can be achieved, then intermodulation-free mixing would be reality.

An ideal linear time-varying resistance cannot be created, but it is possible to find something close to it, such as the channel of an unbiased GaAs MESFET. The channel resistance of a cold MESFET (with no drain bias) can change when a signal is applied to the gate terminal. If the transistor is drain-biased with a very small (or zero) voltage, then the relation between the drain to source current and the gate-source voltage is non-linear. However, drain-source current varies almost linearly with drain-source voltage. This last characteristic permits a very small distortion level compared to that generated by a diode. Since the channel resistance is non-linear with V_{gs}, the LO signal must be applied between these two terminals. The RF signal will be injected through the drain, and now that R_{ds} is almost 'linear' with V_{ds}, very little distortion is generated.

Since the LO signal is modulating the channel conductance, it can be expanded in a Fourier series:

$$G_{ds} = g_o + 2g_1 \cos(\omega_{LO}t) + \ldots \tag{5.110}$$

where ω_{LO} is the LO frequency signal. Taking into account only the first two terms and when load and source are conjugately matched, the conversion losses are given by [5]:

$$I_c \cong \frac{g_0^2}{g_1^2} \tag{5.111}$$

In order to minimise the conversion losses, g_1 must be as large as possible. Thus, the transistor will be biased to maximise the fundamental component of the channel conductance, g_1. This bias point corresponds to that where the channel resistance is most sensitive to bias modulation by the LO. This point is a little above the pinch-off value, where the channel conductance is most non-linear with V_{gs}.

Since drain-gate capacitance is greater for non-biased transistors compared with the same transistor in the saturation region, there is a coupling between these ports. Therefore, LO signal leakage will appear in the drain, increasing the drain-source voltage and the intermodulation level generated. A low DC value at drain port can be obtained by shorting the terminal at LO frequency and its harmonics. Balanced topologies can be used when LO and RF frequency bands are very close. An example of a double-balanced resistive FET mixer is shown in Section 5.9.

5.7 Double-Gate FET Mixers

When designing a mixer with a single gate device, the first problem is how to apply both signals, RF and LO, to the transistor gate. Passive couplers are commonly used in conventional hybrid technology. Nevertheless, for low frequencies, passive coupling is not suitable due to the size restrictions. Using double-gate transistors, this problem can be solved. The advantages of employing these devices are:

- intrinsic separation of signal and local oscillator ports and the possibility of separate matching;
- direct combination of the corresponding powers inside the device.

The operation mode of a double-gate MESFET can be considered equivalent to a cascade connection [6, 7] of two single-gate MESFETs, as can be seen in Figure 5.34. The LO signal is usually injected in the upper gate MESFET and the RF signal is applied in the lower gate MESFET. The main operation regions are shown in Figure 5.35 where the FET1 DC curves are shown with the FET2 DC curves inverse overlapped, but the latter have been plotted as a function of V_{g2}. When a sinusoidal voltage is injected into FET2 gate, three non-linear operating regions can be outlined (Figure 5.35):

1. Low noise mode. Defined by $V_{gs2} < -1$ V.
2. Self-oscillating mode. Defined by -0.5 V. $< V_{gs2} < 1$ V., $V_{gs1} < -1$ V.
3. Image-rejection mode. Defined by 2.5 V. $< V_{gs2} < 3.5$ V. and $V_{gs1} > -1.5$ V.

When the transistors are biased in one of these three modes, some parts of the device act non-linearly causing frequency conversion, while the rest acts as a RF or a IF amplifier.

1. For the first mode, the lower FET is in the linear region, while the upper FET is in the saturation region. The mixing process takes place in the lower FET. The most important non-linear elements are the transconductance g_m and the channel resistance R_{ds}. The upper

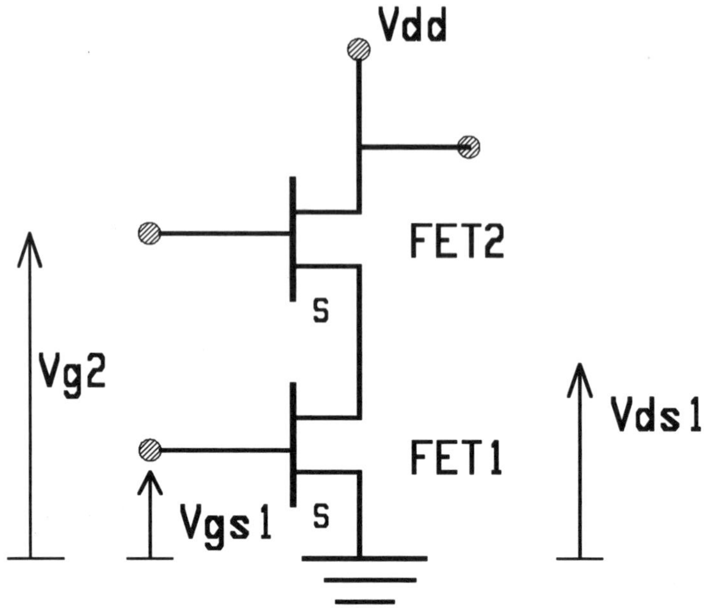

Figure 5.34 Cascade connection for two transistors

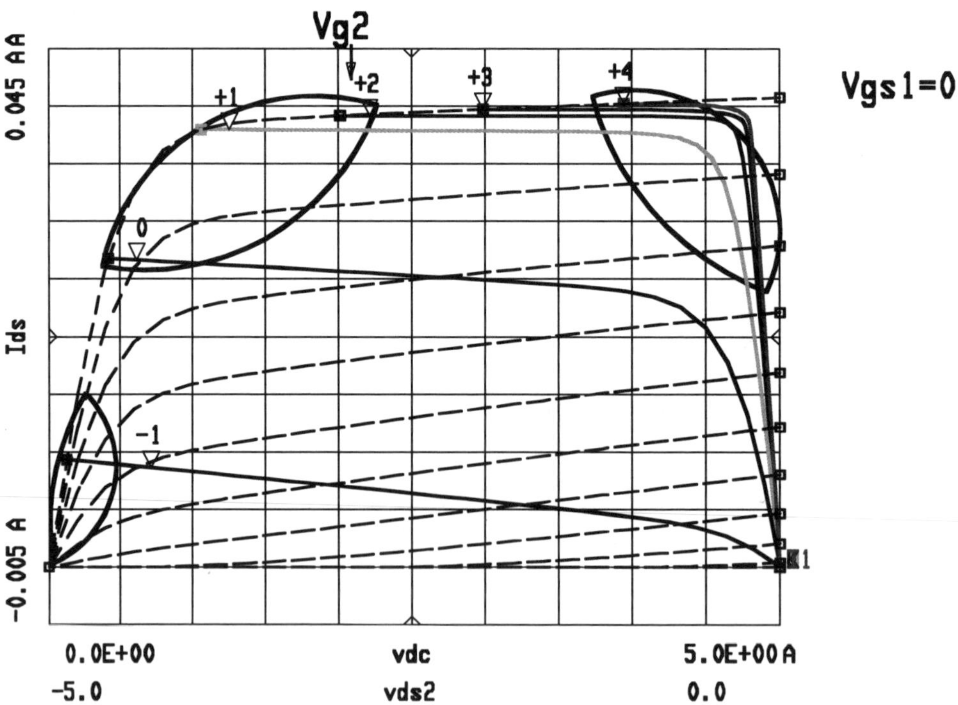

Figure 5.35 I/V Curves of the transistors

FET acts as an IF amplifier. The generated current level is quite low, and this operating mode is therefore suitable for low noise applications.

2. For the second mode, the non-linear elements are the same as for the first mode, although the channel resistance is a more non-linear element. The mixing takes place inside FET1, and FET2 amplifies the IF signal.
3. For the third mode, FET1 is in the saturation region and FET2 in the linear region. The mixing takes place in FET2, while FET1 acts as a RF preamplifier. The main nonlinearities in FET2 are g_m, R_{ds} and C_{gs}.

The second mode is suitable for biasing the transistors in order to get more conversion gain. Both transistors must be biased so that the variation of V_{g2} by the LO signal changes the g_m of FET1 as much as possible. For this operation mode, the upper FET holds in the saturation region during the LO cycle, because its V_{ds} voltage is greater than 1.5 V. The lower transistor is responsible for the mixing. Considering g_m and R_{ds} as the only non-linearities of the transistor, a low frequency simplified analysis can be done. This analysis will allow us to fix the values for both FET's gate DC voltages (V_{g1} and V_{g2}).

For the FET1 (in Figure 5.34) the I_{ds} current can be expressed as:

$$I_{ds} = I_{dss}\left(1 - \frac{V_{gs}}{V_T}\right)^2 (1 + \lambda V_{ds})\left[1 - \left(\frac{\alpha V_{ds}}{3}\right)^3\right] \tag{5.112}$$

The FET1 transconductance is:

$$g_m = \frac{\partial I_{ds}}{\partial V_{gs}} = -\frac{2}{V_T - V_{gs}} I_{ds} \tag{5.113}$$

With $V_{gs} = V_{g1}$ the channel conductance for FET1 is:

$$g_{ds} = \frac{\partial I_{ds}}{\partial V_{ds}} = \left[\frac{\lambda}{1 + \lambda V_{ds}} - \frac{\alpha\left(1 - \frac{\alpha V_{ds}}{3}\right)^2}{1 - \left(1 - \frac{\alpha V_{ds}}{3}\right)^3}\right] I_{ds} \tag{5.114}$$

with $V_{ds} = V_{ds1}$.

FET2 holds in the saturation region during whole the LO cycle, so its current expression can be approximated as:

$$I_{ds} = I_{dss}\left(1 - \frac{V_{gs}}{V_T}\right)^2 (1 + \lambda V_{ds}) \tag{5.115}$$

with $V_{gs} = V_{g2} + V_{OL}\cos(\omega_p t) - V_{ds1}$. Notice that the total voltage at gate two has a DC part (V_{g2}) and a AC part ($V_{OL}\cos(\omega_p t)$). As the drain currents of FET1 and FET2 are the same, we can substitute (5.115) into the g_m and g_{ds} expressions. Hence the large LO signal modulates the transconductance and the output conductance of the FET1. Using the last V_{gs} expression in Equation (5.113), we can represent g_m for FET1 as a function of V_{g1} and V_{g2}, as shown in Figure 5.36.

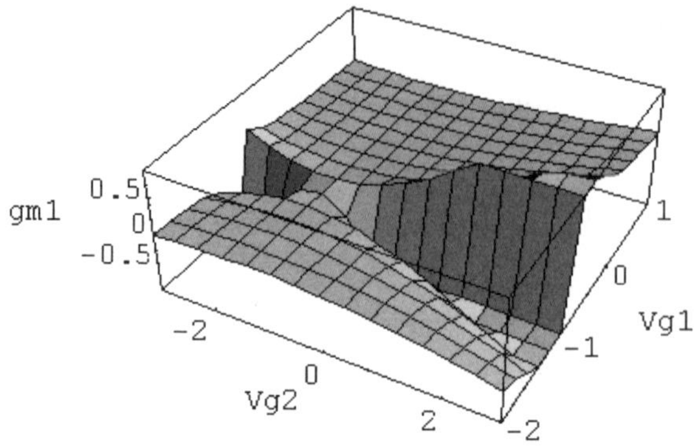

Figure 5.36 Variation of g_{m1} (FET1) versus V_{g1} and V_{g2}

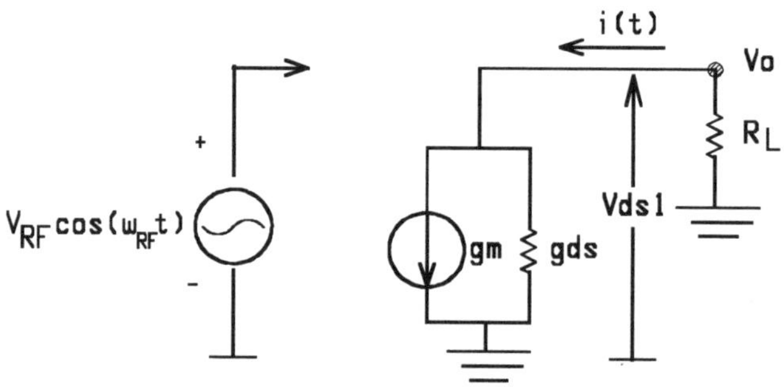

Figure 5.37 Simplified equivalent circuit for the small signal analysis

From this figure, we can get an initial value for V_{g1}. The best value will allow g_m to have the largest excursion when V_{g2} is varied [2]. For V_{g1} values between −0.5 and 0 Volt., the g_m excursion is maximum.

If a radiofrequency small signal voltage drives the gate of the FET1 (Figure 5.37), then the small signal current will be:

$$i(t) = g_m(t)V_{RF}(t) + g_{ds}(t)V_{ds}(t) \tag{5.116}$$

with

$$V_{RF} = \overline{V_{RF}}\cos(\omega_{RF}t)$$

However,

$$V_{ds}(t) = -(\overline{R_{ds}}\,\|R_L)g_m(t)V_{RF}(t)$$

when $\overline{R_{ds}}$ is the R_{ds} time averaged. The IF output voltage will be:

$$V_o(t) = \left(\frac{R_L AB}{V_T^2} - \frac{3R_L ABC}{2V_T^2}\overline{V_{OL}^2} - \frac{R_L 2B^3CA}{V_T^2}\right)(\overline{V_{OL}}\ \overline{V_{RF}})\cos(\omega_{FI}t) \tag{5.117}$$

where $\omega_{FI} = \omega_p - \omega_s$ is the intermediate frequency and A, B and C are:

$$A = \left(\frac{-2I_{dss}(1 - \lambda V_{ds2})}{V_T - V_{g1}} \right) \qquad B = V_T - V_{g2} + V_{dd} - V_{ds2}$$

$$C = \frac{RI_{dss}}{V_T^2} \left[\frac{\lambda}{1 + \lambda(V_{dd} - V_{ds2})} + \frac{\alpha\left(1 - \frac{\alpha(V_{dd} - V_{ds2})}{3}\right)^2}{1 - \left(1 - \frac{\alpha(V_{dd} - V_{ds2})}{3}\right)^3} \right] \tag{5.118}$$

V_{ds2} is the drain-source voltage for the FET2, which can be considered as a constant as this transistor remains in its saturation region throughout the LO cycle. The IF output power will be:

$$P_{out}(\omega_{FI}) = \frac{1}{2}\mathrm{Re}(V_o(\omega_{IF})I_o^*(\omega_{FI})) = \frac{1}{2}|V_o(\omega_{FI})|^2\,\mathrm{Re}(Y_L) \tag{5.119}$$

Substituting (5.117) in the last expression, the intermediate frequency output power is obtained as a function of V_{g2}. The best value for V_{g1} has been chosen in order to maximize the g_m excursion. The variation of IF output power with external bias V_{g2} is shown in Figure 5.38.

From the last figure, it is possible to choose the best value for external bias V_{g2} in order to maximize IF output power.

The expression (5.119) has been calculated considering V_{ds2} constant, in order to obtain a simplified form for IF output power. Nevertheless there is a small change for V_{ds2} when the external bias V_{g1} and V_{g2} are varied. This variation and the influence of the access resistances have been taken into account in more elaborate analyses of the mixer.

For the FET1, the drain source current can be expressed as:

$$I_{ds} = I_{dss}\left(1 - \frac{V_{dsi1}}{V_{T1}}\right)^2 \mathrm{Tanh}(\alpha V_{dsi1}) \tag{5.120}$$

with $V_{gsi1} = V_{g1} - I_{ds}R_s$, $V_{dsi1} = V_{ds1} - I_{ds}(R_s + R_d)$ and $V_{T1} = -1.32$ volt.

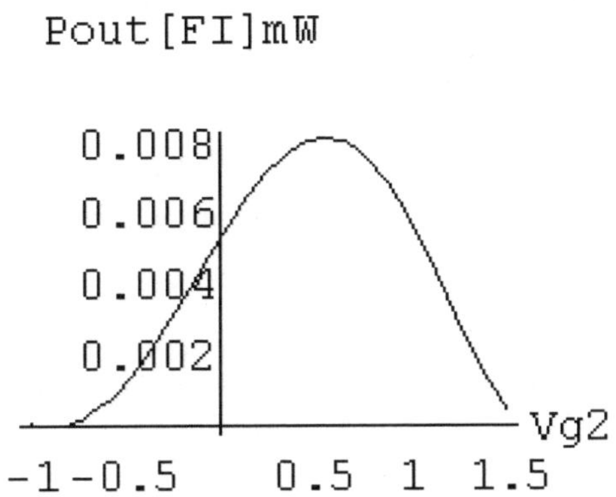

Figure 5.38 IF output power vs V_{g2}

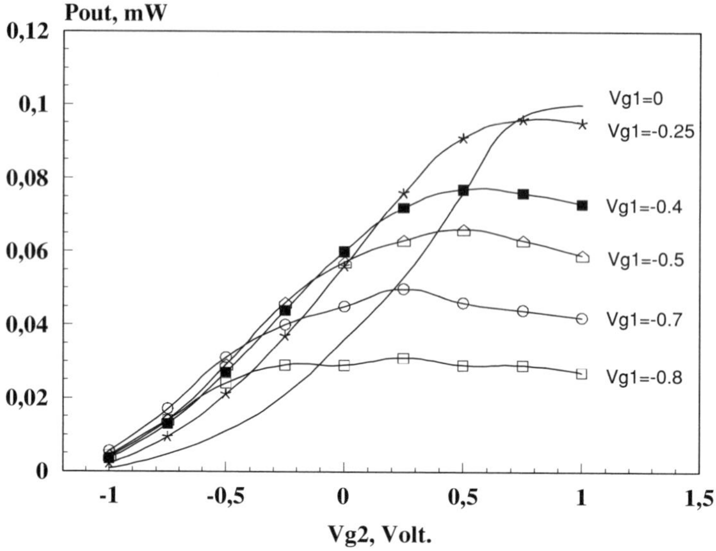

Figure 5.39 IF output power vs V_{g2}

For the FET2, the drain source current can be expressed as:

$$I_{ds} = I_{dss}\left(1 - \frac{V_{gsi2}}{V_{T2}}\right)^2 (1 + \lambda V_{dsi2}) \tag{5.121}$$

with $V_{dsi2} = V_{ds2} - I_{ds}(R_s + R_g)$, $V_{ds2} = V_{dd} - V_{ds1}$, $V_{gsi2} = V_{g2} - I_{ds}R_s - V_{ds1}$ and $V_{T2} = -1.8$ volt.

However, the drain current for FET1 and FET2 are the same, so expressions (5.120) and (5.121) must be identical. Solving the previous system, we can obtain the drain-source voltage for FET1 as a function of the external bias V_{g1} and V_{g2}. Using these values in (5.119), we can plot the IF output power as a function of V_{g2} with V_{g1} as a parameter, as shown in Figure 5.39.

A simplified model of the MESFET transistor for the F20 process of the GEC Marconi foundry has been used to perform all the calculations. The gate width for the transistors used was 300 μm. The parameters used are:

$$\lambda = 0.008 \quad \alpha = 2.52 \quad I_{dss} = 0.0427 \text{ A}$$

From the simplified analysis exposed above, we can get a first design for the mixer. After that, a final adjustment of the circuit can be done with a commercial non-linear simulator.

5.7.1 IF Amplifier

In order to get a greater conversion gain an IF amplifier has been added. A common source amplifier has been used with a matching stage at the input. A common gate transistor has been used as matching stage (Figure 5.40).

The transistor width has been chosen in order to match the output impedance of the mixer to the input impedance of the amplifier, which is inversely proportional to g_m. The value of R_{in} can be fixed with this expression:

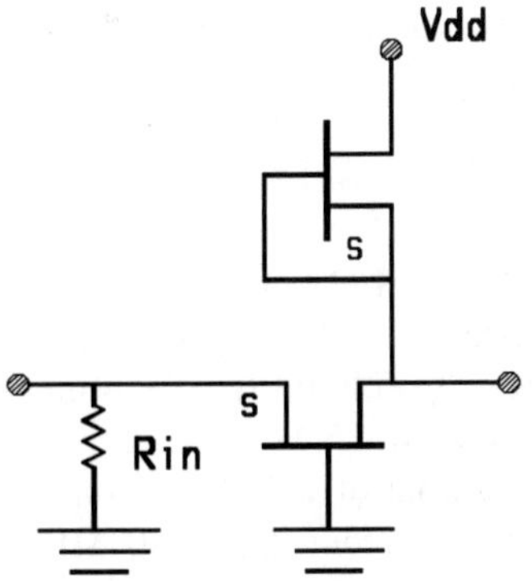

Figure 5.40 IF amplifier matching stage

$$R_{in} = \frac{-V_{gs}}{I_d} \tag{5.122}$$

where I_d corresponds to the saturation drain current: $I_d \cong I_{dss}\left(1 - \frac{V_{gs}}{V_p}\right)^2$

5.7.2 Final Design

Using the initial values obtained in the simplified analysis, the whole mixer has been simulated with the MDS (Microwave Design System) program, from Hewlett-Packard. The Harmonic Balance technique has been used to simulate this mixer because it is considered more suitable for this kind of circuit. From this simulation, load cycles have been found for both transistors (Figure 5.41). I/V DC curves have been added in the same figure. For the upper transistor (FET1), it can be seen that it is biased into the linear region, while FET2 is biased into the saturation region during the whole LO cycle. Thus it is verified that the transistor

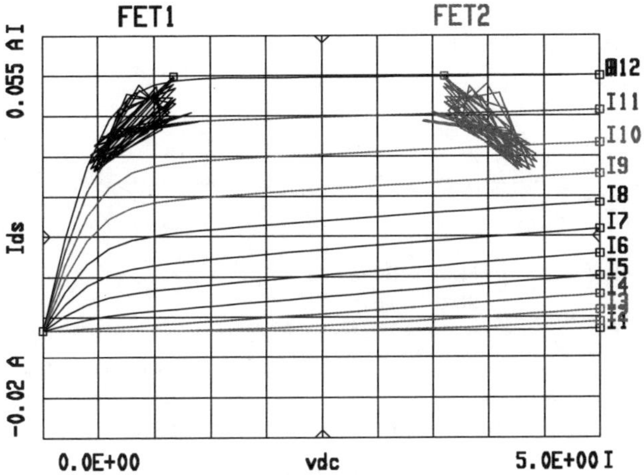

Figure 5.41 Load cycles for both transistors

bias points are the same as the calculated ones in the simplified analysis, which allows us to validate the design procedure.

5.7.3 Mixer Measurements

The circuit has been fabricated in the GEC Marconi foundry, which uses 0.5 μm gate length MESFET transistors. The chip size is 2×2 mm^2. The same design method can be used if a hybrid implementation is desired. This circuit has been measured on wafer with a coplanar probe station (Cascade Microtech). Figure 5.42 shows gain conversion in the IF band between 40 to 460 MHz, for an input RF signal of 1.5 GHz. Simulation results have been added in the same figure showing a good agreement. For a 40 MHz IF signal, gain conversion has been measured. Figure 5.43 shows measured and simulated results for a LO input power of +7 dBm.

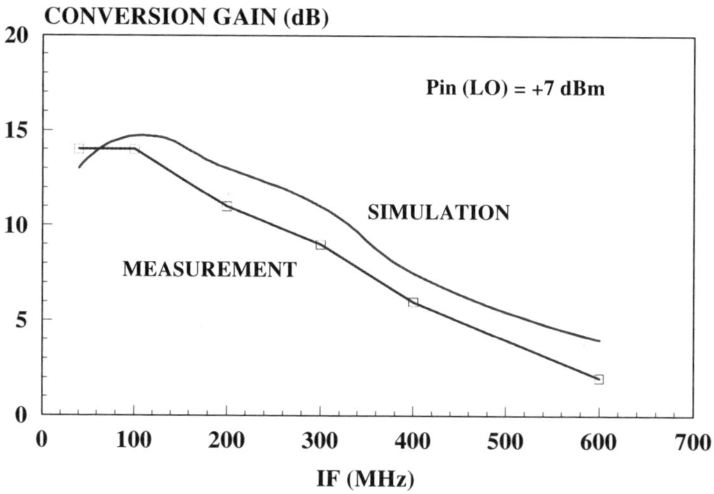

Figure 5.42 Conversion gain vs IF frequency

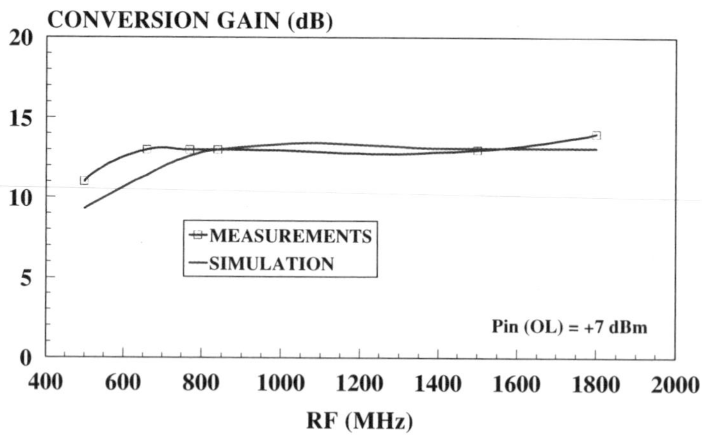

Figure 5.43 Conversion gain vs RF frequency

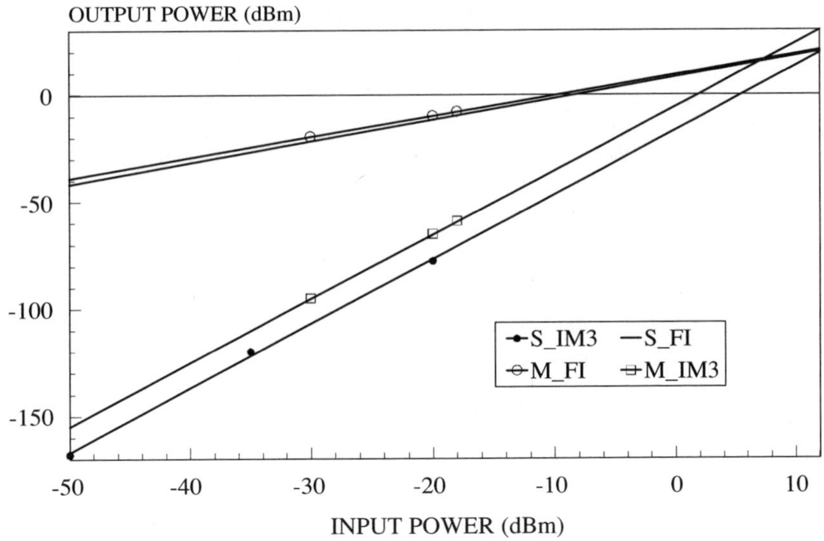

Figure 5.44 Measured and simulated IP3

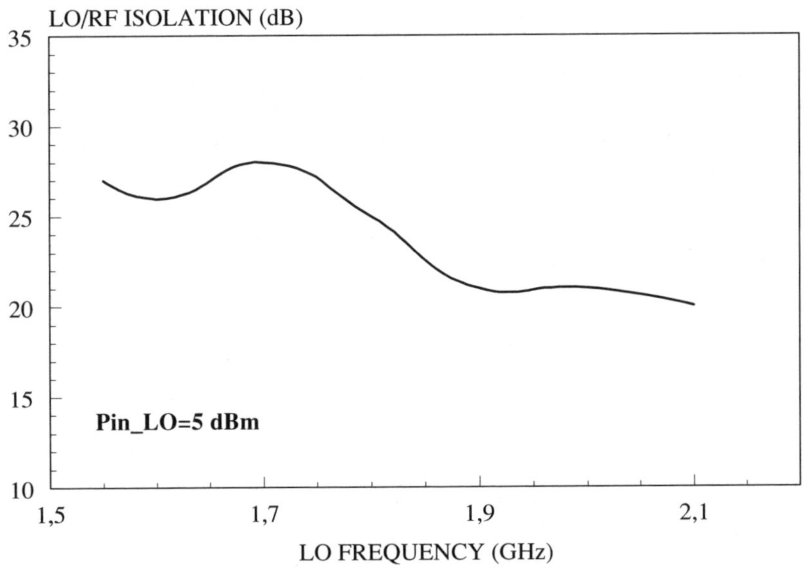

Figure 5.45 LO/RF isolation vs LO frequency

Two tones test has been done to characterise the mixer intermodulation performance. 7 dBm third order interception point (IP3) of input power has been obtained. Measurement and simulation results can be seen in Figure 5.44.

LO/RF isolation has been measured in the LO frequency band, as can be seen in Figure 5.45. More than 20 dB has been obtained, illustrating the inherent isolation between these ports for this topology.

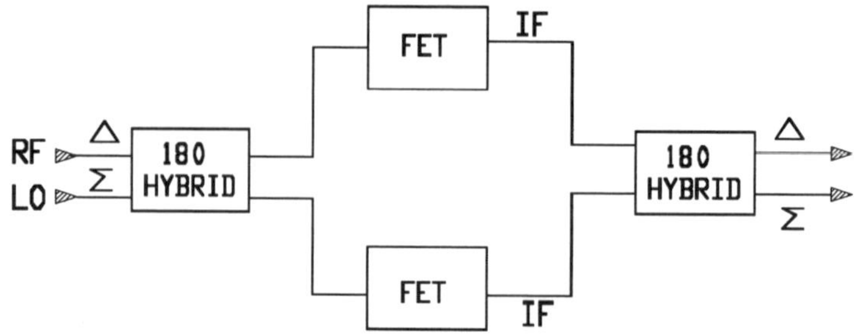

Figure 5.46 180-degree singly balanced FET mixer

5.8 Single-Balanced FET Mixers

Two transistors are required to create a singly balanced mixer. Two single-device mixers can be combined via 90° or 180° hybrids to make a singly balanced mixer. The properties of this kind of mixers are similar to those of the singly balanced diode mixers. In order to add the IF currents, the diodes are placed in opposite senses. However, FETs cannot be reversed, so an IF hybrid is necessary at the output [1] (see Figure 5.46). The design of these hybrids is the same as that of a diode mixer, and must be chosen depending on the frequency band and the implementation type.

In the most general case, RF and LO are applied to the sum (Σ) and difference (Δ) ports, respectively, because in this case, LO signal is cancelled at the delta port. Nevertheless, RF and LO ports can be reversed and the conversion gain and noise figure will be the same, but the spurious-response characteristics will change. Because the IF output is derived from the delta port, the spurious-response rejection properties of the singly balanced FET mixer are the opposite of a singly balanced diode mixer.

Figure 5.47 shows an example of a singly balanced HEMT upconverter. The LO signal is applied to the gates of the transistors thanks to a 180° balun, which is simply a microstrip

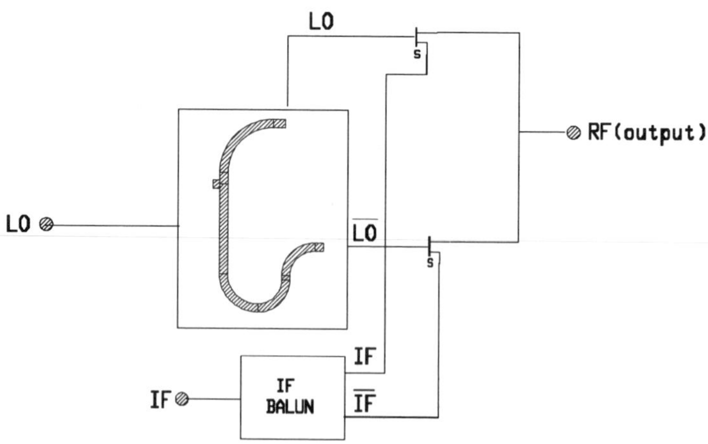

Figure 5.47 Singly balanced HEMT upconverter

T-junction with a electrical length difference of $\lambda/2$ between the two outputs. The IF input signals are applied to the source of the transistors using a 180-degree balun. Lumped elements were used for this balun, since the IF frequency is not high enough.

The RF signal is extracted from the drains of the transistors. Because LO signals are applied in the opposite phase, they are cancelled at the RF output. So good LO/RF isolation is obtained without filters. In order to eliminate the sum frequency, a band-pass filter must be added at the output. Drains of the HEMTs are not biased, so the channel resistance of the transistors causes the mixing. The transistor gate must be biased where the channel resistance is most non-linear with gate-source voltage, normally near the pinch-off value. To optimise the whole upconverter, a non-linear simulator can be used.

5.9 Double-Balanced FET Mixers

Double-balanced mixers make use of four devices for mixing. The LO and RF signals must be applied with a 180° phase shift, so two baluns must be included to provide these phase differences. Double-balanced FET mixers show similar properties to double-balanced diode mixers, although the former presents conversion gain. Since there are four devices, it is not possible to optimise or adjust the whole mixer in the same way as in single-device mixers. On the other hand, a perfect balance between branches is very difficult to achieve in hybrid technology, which provokes a poor performance at high frequencies.

The Gilbert cell is perhaps the best-known double-balanced topology. The basic structure is shown in Figure 5.48. It consists of two differential pairs connected in such a way that the different ports are mutually isolated. RF signals are injected through the gates of the lower transistors, with a 180° phase difference. LO signals are injected through the gates of the upper transistors, with a 180° phase difference as well. The lower transistors are biased in the saturation zone, so they amplify the RF signal. The LO signal is injected through the upper transistors, and this is where the mixing occurs.

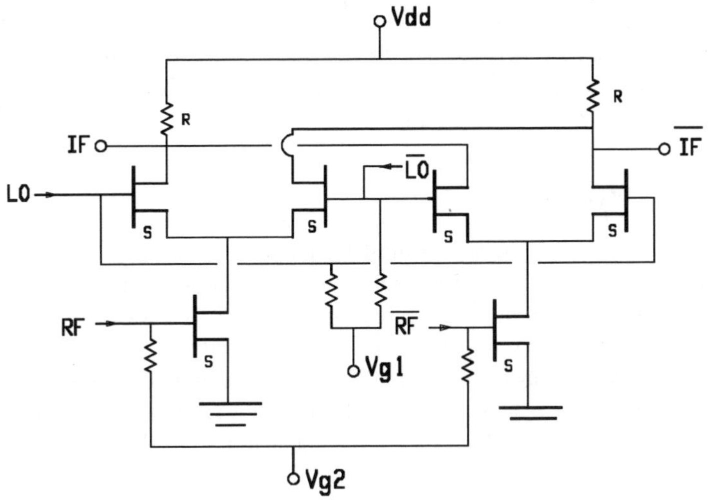

Figure 5.48 Basic structure of a Gilbert cell

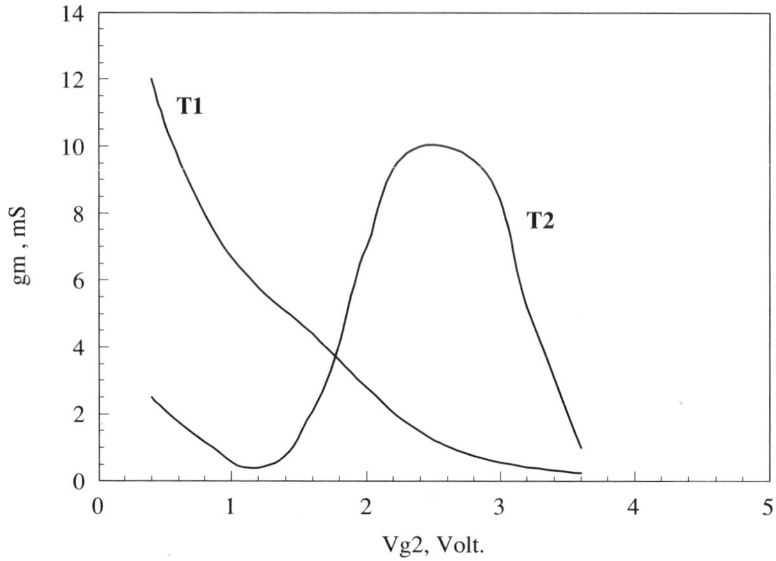

Figure 5.49 Transconductance of T1 and T2 as a function of T2 gate voltage

Figure 5.49 shows the transconductance variation [7] of one of the lower transistors (T1) and one of the upper transistors (T2) with gate voltage of T2.

The gate-source voltage of T1 transistor is chosen to place it in its saturation region. The load line for transistor T2 (I_{ds} vs V_{ds}, when V_{g2} is varied) has been drawn as well as the I/V DC curves for that transistor (Figure 5.50).

M3 can be a good point since V_{g2} corresponds to 1.4 volt. and it corresponds to a non-linear zone which will give rise to a maximum conversion gain.

In order to reduce the DC consumption, the load resistors (*R* in Figure 5.48) were changed to active load (transistors with gate and source short circuited). The gate width of these transistors must be the same as the lower transistors, since the DC current is equal. Figure 5.51 shows the final scheme of the Gilbert cell, which must be optimised using a non-linear simulator (harmonic balance, for instance).

5.10 Harmonic Mixers

A harmonic mixer is a device where a high frequency signal (RF) is mixed with a local oscillator signal, whose frequency is much lower than the former. An output signal is obtained whose frequency is the difference between a harmonic of the local-oscillator and the RF signal. Historically harmonic mixing has been used primarily at the higher millimetre wave frequencies where reliable and stable LO sources are either not available or prohibitively expensive. Sometimes these devices are used in frequency-multiplier design by means of phase loop oscillators (PLOs) or as external mixers of spectrum analysers. Although theoretically any LO harmonic can be used, second and third order are the most common, since the conversion loss is increased with higher orders. Schottky diodes and FETs (MESFET or HEMTS) are the mixing element most used.

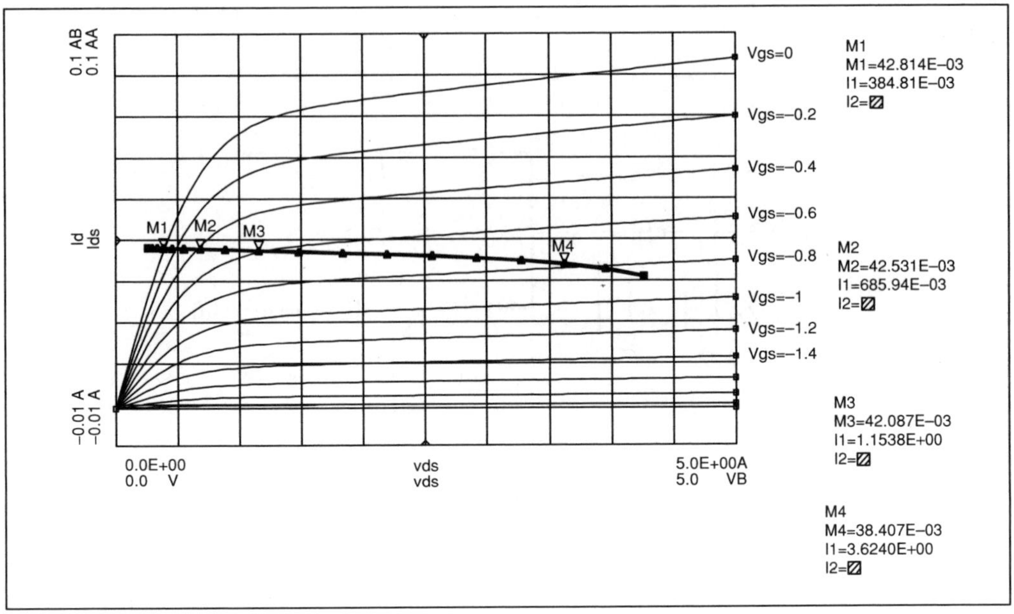

Figure 5.50 I/V DC curves and load line for the T2 transisor

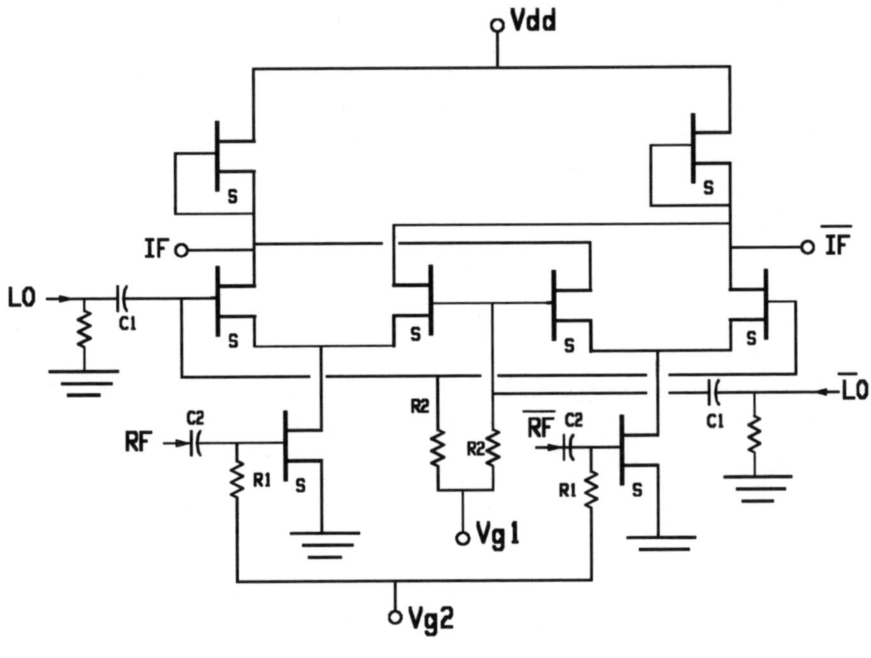

Figure 5.51 Electrical scheme of the mixer

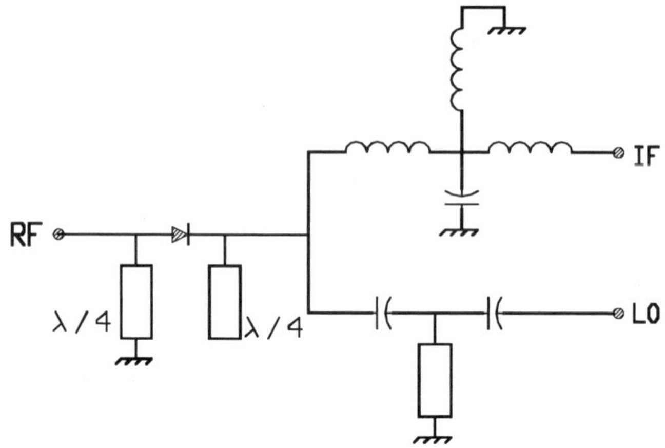

Figure 5.52 Scheme of the harmonic mixer

5.10.1 Single-Device Harmonic Mixers

Harmonic mixers using a single device can obtain the lowest conversion loss and since LO and RF signals are very different, the design is usually quite simple. In order to obtain low conversion loss, fundamental mixing between the signal and LO must be suppressed.

Figure 5.52 shows an example of a harmonic mixer [8] using the seventh harmonic of the LO signal. A Schottky diode was used as the mixing element. On the RF input side, a ($\lambda_R/4$) short-circuited stub allows the signal to pass but stops the IF signal. Similarly, on the LO and IF side the ($\lambda_R/4$) open circuited stub allows the LO and IF to pass but stops the RF signal.

A diplexer is used to inject the LO signal and to extract the IF, realised by high-pass and low-pass filters. An inductance to ground in the low-pass filter is used as the DC return.

The RF input frequency band is 12.105–12.365 GHz and the LO signal is fixed in 1.815 GHz. The IF frequency band is 0.34–0.59 GHz. The circuit was implemented in hybrid technology, using Cuclad 2.17 as substrate, and mounted on a metal case with 3.5 mm connectors to feed the RF, LO and IF signals. The diode used is a silicon Schottky diode, 5082–2774, from Hewlett-Packard. The harmonic mixer was simulated using harmonic balance and the microstrip lines were adjusted with the aid of an electromagnetic simulator (Momentum module from Hewlett-Packard). The conversion loss is shown in Figure 5.53, where a conversion loss of 24 dB is seen across the whole RF frequency band, with 13 dBm LO drive. The isolation between LO and IF ports was better than 47 dB thanks to the diplexer, and more than 50 dB of LO/RF isolation were measured in the RF frequency band. Figure 5.54 shows a photograph of the harmonic mixer.

5.10.2 Balanced Harmonic Mixers

Although single device harmonic mixers can achieve the lowest conversion loss, balanced topologies are commonly used, since many spurious products are rejected. Figure 5.55 shows a generic circuit of a Schottky diode-based subharmonic mixer [9]. It incorporates an

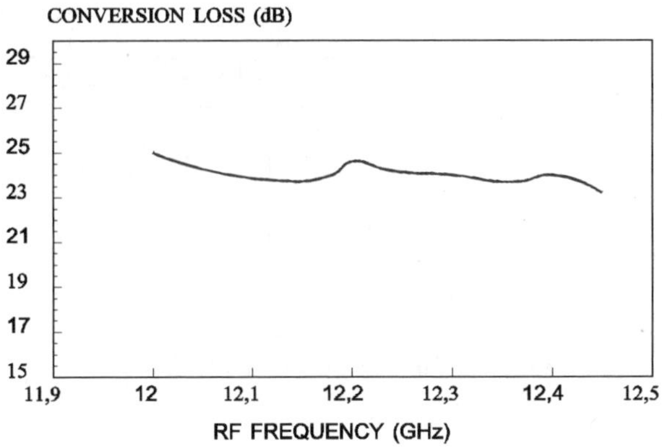

Figure 5.53 Conversion loss of the harmonic mixer

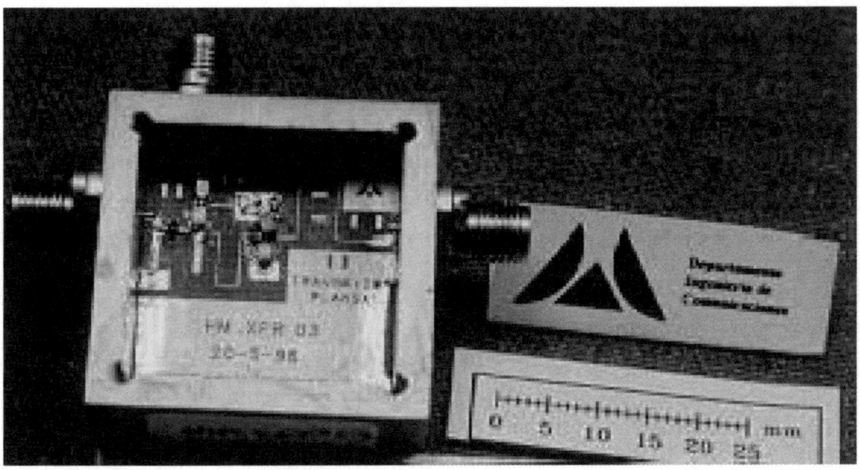

Figure 5.54 A harmonic mixer

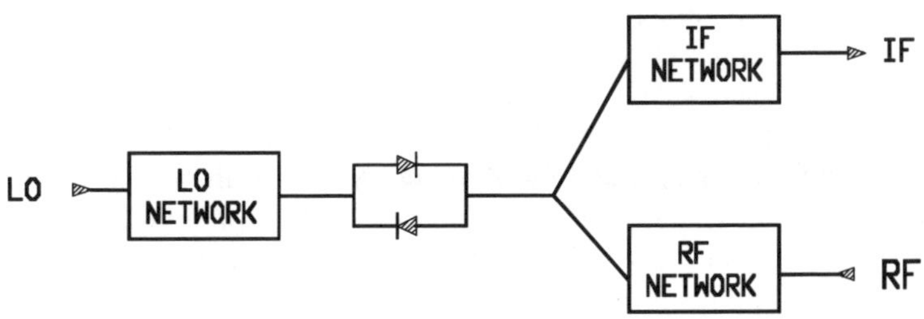

Figure 5.55 Subharmonic diode mixer

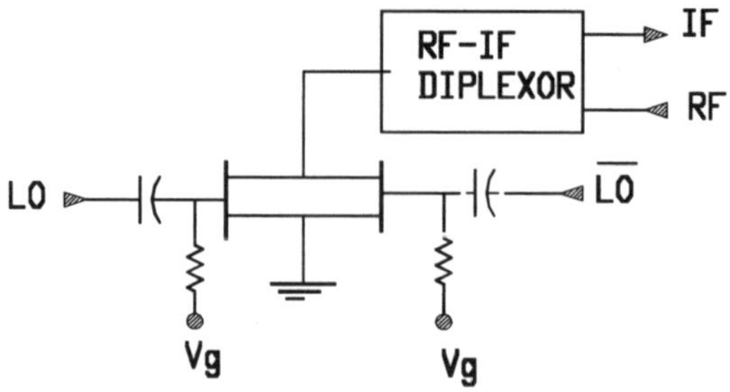

Figure 5.56 Subharmonic FET mixer

anti-parallel diode pair, and the even order mixing products are suppressed ($mf_{RF} \pm nf_{LO}$). Thus the fundamental mixing product, $f_{RF} - f_{LO}$, is quenched, thereby eliminating an additional loss mechanism and interference source.

The anti-parallel diode pair I/V curve is an odd function: $I(V) = \sum_{n=0}^{\infty} C_{2n+1} V^{2n+1}$, and only odd order mixing products are generated at the diode pairs' terminals.

Although the diode has been the most used non-linear element for this kind of mixer, FETs can replace the former as shown in Figure 5.56. The LO signal is applied to the gates with a 180° phase difference. The sources and drains are connected together. RF signal is applied to the drains where the IF signal is also extracted, although the RF signal can be applied to the gates if frequency bands are far enough apart.

FETs can be considered to operate in the passive or active regions. For the former, the conductance, $\partial I_{ds}/\partial V_{ds}$, is the dominant non-linearity, while the transconductance, $\partial I_{ds}/\partial V_{gs}$, is the main non-linearity for the latter. In both cases, the devices are operated near the pinch-off value to achieve the highest conversion gain.

One difference between diode and FET subharmonic mixers is that the FET mixer uses a LO balun, which is not necessary for the diode implementation. This balun represents the major disadvantage of this circuit, especially at low frequencies. Nevertheless when conversion gain is required, balanced FET subharmonic mixer is the best solution.

5.11 Monolithic Mixers

The gallium arsenide monolithic microwave integrated circuit (GaAs MMIC) technology has undergone great development in the past twenty years, achieving a maturity grade similar to that of the silicon technology [10]. A monolithic circuit is one where all components, both passive and active, are incorporated into a single semi-conductor die allowing complete operation by the application of DC and microwave signals. Thus, very little wire-bonding and assembly is required and the size and weight of the circuits are usually much smaller, allowing each subsystem size reductions within the same volume as occupied by a hybrid circuit. Monolithic circuits can be fabricated in large quantity at low cost, allowing their use in consumer applications. Although the prices have been decreasing in the last years, the major limitation is the high cost of prototypes. Only for large quantities, are the prices comparable to the hybrid circuits. Perhaps the major advantage to be gained from monolithic microwave technology will come about when high levels of integration can be produced at affordable costs with acceptable performance. Many circuit functions now available with MMICs would have been impossible to produce using conventional substrate-based hybrid technology. This is particularly true in the case of circuits requiring many different gate-width FETs. Although the hybrid circuits are still being used in many applications, MMICs have begun to make up an important part of many available microwave products.

5.11.1 Characteristics of the Monolithic Medium

Due to the great development in monolithic technology MMICs are being used in many applications. The main reasons for improved performance of MMICs over their hybrid counterparts include the following:

- The assembly interconnects are eliminated, which reduces the parasitics due to bond wires.
- The reliability of the circuits can be improved owing to the much-reduced number of interconnections.
- All the components can be optimised to meet the needs of the circuit performance, without being limited to a discrete catalogue of components.
- The cost of each MMIC does not depend on the number of active elements, as in hybrid technology.
- Highly reproducible performance that provided by batch processing is possible and
- All circuit functions can be integrated.

Nevertheless, there are some disadvantages compared to hybrid circuits. Considerable investment is required in manufacturing facilities and staff in order for a company to produce its own circuits. Although the technology has improved, MMIC prices are only of interest for circuits that will be made in large numbers. On the other hand, the fabrication of the monolithic circuit is the end of its design cycle. No adjustment can be made after the circuit is finished. This fact requires us to model the components of the circuit in a very accurate way. In fact, the designer is limited by the precision of these models, above all by those of the active elements. The modelling of active components is not very different from modelling hybrid circuits. As there is great freedom in adjusting the geometries of FET devices, the modelling of such components can be more difficult.

For mixers, the non-linear performance of the devices that produce the mixing must be characterised in an accurate way, in order to predict the conversion losses of the circuit. This is a difficult task for some devices, so the performance prediction in mixer circuits is more complicated than for other circuits. Passive elements are sometimes measured and their s-parameters stored in a database, which are then used in the design.

From all the chips produced on the same wafer, only a proportion of them will operable due to imperfections in processing. The term yield refers to the number of circuits on a given wafer that deliver acceptable electrical performance. Clearly the cost of a chip is inversely proportional to yield, and thus designers must avoid structures that cannot be produced reliably.

5.11.2 Devices

MMIC mixers can be classified into transistor and diode structures. If GaAs MESFET technology is considered, the diode mixer is made with a FET, using the junction between the gate and the channel. Although the cut-off frequency is higher for the diodes, the width and the geometry of the MESFETs can be modified in order to obtain the desired specifications. Its is nearly impossible to find this degree of freedom with discrete devices in hybrid circuits. The MESFET transistor is commonly used in monolithic technology for frequencies above 1 GHz. Active MESFET mixers offer many advantages over passive ones. This is especially true for double-gate MESFETs, because there is an inherent isolation between the RF and LO ports.

However, there are several drawbacks when designing MESFET mixers. If a diode is used as the non-linear element, it is possible to obtain a good first-order approximation with a linear analysis. But, with the MESFET mixers, the analysis becomes very complicated. Moreover, when there are several transistors, it is necessary to use computer-aided design in order to predict the performance of the mixer.

There are some applications where mixers can be done with FETs but not with diodes. Sometimes, this advantage is due to the compatibility of the FET transistor with the monolithic circuit processes. However, in other cases, it is due to the inherent advantages of the FET over the diode mixers.

More recently HEMTs and HBTs have appeared; the former uses a GaAs' substrate and the latter can use either GaAs or silicon. Both of them have a common characteristic: a heterojunction is used in its construction. These heterojunctions are formed between semiconductors of different compositions and band gaps. The properties of these new devices are superior to those of the MESFET in having a higher cut-off frequency, greater gain, and lower noise figure. As these processes are newer, one might suppose that yield will be lower than in MESFET processes. Much effort has been put in over recent years in order to improve yield of these processes. Nowadays, although the maturity of these technologies is not that of the silicon ICs, we can hope that it will be possible in the near future.

5.11.3 Single-Device FET Mixers

The design procedure used for these mixers is the same as that of the hybrid mixers. Single-device mixers usually have greater conversion gain than balanced topologies and the optimisation method is easier. In order to avoid using a large substrate area, lumped-element circuits can be used when the frequency band is not very high.

Figure 5.57 shows an example of a single-device HEMT mixer [11]. This topology corresponds to a drain mixer, since the LO signal is injected by the drain. RF and LO signals were applied to separate terminals because their frequency bands were very close. As explained in Section 5.6.1.3, when the LO signal is injected through the drain, the transconductance g_m and the channel resistance R_{ds} are modulated and become time-varying functions, with a period equal to that of the LO signal. The maximum conversion gain is given by (see Section 5.6.1.3 for more details):

$$G_m = \frac{|\mu_1|^2}{4\omega_1^2 C_{gs}^2 (R_{gm} + R_s)(R_{dr} + R_s + R_{do})} \tag{5.123}$$

Where $\mu = g_m R_{ds}$ is the voltage amplification factor.

Therefore, the fundamental component of this voltage amplification factor must be maximised in order to get as great a conversion gain as possible. For this purpose it is necessary to find a bias point for the transistor in which μ is more sensitive to the bias modulation by the LO signal. For a drain mixer, the channel resistance non-linearity has the same contribution to the mixing as the tranconductance non-linearity. For this reason, both of them must be taken into account. Figure 5.58 shows G_m versus drain-source voltage with gate-source voltage as a parameter (following Section 5.56).

At the RF input a lumped-elements low-pass filter structure has been used. This structure has similar properties to those of the ladder transformers, but it can be built in a compact way, even at low frequencies. The main difference between this topology and a low pass filter is that the resistances at the terminals can be different, which means that the reflection losses at zero frequency will be reasonable. In this case, lumped elements were used to implement the input and output networks since the working frequency is not very high and these networks would occupy a large substrate area if distributed elements were used. To bias the transistor gate, a lumped-element circuit was designed, also creating a short circuit at IF on the RF port. This circuit down-converts 14–17 GHz RF signals to a 1 GHz IF band with 2.5 dB of conversion gain by using 15 dBm of LO drive. Table 5.1 summarises the measured results of this circuit.

For the LO and the IF signals a diplexer has been designed, since both frequency bands are very distant in frequency and good isolations can be achieved. Lumped elements were used for the same reason than in the RF input network.

This mixer was fabricated by Philips Microwave Limeil (France) using the D02AH process, which uses 0.2 μm of gate length in the HEMT transistors. The chip size is 1.5 mm^2, and a photograph of the mixer is shown in Figure 5.57.

Table 5.1 14–17 GHz MMIC single-ended mixer

RF bandwidth	14–17 GHz
RF bandwidth	1 GHz
NF (SSB)	7.6 dB (15 GHz)
Gain	2.5 dB
Isolation LO to RF	> 23 dB
LO to IF	> 42 dB
Return Losses (LO, RF)	> 10 dB
LO power	15 dBm

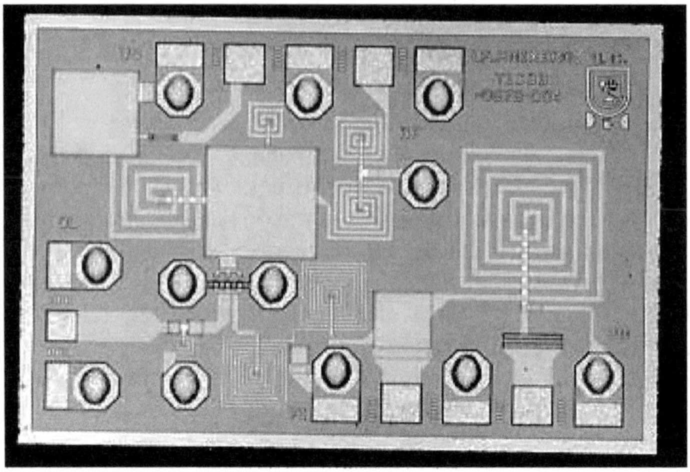

Figure 5.57 The mixer

5.11.4 Single-Balanced FET Mixers

This kind of mixer is widely used in monolithic circuits. Thanks to the homogeneous performance of the components fabricated in the same wafer, the isolations between ports is higher than in hybrid technology. In the latter case balanced mixers often use large passive structures, such as Branch-Line or Rat-Race couplers, to form the baluns. Nevertheless, a large substrate area is required if monolithic implementation is desired. Lumped elements are another option to make these distributed baluns, but they generally present low bandwidth and higher losses.

Figure 5.58 shows an example of singly balanced mixer [12]. Mixing and balun functions are carried out by the same structure. This topology is similar to that of the centre tapped balun. Two transistors in gate-common, source-common structures realise the mixing and the phase shifting, thanks to the way they are connected.

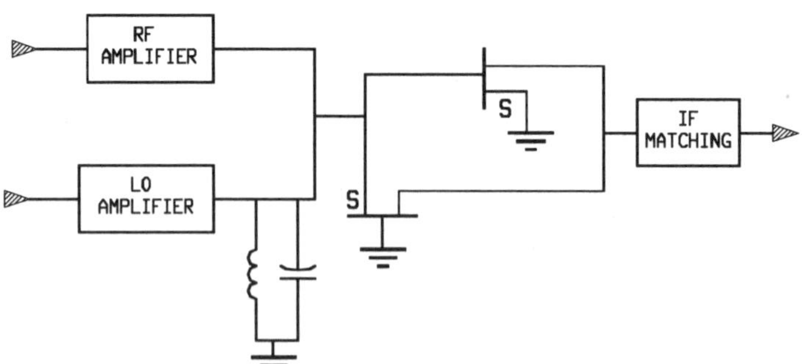

Figure 5.58 Singly balanced FET mixer

For the common gate HEMT, the LO signal is injected between the gate and source ports. In this case, the time-varying transconductance is the dominant contributor to frequency conversion, and the effect of other non-linearities is minimal. The conversion gain is proportional to the fundamental LO-frequency component of such a transconductance waveform $g_m(t)$. In a HEMT transistor, g_m is the maximum when V_{ds} is the maximum, so its drain voltage must remain in the saturation region throughout the LO cycle.

In order to avoid a drop in the drain/source resistance and an increase in gate/source capacitance, it is important to guarantee that V_{ds} does not decrease below the knee voltage. Otherwise, the conversion gain will decrease and the noise figure will increase. For the common-gate transistor, the input impedance is proportional to $1/g_m$. The width of this transistor is chosen to match the combiner output impedance to the common-gate transistor input impedance.

For the common-source transistor, the time-varying transconductance is also the dominant contributor to frequency conversion. The maximum conversion gain is [4]:

$$G_c = \frac{g_1^2 \overline{R_d}}{4\omega_1^2 \bar{C}^2 R_{in}} \tag{5.124}$$

Where $R_{in} = R_g + R_i + R_s$, $\bar{C}$ is the average value of C_{gs} and $\overline{R_d}$ is the average value of R_{ds}. g_1 is the magnitude of the fundamental component of the transconductance waveform. In order to maximise g_1, the transistor must be biased near the pinch-off value. The variation of g_m with the gate voltage for a 300 μm gate width is shown in Figure 5.59. In the same figure the variation of g_1 as a function of the gate voltage was added. It can be seen that maximum value of g_1 corresponds to the device turn-on voltage. From both figures, it can be seen that the g_1 maximum (13 mS) is 1/3.3 of g_m maximum (43 mS). This ratio is very close to the $1/B$ ratio obtained for the ideal case, when the g_m is a step function of gate voltage.

By selecting the gate width of the common-source transistor, it is possible to have the same gain in both devices (common-source and common-gate), and IF and LO signals will be cancelled at the output of both transistors due to the phase shift (ideally 180°) in the common-source transistor. This is true in low frequencies but for higher frequencies, the

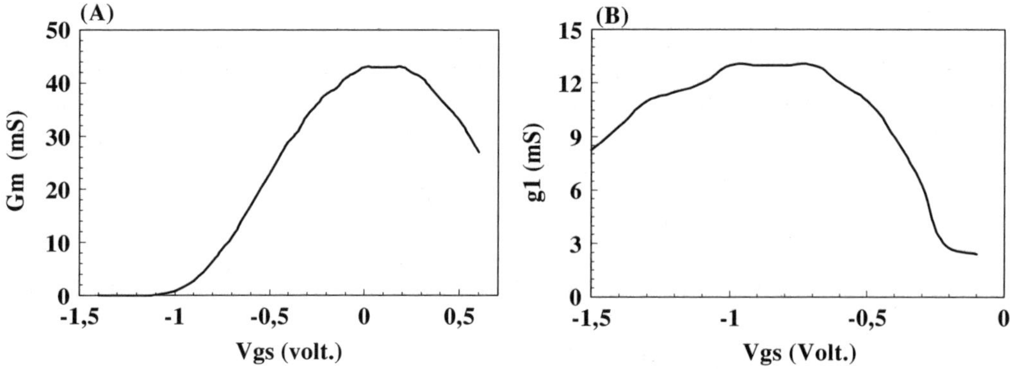

Figure 5.59 (a) g_m vs V_{gs}. (b) g_1 vs V_{gs}

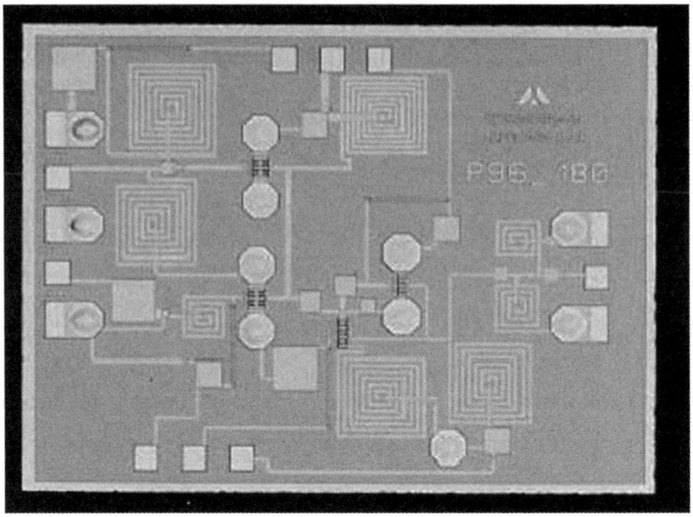

Figure 5.60 The singly balanced FET mixer

phase shift begins to differ from 180°, decreasing the isolation between output and input ports. When LO and RF signals are very close in frequency, it will be easier to optimise the mixer at both frequency bands. Nevertheless, if LO and RF frequencies are very different, that is the case of an upconverter, it will not be possible to have good balance (phase and amplitude) at both frequencies, and filters will be necessary in order to achieve good isolations.

The combiner for LO and RF signals is formed by two amplifiers, providing good isolation between both ports, and furthermore, amplifying both signals. Lumped elements were used to match amplifier input circuits.

The output network can be different depending on whether the mixer is to be a down-converter or an upconverter. For the first option, a matching network is sufficient, but for an upconverter a high-pass filter will be necessary for the reasons explained above.

The mixer of the example up-converts 1.885 GHz IF signal to a 14–14.25 GHz RF band with 4.2 dB of conversion gain by using only 3 dBm LO drive. This mixer was fabricated in Philips Microwave Limeil (France) using the D02AH process, which incorporates 0.2 μm of gate length for the HEMT transistors. Figure 5.60 shows a photograph of the mixer. The chip size is 3 mm^2.

5.11.5 Double-Balanced FET Mixers

When the input and output frequency bands are very close, or for very broadband applications, double-balanced mixers can obtain the best performances. This is especially true in a monolithic implementation, where the dispersion is equal for similar elements. This means that all the transistors in the chip will have the same performance, and the same can be said for the capacitors, inductors, etc. on the chip. This property produces double-balanced

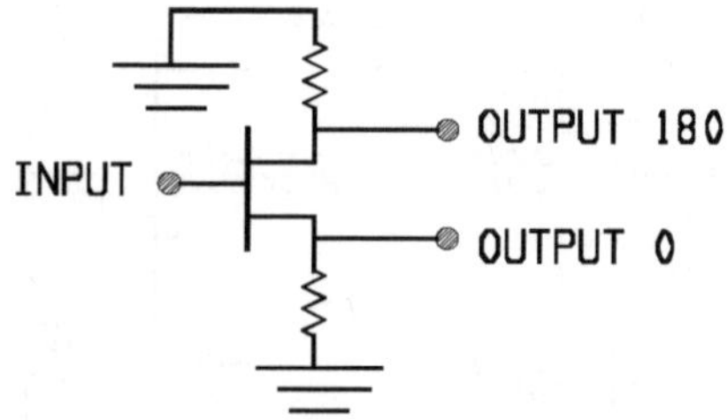

Figure 5.61 Active balun

mixers with better isolations between the ports when they are fabricated in monolithic technology.

The design problem for balanced mixers can be divided into two main areas: the non-linear element and the balun. If a monolithic implementation is desired, the balun dimension is limited by the chip area, especially for low frequencies (below 20 GHz). Thus, active balun or lumped-element transformers are the only viable options. An active balun is shown in Figure 5.61. This circuit uses the known property of a transistor amplifier, where the phase shift between drain and source ports is 180° ideally. This property is only true at low frequencies. When the frequency begins to increase, the phase shift change due to capacitance and inductance affects the transistors.

In a high frequency analysis, and using a unilateral electrical model for the MESFET transistor, we can get:

$$\frac{V_1}{V_2} = \frac{-g_m R_{ds} - \omega^2 R R_{ds} C_{gs} C_{ds} + j\omega R C_{gs}}{g_m R_{ds} - \omega^2 R R_{ds} C_{gs} + j\omega (R_{ds} + R C_{gs})} \tag{5.125}$$

where:

R_{ds}: channel resistance
g_m: transconductance,
R: the load resistance at the output
C_{gs}: the gate-source capacitor
C_{ds}: the drain-source capacitor.

There is other balun topology which can be implemented using monolithic technology, and can eliminate these problems. Two source common amplifiers coupled by their sources to a resistance or current source are used (Figure 5.62). This topology, commonly used in low frequency circuits with bipolar transistors, can be translated to microwaves frequencies using MESFET transistors as amplifier element. In the traditional differential stages, the basic function is to amplify the difference between two input signals. Nevertheless, one of the inputs is connected to ground using a capacitor if the circuit must operate as a balun. The outputs are taken from the transistor drains.

Although this topology obtains better results, when the frequency increases, the phase errors increase as well, which makes the circuit not very useful for microwave balanced mixers. Better results can be obtained if a second stage is connected at the output.

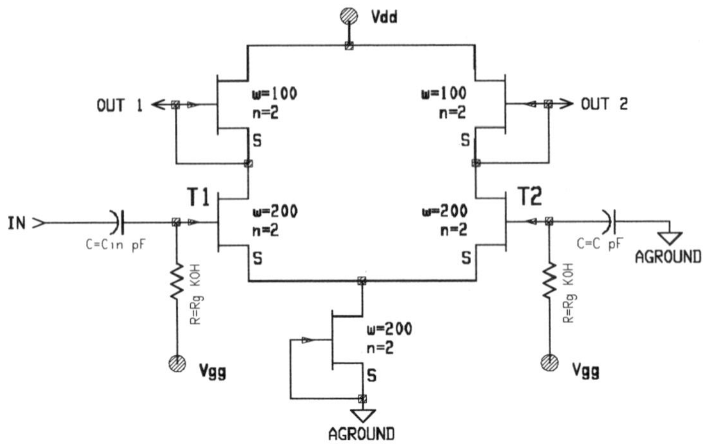

Figure 5.62 Balun topology

By using this last topology as balun, a double-balanced mixer was designed [7, 13]. A MESFET ring was used as mixer and two differential pairs as RF and LO baluns. In order to get low intermodulation levels, cold FETs were used as the mixing element. The channel resistance of a MESFET transistor without biasing can vary when a signal is applied to the gate. For a non-biasing transistor, the drain current is nonlinear with V_{gs}, which allows the channel resistance to change with time. Nevertheless, the drain current is almost linear with V_{ds}. Thanks to this last characteristic the distortion level generated is very low when compared to that generated by a diode. Thus, the LO signal is applied to the gate and the RF signal to the drain.

The IF signal is filtered from the drain for single-ended mixers, but for balanced topologies it is obtained from the source. In both structures (single or balanced), there is a problem if the transistor is biased at zero volts. The capacitance between the drain and gate terminals increases until values close to the C_{gs} capacitance, which would provoke some coupling between LO and RF ports and diminish the isolation between these terminals. Also, if LO signal is coupled to drain port the voltage at this terminal would greater than zero, increasing the intermodulation level. It is possible to avoid these problems by using filters in both ports. However, for broadband applications balanced topologies must be used, as it is shown in the example of the Figure 5.63.

The variation of channel conductance with gate-source voltage, for a 400 μm gate width MESFET, is shown in Figure 5.64. The mixer transistors are biased in the most non-linear region in order to get the lowest conversion losses. As the channel conductance is being modulated by the LO signal, G_{ds} can be developed in a Fourier series:

$$G_{ds} = g_0 + 2g_1 \cos(\omega_{LO} t) + \ldots \tag{5.126}$$

where ω_{LO} is the local oscillator frequency.

g_1 must be maximised in order to get the minimum conversion losses. This means that the transistors will be biased near the pinch-off voltage ($\cong -1.7$ volt.). This gate biasing point is

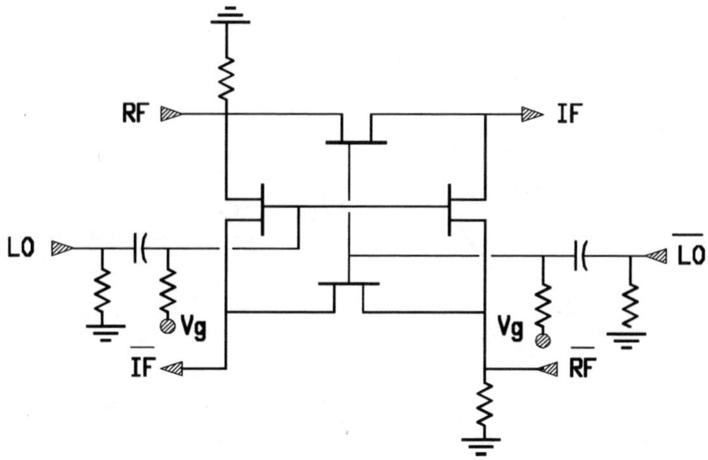

Figure 5.63 Double-balanced FET mixer

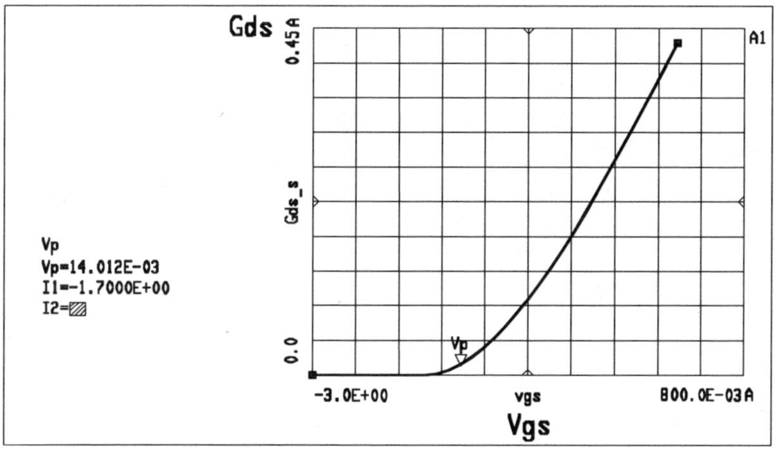

Figure 5.64 G_{ds} versus V_{gs}

the most convenient, since small variations of V_{gs} provoke large changes of G_{ds}. It means that this is the most sensitive biasing point for the channel resistance with respect to gate voltage changes.

In order to get a low distortion level, a RF balun should be designed with a view to not degrading the IM performance of the mixer.

The double-balanced mixer of the example down-converts 1.8 GHz of RF frequency to 40–860 MHz IF band with 2 dB of conversion losses. LO/IF isolation is shown in Figure 5.65 as a function of the transistor gate voltage. As we can see in the figure, more than 40 dB is achieved, showing the good isolation between ports in double-balanced topologies.

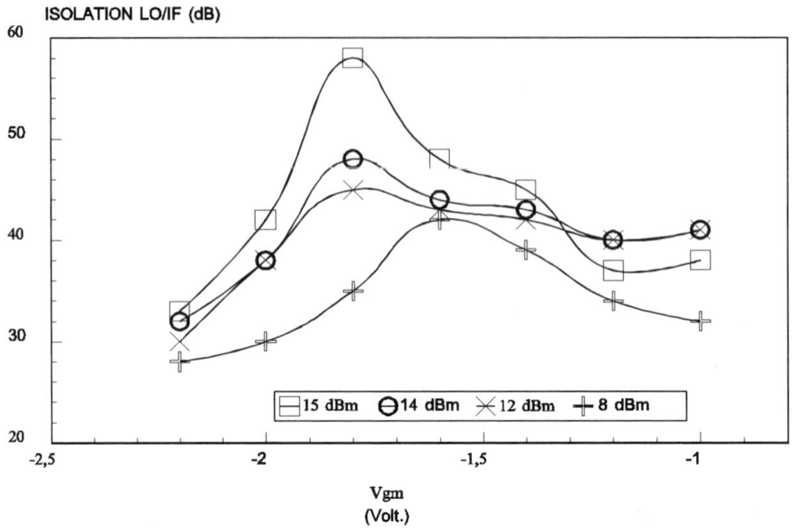

Figure 5.65 LO/IF isolation vs V_{gs}

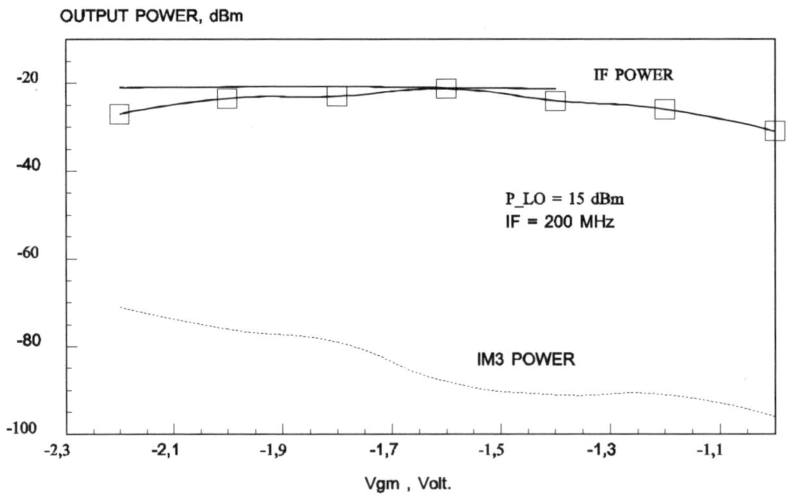

Figure 5.66 Performance of the double-balanced mixer

Low level intermodulation has been measured, as can be seen in Figure 5.66, in the two-tone test, thanks to the cold FET mixer. This downconverter was fabricated by Philips Microwave Limeil (France), using the ER07AD process, which incorporates 0.7 μm of gate length for the MESFET transistors. Figure 5.67 shows a photograph of the whole downconverter, where the chip size is 3 mm^2.

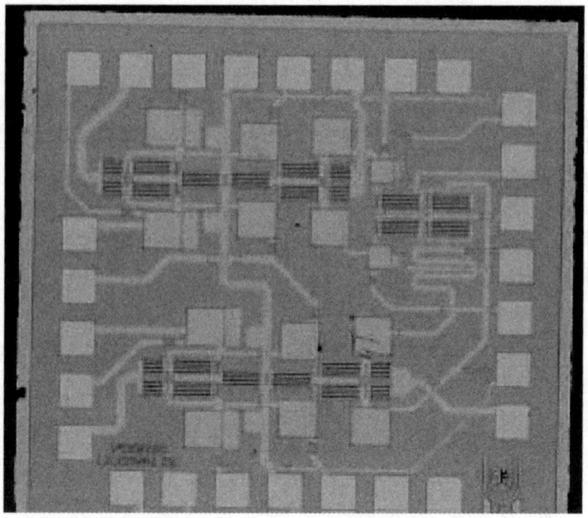

Figure 5.67 The double-balanced mixer

Appendix I

$$\cos^k x = \frac{1}{2^{k-1}} \cdot \left[\sum_{y=0}^{Y} \frac{k!}{(k-y)! \cdot y!} \cdot \cos((k-2y) \cdot x) + b_k \right]$$

where

$$Y = \begin{cases} \dfrac{k-2}{2} & \textit{for k even} \quad \textit{and} \quad k \neq 0 \\ \dfrac{k-1}{2} & \textit{for k odd} \end{cases} \quad \text{and} \quad b_k = \begin{cases} \dfrac{1}{2} \cdot \dfrac{k!}{(k/2!)^2} & \textit{for k even} \quad \textit{and} \quad k \neq 0 \\ 0 & \textit{for k odd} \end{cases}$$

k is a natural number and $k - 2y$ is always greater than zero.

Appendix II: Modified Bessel functions of first species and *n* order

$$I_n(x) = \sum_{k=0}^{\infty} \frac{\left(\dfrac{x}{2}\right)^{n+2\cdot k}}{k! \cdot \Gamma(n+k+1)} \quad \text{Where} \quad \Gamma(n+k+1) = (n+k) \cdot \Gamma(n+k) = (n+k)!$$

and

$$I_0(x) = 1 + \frac{x^2}{2^2} + \frac{x^4}{2^2 \cdot 4^2} + \frac{x^6}{2^2 \cdot 4^2 \cdot 6^2} + \ldots$$

$$I_1(x) = 1 + \frac{x^2}{2^2} + \frac{x^4}{2^2 \cdot 4^2} + \frac{x^6}{2^2 \cdot 4^2 \cdot 6^2} + \ldots$$

$I_0(x)$ Function

x	0	1	2	3	4	5	6	7	8	9
0.	1.000	1.003	1.010	1.023	1.040	1.063	1.092	1.126	1.167	1.213
1.	1.226	1.326	1.394	1.469	1.553	1.647	1.750	1.864	1.990	2.128
2.	2.280	2.446	2.629	2.830	3.049	3.290	3.553	3.842	4.157	4.503
3.	4.881	5.294	5.747	6.243	6.785	7.378	8.028	8.739	9.517	10.37
4.	11.30	12.32	13.44	14.67	16.01	17.48	19.09	20.86	22.79	24.91
5.	27.24	29.79	32.58	35.65	39.01	42.69	46.74	51.17	56.04	61.38
6.	67.23	73.66	80.72	88.46	96.96	106.3	116.5	127.8	140.1	153.7
7.	168.6	185.0	202.9	222.7	244.3	268.2	294.3	323.1	354.7	389.4
8.	427.6	469.5	515.6	566.3	621.9	683.2	750.5	824.4	905.8	995.2
9.	1094	1202	1321	1451	1595	1753	1927	2119	2329	2561

$I_1(x)$ Function

x	0	1	2	3	4	5	6	7	8	9
0.	0.000	0.050	0.100	0.151	0.204	0.257	0.313	0.372	0.433	0.497
1.	0.565	0.637	0.715	0.797	0.886	0.982	1.085	1.196	1.317	1.448
2.	1.591	1.745	1.914	2.098	2.298	2.517	2.755	3.016	3.301	3.613
3.	3.953	4.326	4.734	5.181	5.670	6.206	6.793	7.436	8.140	8.913
4.	9.759	10.69	11.71	12.82	14.05	15.30	16.86	18.48	20.25	22.20
5.	24.34	26.68	29.25	32.08	35.18	38.59	42.33	46.44	50.95	55.90
6.	61.34	67.32	73.89	81.10	89.03	97.74	107.3	117.8	129.4	142.1
7.	156.0	171.4	188.3	206.8	227.2	249.6	274.2	301.3	331.1	363.9
8.	399.9	439.5	483.0	531.0	583.7	641.6	705.4	775.5	852.7	937.5
9.	1031	1134	1247	1371	1508	1658	1824	2006	2207	2428

References

[1] S.A. Maas, *Microwave Mixers*, 2nd edn, Artech House, MA, 1993.

[2] S.A. Maas, *Non Linear Microwave Circuits*, Artech House, MA, 1989.

[3] R.A. Pucel, D. Massé, R. Bera. 'Performance of GaAs MESFET Mixers at X Band', *IEEE MTT*, vol. MTT-24, no. 6, June 1976, pp. 351–360.

[4] G.D. Vendelin, D.M. Pavio, V.L. Rohde, *Microwave Circuit Design*. Artech House, MA, 1979.

[5] S. Balatchev, J.L. Gautier, B. Delacressonnière, 'Using a negative conductance for optimizing the resistive mixers' conversion losses', *Galium Arsenide Applications Symposium*, GAAS 96, 4C5.

[6] C. Tsironis, R. Meier, R. Stahlamann, 'Dual-Gate MESFET Mixers', *IEEE MTT*, vol. MTT-32, no. 3, March 1984, pp. 248–255.

[7] M.L. de la Fuente, 'Diseño de mezcladores de microondas en tecnología monolítica', PhD thesis, Universidad de Cantabria, November 1997.

[8] F. Diaz, A. Herrera, E. Artal, A. Tazón, M.L. de la Fuente, J.M. Zamanillo, F. López, 'Diseño de un PLO Sintetizado en Banda Ku', *URSI XI Simp. Nac.*, Madrid, Sept. 1996.

[9] A. Madjar, 'A novel general approach for the optimum design of microwave and millimeter wave subharmonic mixers', *IEEE MTT*, vol. 44, no. 11, November 1996, pp. 1997–2000.
[10] R. Goyal, *Monolithic Microwave Integrated Circuits*, Artech House, MA, 1989.
[11] M.L. de la Fuente, J. Portilla, E. Artal, 'Low noise Ku-band drain mixer using P-HEMT technology', paper presented at IEEE, 5th International Conference on Electronics, Circuits and Systems, Lisbon, September 1998, pp. 175–178.
[12] M.L. de la Fuente, J. Portilla, J.P. Pascual, E. Artal, 'Low-noise Ku-band MMIC balanced P-HEMT upconverter', *IEEE Solid-State Circuits*, vol. 34 February 1999, pp. 259–263.
[13] M.L. de la Fuente, J.P. Pascual, E. Artal, 'Low intermodulation converter system for TV distribution', paper presented at XII Design of Circuits and Integrated Systems Conference, Seville, Spain, November 1997.

6

Filters

A. Mediavilla

6.1 Introduction

Filter circuits are a key component in any high frequency wireless system. Modern trends have been to try to move away from analogue filtering as much as possible and to implement filtering using digital signal processing wherever possible. However, this is only achievable at lower frequencies where the analogue signal can be successfully transformed into the digital domain. In processing signals at microwave frequencies, the standard approaches to filter designs will still be required for the foreseeable future.

This chapter describes the different types of filter (low pass, high pass, band pass and band stop) and their characteristic responses (Butterworth, Chebyshev, Bessel and Elliptic). To enable a generalised approach to filter design, given any particular specification, the chapter describes how a low pass prototype filter can be transformed into any of the other types of filter and how the frequency response can also be scaled to fit the defined corner frequencies.

6.2 Filter Fundamentals

6.2.1 Two-Port Network Definitions

Let us consider a general microwave two-port network, with generator and load termination R_G and R_L respectively, under sinusoidal operation, as shown in Figure 6.1.

In this case, and assuming that the source generator is E_g, we can define the power content in the network as a function of the terminal voltages and currents as follows:

$$P_{in} = \frac{|E_g|^2}{8R_G} \quad P_{diss} = 1/2 \ \mathrm{Re}\{V_1 \cdot I_1^*\} \quad P_{ref} = P_{in} - P_{diss}$$
$$P_L = 1/2 \ \mathrm{Re}\{V_2 \cdot (-I_2^*)\} = 1/2 \ \frac{|V_2|^2}{R_L} = 1/2 \ R_L \cdot |I_2|^2 \tag{6.1}$$

where P_{in}, P_{diss}, P_{ref} and P_L are the incident or available, dissipated at the input, reflected, and output power. All the voltages and currents in the equations are in phasor form (complex

Microwave Devices, Circuits and Subsystems for Communications Engineering Edited by I. A. Glover, S. R. Pennock and P. R. Shepherd

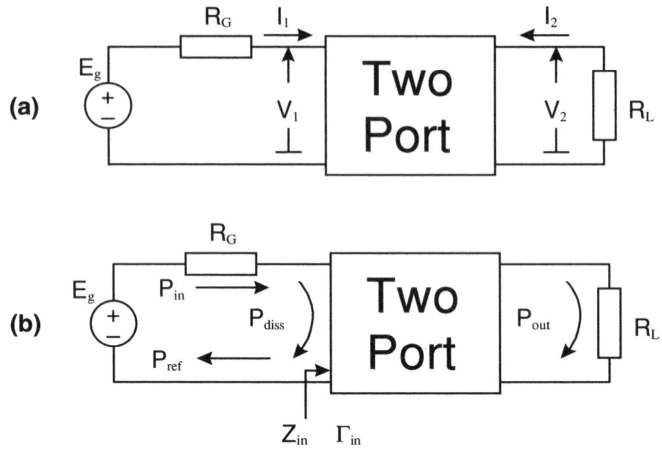

Figure 6.1 Two-port microwave network definitions

values where the modulus is the peak value), and the input impedance of this network is defined as:

$$Z_{in} = V_1/I_1 \tag{6.2}$$

In the same way we can define the concepts of voltage gain, power transfer gain and attenuation:

Voltage gain:

$$Av = V_2/V_1 \tag{6.3}$$

$$Av|_{dB} = 20 \cdot \text{Log}|Av| \tag{6.4}$$

Power transfer gain:

$$G_T = P_L/P_{in} \tag{6.5}$$

$$G_T|_{dB} = 10 \cdot \text{Log}(G_T) \tag{6.6}$$

Network attenuation:

$$Atn = P_{in}/P_L = 1/G_T \tag{6.7}$$

$$Atn|_{dB} = 10 \cdot \text{Log}(Atn)$$

Furthermore, we can introduce the high frequency concepts such as reflection coefficient at the input, return loss and their relationship with the power content and the scattering parameters of the network.

The reflection coefficient Γ_{in} at the input (referred to R_G) is defined as:

$$\Gamma_{in} = \frac{Z_{in} - R_G}{Z_{in} + R_G} \tag{6.8}$$

where its modulus ρ is the ratio between the incident power and the reflected power at the input of the network:

$$\rho = |\Gamma_{in}| = \left|\frac{Z_{in} - R_G}{Z_{in} + R_G}\right| \quad \rho^2 = P_{ref}/P_{in} \tag{6.9}$$

Return loss:

$$R_{loss} = P_{ref}/P_{in} = \rho^2 = 1 - P_{diss}/P_{in} \quad (6.10)$$

Scattering parameter S_{11}:

$$S_{11} = \Gamma_{in} \quad |S_{11}|^2 = \rho^2 = R_{loss} \quad (6.11)$$

Scattering parameter S_{21}:

$$|S_{21}|^2 = P_L/P_{in} = G_T \quad (6.12)$$

If the two-port is a lossless network, the dissipated input power P_{diss} and the output power P_L are the same because inside the network there is no dissipating device such as resistors. In this case, we can write the following modifications:

$$G_T = 1 - \rho^2 \quad Atn = \frac{1}{1 - \rho^2} \quad (6.13)$$

$$R_{loss} = 1 - G_T = 1 - 1/Atn \quad (6.14)$$

6.2.2 Filter Description

Generally speaking, a filter is any passive or active network with a predetermined frequency response in terms of amplitude and phase.They can be classified, depending on their application, as low pass, high pass, band pass and band stop filters, as shown in Figure 6.2.

Low pass and high pass filters are characterised by their cut-off frequency F_c where the transfer gain usually drops to one half (3 dB point). Conversely, band pass and band stop filters are defined by their centre frequency F_0 along with their 3dB bandwidth BW3.

It seems obvious that in a filter design it is not enough to define the cut-off frequencies or 3dB bandwidths. Most applications require a given attenuation at a given frequency – out of

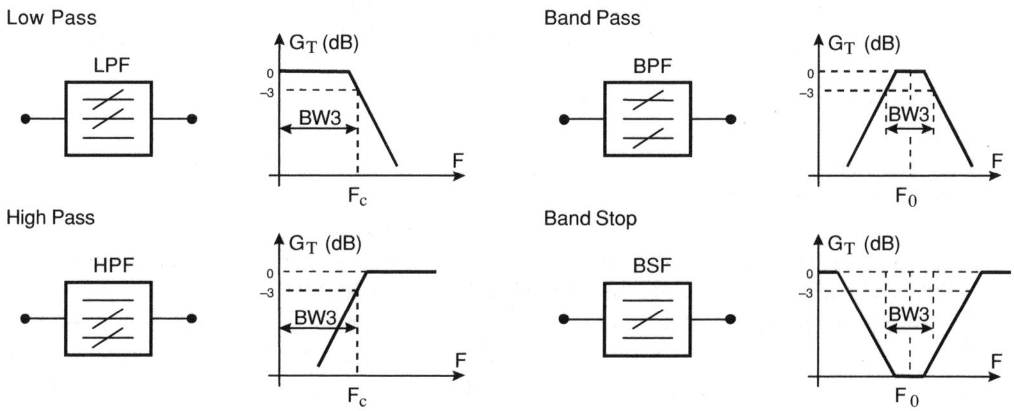

Figure 6.2 Types of filter

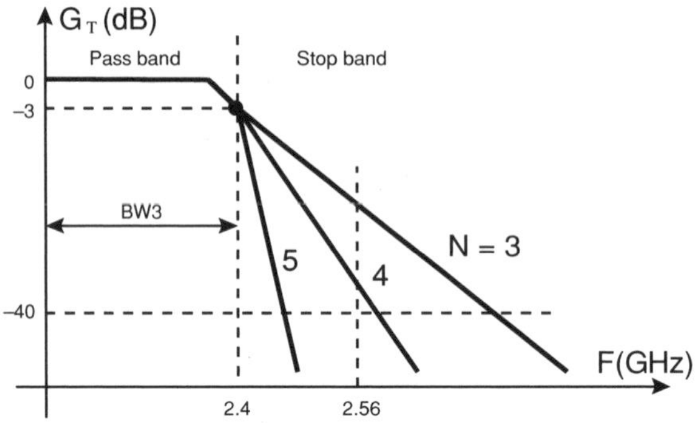

Figure 6.3 Variation of stopband characteristics with filter order number

band – while maintaining the cut-off values and the 3dB bandwidth. This new requirement will condition the filter response and the number of sections (complexity).

In the Figure 6.3 we have several low pass filters having the same cut-off frequency of 2.4 GHz with a different number of sections, N.

As we increase the number of sections N, we have a steeper transition from the passband to the stopband while maintaining the same passband characteristics and cut-off frequency. If our specification at, for example, 2.56 GHz is that the attenuation should be better than 40 dB, it is necessary to use an order $N > 4$.

Apart from the order of a filter, we have another important characteristic, that is the filter response: there are many types of frequency responses, and the choice depends on the application. The most important are:

- Butterworth: maximally flat in amplitude;
- Chebyshev: amplitude passband equi-ripple;
- Bessel: maximally flat in phase;
- Elliptic: amplitude passband and stopband equi-ripple.

The choice of one of the above characteristics depends mainly on the system application where the filter should be used as well as the type of signal that is passing through the filter:

- pure sinusoidal;
- square (train of pulses);
- AM or FM modulation;
- more complex modulations.

In Figure 6.4 we have three different band pass $N = 3$ filters at the same centre frequency $F_0 = 900$ MHz and with the same 3dB bandwidth BW3 = 60 MHz.

The Elliptic response exhibits a more abrupt transition but the stopband has a residual ripple. The Bessel response has a very poor amplitude response but it has an excellent phase behaviour. Finally, Butterworth and Chebyshev filters can be considered as a compromise between the other two. In the passband Chebyshev and Elliptic filters have an equi-ripple response while Bessel and Butterworth filters have a flat behaviour.

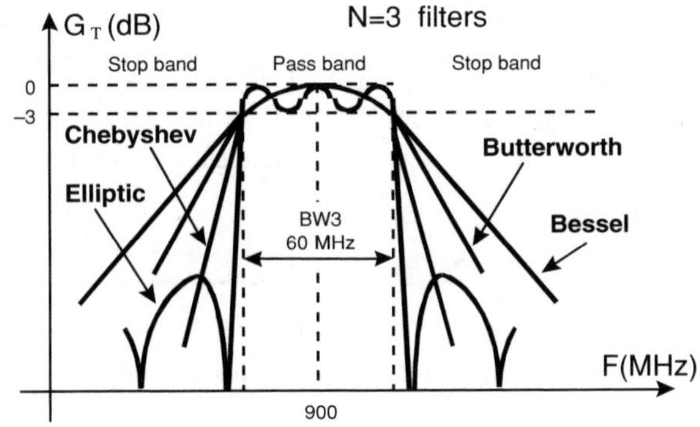

Figure 6.4 Different filter response types

6.2.3 Filter Implementation

All the filters described above, having their own particular characteristics, can be built in very different ways and technologies. The choice mainly depends on the working frequency, filter selectivity and losses in the passband.

In the low frequency range (up to several MHz) most filters are designed on the basis of operational amplifiers: active filters. This is due to the fact that an implementation with LC networks leads to LC values out of commercial range, apart from the design flexibility when using Opamps.

In the radiofrequency band (up to a few GHz) filter synthesis is accomplished by using classical discrete RLC elements, or more recently high Q coaxial resonators. The only inconvenience here are the lumped-element losses when increasing the frequency.

When we want to design up to 30/40 GHz, filter synthesis does not use lumped elements but coaxial, planar transmission line technologies (microstrip, strip, slab, coplanar lines) or waveguide technology (rectangular, circular, ridge, etc.). The discrete L and C elements can be built by using sections of transmission lines when the bandwidth is not too low, or alternatively dielectric resonators for high Q filters.

Finally, at frequencies above 40 GHz (high frequency radio links, radio astronomy, etc.), the losses in coaxial and planar technologies are so high that it is necessary to use waveguide configurations.

6.2.4 The Low Pass Prototype Filter

It is obvious that the mathematical techniques for filter synthesis cannot look for the infinite variety of cut-off frequencies, centre frequencies, 3dB bandwidths that we can imagine, as well as whether the filter is a high pass or stop band, etc.

Consequently, all the filter synthesis techniques and filter shapes refer to the prototype low pass filter (PLPF). This normalised filter shown in Figure 6.5 has a cut-off frequency of 1 rad/sec (0.159 Hz) for a general 3dB attenuation.

We will see in the next section that we can synthesise PLPF with Butterworth, Bessel, Chebyshev, etc. characteristics. We also see how we can scale in frequency this PLPF in

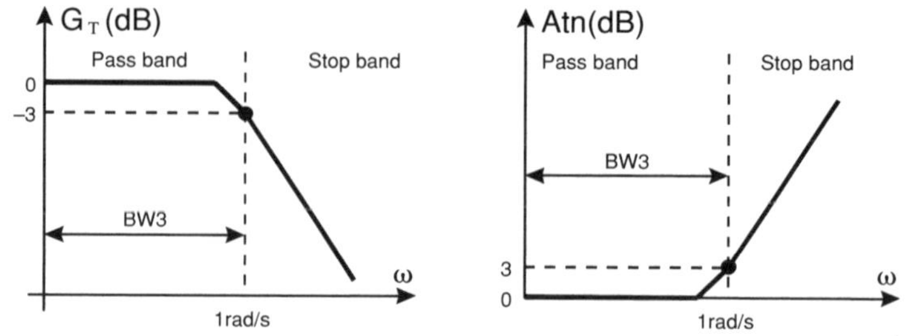

Figure 6.5 Normalised LPF characteristics

order to have LPF at the frequencies of interest, and how to go directly from a PLPF to an HPF, BPF or BSF at a given frequency. Finally, we will see how the actual out of band specifications can be translated into PLPF constraints. In conclusion, our problem is reduced to mathematically synthesising prototype filters PLPF with specified attenuation in the passband and stopband.

6.2.5 The Filter Design Process

Starting from the above considerations, the methods used in filter design can be summarised as follows:

1. The designer has several filter specifications (inband and outband) as well as system and incoming signal constraints. The first step is to select one or two types of filter that could meet the specifications along with the presumed technology.
2. The frequency response specifications are transformed into PLPF specifications with a cut-off frequency of $\omega_c = 1$ rad/s.
3. We compare in the PLPF plane the desired response with the theoretical responses (Bessel, Chebyshev, etc.) and we select the minimum number of sections and the type of filter.
4. We perform frequency scaling and transformations in order to arrive at the specified frequency range.
5. We implement the filter into the specified technology, depending on the frequency range.

6.2.5.1 Filter simulation

The most important measures that a filter designer should have at his or her disposal are obviously the attenuation (or the inverse transfer gain), the return loss in order to see the frequency matching, and more specifically the group delay if the input signal has any kind of modulation. We must remember that the attenuation is directly related to the S_{21} parameter while the return loss is related to the S_{11}.

Although many commercially available simulators have implemented all the above measures, other popular simulators such as PSPICE do not have these facilities: they do not have the concept of scattering parameters. Therefore, it is of interest to arrange the circuit implementation in order to directly measure the key parameters.

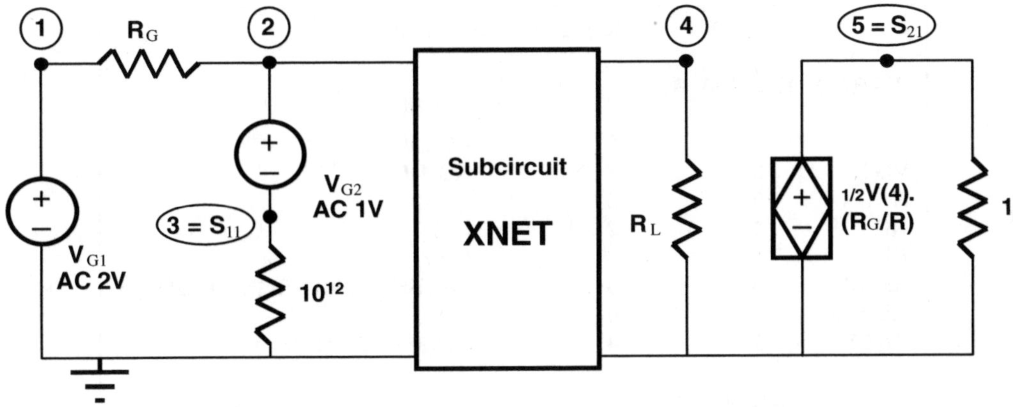

Figure 6.6 PSPICE configuration for generating s-parameters

Figure 6.6 shows a PSPICE configuration able to generate the scattering parameters of a two-port network measured with respect to any R_G and R_L (resistive source and load impedances).

The AC voltage at node 3 is exactly (in modulus and phase) the S_{11} parameter while the same applies for the node 5 and the S_{21} parameter. If we want to calculate S_{22} and S_{12}, we have to turn the two-port network. Figure 6.7 shows the details of the circuit description in PSPICE along with the measurements to be performed.

6.3 Mathematical Filter Responses

As has been stated above, our interest now is to design and to control prototype low pass filters PLPF with different shapes of pass band and stop band attenuation. In this section we will study the commonly used characteristics: Butterworth, Chebyshev, Bessel and Elliptic, although many other characteristics are available in the literature for special applications.

6.3.1 The Butterworth Response

Commonly known as a 'maximally flat' filter, the attenuation has a frequency response where the $N - 1$ first derivatives at DC ($\omega = 0$) are null, N being the order of the filter. This can be written as shown is Equation (6.15):

$$Atn = 1 + (\omega)^{2N} \quad Atn|_{dB} = 10 \cdot \mathrm{Log}[1 + (\omega)^{2N}] \quad N = \text{order} \tag{6.15}$$

From this equation we can immediately see that:

$$\text{at } \omega = 0 \quad Atn = 1 \quad Atn|_{dB} = 0 \quad \text{(independent of the order } N\text{)}$$

$$\text{at } \omega = 1 \quad Atn = 2 \quad Atn|_{dB} = 3 \quad \text{(independent of the order } N\text{)}$$

PSPICE Circuit Description

```
.PARAM  RG=100 RL=50

VG1     1   0   AC 2V
VG2     2   3   AC 1V
RAUX    3   0   1.E12
RG      1   2   {RG}
RL      4   0   {RL}
EAUX    5   0   VALUE={ V(4) * SQRT(RG/RL) }
REAUX   5   0   1.0
XNET    2   4   Name of Subcircuit

.AC  Frequency Sweep
```

PSPICE Measurements with PROBE

```
VM(3)  ----------------------  Amplitude of S11
VP(3)  ----------------------  Phase of S11
VM(5)  ----------------------  Amplitude of S21
VP(5)  ----------------------  Phase of S21
VDB(3)  --------------------  Return Loss (dB)
VDB(5)  --------------------  Transfer Gain (dB)
-VDB(5)  -------------------  Attenuation (dB)
-1/360 * d(VP(5))  ------  Group delay (sec)
```

Figure 6.7 PSPICE circuit description for s-parameter generation

at $\omega \gg 1$ *Atn* grows towards infinity at a ratio of 6*N* dB/octave because the nominal value grows as ω^{2N}.

$$\text{at } \omega = 0 \quad d^n(Atn)/d\omega^n = 0 \quad \text{for } 0 < n < N \quad \text{(maximally flat condition)}$$

Figure 6.8 shows the typical response of these filters as a function of the number of sections *N*.

We can observe that the cut-off frequency is always $\omega_c = 1$ rad/s for an attenuation of 3 dB, independent of *N*. As the order *N* increases, the attenuation in the stopband area grows while the flatness in the pass band area is more evident. This kind of filter is an 'all pole filter' because infinite attenuation is reached with ω going to infinity.

6.3.2 The Chebyshev Response

This characteristic has, for the same order *N*, a higher degree of stop band attenuation than the Butterworth response. This can be accomplished by allowing a certain amount of controlled ripple in the pass band. That is why these kinds of filters are called 'ripple-controlled filters' and, as well as the Butterworth filters, are 'all pole filters'. The attenuation can be written as shown in Equation (6.16):

$$Atn = 1 + K^2 \cdot T_N^2(\omega) \quad Atn|_{dB} = 10 \cdot \text{Log}[1 + K^2 \cdot T_N^2(\omega)] \quad N = \text{order} \tag{6.16}$$

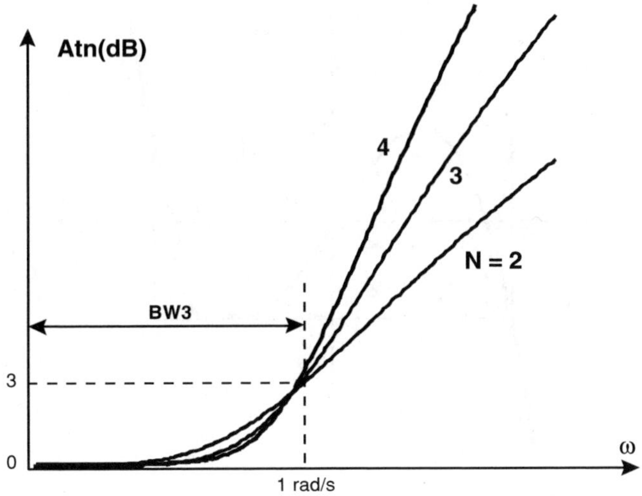

Figure 6.8 Butterworth LPF responses as a function of N

where T_N is the Chebyshev polynomial of degree N:

$T_1(X) = X$ $\quad$ $T_5(X) = 16X^5 - 20X^3 + 5X$

$T_2(X) = 2X^2 - 1$ $\quad$ $T_6(X) = 32X^6 - 48X^4 + 18X^2 - 1$

$T_3(X) = 4X^3 - 3X$ $\quad$ $T_7(X) = 64X^7 - 112X^5 + 56X^3 - 7X$

$T_4(X) = 8X^4 - 8X^2 + 1$ $\quad$ Recurrence: $T_N(X) = 2X \cdot T_{N-1}(X) - T_{N-2}(X)$

Alternatively, we can write:

$$T_N(X) = \text{COS}[\text{N.ACS}(X)] \quad \text{for} \quad |X| \leqslant 1 \tag{6.17}$$

$$T_N(X) = \text{COSH}[\text{N.ACSH}(X)] \quad \text{for} \quad |X| > 1 \tag{6.18}$$

Figure 6.9 shows the Chebyshev polynomials up to $N = 4$.

From Figure 6.9 and the definition we can observe the following properties:

- $T_N(X)$ varies between -1 and $+1$ when the argument X is restricted to the range $(-1, +1)$.
- $T_N(X = 0) = 0$ if N is even.
- $|T_N(X = 0)| = 1$ if N is odd.
- $|T_N(X)|$ grows toward infinity for $|X| > 1$, and the growing rate increases with N.

In the case of the Chebyshev filter response, the argument ω is positive and the equation uses T_N^2. Figure 6.10 shows this behaviour up to $N = 5$.

We can observe that for N odd, $T_N^2(X = 0) = 0$ and we will have $(N - 1)/2$ additional passes through zero in the pass band $(0 < X < 1)$. Conversely, for N even, we have $T_N^2(X = 0) = 1$ and $N/2$ passes through zero.

Finally, if we consider the total equation for the Chebyshev attenuation, Figure 6.11 shows this behaviour for $N = 3$ and $N = 4$.

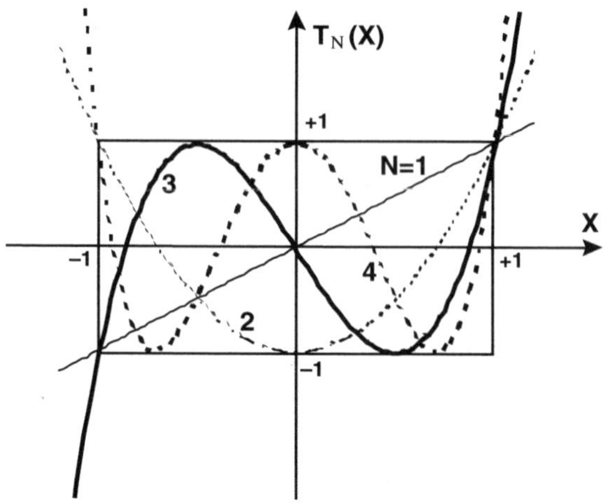

Figure 6.9 Chebyshev polynomials to $N = 4$

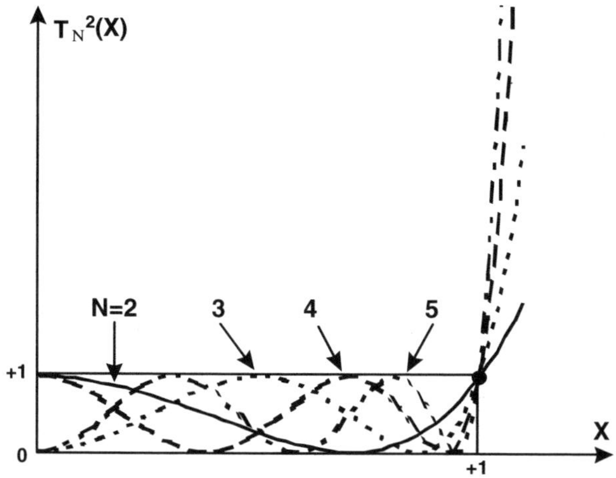

Figure 6.10 Chebyshev filter response up to $N = 5$

From Figure 6.11 and the definition, we can extract the following properties:

- at $\omega = 0$ $Atn = 1$ for N odd and $Atn = 1 + K^2$ for N even
- at $\omega = 1$ $Atn = 1 + K^2$ for N even and odd. $Atn|_{dB} = \text{RdB} = 10 \cdot \text{Log}(1 + K^2)$
- The pass band has a constant ripple from zero to RdB. The pass band up to $\omega = 1$ is called BWR as opposed to BW3 that is the 3dB point
- at $\omega \gg 1$ Atn grows toward infinity as $(K^2/4) \cdot (2\omega)^{2N}$. This means that the growing rate is $6N$ dB/octave but at a given frequency the stop band attenuation is higher than the Butterworth characteristic.

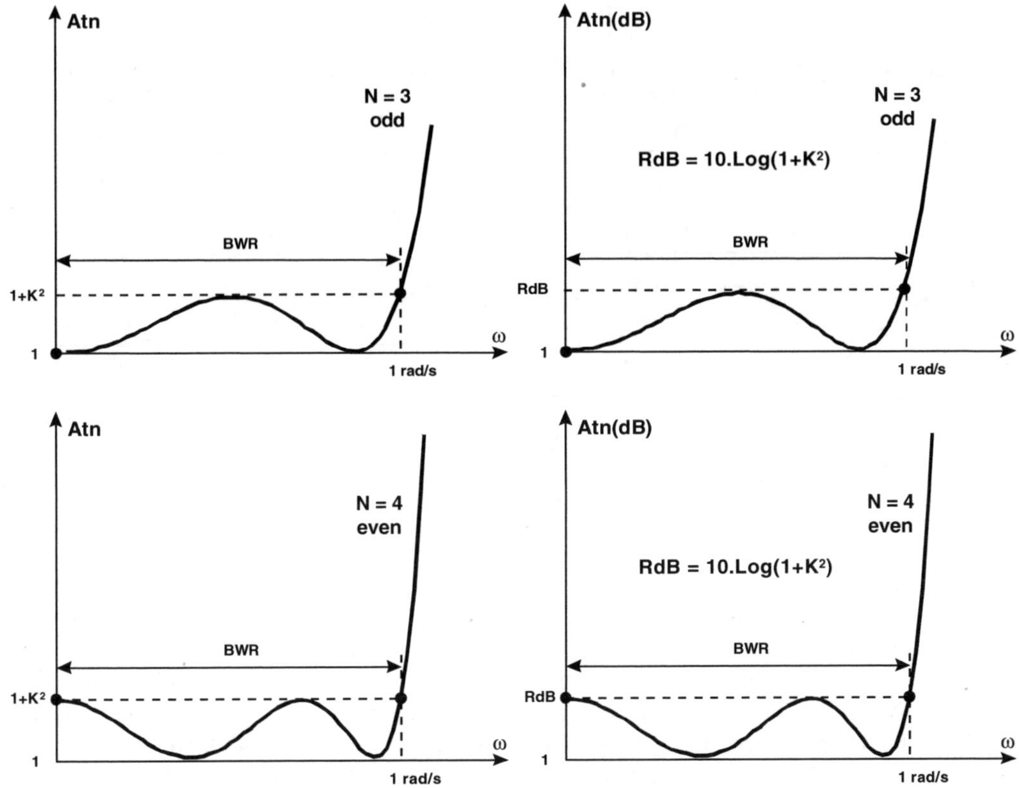

Figure 6.11 Chebyshev filter attenuation for $N = 3$ and 4

In order to normalise this filter, it is necessary to know the frequency for which the attenuation is exactly 3dB ($Atn = 2$). This occurs at a frequency named ω3dB slightly higher than 1 rad/sec and depends on the order N and the ripple factor K:

$$Atn = 2 = 1 + K^2 \cdot T_N^2(\omega 3\text{dB}) \rightarrow T_N(\omega 3\text{dB}) = 1/K$$

and using the property $T_N(X) = \text{COSH[N.ACSH}(X)]$, we can find:

$$\omega 3\text{dB} = \text{COSH}[1/\text{N.ACSH}(1/K)] \tag{6.19}$$

It is now evident that an attenuation function of the form:

$$Atn = 1 + K^2 \cdot T_N^2(\omega 3\text{dB} \cdot \omega) \quad Atn|_{dB} = 10 \cdot \text{Log}[1 + K^2 \cdot T_N^2(\omega 3\text{dB} \cdot \omega)] \tag{6.20}$$

ensures that, for any value of N and K, the attenuation is 3 dB for $\omega = 1$. So we have compatibility with the Butterworth response. Figure 6.12 shows these two responses for a $N = 4$ filter.

In the second case, the maximum frequency for RdB is exactly:

$$\omega\text{RdB} = 1/\omega 3\text{dB} \tag{6.21}$$

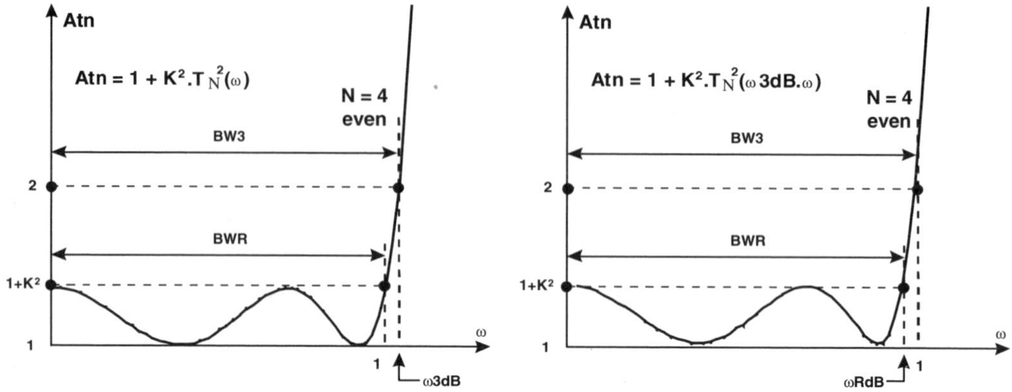

Figure 6.12 Chebyshev LPF responses for $N = 4$

6.3.3 The Bessel Response

The Butterworth characteristic has a good behaviour in amplitude selectivity along with an acceptable phase response. This means that its transient characteristic is normally good for many applications. Conversely, the Chebyshev family (depending on the ripple factor K) offers a very important increase in stop band attenuation but a poor phase response.

The Bessel response has been optimised in order to have a maximally flat phase response in the passband. This means that the amplitude selectivity in the stop band has been seriously degraded, but the filter response under transient operation or complex modulations is optimum.

The generic low pass transfer function that has a constant delay can be written as a ratio of hyperbolic trigonometric functions of the frequency:

$$T(s) = \frac{1}{\sinh(s) + \cosh(s)} \qquad s = j\omega \tag{6.22}$$

The expansion of the above equation is rather tedious and, in an approximate way, we can write the attenuation as:

$$Atn|_{dB} \cong 3 \cdot \omega^2 \quad \text{for any } N \text{ and valid up to } \omega = 2 \tag{6.23}$$

This filter is again an 'all pole filter' and Figure 6.13 shows its frequency response. It is clear that for $\omega > 2$, the selectivity in the stop band increases with N and the linearity of the phase response extends towards higher frequencies. In fact, these kinds of filters are only used when the transient properties are critical.

As the mathematical filter response is not easy to use, Figure 6.14 shows in a graphical form the filter response in the pass band and stop band for N up to 7.

6.3.4 The Elliptic Response

All the characteristics studied above are 'all pole', that is, infinite attenuation (transmission zero) is obtained for infinite ω. In these cases the best frequency selectivity in the stopband is assured by the Chebyshev response.

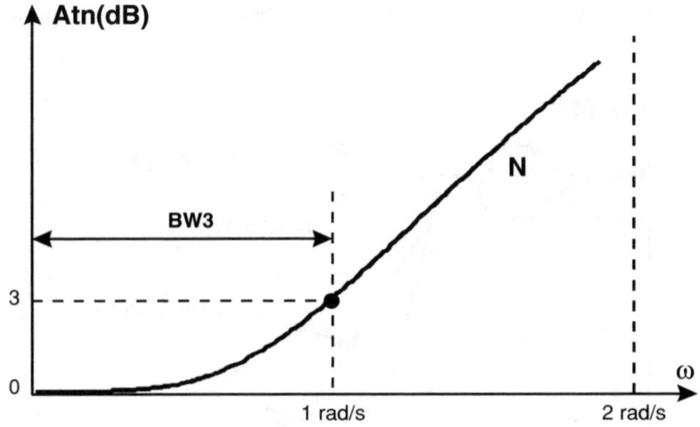

Figure 6.13 Bessel LPF response

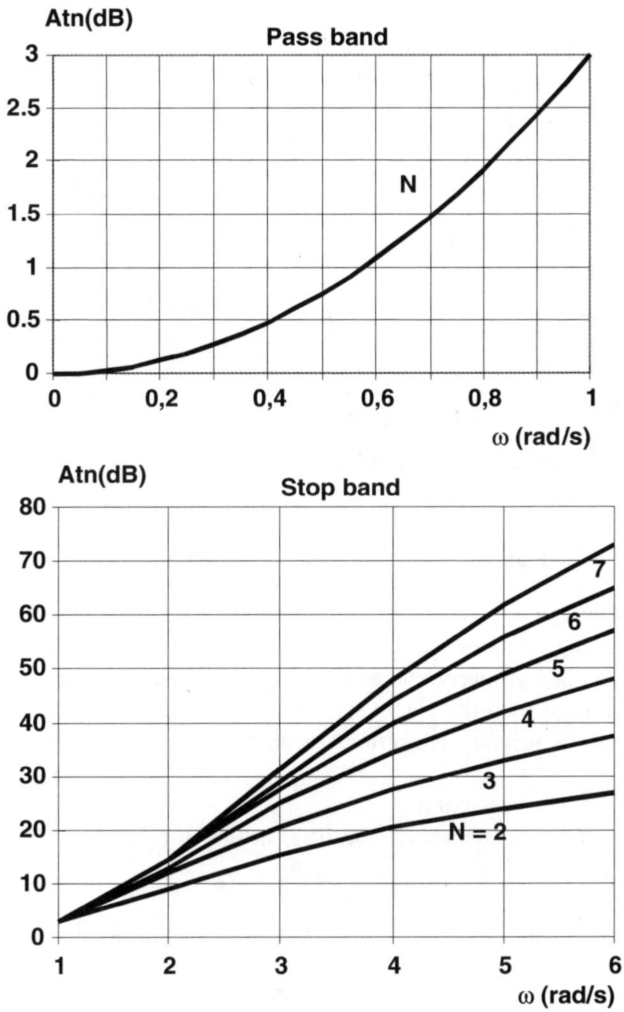

Figure 6.14 Bessel LPF response for N up to 7

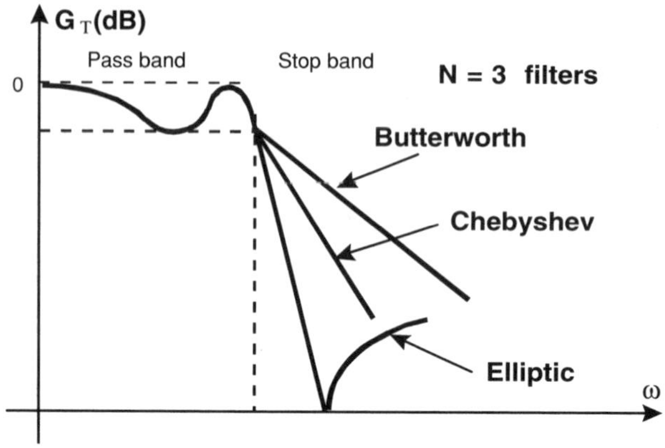

Figure 6.15 Elliptic LPF response compared with the Butterworth and Chebyshev

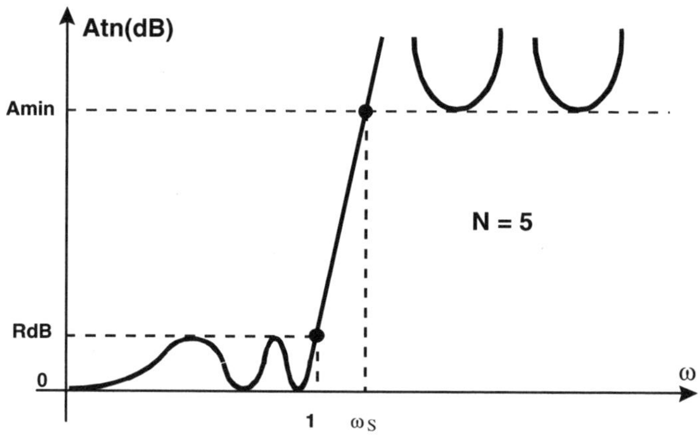

Figure 6.16 Definitions for the Elliptic filter

The Elliptic response allows transmission zeros at controlled finite frequencies. This is accomplished by having a Chebyshev-like ripple RdB in the pass band and an extra ripple in the stop band. This last property means that the transition from the pass band to the stop band will be more abrupt than the Chebyshev response as shown in Figure 6.15.

Figure 6.16 shows the primary definitions of an Elliptic filter where:

- RdB is the ripple in the pass band.
- Amin is the minimum attenuation (ripple like) in the stop band.
- ω_S is the lowest frequency where Amin occurs.

As we can see, the stop band region has transmission zeros at finite frequencies, and the first one occurs at a frequency slightly greater than ω_S. The attenuation of these kinds of filters can be written as:

$$Atn = 1 + K^2 \cdot Z_N^2(\omega) \tag{6.24}$$

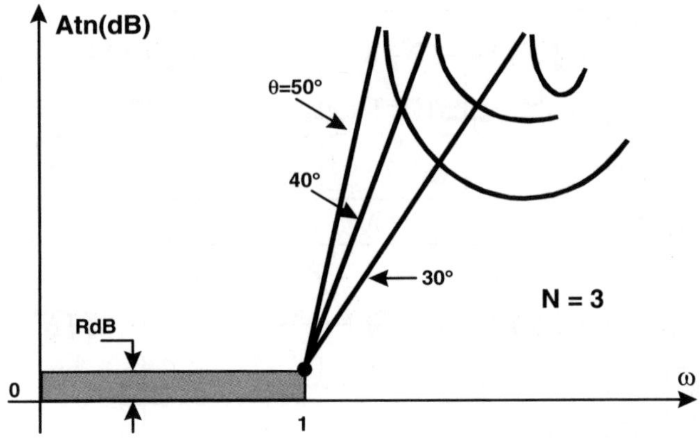

Figure 6.17 Variation of filter response with θ

where K corresponds to the pass band ripple RdB, and Z_N is an Elliptic function of order N:

$$Z_N(\omega) = \frac{W \cdot (A_2^2 - \omega^2) \cdot (A_4^2 - \omega^2) \ldots (A_m^2 - \omega^2)}{(1 - A_2^2\omega^2) \cdot (1 - A_4^2\omega^2) \ldots (1 - A_m^2\omega^2)} \quad N \text{ odd: } m = (N-1)/2 \qquad (6.25)$$

$$Z_N(\omega) = \frac{(A_2^2 - \omega^2) \cdot (A_4^2 - \omega^2) \ldots (A_m^2 - \omega^2)}{(1 - A_2^2\omega^2) \cdot (1 - A_4^2\omega^2) \ldots (1 - A_m^2\omega^2)} \quad N \text{ even: } m = N/2 \qquad (6.26)$$

The transmission zeros are at $A_2, A_4, \ldots A_m$ and the poles are at the symmetric points $1/A_2$, $1/A_4, \ldots, 1/A_m$. This symmetry results in a controlled ripple in the pass band as well as in the stop band. Straightforward calculations give us the zero and pole position as a function of the pass band ripple factor K and the order N.

The last parameter associated with this kind of filter is the modulation angle θ that is defined as:

$$\theta = \text{ASIN}(1/\omega_S) \quad \text{with} \quad 0 < \theta < 90° \qquad (6.27)$$

and is the most popular parameter that defines the filter selectivity. As θ increases, the position of ω_S decreases towards $\omega = 1$, and the transition from the pass band to the stop band is more abrupt with a lower value of Amin. This fact is shown in Figure 6.17. On the other hand, if we have fixed values for θ (or their equivalent ω_S) and the order N, the stop band attenuation Amin can be increased by allowing a higher value of RdB in the pass band.

It is clear that it is not easy to draw the attenuation characteristics of Elliptic filters because we have three independent variables: K, N and Amin. So we will defer this problem to the next sections.

6.4 Low Pass Prototype Filter Design

The actual specifications for a given filter with a particular characteristic are always translated, in a first step, to the concept of low pass prototype filter. The generic ladder network

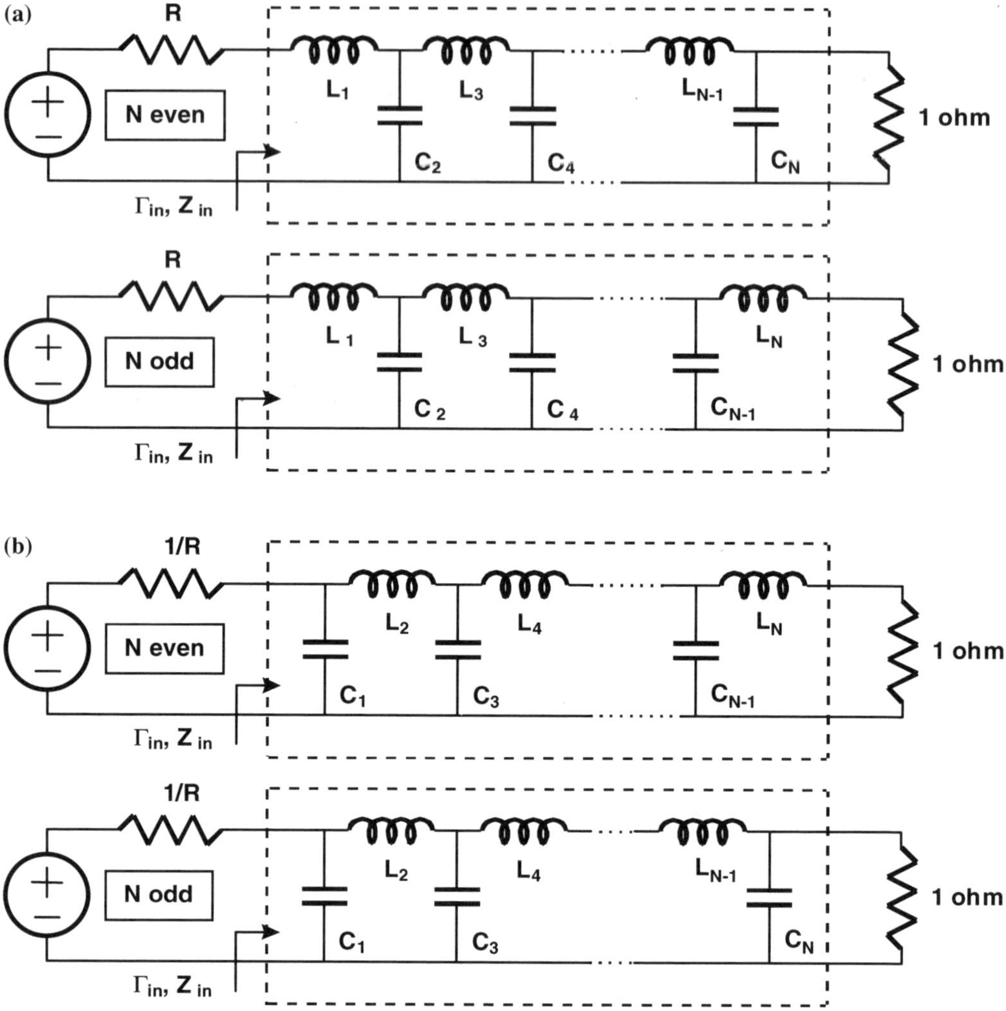

Figure 6.18 Prototype low pass filter definitions

that conforms to this kind of prototype filter is shown in Figure 6.18(a), where the load resistor is always unity while the source resistor is in general R.

The dual network of the above description is shown in Figure 6.18(b), where the element values L_j and C_j for any value of j are the same and the source resistor is inverted. The frequency response of both networks is exactly the same.

If we look for any of the networks shown in Figure 6.18(a), the reflection coefficient at the input, measured with respect to the reference R, is:

$$\Gamma_{in} = \frac{Z_{in} - R}{Z_{in} + R} \qquad \rho = |\Gamma| \tag{6.28}$$

On the other hand, the attenuation of this lossless network can be written as:

$$Atn = P_{in}/P_L = \frac{1}{1-\rho^2} = \frac{1}{1-|(Z_{in}-R)/(Z_{in}+R)|^2} =$$

$$= \frac{1}{1-\dfrac{|Z_{in}-R|^2}{(Z_{in}+R)(Z_{in}^*+R)}} = \ldots \text{ after some manipulation } =$$

$$Atn = 1 + \frac{|Z_{in}-R|^2}{2R\cdot(Z_{in}+Z_{in}^*)} \tag{6.29}$$

in an analogous form, the above expression can be written as a function of the input admittance of the filter:

$$Atn = 1 + \frac{|Y_{in}-G|^2}{2G\cdot(Y_{in}+Y_{in}^*)} \qquad G = 1/R \tag{6.30}$$

The general idea is to fit these expressions to the mathematical filter characteristics such as Butterworth, Chebyshev, etc.

For the above prototype network, the circuit is reduced to that shown in Figure 6.19 at zero frequency (DC).

In this case, $Z_{in} = 1$ and the attenuation is:

$$Atn = 1 + \frac{(1-R)^2}{4R} \quad \text{for } \omega = 0 \text{ (DC)} \tag{6.31}$$

6.4.1 Calculations for Butterworth Prototype Elements

We remember that the maximally flat characteristic in amplitude is:

$$Atn = 1 + (\omega)^{2N} \quad Atn|_{dB} = 10\cdot \text{Log}[1+(\omega)^{2N}] \quad N = \text{order number} \tag{6.32}$$

and must be fitted to the characteristic of the prototype network:

$$Atn = 1 + \frac{|Z_{in}-R|^2}{2R\cdot(Z_{in}+Z_{in}^*)} \tag{6.33}$$

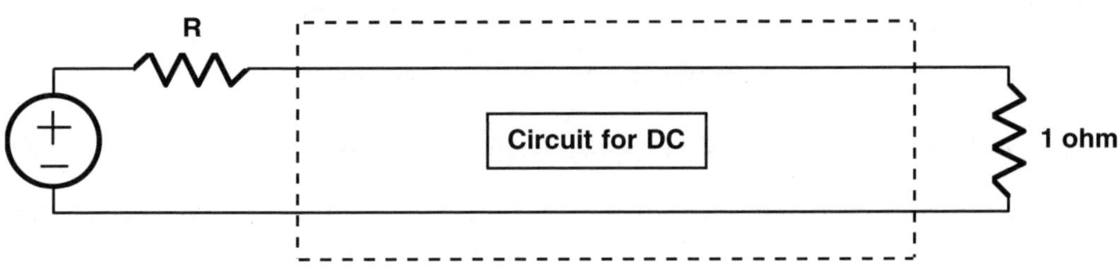

Figure 6.19 PLPF at DC

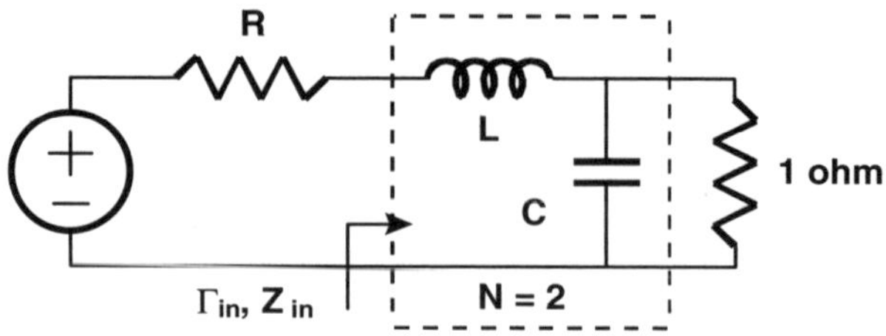

Figure 6.20 PLPF for order 2

So, the problem is to find the reactive element values that make both expressions identical. For clarification purposes, we can develop the calculations for $N = 2$. In this case, the prototype network can be reduced, for example, to Figure 6.20.

The Butterworth attenuation for $N = 2$ is:

$$Atn = 1 + \omega^4 \tag{6.34}$$

and the input impedance of the prototype network is:

$$Z_{in} = jL\omega + \frac{1}{1 + jC\omega} \tag{6.35}$$

After substitution of Z_{in} into the general expression for the attenuation of the prototype network, we have:

$$Atn = 1 + \frac{(1 - R)^2 + \omega^2 \cdot (L^2 + C^2R^2 - 2LC) + \omega^4 \cdot L^2C^2}{4R} \tag{6.36}$$

and using this equation we can observe that at DC ($\omega = 0$) we have:

$$Atn(\omega = 0) = 1 + (1 - R)^2/4R \quad \text{for } R \neq 1 \quad \text{(mismatch at DC)}$$

$$Atn(\omega = 0) = 1 \quad \text{for } R = 1 \quad \text{(matching at DC)}$$

So we can illustrate the attenuation for both cases in Figure 6.21. We observe a linear translation on the attenuation axis.

In the general case, for any R, the attenuation at DC and cut-off are respectively:

$$Atn(\omega = 0) = 1 + (1 - R)^2/4R \quad \text{for any } R \tag{6.37}$$

$$Atn(\omega = 1\text{: cut-off}) = 2 \cdot [1 + (1 - R)^2/4R] \quad \text{for any } R \quad \text{(3 dB point)} \tag{6.38}$$

If we compare the Butterworth attenuation for $N = 2$ and the mathematical expression for the prototype ladder network, we can deduce that the term in ω^2 must be zero:

$$L^2 + C^2R^2 - 2LC = 0 \quad (1° \text{ condition}) \tag{6.39}$$

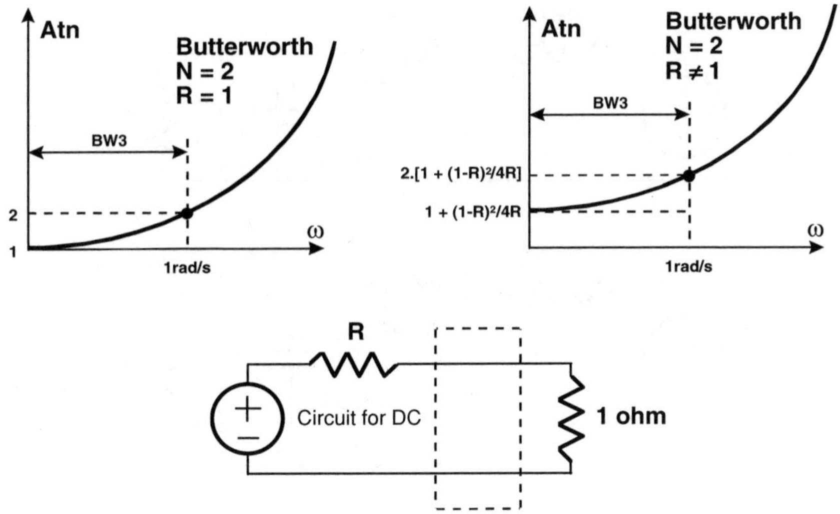

Figure 6.21 Attenuation characteristics of Butterworth PLPF

Furthermore, the attenuation at $\omega = 1$ (cut-off) must be $2 \cdot [1 + (1 - R)^2/4R]$. This means the following identity:

$$2 \cdot [1 + (1 - R)^2/4R] = 1 + \frac{(1 - R)^2}{4R} + \frac{L^2C^2}{4R} \rightarrow LC = 1 + R \text{ (2° condition)} \quad (6.40)$$

From the above two conditions we can deduce the values of L and C:

$$C^2 = \frac{1 + R}{R^2} \cdot [1 \pm (1 - R^2)^{1/2}] \quad L^2 = \frac{(1 + R) \cdot R^2}{[1 \pm (1 - R^2)^{1/2}]} \quad (6.41)$$

As an example, we can deduce the element values for two cases: $R = 1$ and $R = 0.5$:

$R = 1$	$C = 1.414$ F	$L = 1.414$ H	$Atn(\omega = 0) = 1.0$	(0 dB)
			$Atn(\omega = 1) = 2.0$	(3 dB)
$R = 0.5$	$C = 3.346$ F	$L = 0.448$ H	$Atn(\omega = 0) = 1.125$	(0.5 dB)
			$Atn(\omega = 1) = 2.25$	(3.5 dB)

and the two resulting networks are shown in Figure 6.22.

From the equations for L and C we can observe that there is no solution for $R > 1$. It is possible, however, to design filters with generator impedances greater than load impedances. If we look for the filter $R = 0.5$, we can remember that the dual network can be designed as in Figure 6.23.

So all the problem consists of is in generating dual networks. We must remember at this point that if $R \neq 1$ the attenuation at DC is not zero but $[1 + (1 - R)^2/4R]$.

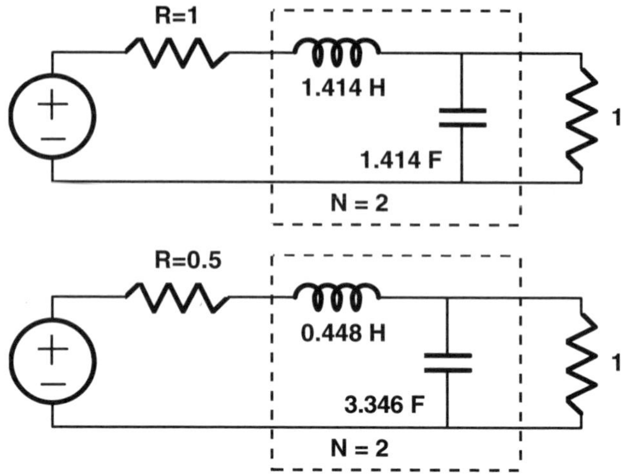

Figure 6.22 Example Butterworth PLPFs

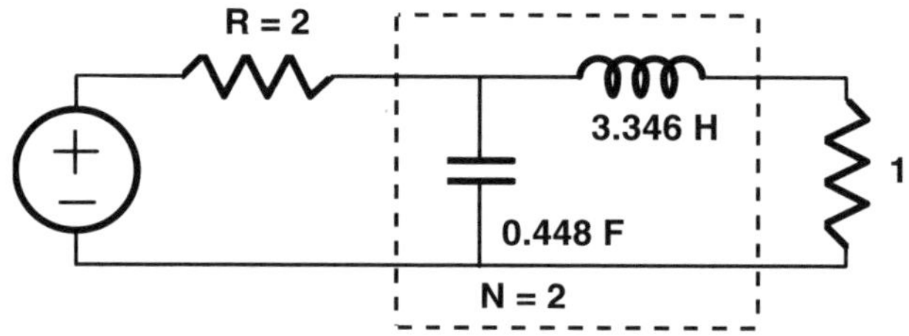

Figure 6.23 Dual network for $R = 0.5$

There are recursive equations for the prototype values for any value of N. In the particular case of $R = 1$ (equality of generator and load impedances) the filter values are symmetrical.

Table 6.1 (see Appendix) shows the prototype element values in Henries and Farads up to $N = 7$ (sufficient for most applications) and the table is organised as follows:

1. If we read the element values using the template shown in the top of the table, the associated network is shown at the upper part of Figure 6.24.
2. If we read the element values using the template shown in the bottom of the table, the associated network is shown in the bottom of Figure 6.24 (dual network).

Obviously, from Table 6.1 we can deduce the two above filters calculated for $N = 2$.

Figure 6.25(a) shows as an example an $N = 4$ Butterworth prototype filter for $R = 1$ (read from the top template of Table 6.1), while Figure 6.25(b) shows its dual network.

In the same way, Figure 6.26(a) shows an $N = 3$ Butterworth prototype filter for $R = 2.5$ (read from the bottom template of Table 6.1), while Figure 6.26(b) shows its dual network.

The attenuation values (in the pass band ($\omega < 1$) or in the stop band ($\omega > 1$) can be easily calculated, at any frequency and for any order N, by using the Butterworth expression for the

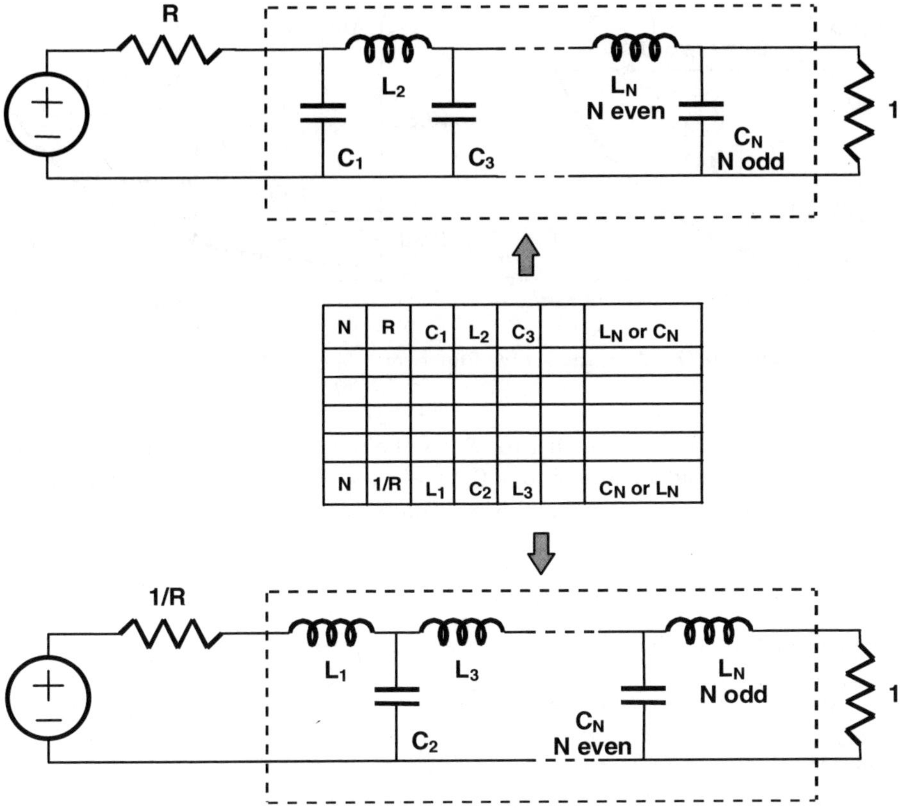

Figure 6.24 Network format for Table 6.1

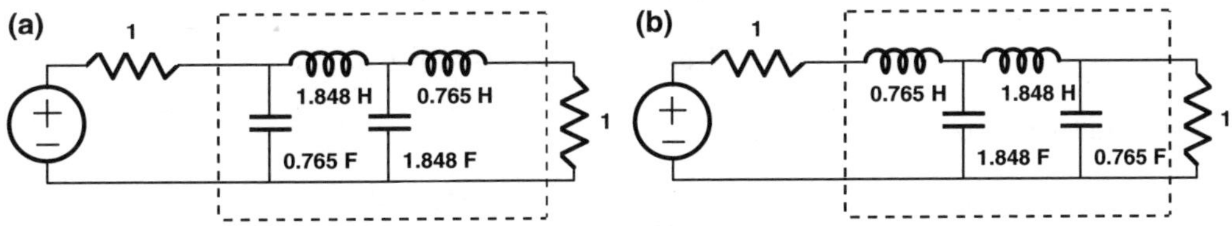

Figure 6.25 Prototype Butterworth LPF with $N = 4$

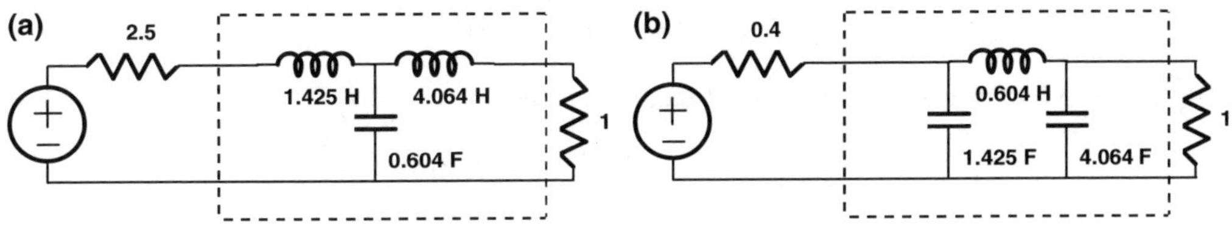

Figure 6.26 Prototype Butterworth LPF with $N = 3$ and $R = 2.5$

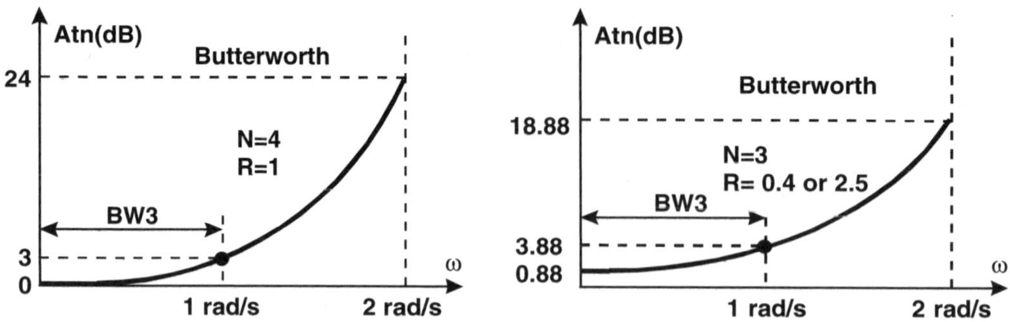

Figure 6.27 Attenuations for Butterworth filters with $N = 3$ and 4

attenuation. The only restriction is that for $R \neq 1$ the theoretical characteristic is shifted by an amount given by the attenuation at $\omega = 0$, that is $1 + (1 - R)^2/4R$. Figure 6.27 shows the predicted attenuations for the $N = 4$ and $N = 3$ filters of the examples.

6.4.2 *Calculations for Chebyshev Prototype Elements*

We remember that the mathematical Chebyshev characteristic is:

$$Atn = 1 + K^2 \cdot TN^2(\omega) \quad \rightarrow \quad \text{for } \omega = 1 \quad Atn = 1 + K^2 \quad \text{(ripple point)}$$

or alternatively:

$$Atn = 1 + K^2 \cdot TN^2(\omega 3\text{dB} \cdot \omega) \quad \rightarrow \quad \text{for } \omega = 1 \quad Atn = 2 \text{ (3dB point)}$$

We are going to perform the same extraction process as Butterworth for $N = 2$, and we can start by using the first mathematical characteristic: attenuation at cut-off is $1 + K^2$:

$$Atn = 1 + K^2 \cdot T_2^2(\omega) = 1 + K^2[1 - 4\omega^2 + 4\omega^4] \tag{6.42}$$

and the ladder network attenuation is:

$$Atn = 1 + \frac{(1 - R)^2 + \omega^2 \cdot (L^2 + C^2R^2 - 2LC) + \omega^4 \cdot L^2C^2}{4R} \tag{6.43}$$

By fitting both equations at $\omega = 0$ we have:

$$1 + K^2 = 1 + (1 - R)^2/4R \tag{6.44}$$

and this means that in this case R is a function of the ripple factor K:

$$R = 2K^2 + 1 \pm [4K^2 \cdot (1 + K^2)]^{1/2} \tag{6.45}$$

if we want the $Atn = 1 + K^2$ at DC. In general, R cannot be unity (equal source and load resistance) for N even. In the case of N odd, we can have equality.

On the other hand, by equalling the general attenuation equations, we have:

$$1 + K^2[1 - 4\omega^2 + 4\omega^4] = 1 + \frac{(1-R)^2}{4R} + \frac{\omega^2 \cdot (L^2 + C^2R^2 - 2LC)}{4R} + \frac{\omega^4 \cdot L^2C^2}{4R} \quad (6.46)$$

or, using the condition for R:

$$-4K^2\omega^2 + 4K^2\omega^4 = \omega^2 \cdot \frac{(L^2 + C^2R^2 - 2LC)}{4R} + \omega^4 \cdot \frac{L^2C^2}{4R} \quad (6.47)$$

that is, we arrive at a system of equations:

$$L^2 + C^2R^2 - 2LC = -16K^2R \quad (1^\circ \text{ condition}) \quad (6.48)$$

$$L^2C^2 = 16K^2R \quad (2^\circ \text{ condition}) \quad (6.49)$$

to calculate L and C.

As an example for this $N = 2$ filter, we can suppose that the allowed ripple in the passband is RdB = 0.1 dB:

$$10 \cdot \text{Log}(1 + K^2) = 0.1 \quad \rightarrow \quad K = 0.15262$$

and, as we have N even, there are two solutions different from the unity for the source resistor R:

$$R_a = 1.35536 \qquad \text{with } R_a = 1/R_b \qquad R_b = 0.73781$$

If our choice is, for example, $R = R_b = 0.73781$, then the solutions for the reactive elements are:

$$R = 0.73781 \quad C = 0.843 \text{ F} \quad L = 0.622 \text{ H} \quad Atn(W = 0) = 1 + K^2 = 0.02329 \text{ (0.1 dB)}$$

$$Atn(W = 1) = 1 + K^2 = 0.02329 \text{ (0.1 dB)}$$

Figure 6.28 shows this filter along with its theoretical response calculated from the Chebyshev formula.

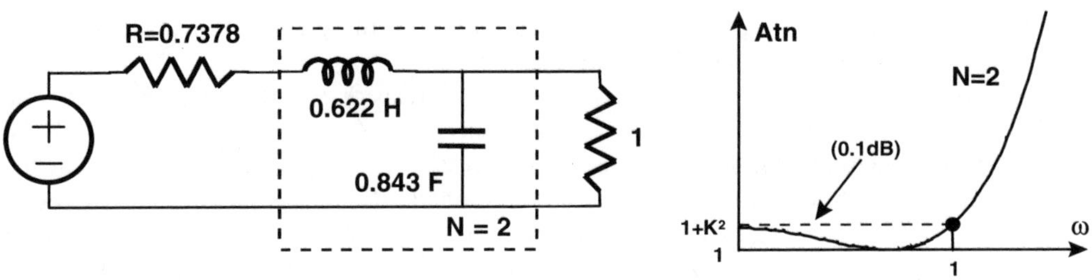

Figure 6.28 Chebyshev PLPF with $N = 2$ and $R = 0.73781$

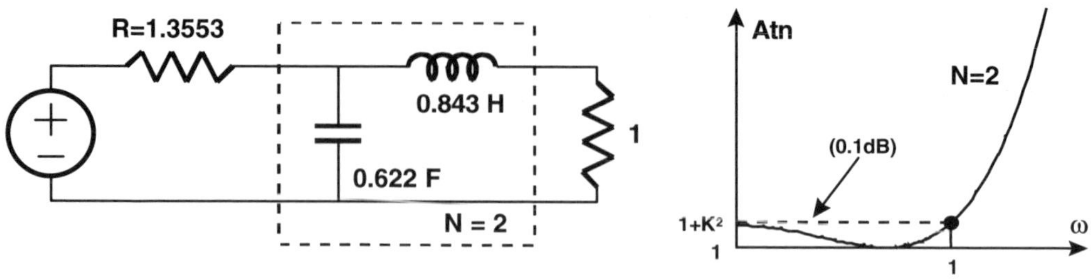

Figure 6.29 Chebyshev PLPF with $N = 2$ and $R = 1.35536$

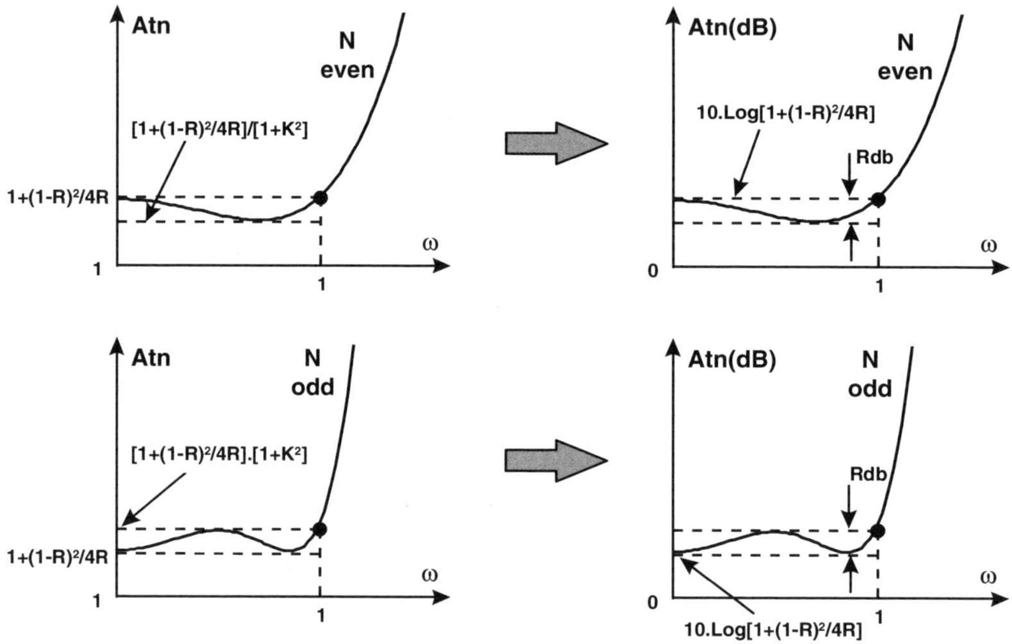

Figure 6.30 Chebyshev PLPF responses with $N = 2$

Conversely, if our choice is $R = R_a = 1.35536$, then the L and C values result in an imaginary form. This means that this ladder network is only possible for $R < 1$. Again we can use the concept of dual network to obtain a filter with these characteristics and with a generator resistance R_a, as shown in Figure 6.29.

If we now try to design the same filter for a source resistor R different to R_a or R_b (the optimum resistors in order to have the ripple value $1 + K^2$ at cut-off), we can still use the system of equations but in this case the Chebyshev characteristics are shifted in the attenuation axis as shown in Figure 6.30 for N even and odd, that is, always the DC point should have $Atn = 1 + (1 - R)^2/4R$ while maintaining the ripple.

As an example, suppose that the desired source resistor is $R = 0.5$. In this case the system of equations gives the following results:

$$R = 0.5 \quad L = 0.288\ \text{H} \quad C = 1.5715\ \text{F}$$

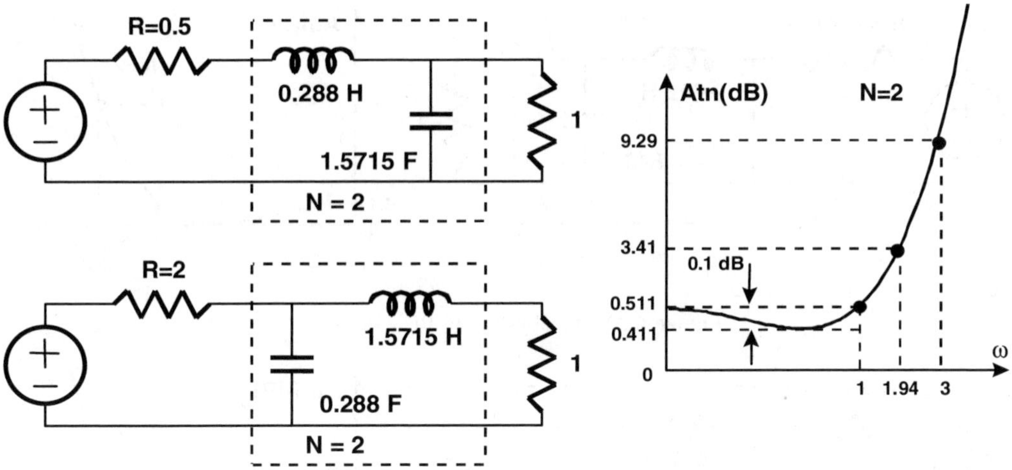

Figure 6.31 Chebyshev PLPF with $N = 2$, $R = 0.5$ and 2.0

and the dual for $R = 1/0.5 = 2$ is:

$$R = 2 \quad L = 1.5715 \text{ H} \quad C = 0.288 \text{ F}$$

in both cases, the DC attenuation is $10 \cdot \text{Log}[1 + (1 - R)^2/4R] = 0.511$ dB.

Figure 6.31 shows both circuits along with the theoretical response. We can observe the curve is shifted by an amount equal to $0.511 - 0.1 = 0.411$ dB.

Using the Chebyshev formulas we can know that $\omega 3\text{dB} = 1.9432$ and the attenuation at this point should be $3 + 0.411 = 3.411$ dB, and the attenuation at say $\omega = 3$ is $8.88 + 0.411 = 9.29$ dB.

In conclusion, Chebyshev filters with N even cannot have the same source and load resistor if we want to have the ripple RdB at cut-off and there is an optimum value of R for this purpose. Conversely, for N odd, the source and load resistors should be the same ($R = 1$) if we want RdB at cut-off. Finally, if we choose any other resistor R different from the above explained, the response of attenuation in decibel will be shifted by an amount given by:

$$10 \cdot \text{Log}[1 + (1 - R)^2/4R] \qquad \text{for } N \text{ odd} \qquad (6.50)$$

$$10 \cdot \text{Log}[1 + (1 - R)^2/4R] - \text{RdB} \qquad \text{for } N \text{ even} \qquad (6.51)$$

Up to now we have designed Chebyshev filters where $\omega = 1$ corresponds to the end of the ripple. In many applications, however, the designer needs to speak in terms of the 3dB points. So it should be interesting to modify the above designs in order to ensure that $\omega = 1$ corresponds to the 3dB point. In this case the attenuation characteristic is given by:

$$Atn = 1 + K^2 \cdot T_N^2(\omega 3\text{dB} \cdot \omega) \qquad (6.52)$$

$$\text{where } \omega 3\text{dB} = \text{COSH}[1/\text{N.ACSH}(1/\text{K})] \qquad (6.53)$$

One can repeat the calculation process by moving the variable ω towards $\omega \cdot \omega 3\text{dB}$. In fact, it is easy to demonstrate that it is enough to multiply the reactive elements L, C calculated in the sense $\omega = 1 \rightarrow$ RdB, by the factor $\omega 3\text{dB}$ to have $\omega = 1 \rightarrow 3\text{dB}$.

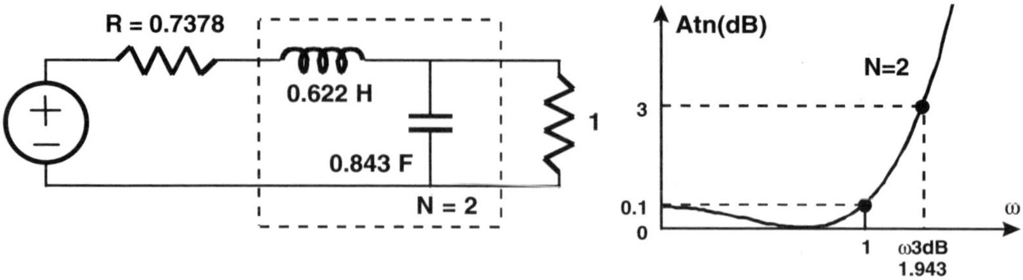

Figure 6.32 Chebyshev PLPF with $N = 2$

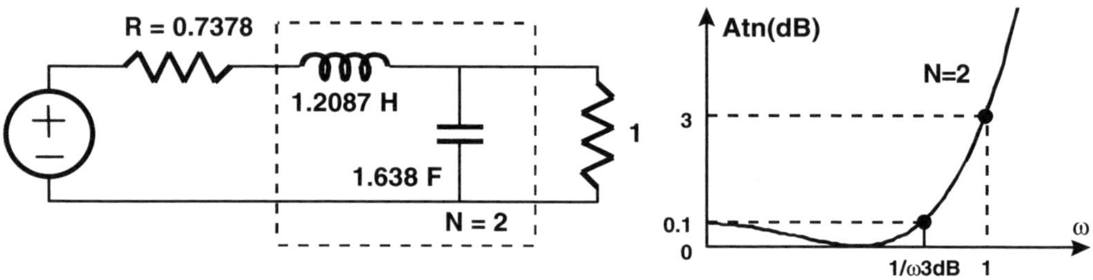

Figure 6.33 Chebyshev LPLF with $N = 2$

As an example, Figure 6.32 repeats the filter design shown in Figure 6.28 for RdB = 0.1, $N = 2$, $R = 0.73781$ and ω3dB = 1.9432 rad/s.

Figure 6.33 shows the modification of the filter for $\omega = 1 \rightarrow$ 3dB.

The same simple process is valid for any other Chebyshev filter.

Table 6.2 (see Appendix) shows the Chebyshev prototype element values in Henries and Farads up to $N = 7$ and for ripples ranging from 0.01 up to 3dB. The elements are calculated in order to have RdB at $\omega = 1$, that is, the R values for N even are the optimum, and unity for N odd. The conventions are the same as for Butterworth filters.

Tables 6.3 to Table 6.7 (see Appendix) show the prototype elements for a variety of source resistors, including the optima, and ripples up to $N = 7$. In these tables the point $\omega = 1$ corresponds to an Atn = 3dB. So it is very easy to translate the reactive elements in this tables in order to have $\omega = 1 \rightarrow$ RdB simply by division of the elements by ω3dB. We must remember at this point that if R is not the optimum one, the theoretical curve is shifted in the attenuation axis.

6.4.3 Calculations for Bessel Prototype Elements

The same approach can be taken for a given Bessel characteristic of order N. As we know, the Bessel characteristic can be approximated by:

$$Atn|_{db} \cong 3 \cdot \omega^2 \quad \text{for any } N \text{ and valid up to } \omega = 2$$

or the graphic format for $\omega > 2$ and N up to 7.

As the theoretical calculations are rather tedious, Table 6.8 (see Appendix) shows the PLPF elements for this response. We must remember that the cut-off $\omega = 1$ corresponds to the 3dB point, and the details are the same as for the Butterworth response.

6.4.4 Calculations for Elliptic Prototype Elements

As these kinds of filters are not 'all pole' filters and they exhibits transmission zeros at finite frequencies, the conventional ladder network is not appropriate in this case. Figure 6.34 shows the conventional Elliptic networks for N up to 7 as well as the number of transmission zeros NZ.

We must remember that these filters have a pass band ripple of RdB such that $\omega = 1 \rightarrow$ RdB, and follow the Chebyshev criteria for the source resistor R. Furthermore, as can be expected, the parallel LC combinations as well as the series LC in the dual network are resonant at the transmission zeros.

The Elliptic PLPF elements are tabulated in Tables 6.9 to 6.16 (see Appendix) as a function of the modulation angle, and for N up to 7. The pass band ripple is 0.01 dB or 0.18 dB which correspond to a reflection coefficient of 0.05 and 0.2 respectively.

Figure 6.35 shows the conventions for reading these tables. The upper network is valid for the upper template and conversely for the dual network shown at the bottom. The values A2, A4, . . . are the position of the transmission zeros.

As an example, Figure 6.36 shows an $N = 3$, RdB = 0.18 Elliptic PLPF having $\omega_s = 2$, Amin = 26.5 dB and the transmission zero at $A2 = 2.27$. The generator and load impedances are unity. Furthermore, Figure 6.37 shows an $N = 4$, RdB = 0.18 Elliptic PLPF having $\omega_s = 3$, Amin = 56.8 dB, $A2 = 3.28$ and $R = 0.66$ using the dual network.

6.5 Filter Impedance and Frequency Scaling

We must remember that when designing prototype low pass filters with specified filter characteristics, the cut-off is at $\omega = 1$ rad/s (0.159 Hz) for an attenuation in general of 3dB, and for a load resistor unity. It is clear that it is not usual to use the above impedance level nor the frequency values, so it is necessary to scale the filter in order to meet actual specifications.

6.5.1 Impedance Scaling

If we know the design of a low pass prototype filter having $R_L = 1$ and any R_G, it should be interesting to derive another filter such that the loading resistors have the actual values while maintaining the same filter shape.

A typical prototype filter ladder network is shown in Figure 6.38, and we will develop the process for a $N = 4$ filter. The extrapolation for any value of N will be clear.

The input impedance of this network can be written as:

$$Z_{in} = j\omega \cdot L1 + \cfrac{1}{j\omega \cdot C2 + \cfrac{1}{j\omega \cdot L3 + \cfrac{1}{j\omega \cdot C4 + 1}}} \tag{6.54}$$

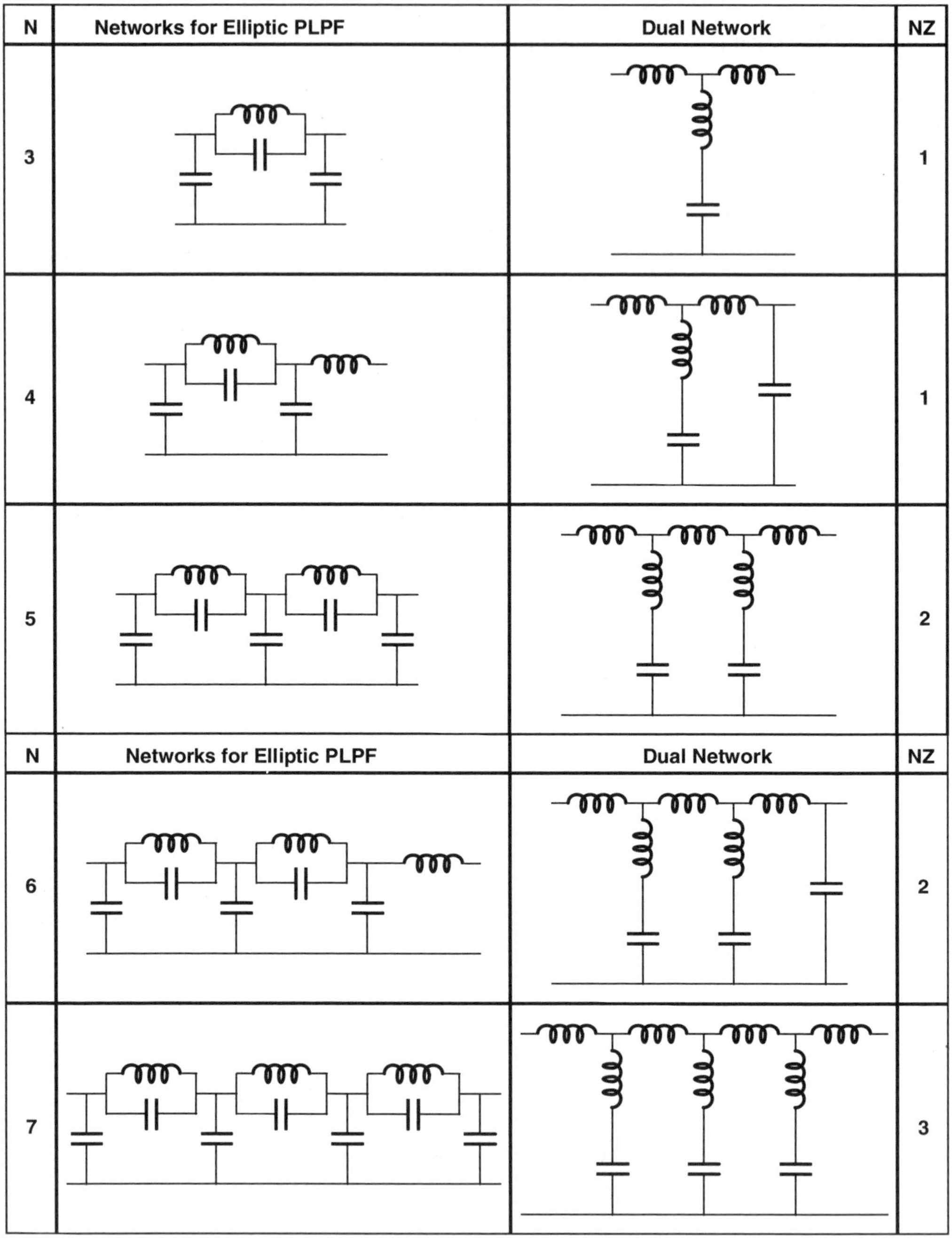

Figure 6.34 Elliptic networks for $N = 3$ to 7

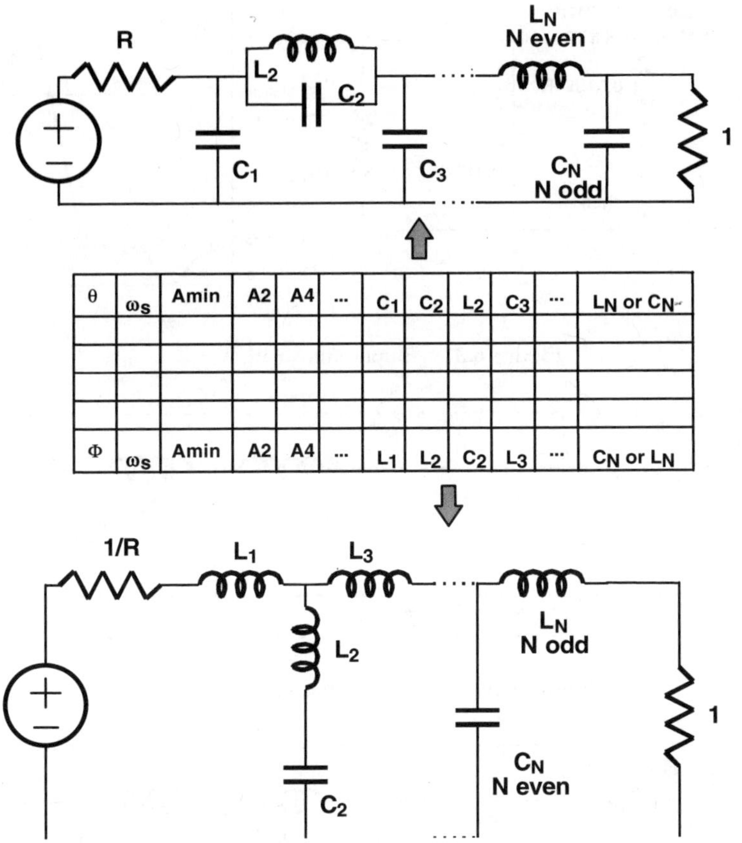

Figure 6.35 Convention for Elliptic filter tables

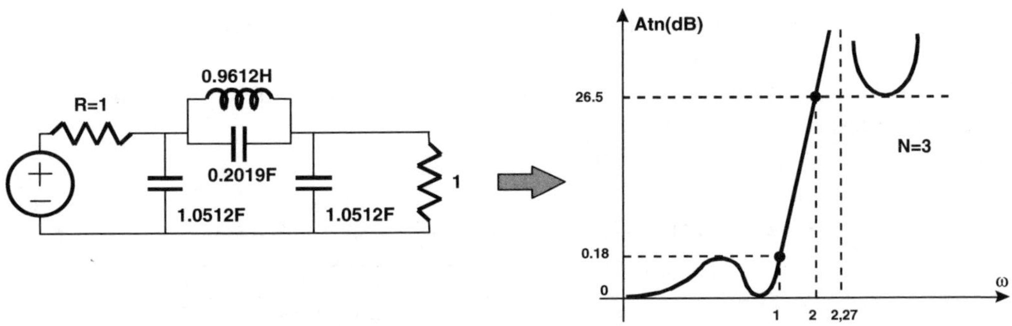

Figure 6.36 Elliptic PLPF with $N = 3$

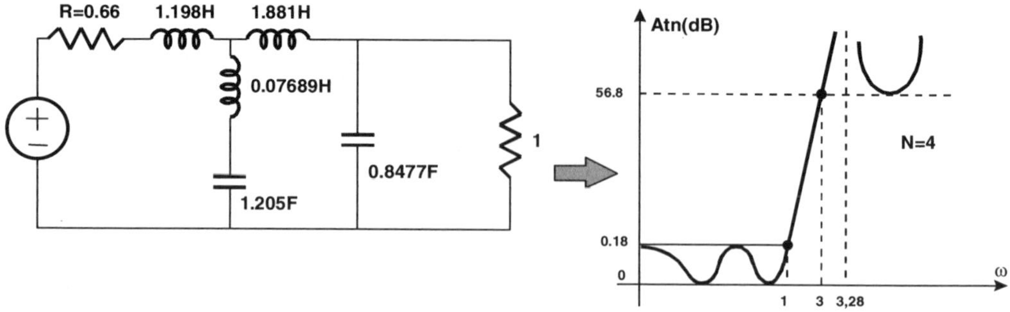

Figure 6.37 Elliptic filter with $N = 4$

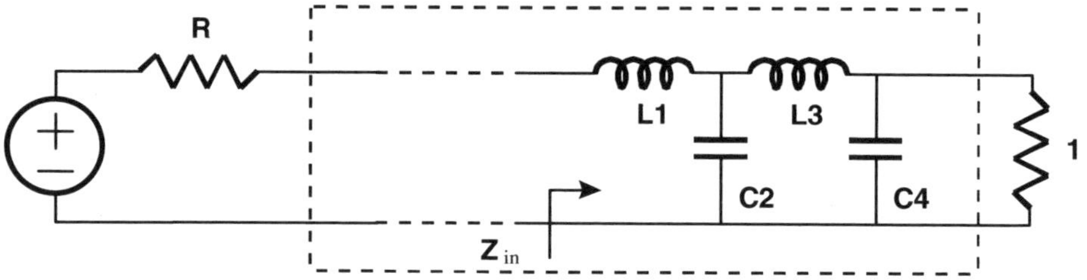

Figure 6.38 Prototype filter ladder network

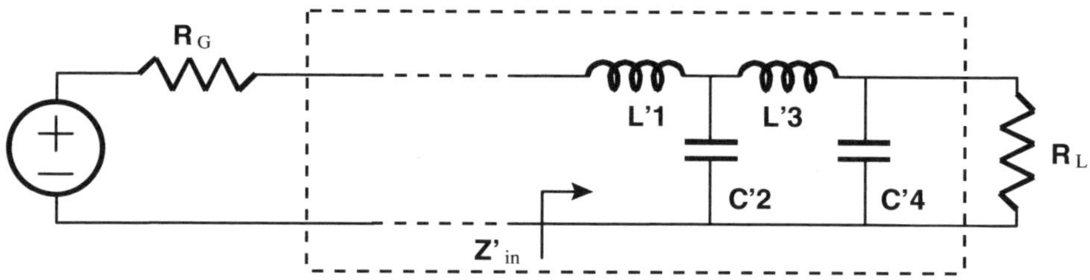

Figure 6.39 Filter network with arbitrary load resistance

Now we have to analyse another network in Figure 6.39 with different reactances and a general load resistor R_L.

For this network the input impedance is:

$$Z'_{in} = j\omega \cdot L'1 + \cfrac{1}{j\omega \cdot C'2 + \cfrac{1}{j\omega \cdot L'3 + \cfrac{R_L}{j\omega \cdot R_L \cdot C'4 + 1}}} \tag{6.55}$$

If we want both networks to have the same frequency response, the following condition should be verified:

$$Z_{in} = Z'_{in}/R_L \tag{6.56}$$

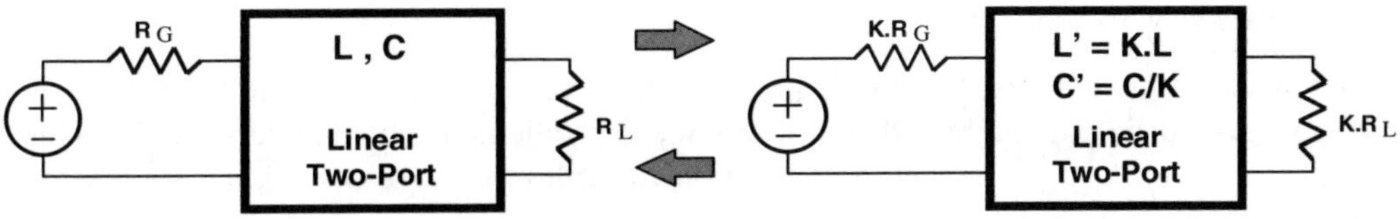

Figure 6.40 Impedance transformation

by applying this condition to both expressions for the input impedance we arrive at:

$$C'2 = C2/R_L \quad L'1 = L1 \cdot R_L \tag{6.57}$$

$$C'4 = C4/R_L \quad L'3 = L3 \cdot R_L \tag{6.58}$$

and this result can be expanded to any number of sections. In general, any linear network maintains its frequency response if all the resistances and inductors are multiplied by a constant and the capacitors are divided by the same constant. This fact is illustrated for our case in Figure 6.40.

Figure 6.41 shows this transformation for a $N = 3$ Butterworth prototype filter when this filter will be used in a 50 ohm system:

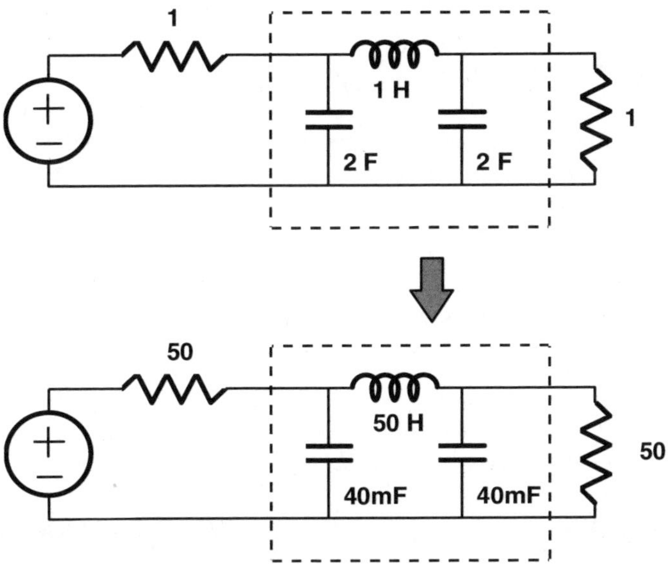

Figure 6.41 Impedance transformation for Butterworth filter

Remember that in the above transformations, the frequency characteristics remain unchanged, that is cut-off $\omega = 1$ corresponds to the 3dB point or the RdB point depending on the filter characteristic.

6.5.2 Frequency Scaling

In practical low pass filters, cut-off frequencies are never $F_C = 0.159$ Hz, and most systems use high pass, band pass and band stop filters with specified cut-offs and bandwidths. Thus, it is necessary to develop frequency translations in order to derive these kinds of practical filters from the well-known prototype low pass filter.

In the following calculations we assume that the load resistor is unity $R_L = 1$ (the impedance scaling is a different problem from frequency scaling and is normally done at the end of the design) and we denote with 'prime' values the prototype low pass filter elements and frequency, while 'unprimed' values are for the scaled filter. The cut-off frequency for the low pass prototype will be $\omega' = 1$ rad/s.

6.5.3 Low Pass to Low Pass Expansion

This expansion is used to derive a new filter where the actual cut-off frequency is ω_C while maintaining the same filter shape:

	Pass band	Attenuation
Prototype filter:	$0 \to \omega' = 1$	$Atn' = Atn'(\omega')$
Actual low pass filter:	$0 \to \omega_C$	$Atn = Atn(\omega)$

Using the following frequency transformation:

$$\omega' = \omega/\omega_C \qquad \begin{array}{l} \omega' = 0 \to \omega = 0 \\ \omega' = 1 \to \omega = \omega_C \end{array} \tag{6.59}$$

both filters will have the same frequency response if the reactances and susceptances in both filters are the same in its own frequency axis:

$$jX'(\omega') = jX(\omega) \quad jB'(\omega') = jB(\omega) \tag{6.60}$$

This fact is depicted in Figure 6.42.

By applying the above statements we have:

$$X' = L' \cdot \omega' \equiv X = L \cdot \omega \quad \to \quad L \cdot \omega_C \cdot \omega' = L \cdot \omega \quad \to \quad L = L'/\omega_C \tag{6.61}$$

$$B' = C' \cdot \omega' \equiv B = C \cdot \omega \quad \to \quad C \cdot \omega_C \cdot \omega' = C \cdot \omega \quad \to \quad C = C'/\omega_C \tag{6.62}$$

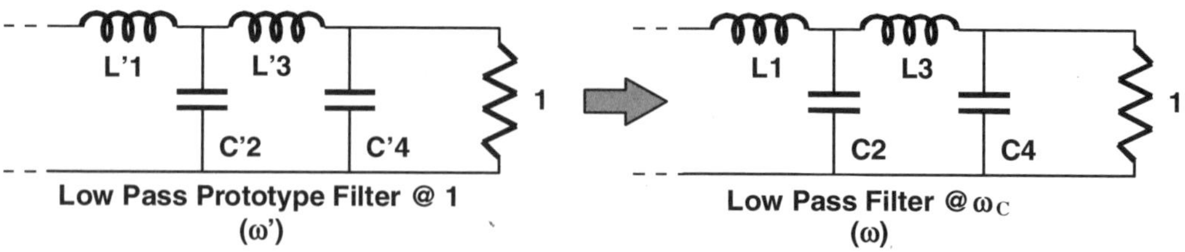

Figure 6.42 LPF frequency transformation

that is:

Prototype	Low Pass		
Inductor: L' →	Inductor: L	with	$L = L'/\omega_C$
Capacitor: C' →	Capacitor: C	with	$C = C'/\omega_C$

In conclusion, it is sufficient to divide all the prototype inductors and capacitors by ω_C to have the same low pass behaviour with a new cut-off frequency ω_C.

Figure 6.43 shows the frequency translation for a $N = 3$ Chebyshev filter with $\omega' = 1 \rightarrow$ RdB. We observe the same shape with an expansion in the frequency axis.

As an example, suppose we have an $N = 3$ Chebyshev prototype filter with $R = 1$ and RdB = 0.1, and we wish to have a low pass filter with a cut-off frequency of 200 MHz such as $F_C \rightarrow$ RdB. Using the tables, the prototype filter is shown in Figure 6.44(a). As $\omega_C = 2 \cdot \pi \cdot 200 \cdot 10^6$, all the inductors and capacitors should be divided by this value. The resulting filter is shown in Figure 6.44(b).

If we want, we can do an impedance transformation at this stage to have a 200 MHz low pass filter in, for example, a 50 ohm system as shown in Figure 6.45(a) and Figure 6.45(b). Remember that 3dB point is at ω3dB = 1.388, that is, at 277.8 MHz.

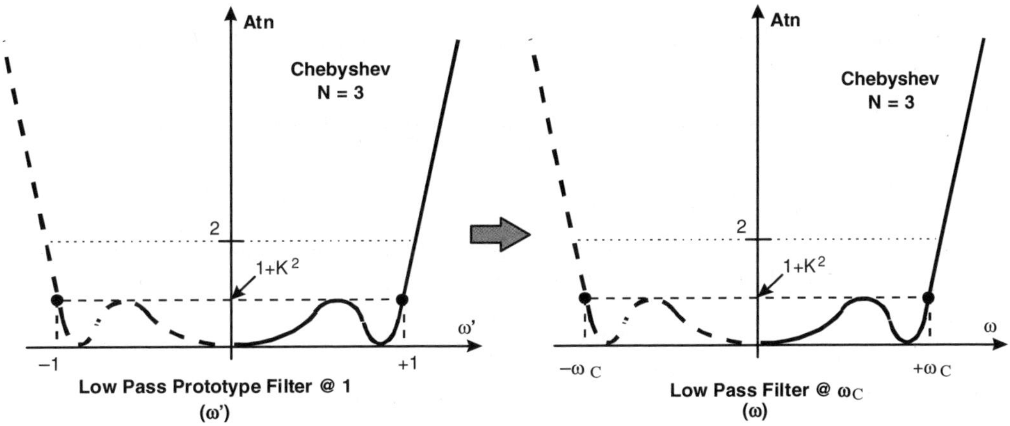

Figure 6.43 Chebyshev filter frequency translation

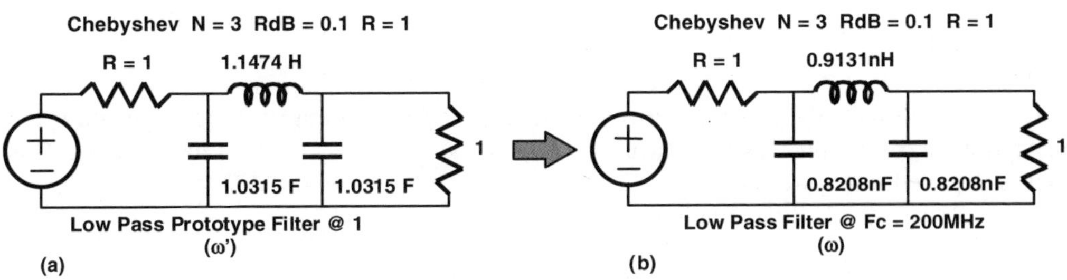

Figure 6.44 Chebyshev filter frequency translation example

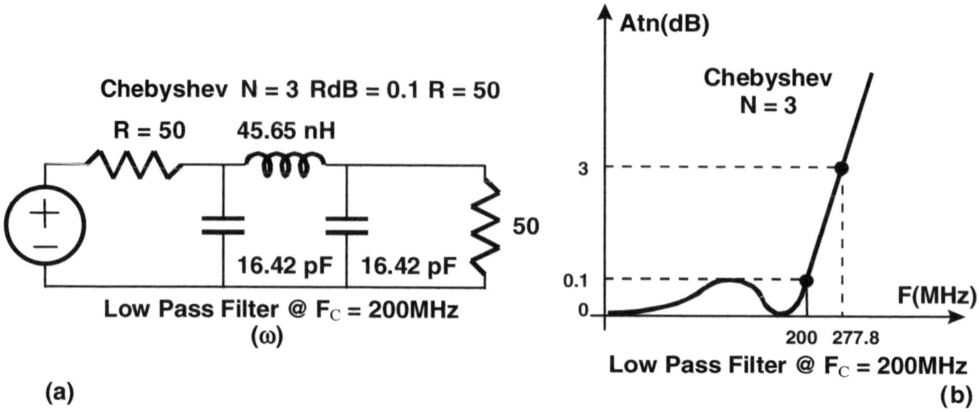

Figure 6.45 LPF frequency and impedance transformation

The expected attenuation at, for example, 600 MHz can be easily calculated. At this frequency we have:

$$\omega' = \omega/\omega_C = F/F_C = 600 \text{ MHz} / 200 \text{ MHz} = 3$$

and using the PLPF Chebyshev characteristic for $N = 3$ and RdB = 0.1 ($K^2 = 0.02329$), we can write:

$$Atn|_{dB} = 10 \cdot \text{Log}[1 + K^2 \cdot T_3^2(\omega')] = 23.6 \text{ dB}$$

6.5.4 Low Pass to High Pass Transformation

This transformation is used to derive a high pass filter where the actual cut-off frequency is ω_C while maintaining the same filter shape as the prototype low pass:

	Pass band	Attenuation
Prototype filter:	$0 \rightarrow \omega' = 1$	$Atn' = Atn'(\omega')$
Actual high pass filter:	$\omega_C \rightarrow \infty$	$Atn = Atn(\omega)$

Using the following frequency transformation:

$$\omega' = -\omega_C/\omega \quad \begin{array}{lll} \omega' = 0 & \rightarrow & \omega = \infty \\ \omega' = +1 & \rightarrow & \omega = -\omega_C \\ \omega' = -1 & \rightarrow & \omega = +\omega_C \end{array} \tag{6.63}$$

both filters will have the same frequency response if we use a high pass topology and if the reactances and susceptances in both filters are the same in their own frequency axis. This fact is depicted in Figure 6.46.

By applying the identities we have:

$$X' = L' \cdot \omega' = L' \cdot (-\omega_C/\omega) \equiv X = -1/C\omega \quad \rightarrow \quad C = 1/(L' \cdot \omega_C) \tag{6.64}$$

$$B' = C' \cdot \omega' = C' \cdot (-\omega_C/\omega) \equiv B = -1/L\omega \quad \rightarrow \quad L = 1/(C' \cdot \omega_C) \tag{6.65}$$

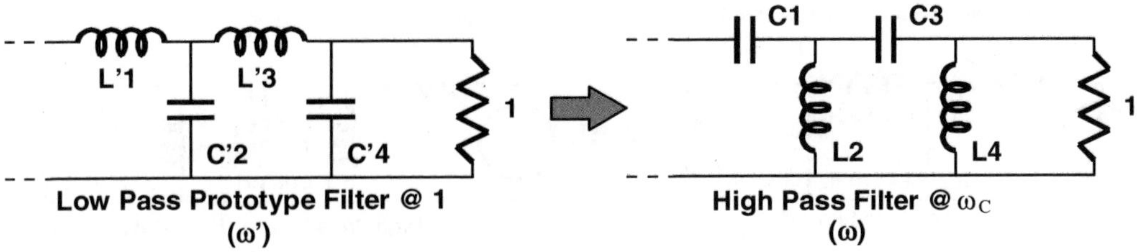

Figure 6.46 LP to HP transformation

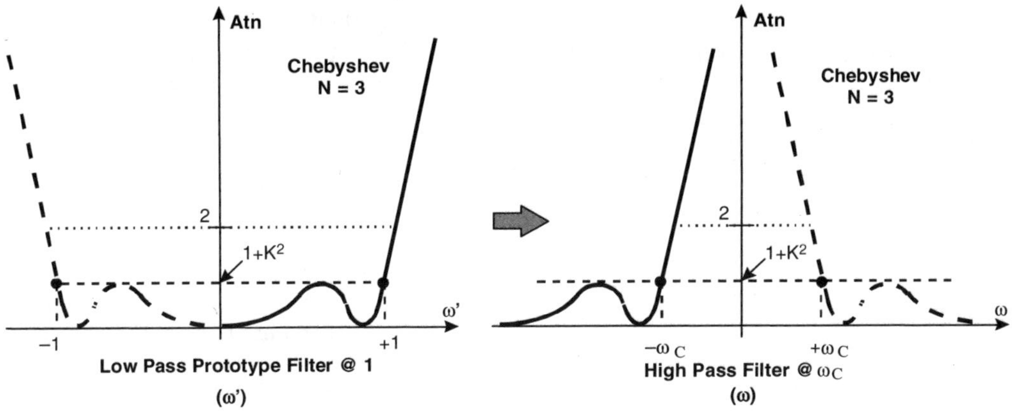

Figure 6.47 LP to HP Chebyshev filter transformation

that is, the inductors translate to capacitors and vice versa.

Prototype	High Pass		
Inductor: L' $\rightarrow$	Capacitor: C	with	$C = 1/(L' \cdot \omega_C)$
Capacitor: C' $\rightarrow$	Inductor: L	with	$L = 1/(C' \cdot \omega_C)$

Figure 6.47 shows this frequency transformation for an $N = 3$ Chebyshev filter with $\omega' = 1$ $\rightarrow$ RdB. We observe the same shape expanded in the frequency axis and symmetrical with respect to $\omega = 0$.

As an example, suppose we want to design an $N = 3$ Chebyshev high pass filter having a ripple of RdB = 0.1 and a 3dB cut-off frequency of 400 MHz. The input and output impedances should be 25 and 50 ohm respectively ($R = 25/50 = 0.5$). Figure 6.48(a) and 6.48(b) show the frequency translation while Figure 6.49(a) and 6.49(b) show the final filter with impedance scaling and the filter response. We should remember that the mismatch due to $R = 0.5$ is 0.511 dB.

The expected attenuation at, for example, 200 MHz can easily be calculated. At this frequency, we have:

$$\omega' = -\omega_C/\omega = -F_C/F = -400 \text{ MHz} / 200 \text{ MHz} = -2$$

$$\omega 3\text{dB} = 1.389.$$

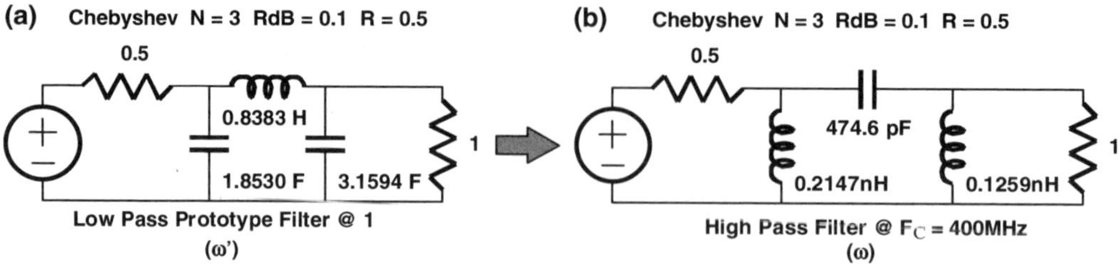

Figure 6.48 LP to HP Chebyshev filter transformation example

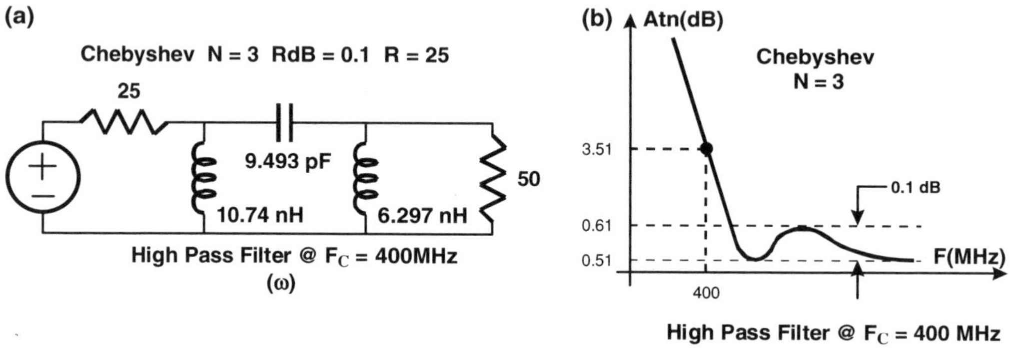

Figure 6.49 Chebyshev filter frequency translation example

Using the PLPF Chebyshev characteristic for $N = 3$ and RdB = 0.1 ($K^2 = 0.02329$) and the ω3dB value, we can write:

$$Atn|_{dB} = 10 \cdot \text{Log}[1 + K^2 \cdot T_3^2(\omega' \cdot \omega 3\text{dB})] = 21.48 \text{ dB} + 0.511 = 21.99 \text{ dB}$$

6.5.5 Low Pass to Band Pass Transformation

This transformation is used to derive a band pass filter where the actual cut-off frequencies are ω_1 and ω_2 while maintaining the same filter shape as the prototype low pass:

	Pass band	Attenuation
Prototype filter:	$0 \rightarrow \omega' = 1$	$Atn' = Atn'(\omega')$
Actual band pass filter:	$\omega_1 \rightarrow \omega_2$	$Atn = Atn(\omega)$

Using the following frequency transformation:

$$\omega' = \frac{\omega_0}{\omega_2 - \omega_1} \cdot [\omega/\omega_0 - \omega_0/\omega] \quad \text{where } \omega_0 = (\omega_1 \cdot \omega_2)^{1/2}$$

or

$$\omega = \omega' \cdot \frac{\omega_2 - \omega_1}{2} \pm \frac{1}{2}[\omega'^2(\omega_2 - \omega_1)^2 + 4\omega_1\omega_2]^{1/2} \tag{6.66}$$

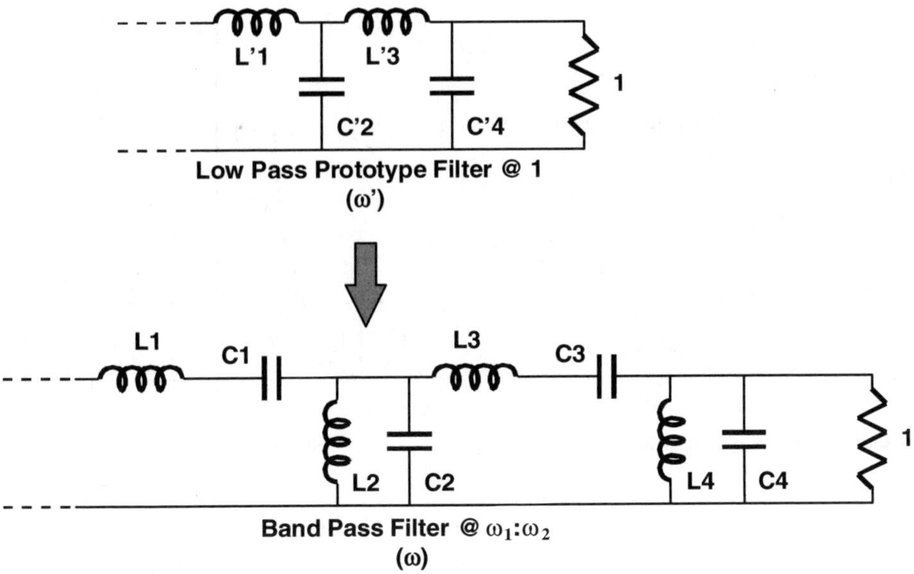

Figure 6.50 LP to BP transformation

then we have:

$$\omega' = 0 \quad \rightarrow \omega = \pm\omega_0$$
$$\omega' = +1 \rightarrow \omega = +\omega_2 \text{ and } \omega = -\omega_1$$
$$\omega' = -1 \rightarrow \omega = -\omega_2 \text{ and } \omega = +\omega_1$$

both filters will have the same frequency response if we use a band pass topology and if the reactances and susceptances in both filters are the same in their own frequency axis. This fact is depicted in Figure 6.50 where each series inductor in the prototype translates to a series LC resonant circuit, while each shunt capacitor moves to a parallel LC resonant network. All the LC resonators (series or parallel) have their resonant frequency at $\omega_0 = (\omega_1 \cdot \omega_2)^{1/2}$.

By applying the reactance identities we have:

Prototype series inductor: $X' = L' \cdot \omega' = L' \cdot \dfrac{\omega_0}{\omega_2 - \omega_1} \cdot [\omega/\omega_0 - \omega_0/\omega]$

Band pass LC series resonant: $X = L\omega - 1/C\omega$

that is,

$$X' = \mathrm{X} \quad \rightarrow \quad L = L'/(\omega_2 - \omega_1) \quad \text{and} \quad C = 1/(L \cdot \omega_0^2)$$

in an analogous form we can derive:

Prototype parallel capacitor: $B' = C' \cdot \omega' = C' \cdot \dfrac{\omega_0}{\omega_2 - \omega_1} \cdot [\omega/\omega_0 - \omega_0/\omega]$

Band pass LC parallel resonant: $B = C\omega - 1/L\omega$

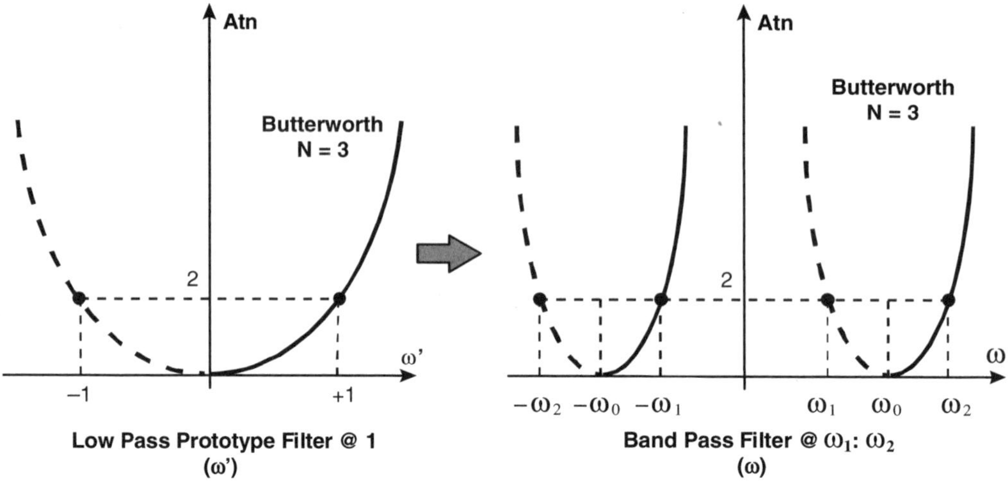

Figure 6.51 Butterworth LP to BP transformation

that is,

$$B' = B \quad \rightarrow \quad C = C'/(\omega_2 - \omega_1) \quad \text{and} \quad L = 1/(C \cdot \omega_0^2)$$

Prototype			Band pass			
Inductor:	L'	$\rightarrow$	Series LC	with	$L = L'/(\omega_2 - \omega_1)$	$C = 1/(L \cdot \omega_0^2)$
Capacitor:	C'	$\rightarrow$	Parallel LC	with	$C = C'/(\omega_2 - \omega_1)$	$L = 1/(C \cdot \omega_0^2)$
	$\omega_0^2 = (\omega_1 \cdot \omega_2)^{1/2} = \text{LC}$					

Figure 6.51 shows this frequency transformation for an $N = 3$ Butterworth filter with $\omega' = 1 \rightarrow$ 3dB. We observe the same shape expanded in the frequency axis as the low pass filter and symmetrical with respect to $\omega = 0$.

As an example we can design an $N = 3$ Butterworth band pass filter where the 3dB points are at $F_1 = 820$ MHz and $F_2 = 900$ MHz and the generator and load impedances are respectively $R_G = 100$ ohm and $R_L = 50$ ohm.

In this case $F_0 = 859.069$ MHz and $R_G/R_L = R = 2$, and we know that this means that in the prototype the attenuation at DC is $Atn(\omega' = 0) = 1 + (1 - R^2)/4R = 1.125$ (0.51 dB) and the attenuation at cut-off is $Atn(\omega' = 1) = 2 * 1.125 = 2.25$ (3.51 dB). If we think about the frequency translation, it is clear that in the actual band pass filter the attenuation at F_0 will be 0.51 dB and the attenuation at F_1 and F_2 will be 3.51 dB.

The prototype filter, from the tables, and the frequency transformation are shown in Figure 6.52 for $R = 2$. Figure 6.53(a) and 6.53(b) show the final filter design, after impedance scaling, and the predicted frequency response fom the Butterworth formula. In this case, we have drawn the transfer gain G_T in dB.

The expected attenuation at, for example, 700 MHz can easily be calculated. At this frequency, we have:

$$\omega' = \frac{\omega_0}{\omega_2 - \omega_1} \cdot [\omega/\omega_0 - \omega_0/\omega] = \frac{F_0}{F_0 - F_1} \cdot [F/F_0 - F_0/F] = -4.4285$$

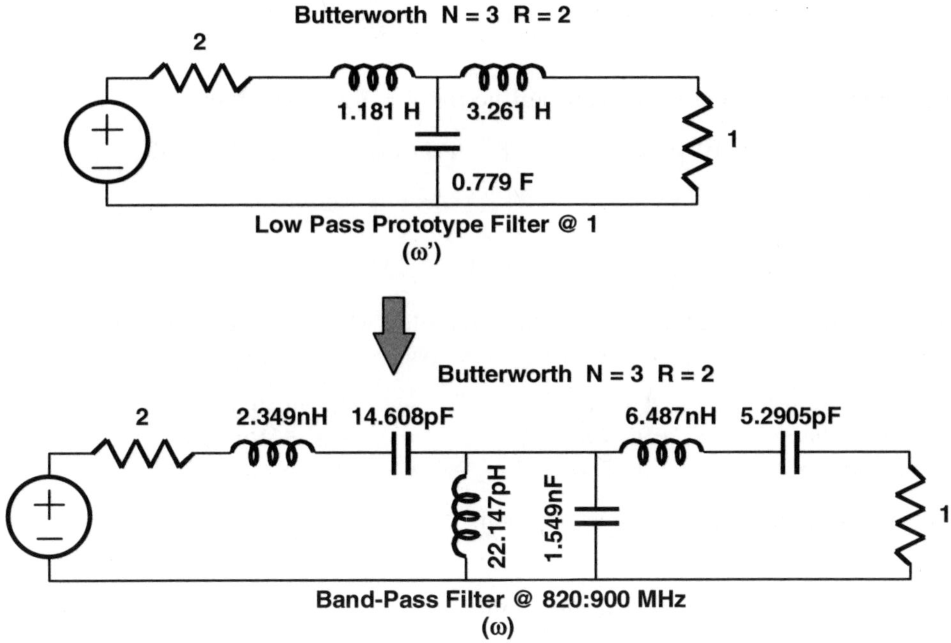

Figure 6.52 Butterworth filter example of LP to BP frequency transformation

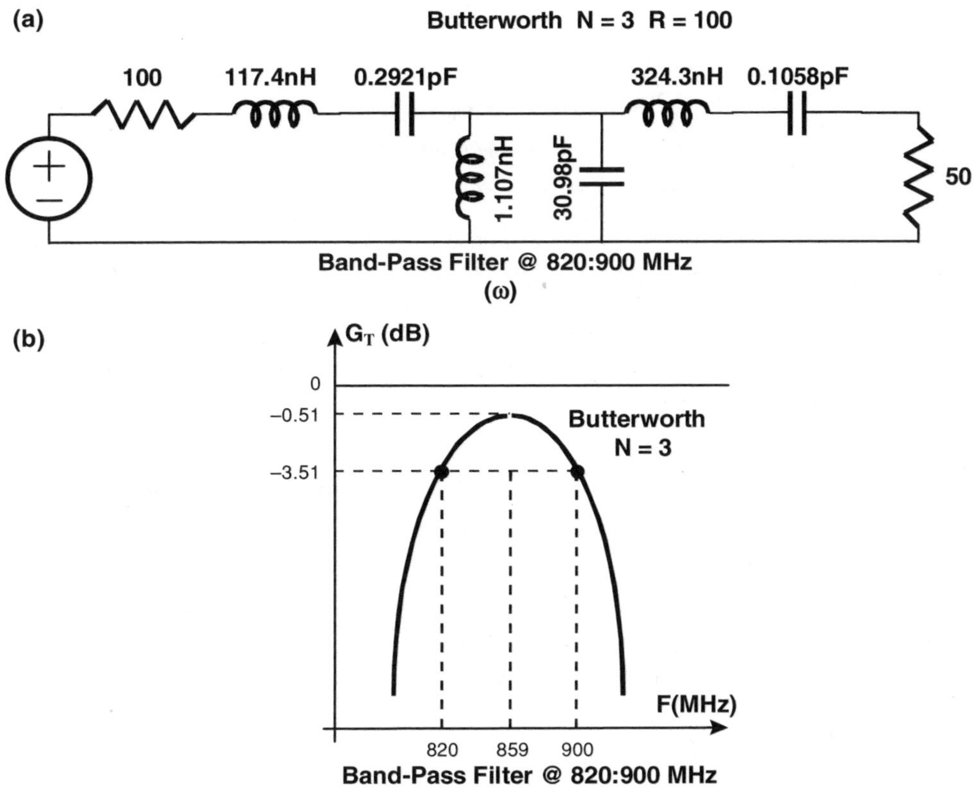

Figure 6.53 Butterworth filter example LP to BP impedance transformation

and using the PLPF Butterworth characteristic for $N = 3$ we can write:

$$Atn|_{dB} = 10 \cdot \text{Log}[1 + (\omega')^{2N}] = 38.77 \text{ dB} + 0.511 = 39.28 \text{ dB}$$

that is $G_T|_{dB} = -39.28$ dB.

6.5.6 Low Pass to Band Stop Transformation

This transformation is used to derive a band stop filter where the actual cut-off frequencies are ω_1 and ω_2 while maintaining the same filter shape as the prototype low pass:

	Pass band	Attenuation
Prototype filter:	$0 \to \omega' = 1$	$Atn' = Atn'(\omega')$
Actual band stop filter:	$0 \to \omega_1$ and $\omega_2 \to inf$	$Atn = Atn(\omega)$

Using the following frequency transformation:

$$\omega' = \frac{\omega_2 - \omega_1}{\omega_0} \cdot \frac{1}{[\omega_0/\omega - \omega/\omega_0]} \quad \text{where } \omega_0 = (\omega_1 \cdot \omega_2)^{1/2}$$

or

$$\omega = (-1/\omega') \cdot \frac{\omega_2 - \omega_1}{2} \pm \frac{1}{2}[(1/\omega')^2 \cdot (\omega_2 - \omega_1)^2 + 4\omega_1\omega_2]^{1/2} \tag{6.67}$$

then we have:

$$\omega' = 0 \quad \to \omega = \pm\, \omega_0$$
$$\omega' = +1 \to \omega = +\omega_1 \text{ and } \omega = -\omega_2$$
$$\omega' = -1 \to \omega = -\omega_1 \text{ and } \omega = +\omega_2$$

both filters will have the same frequency response if we use a band stop topology and if the reactances and susceptances in both filters are the same in ther own frequency axis. This fact is depicted in Figure 6.54 where each series inductor in the prototype translates to a parallel LC resonant circuit, while each shunt capacitor moves to a series LC resonant network. All the LC resonators (series or parallel) have their resonant frequency at $\omega_0 = (\omega_1 \cdot \omega_2)^{1/2}$.

By applying the reactance identities, we have:

Prototype series inductor: $\quad X' = L' \cdot W' = L' \cdot \dfrac{\omega_2 - \omega_1}{\omega_0} \cdot \dfrac{1}{[\omega_0/\omega - \omega/\omega_0]}$

Band stop LC parallel resonant: $\quad X = -1/[C\omega - 1/L\omega]$

that is,

$$X' = X \quad \to \quad L = L' \cdot (\omega_2 - \omega_1)/\omega_0^2 \quad \text{and} \quad C = 1/(L \cdot \omega_0^2)$$

in an analogous form we can derive:

Prototype series inductor: $\quad B' = C' \cdot \omega' = C' \cdot \dfrac{\omega_2 - \omega_1}{\omega_0} \cdot \dfrac{1}{[\omega_0/\omega - \omega/\omega_0]}$

Band stop LC series resonant: $\quad B = -1/[L\omega - 1/C\omega]$

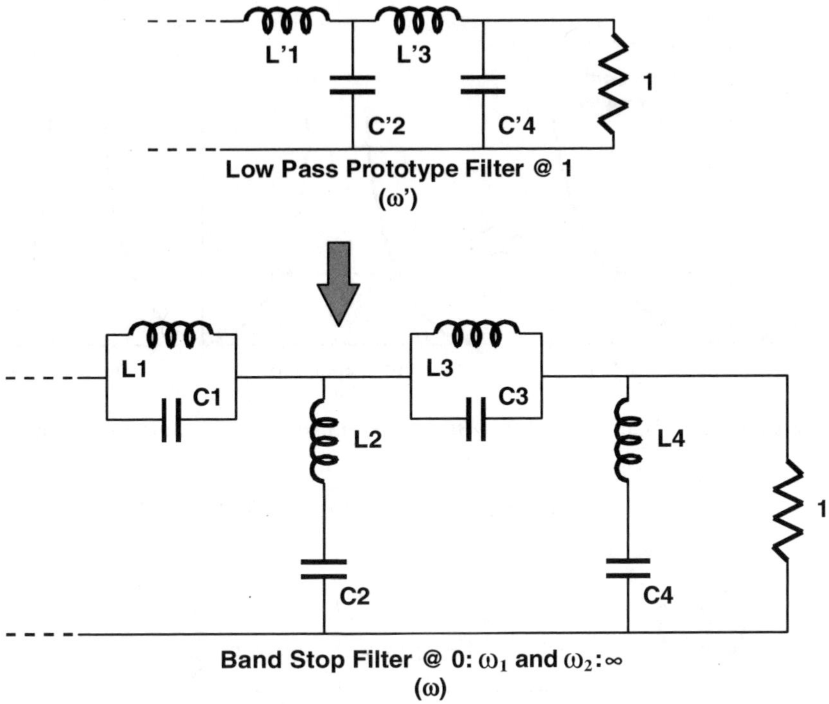

Figure 6.54 Low pass to band stop tranformation

that is,

$$B' = B \quad \rightarrow \quad C = C' \cdot (\omega_2 - \omega_1)/\omega_0^2 \quad \text{and} \quad L = 1/(C \cdot \omega_0^2)$$

Prototype			Band stop			
Inductor:	L'	$\rightarrow$	Parallel LC	with	$L = L' \cdot (\omega_2 - \omega_1)/\omega_0^2$	$C = 1/(L \cdot \omega_0^2)$
Capacitor:	C'	$\rightarrow$	Series LC	with	$C = C' \cdot (\omega_2 - \omega_1)/\omega_0^2$	$L = 1/(C \cdot \omega_0^2)$
	$\omega_0^2 = (\omega_1 \cdot \omega_2)^{1/2} = \mathrm{LC}$					

Figure 6.55 shows this frequency transformation for an $N = 3$ Butterworth filter with $\omega' = 1 \rightarrow$ 3dB. We observe the same shape expanded in the frequency axis as the high pass filter and symmetrical with respect to $\omega = 0$.

As an example, we design an $N = 3$ RdB = 0.1 band stop Chebyshev filter where the Rdb points are at $F_1 = 5$ GHz and $F_2 = 7$ GHz and the generator and load impedances are 50 ohm.

In this case $F_0 = 5.916$ GHz and $R_G/R_L = R = 1$. This means that in the prototype low pass filter the attenuation at DC is 0 dB, the attenuation at cut-off is RdB = 0.1 and the filter is symmetrical. Using the frequency translation the attenuation at F_1 and F_2 will be RdB, the attenuation at F_0 will be infinite and the ripple is maintained when going towards DC and high frequency.

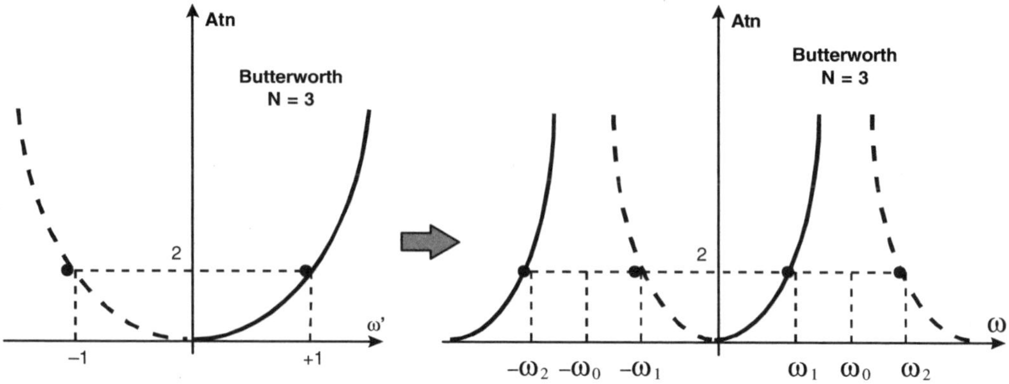

Figure 6.55 Butterworth LP to BS filter example, frequency transformation

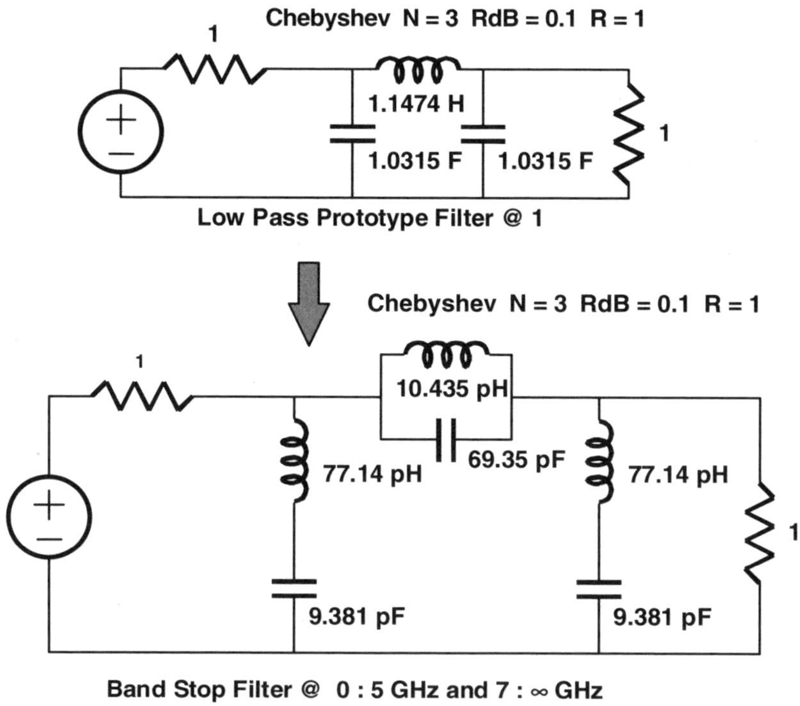

Figure 6.56 Chebyshev LP to BS filter example, frequency translation

The prototype filter, from the tables, and the frequency transformation are shown in Figure 6.56 for $R = 1$. Figures 6.57(a) and 6.57(b) show the final filter design, after impedance scaling, and the predicted frequency response fom the Chebyshev formula.

The expected attenuation at, for example, 6.5 GHz can easily be calculated. At this frequency, we have:

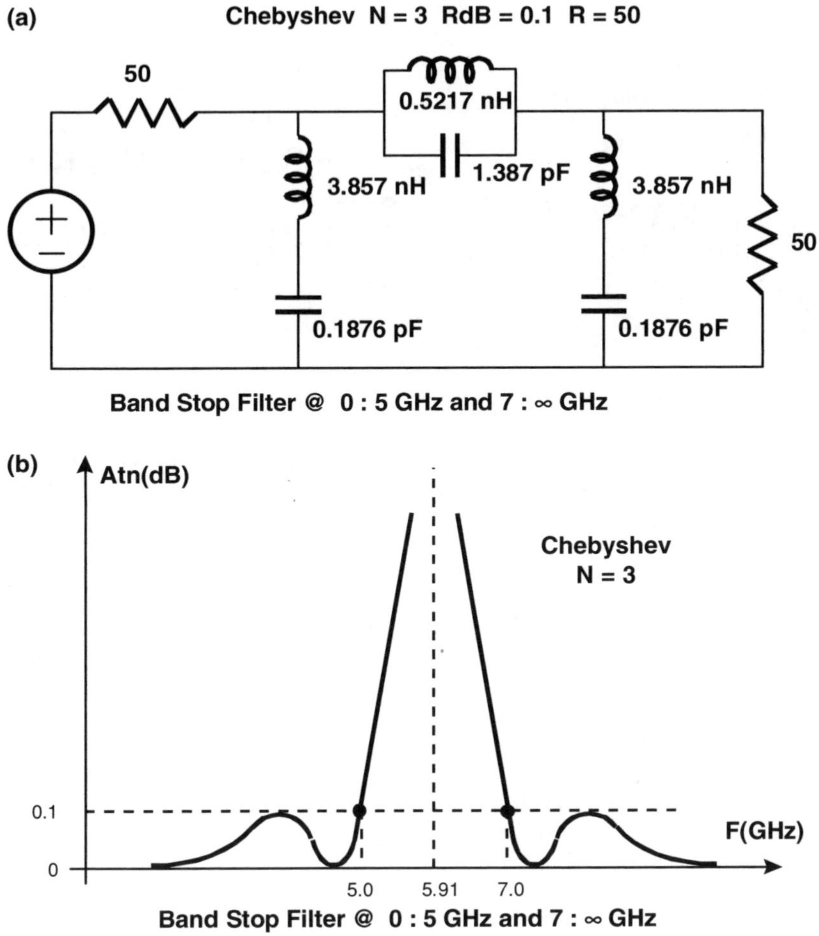

Figure 6.57 Chebyshev LP to BS filter example, impedance transformation

$$\omega' = \frac{F_2 - F_1}{F_0} \cdot \frac{1}{[F_0/F - F/F_0]} = -1.7931$$

and using the PLPF Chebyshev characteristic for $N = 3$ and RdB $= 0.1$ ($K^2 = 0.02329$) we can write:

$$Atn|_{dB} = 10 \cdot \mathrm{Log}[1 + K^2 \cdot T_3^2(\omega')] = 9.18 \text{ dB}$$

6.5.7 Resonant Network Transformations

As we will see in the next section, band pass and band stop translations for Elliptic filters result in networks as shown in Figure 6.58(a), that is, a series resonant circuit at ω_0 in parallel with a parallel LC resonant circuit at the same frequency ω_0.

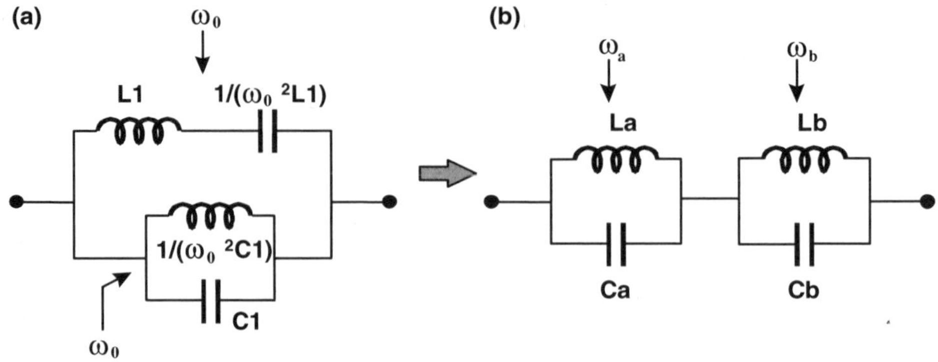

Figure 6.58 Band pass to band stop filter transformation

It should be very interesting to find an equivalent, shown in Figure 6.58(b), where we have a cascade combination of two parallel LC networks. It can easily be shown that both networks are equivalent if:

$$La = \frac{1}{C1 \cdot \omega_0^2 \cdot (H+1)} \qquad Lb = H \cdot La$$

$$Ca = \frac{1}{H \cdot \omega_0^2 \cdot La} \qquad Cb = \frac{1}{\omega_0^2 \cdot La}$$

where $$H = 1 + \frac{1}{2 \cdot L1 \cdot C1 \cdot \omega_0^2} \cdot \{1 + [1 + 4 \cdot \omega_0^2 \cdot L1 \cdot C1]^{1/2}\} \tag{6.68}$$

and the resonant frequencies of the two parallel circuits are respectively:

$$\omega_a = H^{1/2} \cdot \omega_0 \quad \omega_b = H^{-1/2} \cdot \omega_0 \tag{6.69}$$

From the above equations it is easy to see that La,Cb and Lb,Ca resonate at the primitive frequency ω_0.

The same approach can be applied to the circuits shown in Figure 6.59(a) and 6.59(b). In this case, the equations are:

$$Ca = \frac{1}{L1 \cdot \omega_0^2 \cdot (H+1)} \qquad Cb = \frac{1}{\omega_0^2 \cdot La}$$

$$La = \frac{H+1}{H} \cdot L1 \qquad Lb = \frac{1}{\omega_0^2 \cdot Ca}$$

where $$H = 1 + \frac{1}{2 \cdot L1 \cdot C1 \cdot \omega_0^2} \cdot \{1 + [1 + 4 \cdot \omega_0^2 \cdot L1 \cdot C1]^{1/2}\} \tag{6.70}$$

and the resonant frequencies of the two series circuits are respectively:

$$\omega_a = H^{1/2} \cdot \omega_0 \quad \omega_b = H^{-1/2} \cdot \omega_0 \tag{6.71}$$

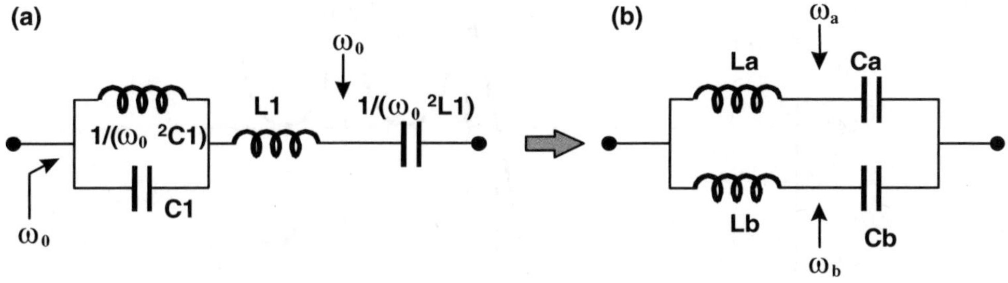

Figure 6.59 Band stop to band pass filter transformation

6.6 Elliptic Filter Transformation

It is clear that all the developed formulas for impedance scaling and frequency transformations are valid here, but there are some details in respect to the position of the transmission zeros that are interesting to note. In order to clarify the situation we will develop the successive transformations starting from an example:

$$N = 5 \text{ Elliptic PLPF with } R = 1,\ \text{RdB} = 0.01,\ \theta = 50°$$

This PLPF has $\omega_s = 1.3054$, $A_{min} = 24.94$ dB and two transmission zeros at $A2 = 1.9480$ and $A4 = 1.3481$. The filter structure, as well as the dual, are shown in Figure 6.60 and the filter response is shown in Figure 6.61. We can observe that the first parallel network resonates at A2 while the second parallel network resonates at A4.

6.6.1 Low Pass Elliptic Translation

In this case, every inductor and capacitor is divided by the new cut-off frequency ω_c. These calculations, along with the filter response, are shown in Figure 6.62 for an LPF having a $F_c = 600$ MHz. Every parallel LC network resonates at ω_c times the zero position:

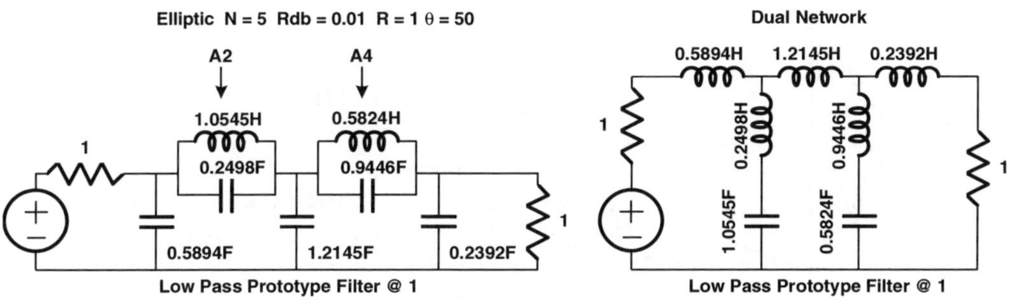

Figure 6.60 Elliptic filter transformation example

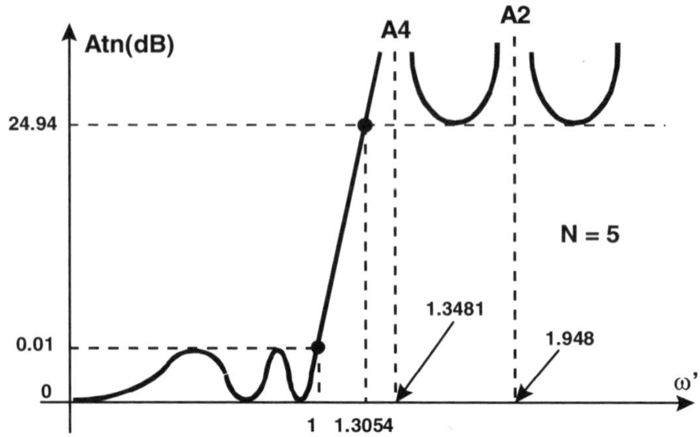

Figure 6.61 Response of filter shown in Figure 6.60

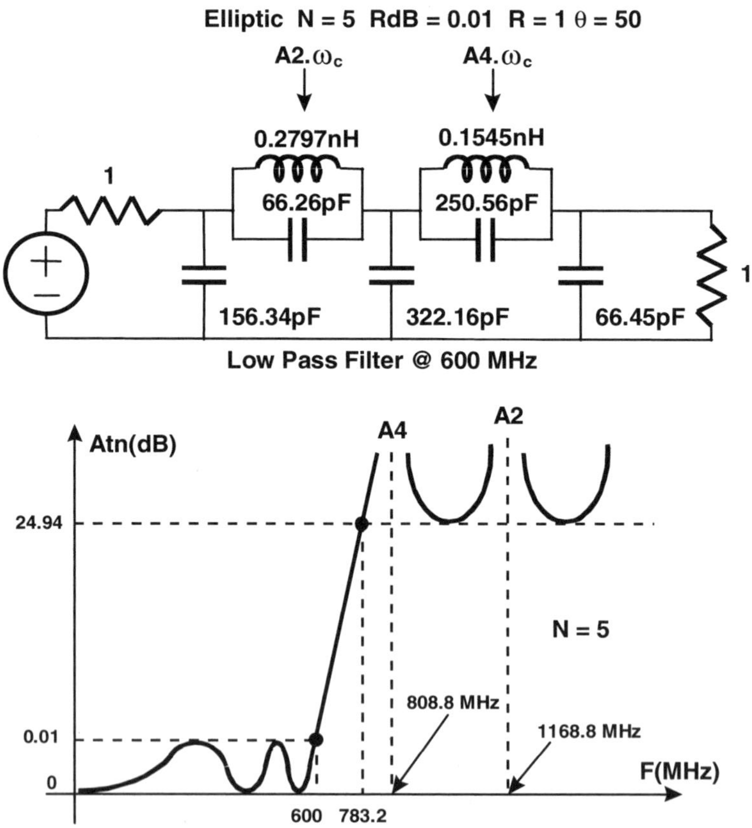

Figure 6.62 Elliptic LP filter transformation

$$\omega' = \omega_s = 1.3054 \quad \rightarrow \quad \omega = 4921.24\ 10^6\ \text{rad/s} \quad \rightarrow \quad F = 783.24\ \text{MHz}$$

$$\omega' = A2 = 1.9480 \quad \rightarrow \quad \omega = 7343.78\ 10^6\ \text{rad/s} \quad \rightarrow \quad F = 1168.8\ \text{MHz}$$

$$\omega' = A4 = 1.3481 \quad \rightarrow \quad \omega = 5082.21\ 10^6\ \text{rad/s} \quad \rightarrow \quad F = 808.8\ \text{MHz}$$

Impedance scaling should be performed at this point.

6.6.2 High Pass Elliptic Translation

Every capacitor translates to an inductor and vice versa. Figure 6.63 shows this transformation for a cut-off frequency of $F_c = 600$ MHz. As before, the position of the transmission zeros follows the rules for the frequency translation:

$$\omega' = \omega_s = 1.3054 \quad \rightarrow \quad F = 459.63\ \text{MHz}$$

$$\omega' = A2 = 1.9480 \quad \rightarrow \quad F = 308.00\ \text{MHz}$$

$$\omega' = A4 = 1.3481 \quad \rightarrow \quad F = 445.07\ \text{MHz}$$

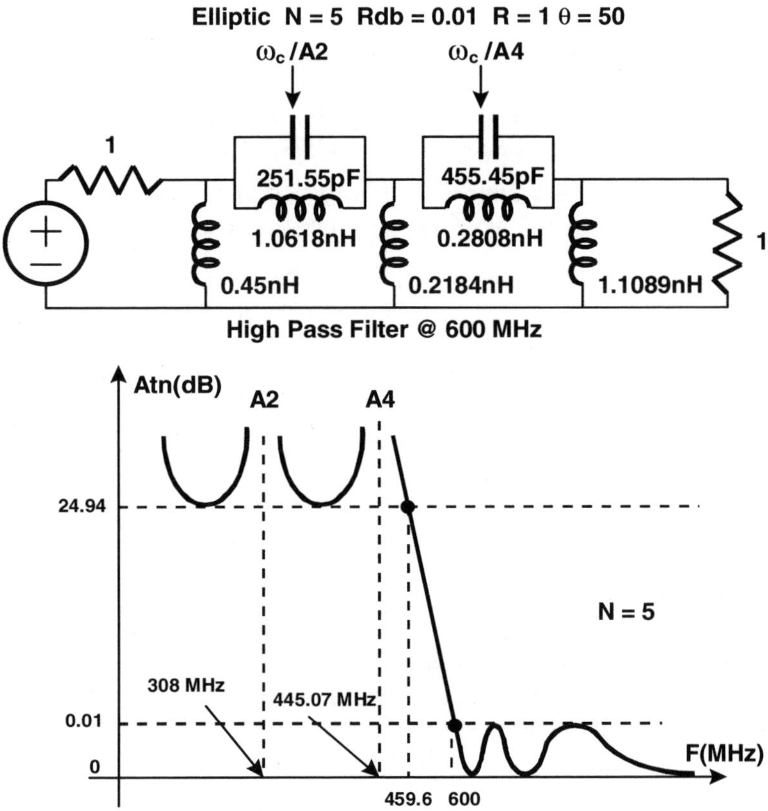

Figure 6.63 Elliptic LP to HP filter transformation

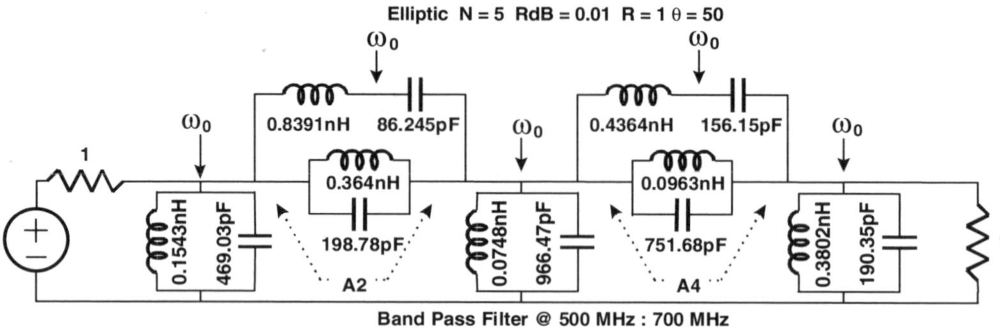

Figure 6.64 Elliptic LP to BP filter transformation

6.6.3 Band Pass Elliptic Translation

Here each parallel capacitor in the PLPF translates to a parallel LC network that resonates at the centre frequency ω_0, and each series inductor translates to a series LC resonant circuit at the same frequency ω_0. If we suppose that F_1 = 500 MHz and F_2 = 700 MHz (F_c = 591.6 MHz), the translated circuit is shown in Figure 6.64, where all the series or parallel resonant circuits have their resonance at ω_0. The band pass translation formulas give:

$$\omega' = \omega_s = 1.3054 \quad \rightarrow \quad F = 736.37 \text{ MHz}$$
$$\rightarrow \quad F = 475.3 \text{ MHz}$$

$$\omega' = A2 = 1.9480 \quad \rightarrow \quad F = 817.65 \text{ MHz} \quad (F_{2a})$$
$$\rightarrow \quad F = 428.05 \text{ MHz} \quad (F_{2b})$$

$$\omega' = A4 = 1.3481 \quad \rightarrow \quad F = 741.58 \text{ MHz} \quad (F_{4a})$$
$$\rightarrow \quad F = 471.96 \text{ MHz} \quad (F_{4b})$$

With this circuit, it is not easy, for example, to adjust the transmission zeros, that is, this network does not meet the requirements of their positions. Using the transformations for the resonant networks we obtain the circuit shown in Figure 6.65 where the parallel resonant circuits in the series branches have their resonant frequencies at the transmission zero.

Furthermore, Figure 6.66 shows the frequency response of this BPF.

6.6.4 Band Stop Elliptic Translation

Here each inductance in the PLPF translates to a parallel LC network that resonates at the centre frequency ω_0, and each capacitor translates to a series LC resonant circuit at the same frequency ω_0. If we suppose as above that F_1 = 500 MHz and F_2 = 700 MHz (F_0 = 591.6 MHz), the translated circuit is shown in Figure 6.67, where all the series or parallel resonant circuits have their resonance at ω_0. The band stop translation formulas give:

$$\omega' = \omega_s = 1.3054 \quad \rightarrow \quad F = 519.89 \text{ MHz}$$
$$\rightarrow \quad F = 673.15 \text{ MHz}$$

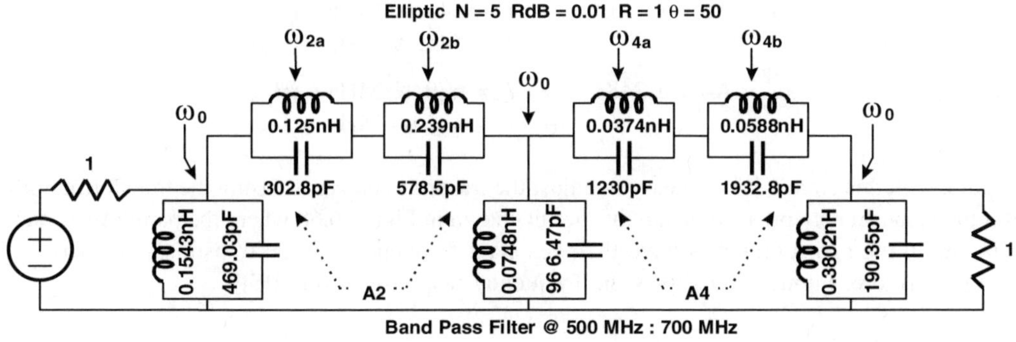

Figure 6.65 Elliptic LP to BP filter transformation including transmission zeros

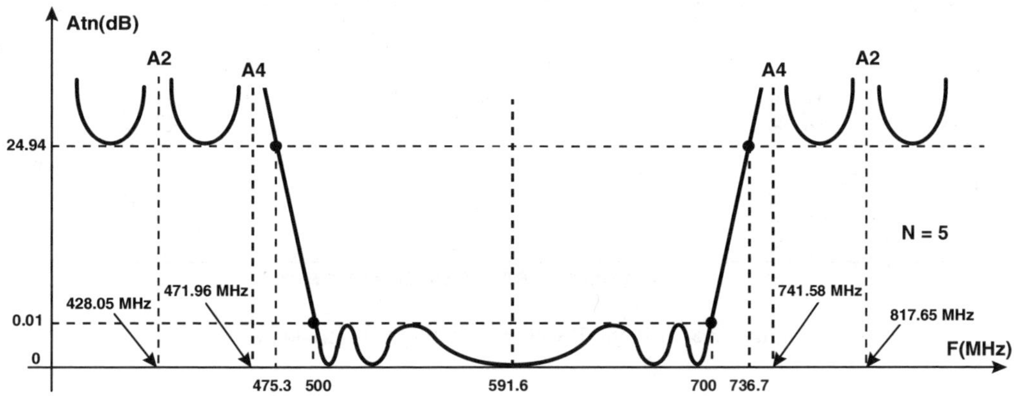

Figure 6.66 Frequency response of Elliptic BP filter

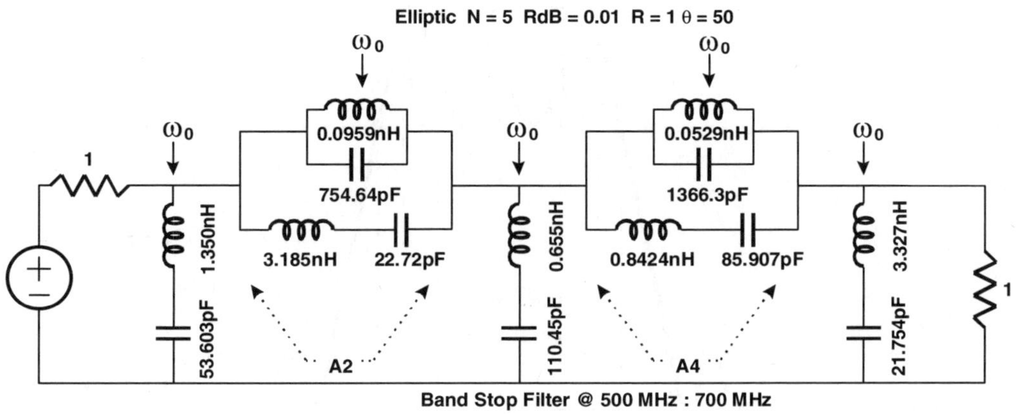

Figure 6.67 Elliptic LP to BS transformation

$$\omega' = A2 = 1.9480 \quad \rightarrow \quad F = 645.16\ \text{MHz} \quad (F_{2a})$$
$$\rightarrow \quad F = 542.49\ \text{MHz} \quad (F_{2b})$$
$$\omega' = A4 = 1.3481 \quad \rightarrow \quad F = 670.41\ \text{MHz} \quad (F_{4a})$$
$$\rightarrow \quad F = 522.06\ \text{MHz} \quad (F_{4b})$$

Again, with this circuit, it is not easy to adjust the transmission zeros. Using the transformations for the resonant networks we obtain the circuit shown in Figure 6.68 where the parallel resonant circuits in the series branches have their resonant frequencies at the transmission zeros.

Furthermore, Figure 6.69 shows the frequency response of this BSF.

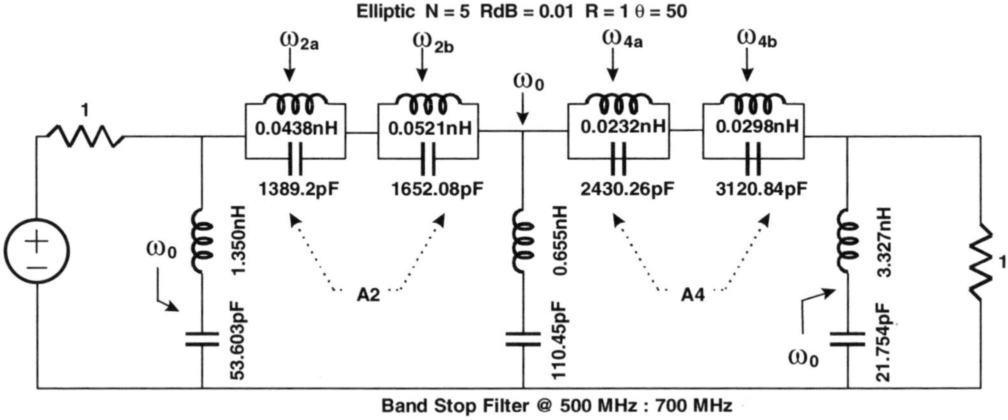

Figure 6.68 Elliptic BS filter including transmission zeros

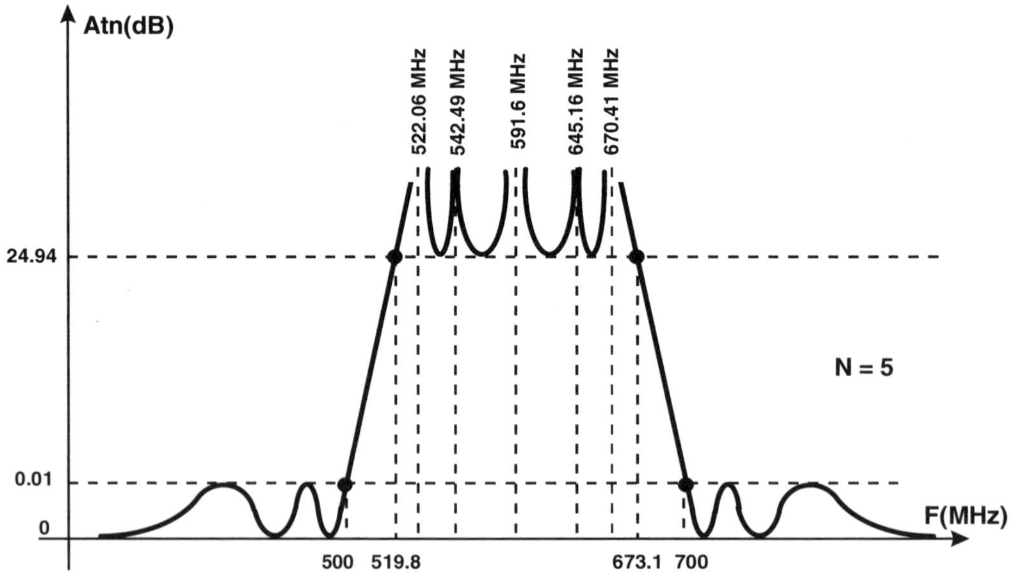

Figure 6.69 Frequency response of Elliptic BS filter

6.7 Filter Normalisation

When we are involved in a filter design process we not only have specifications in the pass band but most of the time also in the stop band. As we know, these stop band requirements will impose the number of sections as well as the filter response. It seems clear now that is imperative to accurately translate the actual filter specifications to PLPF constraints.

For this purpose it is convenient to write here the frequency translation equations:

Prototype Low Pass to Low Pass:

$$\omega' = \omega/\omega_C \tag{6.72}$$

Prototype Low Pass to High Pass:

$$\omega' = -\omega_C/\omega \tag{6.73}$$

Prototype Low Pass to Band Pass:

$$\omega' = \frac{\omega_0}{\omega_2 - \omega_1} \cdot [\omega/\omega_0 - \omega_0/\omega] \quad \text{where} \quad \omega_0 = (\omega_1 \cdot \omega_2)^{1/2}$$

or

$$\omega = \omega' \cdot \frac{\omega_2 - \omega_1}{2} \pm \frac{1}{2}[\omega'^2(\omega_2 - \omega_1)^2 + 4\omega_1\omega_2]^{1/2} \tag{6.74}$$

Prototype Low Pass to Band Stop:

$$\omega' = \frac{\omega_2 - \omega_1}{\omega_0} \cdot \frac{1}{[\omega_0/\omega - \omega/\omega_0]} \quad \text{where} \quad \omega_0 = (\omega_1 \cdot \omega_2)^{1/2}$$

or

$$\omega = (-1/\omega') \cdot \frac{\omega_2 - \omega_1}{2} \pm \frac{1}{2}[(1/\omega')^2 \cdot (\omega_2 - \omega_1)^2 + 4\omega_1\omega_2]^{1/2} \tag{6.75}$$

where ω_C is the cut-off frequency in low pass and high pass filters, while ω_1 and ω_2 are the cut-off frequencies in band pass and band stop filters.

6.7.1 Low Pass Normalisation

In this kind of filter it is common to specify the cut-off frequency F_C (usually the 3dB point) and a minimum of attenuation XdB at a given frequency F_a in the stop band, as shown in Figure 6.70. Taking into account the frequency translation of a low pass filter, it is clear that the frequency F_a has a PLPF response given by:

$$\omega' = \omega_a/\omega_C = F_a/F_C \tag{6.76}$$

and with this response the PLPF should have the same attenuation as the original LPF, as shown in Figure 6.70.

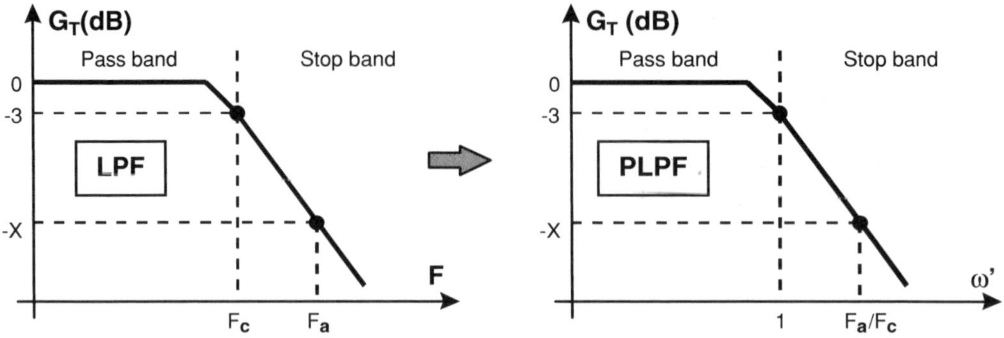

Figure 6.70 LPF normalisation

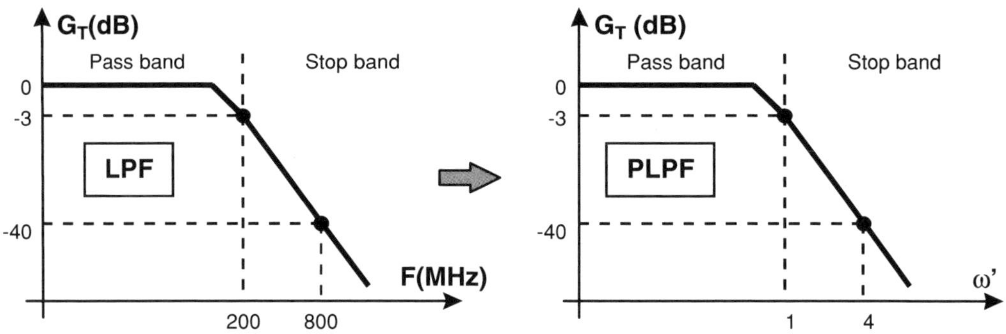

Figure 6.71 Example of LPF normalisation

As an example, suppose we have the following low pass specification:

- cut-off 3dB point at $F_C = 200$ MHz
- 40 dB minimum attenuation at $F_a = 800$ MHz.

The value of ω' is 4 and the PLPF specifications should be:

- cut-off 3dB point at $\omega' = 1$ rad/s
- 40 dB minimum attenuation at $\omega' = 4$ rad/s.

This normalisation is depicted in Figure 6.71. We are now able to work with the PLPF network design, performing the frequency and impedance scaling.

6.7.2 High Pass Normalisation

In this case the specifications are the same as low pass filters, that is, a cut-off frequency F_C and a desired attenuation of XdB at F_a. After the frequency translation equation for HPF, the PLPF response is given by:

$$\omega' = -\omega_C/\omega_a = -F_C/F_a \tag{6.77}$$

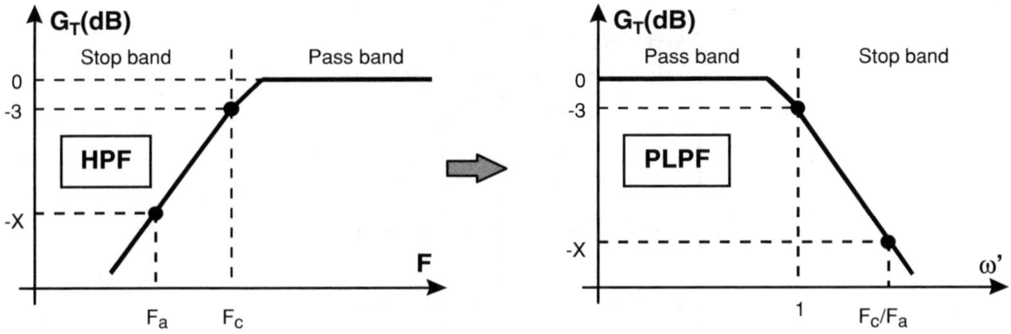

Figure 6.72 HPF transform and normalisation

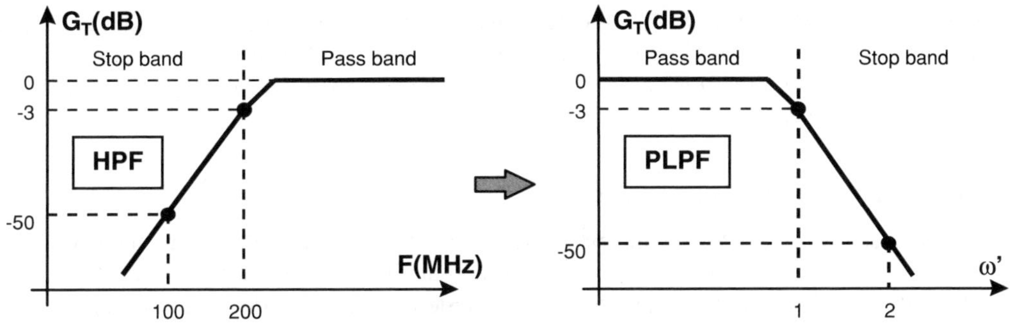

Figure 6.73 Example of HPF normalisation

The PLPF shown in Figure 6.72 should have XdB of attenuation. But this frequency translation is symmetrical with respect to $\omega' = 0$, so it is more convenient to look for a PLPF frequency given by the negative of the above:

$$\omega' = +\omega_C/\omega_a = +F_C/F_a \tag{6.78}$$

As an example, suppose we have the following high pass specification:

- cut-off 3dB point at $F_C = 200$ MHz
- 50 dB minimum attenuation at $F_a = 100$ MHz.

$\omega' = 2$ and consequently the PLPF specification, shown in Figure 6.73, should be:

- cut-off 3dB point at $\omega' = 1$ rad/s
- 50 dB minimum attenuation at $\omega' = 2$ rad/s.

6.7.3 Band Pass Normalisation

In the case of BPF, Figure 6.74, we can distinguish between broadband and narrowband BPF depending on the fractional bandwidth FBW given by:

$$FBW = (F_2 - F_1)/F_0 = BW/F_0 \tag{6.79}$$

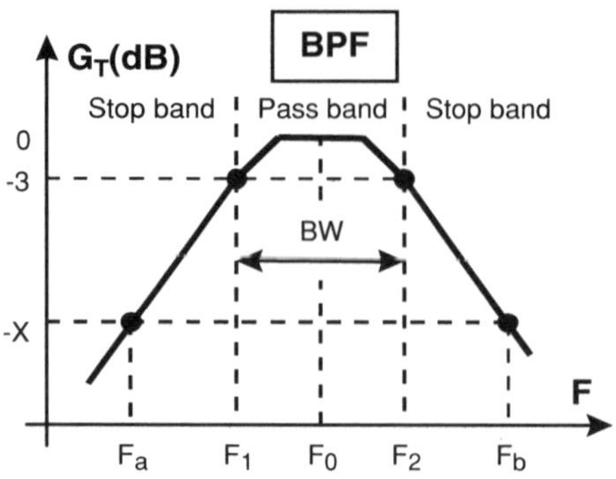

Figure 6.74 BPF response

Filters having a FBW ≥ 0.707 (70.7%) are considered broadband filters. This limit is given by the fact that in this case $F_2 = 2F_1$, that is, we have an octave of bandwidth.

In the general case, this BPF has specifications at the cut-off frequencies F_1, F_2 and at one or two stopband frequencies F_a, F_b.

6.7.3.1 Broadband band pass normalisation

When we deal with a BPF having broadband characteristics, the filter implementation is done by using a cascade connection of an LPF and an HPF as shown in Figure 6.75.

The LPF and HPF sections are normalised independently of the PLPF as shown in Figure 6.76.

As an example suppose we have the following broadband BPF specifications:

- cut-off RdB = 0.1 points at $F_1 = 4.5$ GHz and $F_2 = 10$ GHz
- 40 dB minimum attenuation at $F_a = 2$ GHz and $F_b = 20$ GHz.

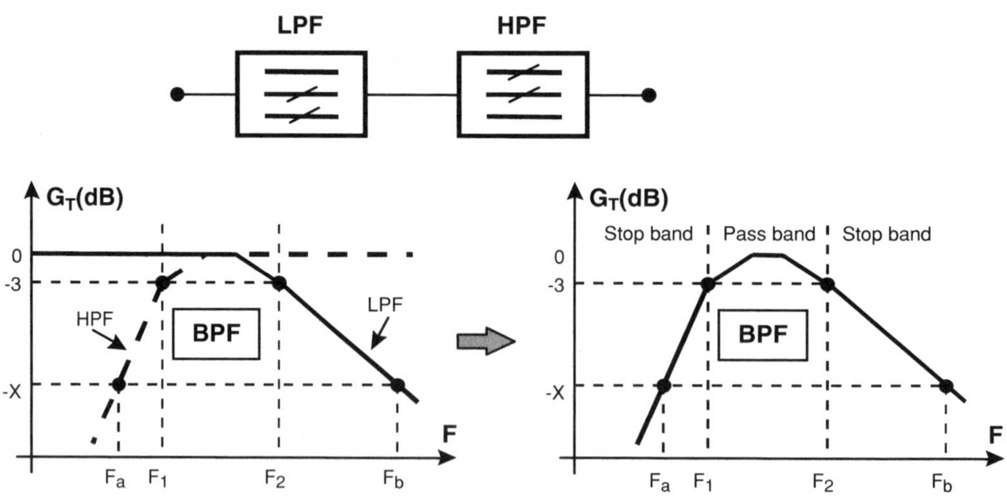

Figure 6.75 BPF generated from LPF and HPF

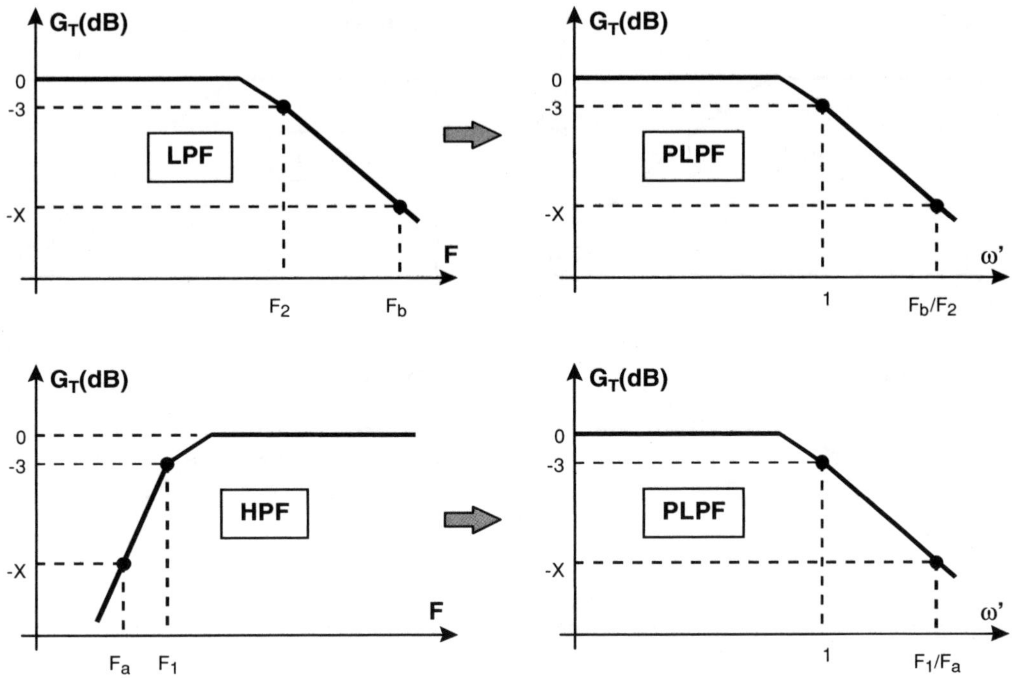

Figure 6.76 LPF and HPF normalisations

Then the PLPF specifications for the LPF section are:

- cut-off RdB = 0.1 at $\omega' = 1$ rad/s
- 40 dB minimum attenuation at $\omega' = 2$ rad/s.

And for the HPF section are:

- cut-off RdB = 0.1 at $\omega' = 1$ rad/s
- 40 dB minimum attenuation at $\omega' = 2.25$ rad/s.

The two normalisations are shown in Figure 6.77. The next step is to design both PLPF and then to perform the two frequency translations.

6.7.3.2 Narrowband band pass normalisation

These BPF have a bandwidth less than one octave and in this case we can directly use the frequency translation properties. It is convenient to take into account that such a frequency translation, Figure 6.78, is symmetrical in a geometric form with respect to $F_0 = (F_1 \cdot F_2)^{1/2}$, that is, a given attenuation of XdB at a given frequency F_x is also encountered at its symmetrical $F_{xs} = F_0^2/F_x$.

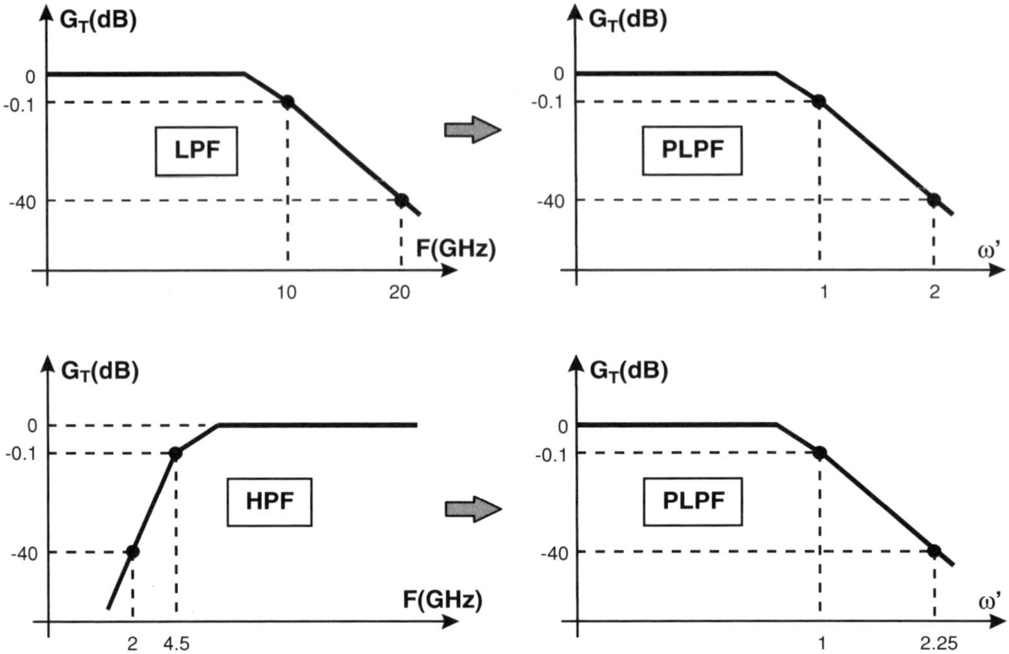

Figure 6.77 Example LPF and HPF normalisations

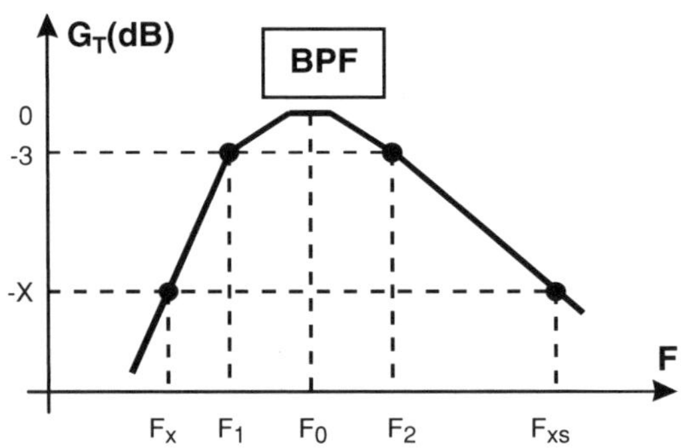

Figure 6.78 Narrowband BPF response

In effect, the PLPF frequency for F_x is:

$$\omega'_x = \frac{F_0}{F_2 - F_1} \cdot [F_x/F_0 - F_0/F_x] \tag{6.80}$$

while the corresponding value for F_{xs} is:

$$\omega'_{xs} = \frac{F_0}{F_2 - F_1} \cdot [F_{xs}/F_0 - F_0/F_{xs}] = \frac{F_0}{F_2 - F_1} \cdot [F_0/F_x - F_x/F_0] = -\omega'_s \tag{6.81}$$

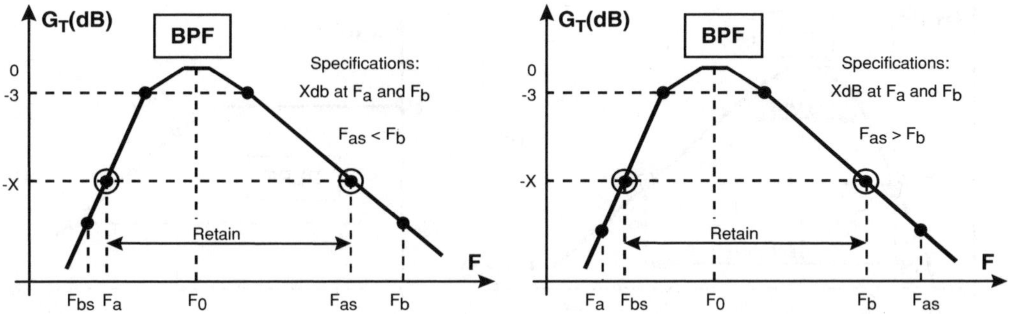

Figure 6.79 Non-symmetric BPF responses

and we know that the attenuation characteristics of the different PLPF are the same if the frequency changes sign.

Most of the time the actual stop band requirements XdB at the two stop band frequencies F_a and F_b are such that F_a and F_b are not symmetrical with respect to F_0. So we have to convert these specifications into a symmetrical form. This is accomplished by obtaining the symmetry $F_{as} = F_0^2/F_a$. If $F_{as} < F_b$ this means that at F_b we will have more than XdB and we will retain the stop band symmetrical bandwidth (F_a : F_{as}). Conversely if $F_{as} > F_b$, then the attenuation at F_b will be less than XdB and we will retain the stop band symmetrical bandwidth (F_{bs} : F_b) because it is clear that in this case $F_a < F_{bs}$. Both cases are shown in Figure 6.79.

As an example we have the following narrowband BPF specifications:

- 3dB bandwidth BW = 300 MHz centered around 1 GHz
- 64 dB minimum attenuation at ± 300 MHz.

In this case $F_1 = 850$ MHz, $F_2 = 1150$ MHz and $F_0 = 988.68$ MHz with a FBW = 0.30 (30%). Furthermore, the stop band frequencies are $F_a = 700$ MHz and $F_b = 1300$ MHz.

The value of F_{as} is 1396.4 MHz, that is, greater than F_b, so the retained stop band requirements are $F_{bs} = 751.9$ MHz and $F_b = 1300$ MHz. Now we have to translate F_{bs} or F_b into the PLPF ω' plane using the equation for the BPF frequency translation. Both results should be the same except for the sign. The resulting value is $\omega' = 1.827$. As a consequence, the PLPF specifications are:

- 3dB cut-off at $\omega' = 1$ rad/s
- 64 dB minimum attenuation at $\omega' = 1.827$ rad/s.

Figure 6.80 shows this normalization.

6.7.4 Band Stop Normalisation

As in the case of BPF, in the BSF – Figure 6.81 – we can distinguish between broadband and narrowband BSF depending on the FBW.

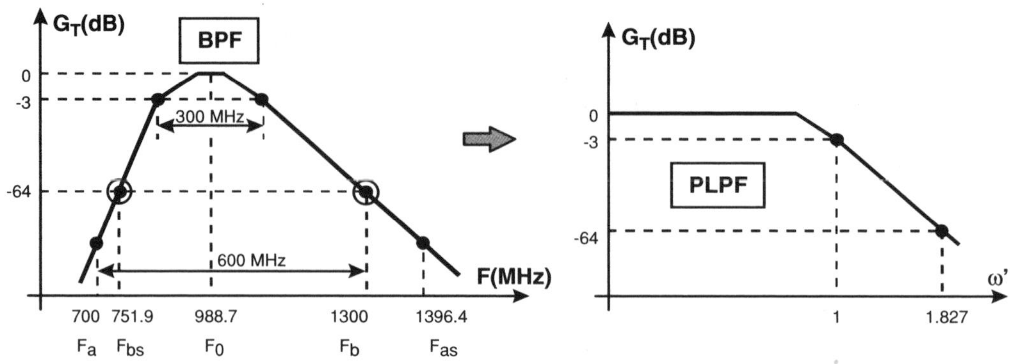

Figure 6.80 Example BPF normalisation

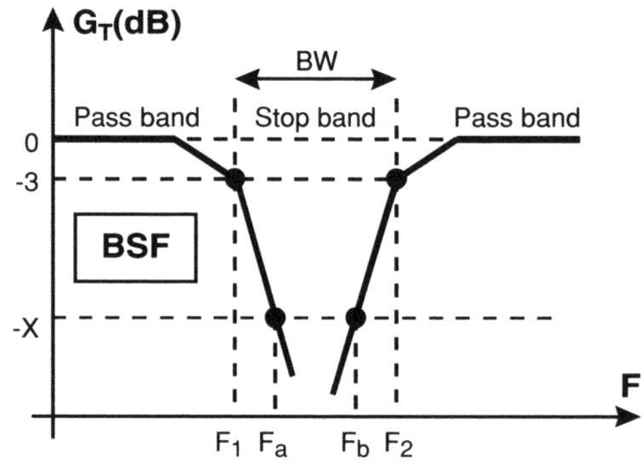

Figure 6.81 BSF definitions

6.7.4.1 Broadband band stop normalisation

In dealing with BPF having broadband characteristics, the filter implementation is done by using a parallel connection of a LPF and a HPF with the output combined as shown in Figure 6.82.

The LPF and HPF sections are normalised independently of the PLPF as shown in Figure 6.83.

As an example, suppose we have the following broadband BSF specifications:

- cut-off 3dB points at $F_1 = 2$ GHz and $F_2 = 8$ GHz
- 40 dB minimum attenuation at $F_a = 3$ GHz and $F_b = 5$ GHz.

Then the PLPF specifications for the LPF section are:

- cut-off 3dB point at $\omega' = 1$ rad/s
- 40 dB minimum attenuation at $\omega' = 1.5$ rad/s.

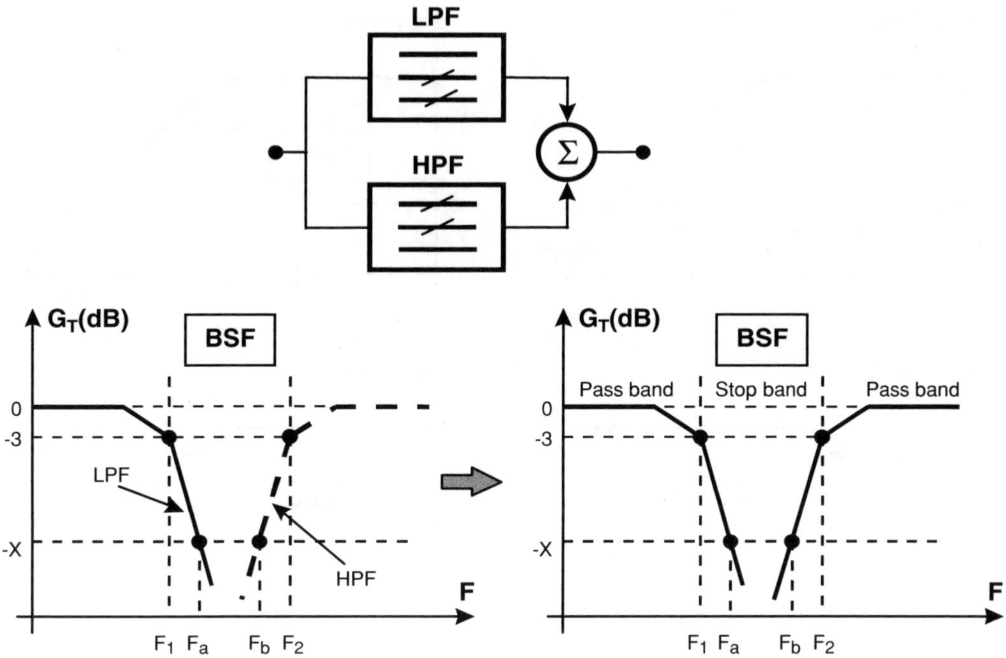

Figure 6.82 BSF realised by combination of LPF and HPF

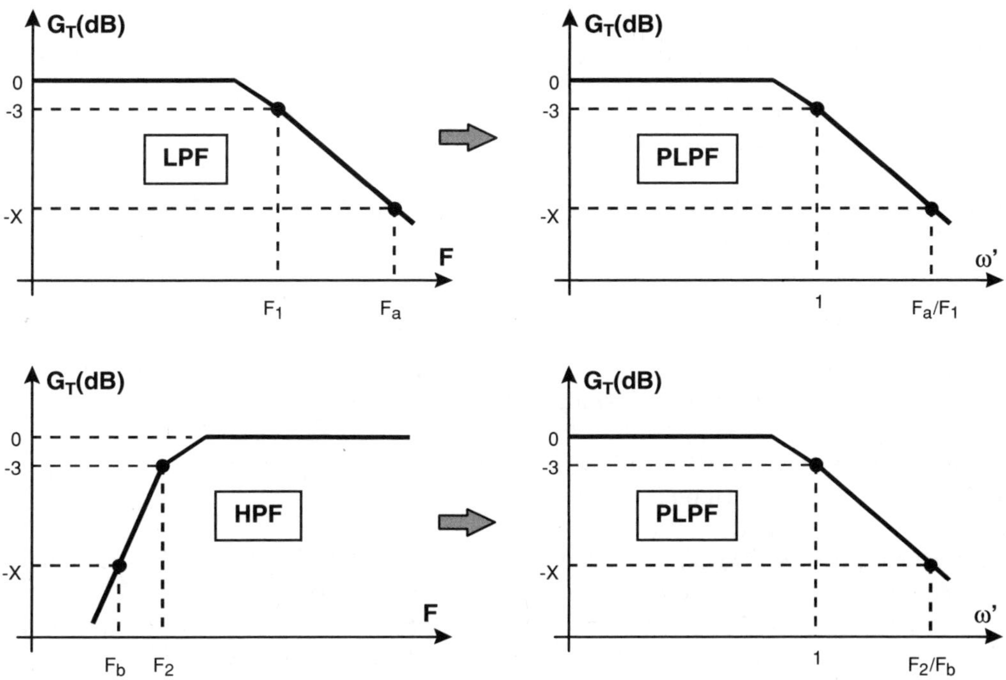

Figure 6.83 LPF and HPF normalisations

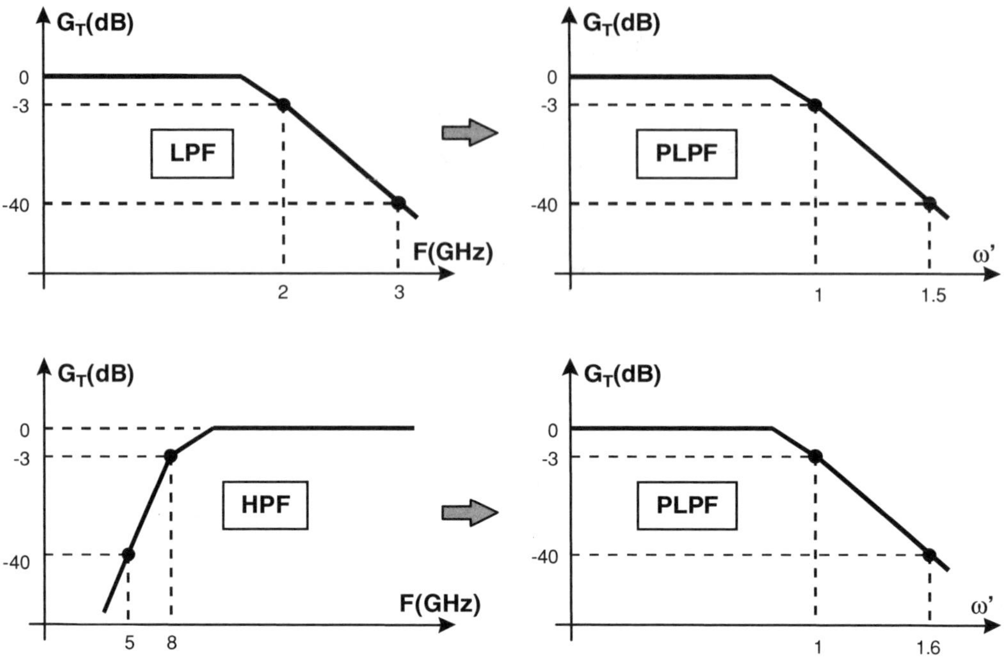

Figure 6.84 Example LPF and HPF normalisations

and for the HPF section are:

- cut-off 3dB point at $\omega' = 1$ rad/s
- 40 dB minimum attenuation at $\omega' = 1.6$ rad/s.

The two normalisations are shown in Figure 6.84. Remember that the two frequency translations are made independently.

6.7.4.2 Narrowband band stop normalisation

All the considerations for BPF are applicable here. So it is easiest to explain with an example. Suppose we have the following BSF specifications:

- 3dB cut-off bandwidth BW = 400 MHz around 1 GHz
- 50 dB minimum attenuation at ± 100 MHz.

Here we have $F_1 = 800$ MHz, $F_2 = 1200$ MHz, $F_0 = 979.8$ MHz with a FBW = 0.4 (40%). The stop band frequencies are $F_a = 900$ MHz and $F_b = 1100$ MHz.

The value of F_{as} is 1066.6 MHz, that is less than F_b, and this means that the presumed attenuation at F_b should be less than 50 dB. So the retained values are $F_{bs} = 872.7$ MHz and $F_b = 1100$ MHz. Now we translate these values into the ω' plane by using the BSF frequency translation. The result is $\omega' = 1.76$ and the PLPF specifications should be:

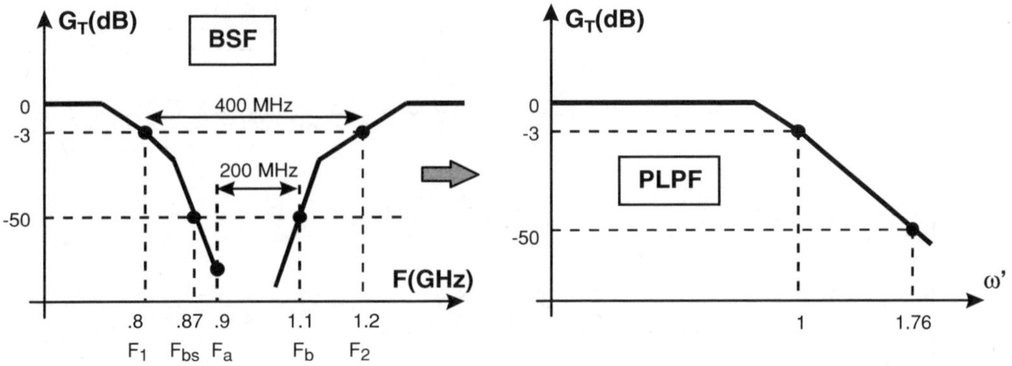

Figure 6.85 Narrowband BSF normalisation

- 3dB cut-off at $\omega' = 1$ rad/s
- 50 dB minimum attenuation at $\omega' = 1.76$ rad/s.

Figure 6.85 shows this normalisation.

Self-assessment Problems

6.1 Design a 3-stage low pass Butterworth filter with a cut-off frequency of 5 GHz for use in a circuit with a normalised load of $R = 1$.

6.2 Design a 4-stage high pass Chebyshev filter with a cut-off frequency of 6 GHz for use in a circuit with a normalised load of $R = 2$, RdB = 0.5.

6.3 Design a 2-stage band pass Bessel filter with cut-off frequencies of 1GHz and 3 GHz for use in a circuit with a normalised load of $R = 1$.

6.4 Design a 3-stage band stop Elliptical filter with cut-off frequencies of 2 GHz and 4 GHz for use in a circuit with a normalised load of $R = 1$, RdB = 0.01.

Appendix Tables for Filter Design

Table 6.1 Low pass prototype filter: Butterworth

N	R	C1	L2	C3	L4	C5	L6	C7
2	1.000	1.414	1.414					
	1.111	1.035	1.835					
	1.250	0.849	2.121					
	1.429	0.697	2.439					
	1.667	0.566	2.828					
	2.000	0.448	3.346					
	2.500	0.342	4.095					
	3.333	0.245	5.313					
	5.000	0.156	0.707					
	10.00	0.074	14.814					
	Inf	1.414	0.707					
3	1.000	1.000	2.000	1.000				
	0.900	0.808	1.633	1.599				
	0.800	0.844	1.384	1.926				
	0.700	0.815	1.165	2.277				
	0.600	1.023	0.965	2.702				
	0.500	1.181	0.779	3.261				
	0.400	1.425	0.604	4.064				
	0.300	1.838	0.440	5.363				
	0.200	2.669	0.284	7.910				
	0.100	5.167	0.138	15.455				
	Inf	1.500	1.333	0.500				
4	1.000	0.765	1.848	1.848	0.765			
	1.111	0.466	1.592	1.744	1.469			
	1.250	0.388	1.695	1.511	1.811			
	1.429	0.325	1.862	1.291	2.175			
	1.667	0.269	2.103	1.082	2.613			
	2.000	0.218	2.452	0.883	3.187			
	2.500	0.169	2.986	0.691	4.009			
	3.333	0.124	3.883	0.507	5.338			
	5.000	0.080	5.684	0.331	7.940			
	10.00	0.039	11.094	0.162	15.642			
	Inf	1.531	1.577	1.082	0.383			
N	1/R	L1	C2	L3	C4	L5	C6	L7

N	R	C1	L2	C3	L4	C5	L6	C7
5	1.000	0.618	1.618	2.000	1.618	0.618		
	0.900	0.442	1.027	1.910	1.756	1.389		
	0.800	0.470	0.866	2.061	1.544	1.738		
	0.700	0.517	0.731	2.285	1.333	2.108		
	0.600	0.586	0.609	2.600	1.126	2.552		
	0.500	0.686	0.496	3.051	0.924	3.133		
	0.400	0.838	0.388	3.736	0.727	3.965		
	0.300	1.094	0.285	4.884	0.537	5.307		
	0.200	1.608	0.186	7.185	0.352	7.935		
	0.100	3.512	0.091	14.095	0.173	15.710		
	Inf	1.545	1.694	1.382	0.894	0.309		
6	1.000	0.518	1.414	1.932	1.932	1.414	0.518	
	1.111	0.289	1.040	1.322	2.054	1.744	1.335	
	1.250	0.245	1.116	1.126	2.239	1.550	1.688	
	1.429	0.207	1.236	0.957	2.499	1.346	2.062	
	1.667	0.173	1.407	0.801	2.858	1.143	2.509	
	2.000	0.141	1.653	0.654	3.369	0.942	3.094	
	2.500	0.111	2.028	0.514	4.141	0.745	3.931	
	3.333	0.082	2.656	0.379	5.433	0.552	5.280	
	5.000	0.054	3.917	0.248	8.020	0.363	7.922	
	10.00	0.026	7.705	0.122	15.786	0.179	15.738	
	Inf	1.553	1.759	1.553	1.202	0.758	0.259	
7	1.000	0.445	1.247	1.802	2.000	1.802	1.247	0.445
	0.900	0.299	0.711	1.404	1.489	2.125	1.727	1.296
	0.800	0.322	0.606	1.517	1.278	2.334	1.546	1.652
	0.700	0.357	0.515	1.688	1.091	2.618	1.350	2.028
	0.600	0.408	0.432	1.928	0.917	3.005	1.150	2.477
	0.500	0.480	0.354	2.273	0.751	3.553	0.951	3.064
	0.400	0.590	0.278	2.795	0.592	4.380	0.754	3.904
	0.300	0.775	0.206	3.671	0.437	5.761	0.560	5.258
	0.200	1.145	0.135	5.427	0.287	8.526	0.369	7.908
	0.100	2.257	0.067	10.700	0.142	16.822	0.182	15.748
	Inf	1.558	1.799	1.659	1.397	1.055	0.656	0.223
N	1/R	L1	C2	L3	C4	L5	C6	L7

Note: ($\omega = 1$ rad/s for Atn = 3dB)

Table 6.2 Low pass prototype filter: Chebyshev

N	RdB	R	C1	L2	C3	L4	C5
2	0.01	1.1007	0.4077	0.4488			
	0.1	1.3554	0.6220	0.8430			
	0.2	1.5386	0.6745	1.0378			
	0.5	1.9841	0.7071	1.4029			
	1.0	2.6599	0.6850	1.8219			
	2.0	4.0957	0.6075	2.4881			
	3.0	5.8095	0.5339	3.1013			
3	0.01	1.0000	0.6291	0.9702	0.6291		
	0.1	1.0000	1.0315	1.1474	1.0315		
	0.2	1.0000	1.2275	1.1525	1.2275		
	0.5	1.0000	1.5963	1.0967	1.5963		
	1.0	1.0000	2.0236	0.9941	2.0236		
	2.0	1.0000	2.7107	0.8327	2.7107		
	3.0	1.0000	3.3487	0.7117	3.3487		
4	0.01	1.1007	0.6476	1.3212	1.2003	0.7128	
	0.1	1.3554	0.8180	1.7703	1.3061	1.1088	
	0.2	1.5386	0.8468	1.9761	1.2844	1.3028	
	0.5	1.9841	0.8419	2.3661	1.1926	1.6703	
	1.0	2.6599	0.7892	2.8311	1.0644	2.0991	
	2.0	4.0957	0.6819	3.6063	0.8806	2.7925	
	3.0	5.8095	0.5920	4.3471	0.7483	3.4389	
5	0.01	1.0000	0.7563	1.3049	1.5773	1.3049	0.7563
	0.1	1.0000	1.1468	1.3712	1.9750	1.3712	1.1468
	0.2	1.0000	1.3394	1.3370	2.1660	1.3370	1.3394
	0.5	1.0000	1.7058	1.2296	2.5408	1.2296	1.7058
	1.0	1.0000	2.1349	1.0911	3.0009	1.0911	2.1349
	2.0	1.0000	2.8310	0.8985	3.7827	0.8985	2.8310
	3.0	1.0000	3.4817	0.7618	4.5381	0.7618	3.4817
N	RdB	1/R	L1	C2	L3	C4	L5

N	RdB	R	C1	L2	C3	L4	C5	L6	C7	C8	L9
6	0.01	1.1007	0.7098	1.4970	1.5350	1.6896	1.3600	0.7813			
	0.1	1.3554	0.8618	1.9029	1.5170	2.0562	1.4039	1.1681			
	0.2	1.5386	0.8838	2.0974	1.4555	2.2394	1.3632	1.3598			
	0.5	1.9841	0.8696	2.4758	1.3137	2.6064	1.2479	1.7254			
	1.0	2.6599	0.8101	2.9367	1.1518	3.0634	1.1041	2.1546			
	2.0	4.0957	0.6964	3.7151	0.9393	3.8467	0.9071	2.8521			
	3.0	5.8095	0.6033	4.4641	0.7929	4.6061	0.7685	3.5045			
7	0.01	1.0000	0.7969	1.3924	1.7481	1.6331	1.7481	1.3924	0.7969		
	0.1	1.0000	1.1811	1.4228	2.0966	1.5733	2.0966	1.4228	1.1811		
	0.2	1.0000	1.3722	1.3781	2.2756	1.5001	2.2756	1.3781	1.3722		
	0.5	1.0000	1.7372	1.2583	2.6381	1.3444	2.6381	1.2583	1.7372		
	1.0	1.0000	2.1664	1.1116	3.0934	1.1736	3.0934	1.1116	2.1664		
	2.0	1.0000	2.8655	0.9119	3.8780	0.9535	3.8780	0.9119	2.8655		
	3.0	1.0000	3.5182	0.7723	4.6386	0.8039	4.6386	0.7723	3.5182		
8	0.01	1.1007	0.7333	1.5554	1.6193	1.8529	1.6833	1.7824	1.4130	0.8072	
	0.1	1.3554	0.8778	1.9444	1.5640	2.1699	1.6010	2.1199	1.4346	1.1897	
	0.2	1.5386	0.8972	2.1349	1.4925	2.3413	1.5217	2.2963	1.3875	1.3804	
	0.5	1.9841	0.8796	2.5093	1.3389	2.6964	1.3590	2.6564	1.2647	1.7451	
	1.0	2.6599	0.8175	2.9685	1.1696	3.1488	1.1839	3.1107	1.1161	2.1744	
	2.0	4.0957	0.7016	3.7477	0.9510	3.9335	0.9605	3.8948	0.9151	2.8733	
	3.0	5.8095	0.6073	4.4990	0.8018	4.6990	0.8089	4.6575	0.7745	3.5277	
9	0.01	1.0000	0.8144	1.4270	1.8043	1.7125	1.9057	1.7125	1.8043	1.4270	0.8144
	0.1	1.0000	1.1950	1.4425	2.1345	1.6167	2.2053	1.6167	2.1345	1.4425	1.1956
	0.2	1.0000	1.3860	1.3938	2.3093	1.5340	2.3728	1.5340	2.3093	1.3938	1.3860
	0.5	1.0000	1.7504	1.2690	2.6678	1.3673	2.7239	1.3673	2.6678	1.2690	1.7504
	1.0	1.0000	2.1797	1.1192	3.1215	1.1897	3.1747	1.1897	3.1215	1.1192	2.1797
	2.0	1.0000	2.8790	0.9171	3.9056	0.9643	3.9598	0.9643	3.9056	0.9171	2.8790
	3.0	1.0000	3.5340	0.7760	4.6692	0.8118	4.7272	0.8118	4.6692	0.7760	3.5340
N	RdB	1/R	L1	C2	L3	C4	L5	C6	L7	C8	L9

Note: ($\omega = 1$ rad/s for Atn = RdB)

Table 6.3 Low pass prototype filter: Chebyshev (RdB = 0.01)

N	R	C1	L2	C3	L4	C5	L6	C7
2	1.1007	1.3472	1.4829					
	1.1111	1.2472	1.5947					
	1.2500	0.9434	1.9974					
	1.4286	0.7591	2.3442					
	1.6667	0.6091	2.7496					
	2.0000	0.4791	3.2772					
	2.5000	0.3634	4.0328					
	3.3333	0.2590	5.2546					
	5.0000	0.1642	7.6498					
	10.000	0.0781	14.749					
	Inf	1.4118	0.7415					
3	1.0000	1.1811	1.8214	1.1811				
	0.9000	1.0917	1.6597	1.4802				
	0.8000	1.0969	1.4431	1.8057				
	0.7000	1.1600	1.2283	2.1653				
	0.6000	1.2737	1.0236	2.5984				
	0.5000	1.4521	0.8294	3.1644				
	0.4000	1.7340	0.6452	3.9742				
	0.3000	2.2164	0.4704	5.2800				
	0.2000	3.1934	0.3047	7.8338				
	0.1000	6.1411	0.1479	15.390				
	Inf	1.5012	1.4330	0.5905				
4	1.1000	0.9500	1.9382	1.7608	1.0457			
	1.1111	0.8539	1.9460	1.7439	1.1647			
	1.2500	0.6182	2.0749	1.5417	1.6170			
	1.4286	0.4948	2.2787	1.3336	2.0083			
	1.6667	0.3983	2.5709	1.1277	2.4611			
	2.0000	0.3156	2.9943	0.9260	3.0448			
	2.5000	0.2418	3.6406	0.7293	3.8746			
	3.3333	0.1744	4.7274	0.5379	5.2085			
	5.0000	0.1121	6.9102	0.3523	7.8126			
	10.000	0.0541	13.469	0.1729	15.510			
	Inf	1.5287	1.6939	1.3122	0.5229			
5	1.0000	0.9766	1.6849	2.0366	1.6849	0.9766		
	0.9000	0.8798	1.4558	2.1738	1.6412	1.2739		
	0.8000	0.8769	1.2350	2.3785	1.4991	1.6066		
	0.7000	0.9263	1.0398	2.6582	1.3228	1.9772		
	0.6000	1.0191	0.8626	3.0408	1.1345	2.4244		
	0.5000	1.1658	0.6985	3.5835	0.9421	3.0092		
	0.4000	1.3983	0.5442	4.4027	0.7491	3.8453		
	0.3000	1.7966	0.3982	5.7721	0.5573	5.1925		
	0.2000	2.6039	0.2592	8.5140	0.3679	7.8257		
	0.1000	5.0406	0.1266	16.741	0.1819	15.613		
	Inf	1.5466	1.7950	1.6449	1.2365	0.4883		
6	1.1007	0.8514	1.7956	1.8411	2.0266	1.6312	0.9372	
	1.1111	0.7597	1.7817	1.7752	2.0941	1.6380	1.0533	
	1.2500	0.5445	1.8637	1.4886	2.4025	1.5067	1.5041	
	1.4286	0.4355	2.0383	1.2655	2.7346	1.3318	1.8987	
	1.6667	0.3509	2.2978	1.0607	3.1671	1.1451	2.3568	
	2.0000	0.2786	2.6781	0.8671	3.7683	0.9536	2.9483	
	2.5000	0.2139	3.2614	0.6816	4.6673	0.7606	3.7899	
	3.3333	0.1547	4.2448	0.5028	6.1631	0.5676	5.1430	
	5.0000	0.0997	6.2227	0.3299	9.1507	0.3760	7.7852	
	10.000	0.0483	12.171	0.1623	18.105	0.1865	15.595	
	Inf	1.5510	1.8471	1.7897	1.5976	1.1904	0.4686	
7	1.0000	0.9127	1.5947	2.0021	1.8704	2.0021	1.5947	0.9127
	0.9000	0.8157	1.3619	2.0886	1.7217	2.2017	1.5805	1.2060
	0.8000	0.8111	1.1504	2.2618	1.5252	2.4647	1.4644	1.5380
	0.7000	0.8567	0.9673	2.5158	1.3234	2.8018	1.3066	1.9096
	0.6000	0.9430	0.8025	2.8720	1.1237	3.2496	1.1310	2.3592
	0.5000	1.0799	0.6502	3.3822	0.9276	3.8750	0.9468	2.9478
	0.4000	1.2971	0.5072	4.1563	0.7350	4.8115	0.7584	3.7900
	0.3000	1.6692	0.3716	5.4540	0.5459	6.3703	0.5682	5.1476
	0.2000	2.4235	0.2423	8.0565	0.3604	9.4844	0.3776	7.8019
	0.1000	4.7006	0.1186	15.872	0.1784	18.818	0.1879	15.652
	Inf	1.5593	1.8671	1.8657	1.7651	1.5633	1.1610	0.4564
N	1/R	L1	C2	L3	C4	L5	C6	L7

Note: ($\omega = 1$ rad/s for Atn = 3dB)

Table 6.4 Low pass prototype filter: Chebyshev (RdB = 0.1)

N	R	C1	L2	C3	L4	C5	L6	C7
2	1.3554	1.2087	1.6382					
	1.4286	0.9771	1.9824					
	1.6667	0.7326	2.4885					
	2.0000	0.5597	3.0538					
	2.5000	0.4169	3.8275					
	3.3333	0.2933	5.0502					
	5.0000	0.1841	7.4257					
	10.000	0.0868	14.433					
	Inf	1.3911	0.8191					
3	1.0000	1.4328	1.5937	1.4328				
	0.9000	1.4258	1.4935	1.6219				
	0.8000	1.4511	1.3557	1.8711				
	0.7000	1.5210	1.1927	2.1901				
	0.6000	1.6475	1.0174	2.6026				
	0.5000	1.8530	0.8383	3.1594				
	0.4000	2.1857	0.6603	3.9675				
	0.3000	2.7630	0.4860	5.2788				
	0.2000	3.9418	0.3172	7.8503				
	0.1000	7.5121	0.1549	15.466				
	Inf	1.5133	1.5090	0.7164				
4	1.3554	0.9924	2.1476	1.5845	1.3451			
	1.4286	0.7789	2.3480	1.4292	1.7001			
	1.6667	0.5764	2.7304	1.1851	2.2425			
	2.0000	0.4398	3.2269	0.9672	2.8563			
	2.5000	0.3288	3.9605	0.7599	3.6976			
	3.3333	0.2329	5.1777	0.5602	5.0301			
	5.0000	0.1475	7.6072	0.3670	7.6143			
	10.000	0.0704	14.887	0.1802	15.229			
	Inf	1.5107	1.7682	1.4550	0.6725			
5	1.0000	1.3013	1.5559	2.2411	1.5559	1.3013		
	0.9000	1.2845	1.4329	2.3794	1.4878	1.4883		
	0.8000	1.2998	1.2824	2.5819	1.3815	1.7384		
	0.7000	1.3580	1.1170	2.8679	1.2437	2.0621		
	0.6000	1.4694	0.9469	3.2688	1.0846	2.4835		
	0.5000	1.6535	0.7777	3.8446	0.9126	3.0548		
	0.4000	1.9538	0.6119	4.7193	0.7333	3.8861		
	0.3000	2.4765	0.4509	6.1861	0.5503	5.2373		
	0.2000	3.5457	0.2950	9.1272	0.3659	7.8890		
	0.1000	6.7870	0.1447	17.957	0.1820	15.745		
	Inf	1.5613	1.8069	1.7659	1.4173	0.6507		
6	1.3554	0.9419	2.0797	1.6581	2.2473	1.5344	1.2767	
	1.4286	0.7347	2.2492	1.4537	2.5437	1.4051	1.6293	
	1.6667	0.5422	2.6003	1.1830	3.0641	1.1850	2.1739	
	2.0000	0.4137	3.0679	0.9575	3.7119	0.9794	2.7936	
	2.5000	0.3095	3.7652	0.7492	4.6512	0.7781	3.6453	
	3.3333	0.2195	4.9266	0.5514	6.1947	0.5795	4.9962	
	5.0000	0.1393	7.2500	0.3613	9.2605	0.3835	7.6184	
	10.000	0.0666	14.220	0.1777	18.427	0.1901	15.350	
	Inf	1.5339	1.8838	1.8306	1.7485	1.3937	0.6383	
7	1.0000	1.2615	1.5196	2.2392	1.6804	2.2392	1.5196	1.2615
	0.9000	1.2422	1.3946	2.3613	1.5784	2.3966	1.4593	1.4472
	0.8000	1.2550	1.2449	2.5481	1.4430	2.6242	1.3619	1.6967
	0.7000	1.3100	1.0826	2.8192	1.2833	2.9422	1.2326	2.0207
	0.6000	1.4170	0.9169	3.2052	1.1092	3.3841	1.0807	2.4437
	0.5000	1.5948	0.7529	3.7642	0.9276	4.0150	0.9142	3.0182
	0.4000	1.8853	0.5926	4.6179	0.7423	4.9702	0.7384	3.8552
	0.3000	2.3917	0.4369	6.0535	0.5557	6.5685	0.5569	5.2167
	0.2000	3.4278	0.2862	8.9371	0.3692	9.7697	0.3723	7.8901
	0.1000	6.5695	0.1405	17.603	0.1838	19.376	0.1862	15.813
	Inf	1.5748	1.8577	1.9210	1.8270	1.7340	1.3786	0.6307
N	1/R	L1	C2	L3	C4	L5	C6	L7

Note: ($\omega = 1$ rad/s for Atn = 3dB)

Table 6.5 Low pass prototype filter: Chebyshev (RdB = 0.25)

N	R	C1	L2	C3	L4	C5	L6	C7
2	2	0.6552	2.7632					
	3	0.3740	4.3118					
	4	0.2637	5.7389					
	8	0.1215	11.259					
	Inf	1.3584	0.8902					
3	1	1.6325	1.4360	1.6325				
	0.5	3.2663	1.0775	1.6325				
	0.333	4.8988	0.9572	1.6325				
	0.25	6.5326	0.8971	1.6325				
	0.125	13.064	0.8081	1.6325				
	Inf	1.5348	1.5285	0.8169				
4	2	0.6747	3.6860	1.0247	1.8806			
	3	0.4149	6.2744	0.7682	2.1302			
	4	0.3020	8.8161	0.6667	2.2533			
	8	0.1448	19.020	0.5334	2.4516			
	Inf	1.4817	1.8213	1.5068	0.7853			
5	1	1.5046	1.4436	2.4050	1.4436	1.5046		
	0.5	3.0103	0.7218	3.6080	1.4436	1.5046		
	0.333	4.5149	0.4812	4.8100	1.4436	1.5046		
	0.25	6.0196	0.3615	6.0130	1.4436	1.5046		
	0.125	12.040	0.1807	10.823	1.4436	1.5046		
	Inf	1.5765	1.7822	1.8225	1.4741	0.7523		
6	2	0.6867	3.2074	0.9308	3.8102	1.2163	1.7088	
	3	0.4330	5.0976	0.5392	6.0963	1.0804	1.8393	
	4	0.3173	6.9486	0.3821	8.2530	1.0221	1.8987	
	8	0.1539	14.310	0.1762	16.719	0.9393	1.9868	
	Inf	1.5060	1.9221	1.8191	1.8329	1.4721	0.7610	
7	1	1.5120	1.4169	2.4535	1.5350	2.4535	1.4169	1.5120
	0.5	3.024	0.7085	4.9069	1.1515	2.4535	1.4169	1.5120
	0.333	4.5361	0.4723	7.3596	1.0230	2.4535	1.4169	1.5120
	0.25	6.0471	0.3542	9.8120	0.9593	2.4535	1.4169	1.5120
	0.125	12.095	0.1776	19.625	0.8631	2.4535	1.4169	1.5120
	Inf	1.6009	1.8287	1.9666	1.8234	1.8266	1.4629	0.7555
N	1/R	L1	C2	L3	C4	L5	C6	L7

Note: ($\omega = 1$ rad/s for Atn = 3dB)

Table 6.6 Low pass prototype filter: Chebyshev (RdB = 0.5)

N	R	C1	L2	C3	L4	C5	L6	C7
2	1.9841	0.9827	1.9497					
	2.0000	0.9086	2.1030					
	2.5000	0.5635	3.1647					
	3.3333	0.3754	4.4111					
	5.0000	0.2282	6.6995					
	10.000	0.1052	13.322					
	Inf	1.3067	0.9748					
3	1.0000	1.8636	1.2804	1.8636				
	0.9000	1.9175	1.2086	2.0255				
	0.8000	1.9965	1.1203	2.2368				
	0.7000	2.1135	1.0149	2.5172				
	0.6000	2.2889	0.8937	2.8984				
	0.5000	2.5571	0.7592	3.4360				
	0.4000	2.9854	0.6146	4.2416				
	0.3000	3.7292	0.4633	5.5762				
	0.2000	5.2543	0.3087	8.2251				
	0.1000	9.8899	0.1534	16.118				
	Inf	1.5720	1.5179	0.9318				
4	1.9841	0.9202	2.5864	1.3036	1.8258			
	2.0000	0.8452	2.7198	1.2383	1.9849			
	2.5000	0.5162	3.7659	0.8693	3.1205			
	3.3333	0.3440	5.1196	0.6208	4.4790			
	5.0000	0.2100	7.7076	0.3996	6.9874			
	10.000	0.0975	15.352	0.1940	14.262			
	Inf	1.4361	1.8888	1.5211	0.9129			
5	1.0000	1.8068	1.3025	2.6914	1.3025	1.8068		
	0.9000	1.8540	1.2220	2.8478	1.2379	1.9701		
	0.8000	1.9257	1.1261	3.0599	1.1569	2.1845		
	0.7000	2.0347	1.0150	3.3525	1.0582	2.4704		
	0.6000	2.2006	0.8901	3.7651	0.9420	2.8609		
	0.5000	2.4571	0.7537	4.3672	0.8098	3.4137		
	0.4000	2.8692	0.6091	5.2960	0.6640	4.2447		
	0.3000	3.5877	0.4590	6.8714	0.5075	5.6245		
	0.2000	5.0639	0.3060	10.054	0.3430	8.3674		
	0.1000	9.5560	0.1525	19.646	0.1731	16.547		
	Inf	1.6299	1.7400	1.9217	1.5138	0.9034		
6	1.9841	0.9053	2.5774	1.3675	2.7133	1.2991	1.7961	
	2.0000	0.8303	2.7042	1.2912	2.8721	1.2372	1.9557	
	2.5000	0.5056	3.7219	0.8900	4.1092	0.8808	3.1025	
	3.3333	0.3370	5.0554	0.6323	5.6994	0.6348	4.4810	
	5.0000	0.2059	7.6145	0.4063	8.7319	0.4121	7.0310	
	10.000	0.0958	15.186	0.1974	17.681	0.2017	14.432	
	Inf	1.4618	1.9799	1.7803	1.9253	1.5077	0.8981	
7	1.0000	1.7896	1.2961	2.7177	1.3848	2.7177	1.2961	1.7896
	0.9000	1.8348	1.2146	2.8691	1.3080	2.8829	1.2335	1.9531
	0.8000	1.9045	1.1182	3.0761	1.2149	3.1071	1.1546	2.1681
	0.7000	2.0112	1.0070	3.3638	1.1050	3.4163	1.0582	2.4554
	0.6000	2.1744	0.8824	3.7717	0.9786	3.8524	0.9441	2.8481
	0.5000	2.4275	0.7470	4.3695	0.8377	4.4886	0.8137	3.4050
	0.4000	2.8348	0.6035	5.2947	0.6846	5.4698	0.6690	4.2428
	0.3000	3.5456	0.4548	6.8674	0.5221	7.1341	0.5129	5.6350
	0.2000	5.0070	0.3034	10.049	0.3524	10.496	0.3478	8.4041
	0.1000	9.4555	0.1513	19.649	0.1778	20.631	0.1761	16.665
	Inf	1.6464	1.7772	2.0306	1.7892	1.9239	1.5034	0.8948
N	1/R	L1	C2	L3	C4	L5	C6	L7

Note: ($\omega = 1$ rad/s for Atn = 3dB)

Table 6.7 Low pass prototype filter: Chebyshev (RdB = 1)

N	R	C1	L2	C3	L4	C5	L6	C7
2	3	0.5723	3.1317					
	4	0.3653	4.6002					
	8	0.1571	9.6582					
	Inf	1.2128	1.1093					
3	1	2.2160	1.0883	2.2160				
	0.5	4.4309	0.8168	2.2160				
	0.333	6.6469	0.7259	2.2160				
	0.25	8.8619	0.6799	2.2160				
	0.125	17.725	0.6120	2.2160				
	Inf	1.6522	1.4595	1.1080				
4	3	0.6529	4.4110	0.8140	2.5346			
	4	0.4517	7.0825	0.6118	2.8484			
	8	0.2085	17.164	0.4275	3.2811			
	Inf	1.3499	2.0102	1.4879	1.1057			
N	1/R	L1	C2	L3	C4	L5	C6	L7

N	R	C1	L2	C3	L4	C5	L6	C7
5	1	2.2072	1.1279	3.1025	1.1279	2.2072		
	0.5	4.4144	0.5645	4.6532	1.1279	2.2072		
	0.333	6.6216	0.3763	6.2050	1.1279	2.2072		
	0.25	8.8288	0.2822	7.7557	1.1279	2.2072		
	0.125	17.656	0.1406	13.961	1.1279	2.2072		
	Inf	1.7213	1.6448	2.0614	1.4928	1.1031		
6	3	0.6785	3.8725	0.7706	4.7107	0.9692	2.4060	
	4	0.4810	5.6441	0.4759	7.3511	0.8494	2.5820	
	8	0.2272	12.310	0.1975	16.740	0.7256	2.7990	
	Inf	1.3775	2.0969	1.6896	2.0744	1.4942	1.1022	
7	1	2.2043	1.1311	3.1472	1.1942	3.1472	1.1311	2.2043
	0.5	4.4075	0.5656	6.2934	0.8951	3.1472	1.1311	2.2043
	0.333	6.6118	0.3774	9.4406	0.7955	3.1472	1.1311	2.2043
	0.25	8.8151	0.2828	12.588	0.7466	3.1472	1.1311	2.2043
	0.125	17.631	0.1414	25.175	0.6714	3.1472	1.1311	2.2043
	Inf	1.7414	1.6774	2.1554	1.7028	2.0792	1.4943	1.1016
N	1/R	L1	C2	L3	C4	L5	C6	L7

Note: ($\omega = 1$ rad/s for Atn = 3dB)

Table 6.8 Low pass prototype filter: Bessel

N	R	C1	L2	C3	L4	C5	L6	C7
2	1.000	0.5755	2.1478					
	1.111	0.5084	2.3097					
	1.250	0.4433	2.5096					
	1.429	0.3801	2.7638					
	1.667	0.3191	3.0993					
	2.000	0.2601	3.5649					
	2.500	0.2032	4.2577					
	3.333	0.1486	5.4050					
	5.000	0.0965	7.6876					
	10.00	0.0469	14.510					
	Inf	1.3617	0.4539					
3	1.000	0.3374	0.9705	2.2034				
	0.900	0.3708	0.8650	2.3745				
	0.800	0.4124	0.7609	2.5867				
	0.700	0.4657	0.6584	2.8575				
	0.600	0.5365	0.5576	3.2159				
	0.500	0.6353	0.4587	3.7144				
	0.400	0.7829	0.3618	4.4573				
	0.300	1.0283	0.2673	6.6888				
	0.200	1.5171	0.1752	8.1403				
	0.100	2.9825	0.0860	15.470				
	Inf	1.4631	0.8427	0.2926				
4	1.000	0.2334	0.6725	1.0815	2.2404			
	1.111	0.2085	0.7423	0.9670	2.4143			
	1.250	0.1839	0.8292	0.8534	2.6304			
	1.429	0.1596	0.9406	0.7410	2.9066			
	1.667	0.1356	1.0886	0.6299	3.2727			
	2.000	0.1120	1.2952	0.5202	3.7824			
	2.500	0.0887	1.6040	0.4120	4.5430			
	3.333	0.0658	2.1174	0.3056	5.8048			
	5.000	0.0434	3.1416	0.2013	8.3185			
	10.00	0.0214	6.2086	0.0993	15.837			
	Inf	1.5012	0.9781	0.6127	0.2114			
5	1.000	0.1743	0.5072	0.8040	1.1110	2.2582		
	0.900	0.1926	0.4542	0.8894	0.9945	2.4328		
	0.800	0.2154	0.4016	0.9959	0.8789	2.6497		
	0.700	0.2447	0.3494	1.1323	0.7642	2.9272		
	0.600	0.2836	0.2977	1.3138	0.6506	3.2952		
	0.500	0.3380	0.2465	1.5672	0.5382	3.8077		
	0.400	0.4194	0.1958	1.9464	0.4270	4.5731		
	0.300	0.5548	0.1457	2.5768	0.3174	5.8433		
	0.200	0.8251	0.0964	3.8352	0.2095	8.3747		
	0.100	1.6349	0.0478	7.6043	0.1036	15.949		
	Inf	1.5125	1.0232	0.7531	0.4729	0.1618		
6	1.000	0.1365	0.4002	0.6392	0.8538	1.1126	2.2645	
	1.111	0.1223	0.4429	0.5732	0.9456	0.9964	2.4388	
	1.250	0.1082	0.4961	0.5076	1.0600	0.8810	2.6554	
	1.429	0.0943	0.5644	0.4424	1.2069	0.7665	2.9325	
	1.667	0.0804	0.6553	0.3775	1.4022	0.6530	3.3001	
	2.000	0.0666	0.7824	0.3131	1.6752	0.5405	3.8122	
	2.500	0.0530	0.9725	0.2492	2.0837	0.4292	4.5770	
	3.333	0.0395	1.2890	0.1859	2.7633	0.3193	5.8467	
	5.000	0.0261	1.9209	0.1232	4.1204	0.2110	8.3775	
	10.00	0.0130	3.8146	0.0612	8.1860	0.1045	15.951	
	Inf	1.5124	1.0329	0.8125	0.6072	0.3785	0.1287	
7	1.000	0.1106	0.3259	0.5249	0.7020	0.8690	1.1052	2.2659
	0.900	0.1224	0.2923	0.5815	0.6302	0.9630	0.9899	2.4396
	0.800	0.1372	0.2589	0.6521	0.5586	1.0803	0.8754	2.6556
	0.700	0.1562	0.2257	0.7428	0.4873	1.2308	0.7618	2.9319
	0.600	0.1815	0.1927	0.8634	0.4163	1.4312	0.6491	3.2984
	0.500	0.2168	0.1599	1.0321	0.3457	1.7111	0.5374	3.8090
	0.400	0.2698	0.1274	1.2847	0.2755	2.1304	0.4269	4.5718
	0.300	0.3579	0.0951	1.7051	0.2058	2.8280	0.3177	5.8380
	0.200	0.5338	0.0630	2.5448	0.1365	4.2214	0.2100	8.3623
	0.100	1.0612	0.0313	5.0616	0.0679	8.3967	0.1040	15.917
	Inf	1.5087	1.0293	0.8345	0.6752	0.5031	0.3113	0.1054
N	1/R	L1	C2	L3	C4	L5	C6	L7

Note: ($\omega = 1$ rad/s for Atn = 3dB)

Table 6.9 Low pass prototype filter: Elliptic

θ	ωs	Amin	Ω_2	C1	C2	L2	C3
1	57.2987	103.56	66.1616	0.6395	0.0002	0.9786	0.6395
2	28.6537	85.50	33.0839	0.6390	0.0009	0.9776	0.6390
3	19.1073	74.93	22.0595	0.6381	0.0021	0.9761	0.6381
4	14.3356	67.43	16.5483	0.6370	0.0037	0.9739	0.6370
5	11.4737	61.61	13.2424	0.6354	0.0059	0.9711	0.6354
6	9.5668	56.85	11.0392	0.6336	0.0085	0.9676	0.6336
7	8.2055	52.82	9.4661	0.6314	0.0116	0.9636	0.6314
8	7.1853	49.33	8.2868	0.6289	10.0152	0.9589	0.6289
9	6.3925	46.25	7.3700	0.6261	0.0193	0.9536	0.6261
10	5.7588	43.49	6.6370	0.6229	0.0240	0.9477	0.6229
11	5.2408	41.00	6.0377	0.6194	0.0291	0.9411	0.6194
12	4.8097	38.71	5.5386	0.6155	0.0349	0.9339	0.6155
13	4.4454	36.61	5.1166	0.6113	0.0412	0.9261	0.6113
14	4.1336	34.66	4.7552	0.6068	0.0482	0.9177	0.6068
15	3.8637	32.85	4.4423	0.6020	0.0558	0.9087	0.6020
16	3.6280	31.14	4.1688	0.5968	0.0640	0.8991	0.5968
17	3.4203	29.54	3.9277	0.5913	0.0729	0.8888	0.5913
18	3.2361	28.03	3.7137	0.5855	0.0826	0.8780	0.5855
19	3.0716	26.60	3.5224	0.5793	0.0930	0.8665	0.5793
20	2.9238	25.24	3.3505	0.5728	0.1043	0.8545	0.5728
21	2.7904	23.95	3.1951	0.5661	0.1164	0.8418	0.5661
22	2.6695	22.71	3.0541	0.5590	0.1294	0.8286	0.5590
23	2.5593	21.53	2.9256	0.5515	0.1434	0.8148	0.5515
24	2.4586	20.40	2.8079	0.5438	0.1585	0.8004	0.5438
25	2.3662	19.31	2.6999	0.5358	0.1747	0.7855	0.5358
26	2.2812	18.27	2.6003	0.5275	0.1921	0.7700	0.5275
27	2.2027	17.26	2.5083	0.5189	0.2108	0.7540	0.5189
28	2.1301	16.30	2.4231	0.5100	0.2309	0.7375	0.5100
29	2.0627	15.37	2.3438	0.5009	0.2526	0.7205	0.5009
30	2.0000	14.47	2.2701	0.4915	0.2760	0.7031	0.4915
31	1.9416	13.61	2.2012	0.4819	0.3012	0.6852	0.4819
32	1.8871	12.77	2.1368	0.4720	0.3284	0.6669	0.4720
θ	ωs	Amin	Ω_2	L1	L2	C2	L3

Note: (N = 3, R = 1, RdB = 0.01)

Table 6.10 Low pass prototype filter: Elliptic

θ	ωs	Amin	A2	C1	C2	L2	C3
1	57.2987	115.77	66.1616	1.1893	0.0002	1.1540	1.1893
2	28.6537	97.70	33.0839	1.1889	0.0008	1.1533	1.1889
3	19.1073	87.13	22.0595	1.1881	0.0018	1.1522	1.1881
4	14.3356	79.63	16.5483	1.1870	0.0032	1.1507	1.1870
5	11.4737	73.81	13.2424	1.1856	0.0050	1.1488	1.1856
6	9.5668	69.05	11.0392	1.1839	0.0072	1.1464	1.1839
7	8.2055	65.03	9.4661	1.1819	0.0098	1.1436	1.1819
8	7.1853	61.54	8.2868	1.1796	0.0128	1.1404	1.1796
9	6.3925	58.46	7.3700	1.1770	0.0162	1.1367	1.1770
10	5.7588	55.70	6.6370	1.1740	0.0200	1.1326	1.1740
11	5.2408	53.20	6.0377	1.1708	0.0243	1.1281	1.1708
12	4.8097	50.92	5.5386	1.1672	0.0290	1.1231	1.1672
13	4.4454	48.82	5.1166	1.1634	0.0342	1.1177	1.1634
14	4.1336	46.87	4.7552	1.1592	0.0398	1.1119	1.1592
15	3.8637	45.05	4.4423	1.1547	0.0458	1.1057	1.1547
16	3.6280	43.35	4.1688	1.1500	0.0524	1.0990	1.1500
17	3.4203	41.75	3.9277	1.1449	0.0594	1.0919	1.1449
18	3.2361	40.23	3.7137	1.1395	0.0669	1.0844	1.1395
19	3.0716	38.80	3.5224	1.1338	0.0749	1.0764	1.1338
20	2.9238	37.44	3.3505	1.1278	0.0834	1.0681	1.1278
21	2.7904	36.14	3.1951	1.1215	0.0925	1.0593	1.1215
22	2.6695	34.90	3.0541	1.1149	0.1021	1.0500	1.1149
23	2.5593	33.71	2.9256	1.1080	0.1123	1.0404	1.1080
24	2.4586	32.57	2.8079	1.1008	0.1231	1.0303	1.1008
25	2.3662	31.47	2.6999	1.0933	0.1345	1.0199	1.0933
26	2.2812	30.41	2.6003	1.0855	0.1466	1.0090	1.0855
27	2.2027	29.39	2.5083	1.0773	0.1593	0.9976	1.0773
28	2.1301	28.41	2.4231	1.0689	0.1728	0.9859	1.0689
29	2.0627	27.45	2.3438	1.0602	0.1869	0.9738	1.0602
30	2.0000	26.53	2.2701	1.0512	0.2019	0.9612	1.0512
31	1.9416	25.63	2.2012	1.0420	0.2176	0.9483	1.0420
32	1.8871	24.76	2.1368	1.0324	0.2343	0.9349	1.0324
33	1.8361	23.92	2.0765	1.0225	0.2518	0.9212	1.0225
34	1.7883	23.09	2.0199	1.0123	0.2702	0.9070	1.0123
35	1.7434	22.29	1.9666	1.0019	0.2897	0.8925	1.0019
36	1.7013	21.51	1.9165	0.9912	0.3103	0.8776	0.9912
37	1.6616	20.74	1.8692	0.9802	0.3320	0.8623	0.9802
38	1.6243	20.00	1.8245	0.9689	0.3549	0.8466	0.9689
39	1.5890	19.27	1.7823	0.9573	0.3791	0.8305	0.9573
40	1.5557	18.56	1.7423	0.9455	0.4047	0.8141	0.9455
41	1.5243	17.86	1.7044	0.9334	0.4318	0.7973	0.9334
42	1.4945	17.18	1.6684	0.9210	0.4605	0.7801	0.9210
43	1.4663	16.52	1.6343	0.9084	0.4909	0.7627	0.9084
44	1.4396	15.86	1.6018	0.8955	0.5232	0.7448	0.8955
45	1.4142	15.22	1.5710	0.8823	0.5576	0.7267	0.8823
46	1.3902	14.60	1.5415	0.8689	0.5942	0.7082	0.8689
47	1.3673	13.98	1.5135	0.8553	0.6331	0.6895	0.8553
48	1.3456	13.38	1.4868	0.8415	0.6747	0.6705	0.8415
49	1.3250	12.79	1.4613	0.8274	0.7192	0.6511	0.8274
50	1.3054	12.22	1.4369	0.8131	0.7668	0.6316	0.8131
θ	ωs	Amin	A2	L1	L2	C2	L3

Note: (N = 3, R = 1, RdB = 0.18)

Table 6.11 Low pass prototype filter: Elliptic

θ	ωs	Amin	A2	C1	C2	L2	C3	L4
6	10.350843	88.5	11.367741	0.7174	0.006461	1.198	1.330	0.6549
7	8.876727	83.1	9.747389	0.7154	0.008813	1.194	1.329	0.6552
8	7.771760	78.5	8.532615	0.7130	0.01154	1.190	1.327	0.6555
9	6.912894	74.4	7.588226	0.7103	0.01464	1.186	1.325	0.6558
10	6.226301	70.7	6.833109	0.7073	0.01814	1.181	1.323	0.6561
11	5.664999	67.3	6.215646	0.7040	0.02202	1.176	1.321	0.6565
12	5.197666	64.3	5.701423	0.7003	0.02630	1.170	1.318	0.6569
13	4.802620	61.5	5.266618	0.6963	0.03100	1.163	1.316	0.6574
14	4.464371	58.9	4.894214	0.6920	0.03612	1.156	1.313	0.6579
15	4.171563	56.5	4.571732	0.6874	0.04166	1.148	1.310	0.6584
16	3.915678	54.2	4.289813	0.6824	0.04766	1.140	1.306	0.6590
17	3.690200	52.1	4.041300	0.6771	0.05411	1.132	1.303	0.6596
18	3.490065	50.1	3.820626	0.6715	0.06103	1.122	1.299	0.6603
19	3.311272	48.1	3.623399	0.6655	0.06845	1.113	1.295	0.6610
20	3.150622	46.3	3.446101	0.6592	0.07637	1.103	1.291	0.6617
21	3.005526	44.6	3.285888	0.6526	0.08482	1.092	1.286	0.6624
22	2.873864	42.9	3.140431	0.6456	0.09383	1.081	1.282	0.6632
23	2.753885	41.3	3.007807	0.6383	0.1034	1.069	1.277	0.6641
24	2.644133	39.8	2.886413	0.6306	0.1136	1.057	1.272	0.6649
25	2.543380	38.4	2.774903	0.6226	0.1244	1.044	1.267	0.6658
26	2.450592	37.0	2.672139	0.6143	0.1359	1.030	1.262	0.6668
27	2.364885	35.6	2.577149	0.6055	0.1481	1.017	1.256	0.6677
28	2.285502	34.3	2.489103	0.5964	0.1611	1.002	1.250	0.6687
29	2.211792	33.0	2.407283	0.5870	0.1748	0.9872	1.244	0.6698
30	2.143189	31.8	2.331070	0.5772	0.1894	0.9717	1.238	0.6708
31	2.079202	30.6	2.259921	0.5670	0.2049	0.9558	1.232	0.6719
32	2.019399	29.4	2.193363	0.5564	0.2213	0.9393	1.226	0.6730
33	1.963403	28.3	2.130982	0.5455	0.2388	0.9223	1.219	0.6742
34	1.910879	27.2	2.072410	0.5341	0.2573	0.9048	1.212	0.6753
35	1.861534	26.1	2.017322	0.5224	0.2771	0.8868	1.205	0.6765
36	1.815103	25.1	1.965429	0.5103	0.2982	0.8683	1.198	0.6777
37	1.771354	24.0	1.916475	0.4978	0.3206	0.8492	1.191	0.6789
38	1.730076	23.0	1.870229	0.4848	0.3446	0.8297	1.184	0.6801
39	1.691083	22.1	1.826485	0.4715	0.3702	0.8098	1.177	0.6813
40	1.654204	21.1	1.785057	0.4577	0.3976	0.7893	1.169	0.6825
θ	ωs	Amin	A2	L1	L2	C2	L3	C4

Note: (N = 4, R = 1.1, RdB = 0.01)

Table 6.12 Low pass prototype filter: Elliptic

θ	ωs	Amin	A2	C1	C2	L2	C3	L4
6	10.35084	100.7	11.36774	1.260	0.006028	1.284	1.932	0.8431
7	8.876727	95.3	9.747390	1.258	0.008216	1.281	1.930	0.843
8	7.771760	90.7	8.532615	1.255	0.01074	1.278	1.928	0.8440
9	6.912894	86.6	7.588226	1.253	0.01362	1.275	1.926	0.8442
10	6.226301	82.9	6.833109	1.250	0.01685	1.271	1.924	0.8443
11	5.664999	79.6	6.215646	1.247	0.02043	1.267	1.921	0.8445
12	5.197666	76.5	5.701423	1.243	0.02436	1.263	1.918	0.8448
13	4.802620	73.7	5.266618	1.239	0.02866	1.258	1.915	0.8450
14	4.464371	71.1	4.894214	1.235	0.03333	1.253	1.912	0.8453
15	4.171563	68.7	4.571731	1.231	0.03837	1.247	1.908	0.8456
16	3.915678	66.4	4.289813	1.226	0.04380	1.241	1.904	0.8459
17	3.690200	64.3	4.041300	1.221	0.04961	1.234	1.900	0.8462
18	3.490065	62.3	3.820626	1.216	0.05581	1.227	1.895	0.8465
19	3.311272	60.3	3.623399	1.210	0.06242	1.220	1.891	0.8469
20	3.150622	58.5	3.446101	1.204	0.06944	1.213	1.886	0.8473
21	3.005526	56.8	3.285888	1.198	0.07689	1.205	1.881	0.8477
22	2.873864	55.1	3.140431	1.191	0.08476	1.196	1.875	0.8481
23	2.753885	53.6	3.007807	1.184	0.09309	1.187	1.870	0.8485
24	2.644133	52.0	2.886413	1.177	0.1019	1.178	1.864	0.8490
25	2.543380	50.6	2.774903	1.169	0.1111	1.169	1.858	0.8494
26	2.450592	49.2	2.672139	1.161	0.1209	1.159	1.851	0.8499
27	2.364885	47.8	2.577149	1.153	0.1311	1.148	1.845	0.8505
28	2.285502	46.5	2.489103	1.145	0.1419	1.138	1.838	0.8510
29	2.211792	45.2	2.407283	1.136	0.1532	1.126	1.831	0.8516
30	2.143189	44.0	2.331070	1.127	0.1651	1.115	1.824	0.8521
31	2.079202	42.8	2.259921	1.117	0.1775	1.103	1.816	0.8527
32	2.019399	41.6	2.193363	1.108	0.1906	1.091	1.808	0.8533
33	1.963403	40.5	2.130982	1.097	0.2043	1.078	1.800	0.8540
34	1.910879	39.4	2.072410	1.087	0.2186	1.065	1.792	0.8546
35	1.861534	38.3	2.017322	1.076	0.2337	1.051	1.784	0.8553
36	1.815103	37.2	1.965429	1.065	0.2495	1.038	1.775	0.8560
37	1.771354	36.2	1.916475	1.054	0.2661	1.023	1.766	0.8567
38	1.730076	35.2	1.870229	1.042	0.2835	1.009	1.757	0.8574
39	1.691083	34.2	1.826485	1.030	0.3017	0.9936	1.748	0.8581
40	1.654204	33.3	1.785057	1.017	0.3208	0.9782	1.738	0.8589
θ	ωs	Amin	A2	L1	L2	C2	L3	C4

Note: (N = 4, R = 1.5, RdB = 0.18)

Table 6.13 Low pass prototype filter: Elliptic

θ	ωs	Amin	A2	A4	C1	C2	L2	C3	C4	L4	C5
2	28.6537	167.86	48.7389	30.1274	0.7661	0.0003	1.3099	1.5877	0.0008	1.3091	0.7656
3	19.1073	150.25	32.4927	20.0893	0.7658	0.0007	1.3095	1.5868	0.0018	1.3075	0.7646
4	14.3356	137.74	24.3697	15.0716	0.7654	0.0012	1.3088	1.5855	0.0033	1.3054	0.7633
5	11.4737	128.04	19.4959	12.0620	0.7648	0.0020	1.3080	1.5839	0.0052	1.3026	0.7615
6	9.5668	120.11	16.2468	10.0565	0.7641	0.0029	1.3070	1.5820	0.0076	1.2993	0.7594
7	8.2055	113.40	13.9260	8.6247	0.7632	0.0039	1.3058	1.5796	0.0103	1.2953	0.7569
8	7.1853	107.59	12.1854	7.5516	0.7623	0.0051	1.3044	1.5770	0.0135	1.2907	0.7540
9	6.3925	102.45	10.8316	6.7175	0.7612	0.0065	1.3028	1.5739	0.0172	1.2855	0.7507
10	5.7588	97.86	9.7486	6.0507	0.7600	0.0080	1.3011	1.5706	0.0213	1.2797	0.7470
11	5.2408	93.69	8.8625	5.5057	0.7586	0.0098	1.2991	1.5669	0.0259	1.2733	0.7429
12	4.8097	89.89	8.1241	5.0520	0.7572	0.0116	1.2970	1.5628	0.0309	1.2663	0.7384
13	4.4454	86.39	7.4993	4.6684	0.7556	0.0137	1.2947	1.5584	0.0364	1.2586	0.7335
14	4.1336	83.14	6.9638	4.3401	0.7538	0.0159	1.2922	1.5536	0.0424	1.2504	0.7283
15	3.8637	80.11	6.4997	4.0559	0.7519	0.0183	1.2895	1.5485	0.0489	1.2416	0.7226
16	3.6280	77.27	6.0936	3.8076	0.7499	0.0209	1.2866	1.5431	0.0559	1.2321	0.7165
17	3.4203	74.60	5.7353	3.5888	0.7478	0.0236	1.2836	1.5374	0.0635	1.2221	0.7101
18	3.2361	72.08	5.4168	3.3946	0.7455	0.0266	1.2803	1.5313	0.0716	1.2115	0.7032
19	3.0716	69.69	5.1318	3.2212	0.7431	0.0297	1.2768	1.5249	0.0802	1.2002	0.6959
20	2.9238	67.41	4.8753	3.0654	0.7406	0.0330	1.2732	1.5182	0.0895	1.1884	0.6883
21	2.7904	65.25	4.6433	2.9246	0.7379	0.0365	1.2694	1.5112	0.0994	1.1760	0.6802
22	2.6695	63.18	4.4323	2.7970	0.7350	0.0402	1.2653	1.5038	0.1099	1.1630	0.6717
23	2.5593	61.20	4.2397	2.6807	0.7321	0.0441	1.2611	1.4962	0.1210	1.1494	0.6628
24	2.4586	59.29	4.0631	2.5743	0.7290	0.0482	1.2567	1.4882	0.1329	1.1353	0.6534
25	2.3662	57.46	3.9007	2.4767	0.7257	0.0524	1.2520	1.4800	0.1454	1.1205	0.6437
26	2.2812	55.70	3.7507	2.3868	0.7223	0.0569	1.2472	1.4715	0.1588	1.1052	0.6335
27	2.2027	54.00	3.6119	2.3038	0.7187	0.0617	1.2421	1.4627	0.1729	1.0893	0.6229
28	2.1301	52.35	3.4829	2.2270	0.7150	0.0666	1.2369	1.4537	0.1879	1.0729	0.6118
29	2.0627	50.76	3.3629	2.1556	0.7112	0.0718	1.2314	1.4444	0.2038	1.0559	0.6003

θ	ωs	Amin	A2	A4	L1	L2	C2	L3	L4	C4	L5
30	2.0000	49.22	3.2508	2.0892	0.7072	0.0772	1.2257	1.4348	0.2206	1.0383	0.5884
31	1.9416	47.72	3.1460	2.0274	0.7030	0.08281	1.2198	1.4250	0.2384	1.0201	0.5760
32	1.8871	46.27	3.0476	1.9695	0.6987	0.08871	1.2136	1.4150	0.2574	1.0015	0.5631
33	1.8361	44.85	2.9553	1.9154	0.6942	0.09481	1.2073	1.4048	0.2774	0.9822	0.5498
34	1.7883	43.47	2.8683	1.8646	0.6896	0.10121	1.2007	1.3943	0.2988	0.9625	0.5360
35	1.7434	42.13	2.7864	1.8170	0.6847	0.1078	1.1938	1.3837	0.3214	0.9422	0.5217
36	1.7013	40.81	2.7089	1.7722	0.6798	0.1148	1.1867	1.3729	0.3455	0.9214	0.5070
37	1.6616	39.53	2.6356	1.7299	0.6746	0.1220	1.1794	1.3619	0.3712	0.9001	0.4917
38	1.6243	38.28	2.5662	1.6901	0.6693	0.1295	1.1717	1.3508	0.3985	0.8782	0.4759
39	1.5890	37.05	2.5003	1.6525	0.6637	0.1374	1.1638	1.3396	0.4278	0.8559	0.4595
40	1.5557	35.85	2.4377	1.6170	0.6580	0.1456	1.1556	1.3282	0.4590	0.8331	0.4426
41	1.5243	34.67	2.3781	1.5833	0.6521	0.1541	1.1472	1.3168	0.4924	0.8099	0.4252
42	1.4945	33.52	2.3213	1.5515	0.6460	0.1630	1.1384	1.3053	0.5283	0.7862	0.4072
43	1.4663	32.38	2.2672	1.5213	0.6397	0.1722	1.1292	1.2937	0.5669	0.7620	0.3885
44	1.4396	31.27	2.2154	1.4926	0.6332	0.1819	1.1198	1.2822	0.6085	0.7375	0.3693
45	1.4142	30.17	2.1660	1.4654	0.6265	0.1920	1.1099	1.2706	0.6535	0.7125	0.3494
46	1.3902	29.09	2.1187	1.4396	0.6195	0.2025	1.0997	1.2591	0.7022	0.6871	0.3288
47	1.3673	28.03	2.0733	1.4150	0.6124	0.2135	1.0891	1.2478	0.7550	0.6614	0.3075
48	1.3456	26.99	2.0299	1.3916	0.6050	0.2251	1.0780	1.2365	0.8126	0.6354	0.2855
49	1.3250	25.95	1.9881	1.3693	0.5973	0.2372	1.0665	1.2254	0.8756	0.6090	0.2628
50	1.3054	24.94	1.9480	1.3481	0.5894	0.2498	1.0545	1.2145	0.9446	0.5824	0.2392
51	1.2868	23.93	1.9095	1.3279	0.5813	0.2632	1.0420	1.2039	1.0206	0.5556	0.2147
52	1.2690	22.94	1.8724	1.3087	0.5729	0.2772	1.0289	1.1937	1.1047	0.5285	0.1894
53	1.2521	21.96	1.8366	1.2903	0.5642	0.2920	1.0152	1.1838	1.1980	0.5013	0.1631
54	1.2361	20.99	1.8021	1.2728	0.5552	0.3076	1.0008	1.1744	1.3020	0.4740	0.1357
55	1.2208	20.04	1.7689	1.2561	0.5460	0.3242	0.9858	1.1656	1.4187	0.4467	0.1073
56	1.2062	19.09	1.7368	1.2402	0.5364	0.3417	0.9700	1.1575	1.5502	0.4194	0.0777
57	1.1924	18.15	1.7057	1.2250	0.5265	0.3605	0.9533	1.1501	1.6993	0.3921	0.0468
58	1.1792	17.23	1.6757	1.2104	0.5163	0.3805	0.9358	1.1437	1.8693	0.3651	0.0145

Note: (N = 5, R = 1, RdB = 0.01)

Table 6.14 Low pass prototype filter: Elliptic

θ	ωs	Amin	A2	A4	C1	C2	L2	C3	C4	L4	C5
2	28.6537	180.07	48.7389	30.1274	1.30163	0.00031	1.34523	2.12770	0.00082	1.34459	1.30112
3	19.1073	162.45	32.4927	20.0893	1.30130	0.00070	1.34483	2.12660	0.00184	1.34339	1.30016
4	14.3356	149.95	24.3697	15.0716	1.30084	0.00125	1.34426	2.12507	0.00328	1.34170	1.29881
5	11.4737	140.25	19.4959	12.0620	1.30024	0.00196	1.34353	2.12311	0.00513	1.33954	1.29708
6	9.5668	132.32	16.2468	10.0565	1.29951	0.00282	1.34264	2.12070	0.00740	1.33689	1.29496
7	8.2055	125.61	13.9260	8.6247	1.29865	0.00384	1.34159	2.11786	0.01008	1.33376	1.29246
8	7.1853	119.80	12.1854	7.5516	1.29766	0.00502	1.34037	2.11459	0.01318	1.33015	1.28957
9	6.3925	114.66	10.8316	6.7175	1.29653	0.00637	1.33899	2.11088	0.01671	1.32607	1.28630
10	5.7588	110.06	9.7486	6.0507	1.29527	0.00787	1.33744	2.10675	0.02067	1.32150	1.28264
11	5.2408	105.90	8.8625	5.5057	1.29387	0.00953	1.33573	2.10217	0.02506	1.31646	1.27859
12	4.8097	102.10	8.1241	5.0520	1.29234	0.01136	1.33386	2.09717	0.02989	1.31094	1.27417
13	4.4454	98.59	7.4993	4.6684	1.29067	0.01335	1.33182	2.09172	0.03516	1.30495	1.26936
14	4.1336	95.34	6.9638	4.3401	1.28887	0.01551	1.32961	2.08588	0.04089	1.29848	1.26416
15	3.8637	92.32	6.4997	4.0559	1.28693	0.01783	1.32724	2.07959	0.04707	1.29154	1.25858
16	3.6280	89.48	6.0936	3.8076	1.28485	0.02033	1.32470	2.07288	0.05371	1.28413	1.25261
17	3.4203	86.81	5.7353	3.5888	1.28263	0.02300	1.32199	2.06574	0.06084	1.27625	1.24627
18	3.2361	84.29	5.4168	3.3946	1.28027	0.02584	1.31911	2.05819	0.06844	1.26790	1.23953
19	3.0716	81.89	5.1318	3.2212	1.27778	0.02885	1.31607	2.05021	0.07655	1.25909	1.23241
20	2.9238	79.62	4.8753	3.0654	1.27514	0.03205	1.31285	2.04182	0.08515	1.24981	1.22491
21	2.7904	77.46	4.6433	2.9246	1.27236	0.03542	1.30945	2.03301	0.09428	1.24007	1.21703
22	2.6695	75.39	4.4323	2.7970	1.26943	0.03898	1.30589	2.02379	0.10393	1.22987	1.20876
23	2.5593	73.40	4.2397	2.6807	1.26636	0.04272	1.30215	2.01416	0.11414	1.21921	1.20010
24	2.4586	71.50	4.0631	2.5743	1.26314	0.04666	1.29823	2.00412	0.12490	1.20809	1.19107
25	2.3662	69.67	3.9007	2.4767	1.25978	0.05079	1.29413	1.99368	0.13625	1.19652	1.18164
26	2.2812	67.91	3.7507	2.3868	1.25262	0.05511	1.28985	1.98283	0.14819	1.18450	1.17183
27	2.2027	66.21	3.6119	2.3038	1.25259	0.05963	1.28540	1.97159	0.16075	1.17203	1.16164
28	2.1301	64.56	3.4829	2.2270	1.24877	0.06436	1.28075	1.95995	0.17396	1.15911	1.15106
29	2.0627	62.97	3.3629	2.1556	1.22480	0.06930	1.27592	1.94792	0.18783	1.14576	1.14010
30	2.0000	61.43	3.2508	2.0892	1.24067	0.07446	1.27091	1.93550	0.20239	1.13196	1.12874

θ	ωs	Amin	A2	A4	L1	L2	C2	L3	L4	C4	L5
31	1.9416	59.93	3.1460	2.0274	1.23638	0.07983	1.26570	1.92270	0.21768	1.11772	1.11700
32	1.8871	58.47	3.0476	1.9695	1.23192	0.08543	1.26030	1.90952	0.23371	1.10305	1.10487
33	1.8361	57.06	2.9553	1.9154	1.22731	0.09126	1.25470	1.89596	0.25054	1.08795	1.09235
34	1.7883	55.68	2.8683	1.8846	1.22252	0.09732	1.24890	1.88203	0.26819	1.07242	1.07944
35	1.7434	54.33	2.7864	1.8170	1.21757	0.10363	1.24290	1.86773	0.28671	1.05648	1.06614
36	1.7013	53.02	2.7089	1.7722	1.21244	0.11019	1.23669	1.85307	0.30614	1.04011	1.05244
37	1.6616	51.74	2.6356	1.7299	1.20714	0.11701	1.23028	1.83806	0.32654	1.02332	1.03835
38	1.6243	50.49	2.5662	1.6901	1.20166	0.12410	1.22364	1.82269	0.34795	1.00613	1.02386
39	1.5890	49.26	2.5003	1.6525	1.19600	0.13146	1.21679	1.80698	0.37044	0.98853	1.00897
40	1.5557	48.06	2.4377	1.6170	1.19015	0.13911	1.20971	1.79093	0.39408	0.97053	0.99368
41	1.5243	46.88	2.3781	1.5833	1.18411	0.14706	1.20241	1.77455	0.41894	0.95213	0.97798
42	1.4945	45.72	2.3213	1.5515	1.17787	0.15532	1.19486	1.75784	0.44510	0.93335	0.96187
43	1.4663	44.59	2.2672	1.5213	1.17144	0.16389	1.18708	1.74081	0.47265	0.91417	0.94535
44	1.4396	43.47	2.2154	1.4926	1.16480	0.17280	1.17904	1.72347	0.50170	0.89462	0.92841
45	1.4142	42.38	2.1660	1.4654	1.15794	0.18206	1.17075	1.70583	0.53236	0.87470	0.91105
46	1.3902	41.30	2.1187	1.4396	1.15088	0.19169	1.16219	1.68789	0.56476	0.85441	0.89326
47	1.3673	40.23	2.0733	1.4150	1.14359	0.20169	1.15336	1.66967	0.59903	0.83376	0.87504
48	1.3456	39.19	2.0299	1.3916	1.13607	0.21210	1.14425	1.65117	0.63534	0.81276	0.85638
49	1.3250	38.15	1.9881	1.3693	1.12831	0.22293	1.13484	1.63241	0.67386	0.79141	0.83727
50	1.3054	37.13	1.9480	1.3481	1.12031	0.23421	1.12513	1.61339	0.71481	0.76973	0.81771
51	1.2868	36.12	1.9095	1.3279	1.11206	0.24596	1.11509	1.59413	0.75841	0.74773	0.79768
52	1.2690	35.13	1.8724	1.3087	1.10354	0.25821	1.10473	1.57465	0.80492	0.72541	0.77717
53	1.2521	34.14	1.8366	1.2903	1.09476	0.27099	1.09401	1.55494	0.85465	0.70278	0.75619
54	1.2361	33.17	1.0821	1.2728	1.08569	0.28433	1.08293	1.53504	0.90794	0.67986	0.73470
55	1.2208	32.20	1.7689	1.2561	1.07633	0.29828	1.07147	1.51496	0.96518	0.65667	0.71270
56	1.2062	31.25	1.7368	1.2402	1.06666	0.31288	1.05960	1.49471	1.02684	0.63320	0.69016
57	1.1924	30.30	1.7057	1.2250	1.05668	0.32817	1.04731	1.47431	1.09344	0.60949	0.66709
58	1.1792	29.36	1.6757	1.2104	1.04636	0.34422	1.03456	1.45379	1.16561	0.58554	0.64344
59	1.1666	28.42	1.6467	1.1966	1.03570	0.36109	1.02134	1.43317	1.24407	0.56138	0.61920
60	1.1547	27.49	1.6185	1.1834	1.02467	0.37885	1.00760	1.41247	1.32969	0.53702	0.59435

Note: (N = 5, R = 1, RdB = 0.18)

Table 6.15 Low pass prototype filter: Elliptic

θ	ωs	Amin	A2	A4	C1	C2	L2	C3	C4	L4	C5	L6
16	3.751039	112.5	5.452491	3.888329	1.299	0.0250	1.344	2.142	0.0468	1.412	2.017	0.8828
17	3.535748	109.3	5.133037	3.664543	1.296	0.0283	1.341	2.135	0.0530	1.405	2.012	0.8830
18	3.344698	106.3	4.849152	3.465915	1.293	0.0318	1.337	2.126	0.0596	1.397	2.006	0.8831
19	3.174064	103.4	4.595218	3.288476	1.290	0.0355	1.333	2.118	0.0666	1.389	2.000	0.8833
20	3.020785	100.7	4.366743	3.129050	1.286	0.0395	1.328	2.108	0.0740	1.380	1.993	0.8835
21	2.882384	98.1	4.160091	2.985065	1.283	0.0436	1.324	2.099	0.0818	1.371	1.987	0.8837
22	2.756834	95.6	3.972284	2.854418	1.279	0.0480	1.319	2.089	0.0901	1.362	1.979	0.8839
23	2.642462	93.3	3.800865	2.735370	1.275	0.0527	1.314	2.078	1.0989	1.352	1.972	0.8841
24	2.537873	91.0	3.643786	2.626475	1.270	0.0576	1.309	2.067	0.1081	1.341	1.964	0.8843
25	2.441895	88.8	3.499325	2.526516	1.266	0.0627	1.303	2.055	0.1177	1.331	1.956	0.8845
26	2.353536	86.7	3.366027	2.434463	1.261	0.0680	1.297	2.043	0.1279	1.320	1.948	0.8848
27	2.271953	84.6	3.242651	2.349441	1.256	0.0736	1.291	2.031	0.1385	1.308	1.939	0.8850
28	2.196422	82.6	3.128134	2.270699	1.251	0.0795	1.285	2.018	0.1497	1.296	1.930	0.8853
29	2.126320	80.7	3.021559	2.197588	1.246	0.0857	1.279	2.005	0.1613	1.284	1.921	0.8855
30	2.061105	78.9	2.922132	2.129549	1.240	0.0921	1.272	1.991	0.1735	1.271	1.911	0.8858
31	2.000308	77.1	2.829162	2.066092	1.235	0.0988	1.265	1.977	0.1863	1.257	1.901	0.8861
32	1.943517	75.3	2.742042	2.006790	1.229	0.1057	1.258	1.962	0.1996	1.244	1.891	0.8864
33	1.890370	73.6	2.660241	1.951268	1.223	0.1130	1.250	1.947	0.2136	1.230	1.881	0.8867
34	1.840548	72.0	2.583290	1.899195	1.216	0.1206	1.243	1.931	0.2281	1.215	1.870	0.8870
35	1.793769	70.4	2.510772	1.850277	1.210	0.1285	1.235	1.915	0.2433	1.200	1.859	0.8873
36	1.749781	68.8	2.442318	1.804254	1.203	0.1367	1.226	1.899	0.2592	1.185	1.847	0.8877
37	1.708362	67.3	2.377598	1.760893	1.196	0.1452	1.218	1.882	0.2758	1.169	1.835	0.8880
38	1.669312	65.8	2.316318	1.719987	1.189	0.1541	1.209	1.864	0.2931	1.153	1.823	0.8884
39	1.632449	64.3	2.258212	1.681350	1.181	0.1634	1.200	1.847	0.3112	1.137	1.811	0.8887
40	1.597615	62.8	2.203043	1.644814	1.174	0.1730	1.191	1.828	0.3301	1.120	1.798	0.8891
41	1.564662	61.4	2.150595	1.610227	1.166	0.1830	1.181	1.810	0.3498	1.103	1.785	0.8895
42	1.533460	60.0	2.100673	1.577454	1.158	0.1934	1.172	1.791	0.3704	1.085	1.771	0.8898
43	1.503888	58.7	2.053102	1.546370	1.149	0.2043	1.161	1.771	0.3920	1.067	1.758	0.8902
44	1.475840	57.3	2.007720	1.516862	1.141	0.2155	1.151	1.751	0.4145	1.049	1.744	0.8906
45	1.449216	56.0	1.964382	1.488829	1.132	0.2272	1.140	1.731	0.4381	1.030	1.729	0.8910
46	1.423927	54.7	1.922953	1.462178	1.123	0.2394	1.130	1.710	0.4628	1.011	1.715	0.8915

θ	ωs	Amin	A2	A4	L1	L2	C2	L3	L4	C4	L5	C6
47	1.399891	53.4	1.883312	1.436822	1.113	0.2521	1.118	1.689	0.4888	0.9910	1.700	0.8919
48	1.377032	52.2	1.845347	1.412684	1.103	0.2653	1.107	1.668	0.5160	0.9711	1.684	0.8923
49	1.355282	50.9	1.808954	1.389693	1.093	0.2791	1.095	1.646	0.5446	0.9508	1.669	0.8928
50	1.334577	49.7	1.774040	1.367782	1.083	0.2935	1.083	1.623	0.5747	0.9302	1.653	0.8932
51	1.314859	48.5	1.740516	1.346891	1.073	0.3084	1.070	1.600	0.6063	0.9092	1.637	0.8937
52	1.296076	47.3	1.708301	1.326965	1.062	0.3241	1.057	1.577	0.6397	0.8878	1.620	0.8942
53	1.278176	46.1	1.677322	1.307952	1.050	0.3404	1.044	1.554	0.6749	0.8661	1.603	0.8946
54	1.261116	45.0	1.647510	1.289805	1.039	0.3574	1.031	1.530	0.7122	0.8440	1.586	0.8951
55	1.244853	43.8	1.618799	1.272479	1.027	0.3752	1.017	1.506	0.7517	0.8216	1.568	0.8956
56	1.229348	42.7	1.591131	1.255935	1.015	0.3939	1.003	1.481	0.7936	0.7989	1.551	0.8961
57	1.214564	41.5	1.564449	1.240135	1.002	0.4135	0.9881	1.456	0.8382	0.7758	1.532	0.8966
58	1.200469	40.4	1.538703	1.225044	0.9894	0.4340	0.9732	1.431	0.8857	0.7523	1.514	0.8971
59	1.187032	39.3	1.513843	1.210630	0.9760	0.4556	0.9578	1.405	0.9365	0.7286	1.495	0.8976
60	1.174224	38.1	1.489825	1.196863	0.9623	0.4783	0.9420	1.379	0.9909	0.7045	1.476	0.8981
61	1.162017	37.0	1.466607	1.183715	0.9481	0.5022	0.9258	1.353	1.049	0.6801	1.456	0.8987
62	1.150388	35.9	1.444148	1.171161	0.9335	0.5274	0.9091	1.326	1.112	0.6554	1.436	0.8992
63	1.139313	34.8	1.422411	1.159176	0.9184	0.5541	0.8920	1.299	1.181	0.6304	1.416	0.8997
64	1.128771	33.7	1.401362	1.147737	0.9028	0.5824	0.8743	1.272	1.255	0.6051	1.395	0.9002
65	1.118742	32.6	1.380967	1.136826	0.8867	0.6125	0.8562	1.244	1.335	0.5795	1.374	0.9008
66	1.109208	31.5	1.361196	1.126421	0.8700	0.6445	0.8374	1.216	1.424	0.5536	1.352	0.9013
67	1.100151	30.4	1.342017	1.116505	0.8528	0.6787	0.8182	1.188	1.521	0.5274	1.330	0.9018
68	1.091555	29.3	1.323405	1.107063	0.8349	0.7153	0.7982	1.160	1.629	0.5010	1.308	0.9023
69	1.083407	28.2	1.305331	1.098078	0.8163	0.7547	0.7777	1.131	1.748	0.4744	1.285	0.9028
70	1.075391	27.1	1.287771	1.089536	0.7970	0.7972	0.7564	1.102	1.883	0.4475	1.261	0.9032
71	1.068397	26.0	1.270700	1.081425	0.7769	0.8433	0.7344	1.073	2.034	0.4204	1.237	0.9037
72	1.061511	24.9	1.254065	1.073732	0.7560	0.8936	0.7116	1.044	2.206	0.3931	1.213	0.9040
73	1.055024	23.7	1.237933	1.066446	0.7341	0.9487	0.6878	1.015	2.405	0.3657	1.188	0.9044
74	1.048925	22.6	1.222193	1.059558	0.7112	1.010	0.6631	0.9860	2.634	0.3381	1.162	0.9047
75	1.043207	21.5	1.206854	1.053059	0.6872	1.077	0.6374	0.9568	2.905	0.3105	1.135	0.9049
76	1.037860	20.3	1.191893	1.046940	0.6620	1.153	0.6104	0.9278	3.226	0.2828	1.107	0.9050

Note: (N = 6, R = 1.5, RdB = 0.18)

Table 6.16 Low pass prototype filter: Elliptic

θ	ωs	Amin	A2	A4	A6	C1	C2	L2	C3	C4	L4	C5	C6	L6	C7
26	2.281172	105.4	5.038750	2.333900	2.859592	1.310	0.0290	1.358	2.100	0.1353	1.357	2.049	0.0995	1.281	1.247
27	2.202689	103.0	4.848897	2.253156	2.756829	1.308	0.0314	1.355	2.089	0.1465	1.345	2.034	0.1034	1.272	1.240
28	2.130054	100.7	4.672457	2.178409	2.661529	1.306	0.0339	1.353	2.078	0.1582	1.332	2.019	0.1117	1.263	1.233
29	2.062665	98.5	4.508037	2.109040	2.572921	1.304	0.0364	1.350	2.066	0.1704	1.319	2.003	0.1204	1.254	1.226
30	2.000000	96.3	4.354434	2.044515	2.490337	1.302	0.0391	1.347	2.054	0.1833	1.305	1.987	0.1295	1.245	1.218
31	1.941604	94.2	4.210595	1.984368	2.413194	1.299	0.0420	1.344	2.042	0.1966	1.292	1.970	0.1390	1.235	1.210
32	1.887080	92.2	4.075602	1.928190	2.340984	1.297	0.0449	1.341	2.029	0.2106	1.277	1.952	0.1490	1.225	1.202
33	1.836078	90.2	3.948647	1.875623	2.273259	1.294	0.0479	1.338	2.016	0.2252	1.262	1.934	0.1593	1.214	1.193
34	1.788292	88.3	3.829016	1.826351	2.209625	1.292	0.0511	1.335	2.002	0.2404	1.247	1.916	0.1702	1.204	1.184
35	1.743447	86.4	3.716076	1.780095	2.149731	1.289	0.0544	1.332	1.988	0.2562	1.232	1.897	0.1815	1.193	1.175
36	1.701302	84.6	3.609267	1.736606	2.093268	1.286	0.0578	1.328	1.973	0.2727	1.216	1.878	0.1932	1.181	1.165
37	1.661640	82.8	3.508087	1.695662	2.039957	1.283	0.0614	1.324	1.959	0.2900	1.199	1.858	0.2055	1.169	1.155
38	1.624269	81.0	3.412086	1.657065	1.989552	1.280	0.0650	1.321	1.943	0.3079	1.183	1.837	0.2183	1.157	1.145
39	1.589016	79.3	3.320862	1.620638	1.941830	1.277	0.0689	1.317	1.928	0.3267	1.165	1.817	0.2317	1.145	1.135
40	1.555724	77.6	3.234050	1.586220	1.896591	1.274	0.0728	1.313	1.912	0.3462	1.148	1.795	0.2456	1.132	1.124
41	1.524253	76.0	3.151325	1.553668	1.853653	1.270	0.0770	1.308	1.895	0.3666	1.130	1.773	0.2601	1.119	1.113
42	1.494477	74.3	3.072388	1.522851	1.812855	1.267	0.0812	1.304	1.879	0.3879	1.112	1.751	0.2753	1.105	1.102
43	1.466279	72.8	2.996969	1.493651	1.774048	1.263	0.0857	1.300	1.862	0.4101	1.093	1.728	0.2911	1.092	1.090
44	1.439557	71.2	2.924824	1.465961	1.737098	1.259	0.0903	1.295	1.844	0.4332	1.074	1.705	0.3076	1.077	1.078
45	1.414214	69.7	2.855727	1.439683	1.701881	1.255	0.0950	1.290	1.826	0.4575	1.055	1.682	0.3248	1.063	1.066
46	1.390164	68.2	2.789476	1.414728	1.668286	1.251	0.1000	1.285	1.808	0.4828	1.035	1.657	0.3428	1.048	1.053
47	1.367327	66.7	2.725881	1.391016	1.636211	1.247	0.1051	1.280	1.789	0.5093	1.015	1.633	0.3617	1.033	1.040
48	1.345633	65.2	2.664770	1.368471	1.605563	1.243	0.1105	1.275	1.770	0.5370	0.9944	1.608	0.3814	1.017	1.027
49	1.325013	63.7	2.605984	1.347026	1.576255	1.238	0.1160	1.269	1.751	0.5661	0.9736	1.583	0.4020	1.001	1.013
50	1.305407	62.3	2.549377	1.326618	1.548208	1.234	0.1217	1.264	1.731	0.5965	0.9525	1.557	0.4235	0.9850	0.9992

θ	ωs	Amin	A2	A4	A6	L1	L2	C2	L3	L4	C4	L5	L6	C6	L7
51	1.286760	60.9	2.494813	1.307190	1.521349	1.229	0.1277	1.258	1.711	0.6286	0.9310	1.531	0.4462	0.9684	0.9848
52	1.269018	59.5	2.442167	1.288687	1.495612	1.224	0.1339	1.252	1.690	0.6622	0.9093	1.504	0.4699	0.9514	0.9699
53	1.252136	58.1	2.391323	1.271063	1.470934	1.219	0.1404	1.246	1.669	0.6977	0.8872	1.477	0.4948	0.9340	0.9547
54	1.236068	56.8	2.342170	1.254270	1.447259	1.213	0.1471	1.239	1.648	0.7351	0.8648	1.450	0.5211	0.9163	0.9391
55	1.220775	55.4	2.294610	1.238269	1.424533	1.208	0.1541	1.232	1.626	0.7745	0.8420	1.422	0.5487	0.8981	0.9230
56	1.206218	54.1	2.248546	1.223020	1.402707	1.202	0.1614	1.225	1.604	0.8163	0.8190	1.394	0.5778	0.8796	0.9065
57	1.192363	52.7	2.203891	1.208487	1.381735	1.196	0.1690	1.218	1.581	0.8605	0.7957	1.365	0.6085	0.8607	0.8896
58	1.179178	51.4	2.160560	1.194638	1.361575	1.190	0.1770	1.211	1.558	0.9075	0.7721	1.336	0.6411	0.8414	0.8722
59	1.166633	50.1	2.118476	1.181442	1.342188	1.183	0.1853	1.203	1.535	0.9576	0.7482	1.307	0.6755	0.8217	0.8543
60	1.154701	48.8	2.077565	1.168869	1.323537	1.177	0.1939	1.195	1.511	1.011	0.7240	1.278	0.7121	0.8016	0.8360
61	1.143354	47.5	2.037756	1.156895	1.305587	1.170	0.2030	1.186	1.487	1.068	0.6995	1.248	0.7510	0.7811	0.8171
62	1.132570	46.2	1.998983	1.145494	1.288307	1.163	0.2125	1.177	1.463	1.129	0.6748	1.218	0.7925	0.7602	0.7976
63	1.122326	44.9	1.961181	1.134644	1.271668	1.155	0.2225	1.168	1.438	1.195	0.6498	1.188	0.8369	0.7389	0.7776
64	1.112602	43.7	1.924292	1.124323	1.255641	1.147	0.2331	1.159	1.412	1.267	0.6245	1.157	0.8845	0.7171	0.7570
65	1.103378	42.4	1.888255	1.114512	1.240200	1.139	0.2441	1.149	1.386	1.344	0.5990	1.126	0.9357	0.6949	0.7357
66	1.094636	41.1	1.853014	1.105192	1.225322	1.130	0.2559	1.138	1.360	1.428	0.5732	1.095	0.9909	0.6722	0.7138
67	1.086360	39.8	1.818515	1.096346	1.210984	1.121	0.2682	1.127	1.333	1.520	0.5472	1.064	1.051	0.6490	0.6911
68	1.078535	38.5	1.784703	1.087959	1.197165	1.112	0.2814	1.116	1.306	1.622	0.5209	1.032	1.116	0.6254	0.6676
69	1.071145	37.2	1.751526	1.080016	1.183845	1.101	0.2953	1.104	1.278	1.734	0.4945	1.001	1.187	0.6013	0.6433
78	1.022341	25.1	1.472529	1.026592	1.083849	0.9782	0.4841	0.9527	1.004	3.822	0.2483	0.7148	2.368	0.3595	0.3710
79	1.018717	23.6	1.442574	1.022499	1.074724	0.9588	0.5177	0.9282	0.9699	4.337	0.2205	0.6841	2.628	0.3295	0.3316
80	1.015427	22.1	1.412537	1.018751	1.065966	0.9376	0.5562	0.9011	0.9356	4.994	0.1929	0.6540	2.946	0.2987	0.2892
81	1.012465	20.6	1.382299	1.015345	1.057569	0.9142	0.6011	0.8707	0.9006	5.858	0.1656	0.6248	3.346	0.2672	0.2431
82	1.009828	18.9	1.351718	1.012276	1.049533	0.8881	0.6545	0.8363	0.8648	7.036	0.1387	0.5968	3.863	0.2350	0.1926

Note: (N = 7, R = 1, RdB = 0.18)

7

Oscillators, Frequency Synthesisers and PLL Techniques

E. Artal, J. P. Pascual and J. Portilla

7.1 Introduction

In any communications system, there must be some source of the microwave signal – these sources are generally termed oscillators. There are a number of parameters which have to be considered when specifying a particular source, the most important being the frequency, power, frequency stability, noise content and harmonic content.

This chapter describes the fundamentals of oscillators and frequency synthesis including simple realisations based on diodes and transistors as well as more advanced concepts such as voltage control of the frequency and stabilisation of the frequency using phase-locking techniques.

7.2 Solid State Microwave Oscillators

7.2.1 Fundamentals

Oscillators are one of the main critical subsystems in microwave converters. The common type of oscillator is based on a semiconductor active device such as a transistor or a diode embedded in a frequency selective passive circuit. This kind of oscillator is used in practically all the low or medium power microwave systems. The semiconductor device works like an energy exchange system: the radio frequency power delivered by the device to a load is obtained from a DC power supply according to a specific bias point of operation.

There are several ways or approaches to analyse or to design microwave oscillators. One possibility is to study the oscillators from the point of view of amplifiers with positive feedback. This approach is a valuable tool in the case of low frequency oscillators but it is not very useful in microwave applications. A more powerful and practical way for microwave oscillator analysis and design is the negative resistance method. It is based on the characterisation of the active device as a one port component with its real part of impedance (or admittance) less than zero ohms at the operation frequency. A very simplified scheme of an oscillator following this approach is shown in Figure 7.1.

Microwave Devices, Circuits and Subsystems for Communications Engineering Edited by I. A. Glover, S. R. Pennock and P. R. Shepherd

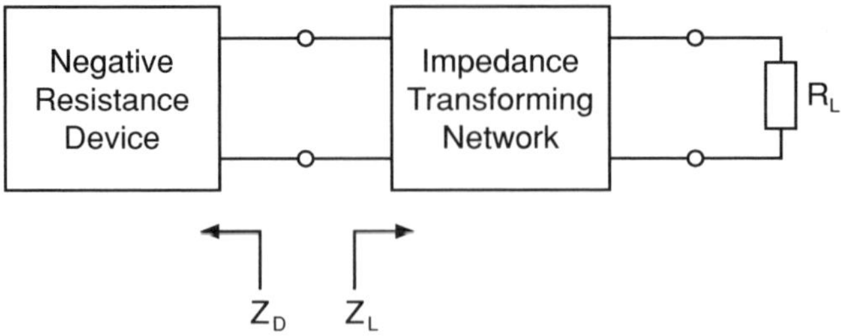

Figure 7.1 Basic scheme of a microwave solid state oscillator

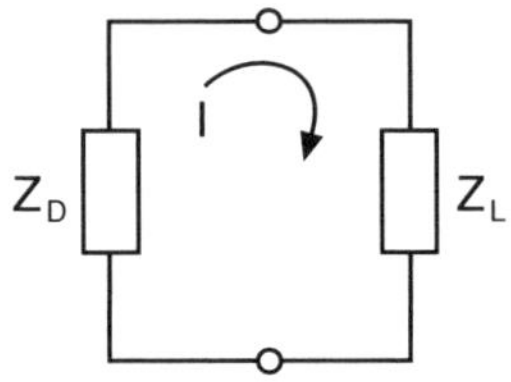

Figure 7.2 Simplified electrical circuit of a microwave oscillator

In Figure 7.1 the oscillator circuit has two separate sections, on the left the active part of the circuit is composed of the active device (negative resistance device), on the right the passive part of the circuit is composed of a frequency selective network including the load. The active device can be a negative resistance diode or a transistor fed back by a suitable passive network. The active part of the oscillator has non-linear behaviour, that is, its impedance depends on the amplitude of the signal. The amplitude can be of the current or the voltage on the active device terminal's plane. The passive part of the device, the impedance transforming network, is supposed to be linear, that is, its behaviour is not dependent on the amplitude of the signal. In order to obtain the maximum radio frequency power delivered to the load, it can be assumed that the linear network is a lossless network. In that case, the power delivered by the active device is fully delivered to the load. In general microwave systems the load is resistive and it is very often equal to the reference impedance system. A standard value of this load is 50 ohm. A first analysis of the oscillator can be made from the simplified electrical circuit shown in Figure 7.2.

The non-linear side of the oscillator is represented by the impedance Z_D (device impedance) and the linear part is represented by the impedance Z_L (load impedance). This last term must be understood as the load directly seen by the active device terminals. Assuming that oscillations are present in the circuit, the Kirchhoff law imposes the next equation:

$$Z_D I = -Z_L I \tag{7.1}$$

where I is the radio frequency current amplitude. This last equation can be now expressed as follows:

$$(Z_D + Z_L)I = 0 \tag{7.2}$$

As the current amplitude I cannot be zero, a necessary condition of oscillation is:

$$Z_D + Z_L = 0 \tag{7.3}$$

By solving Equation (7.3) in real and imaginary parts, the oscillation point is obtained which is defined by the oscillation frequency f_0 and oscillation amplitude I_0. The practical problem is that it is not easy to have a complete knowledge of the frequency and amplitude behaviour of the magnitudes in Equation (7.3). In practice, there are some assumptions that can avoid requiring the total knowledge of the active device behaviour. Moreover, the condition of Equation (7.3) is not sufficient to guarantee the operation of the oscillator. There is an additional condition that ensures that the oscillation point is a stable one. This condition will be discussed later. The next example is intended to clarify some aspects of the oscillator operation.

7.2.1.1 An IMPATT oscillator

An IMPATT diode is a special semiconductor pn junction whose impedance has a negative resistance behaviour in a particular frequency range. This negative resistance comes from the combination of a charge carrier avalanche (electrons and holes), due to an inverse voltage bias of the pn junction, and a transit time effect of these carriers through the semiconductor. A simplified equivalent circuit of the IMPATT diode, only valid for RF purposes, is shown in Figure 7.3.

The negative resistance is non-linearly dependent on the RF current amplitude. The capacitance can be approximated by the pn junction capacitance value under the reverse bias condition. A typical negative resistance dependence is shown in Figure 7.4.

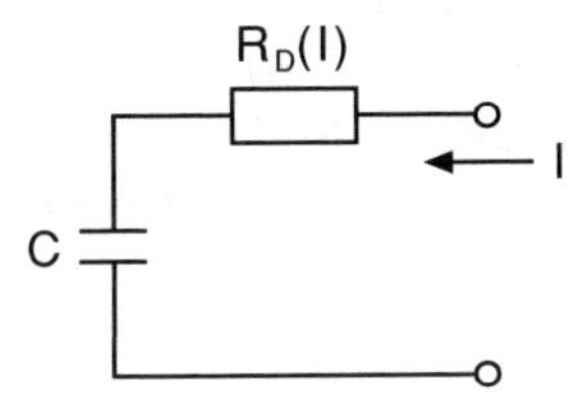

Figure 7.3 Equivalent circuit of an IMPATT diode

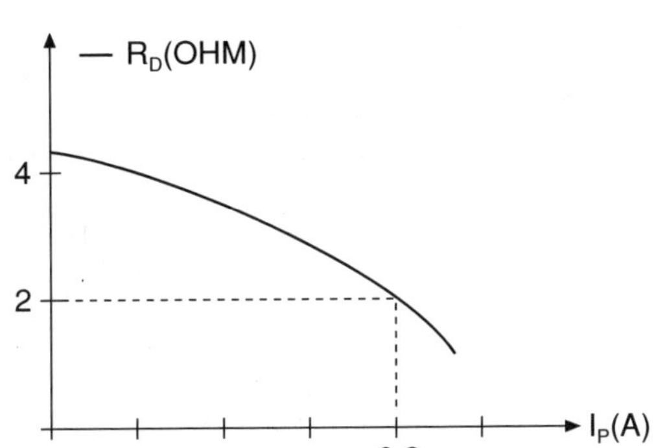

Figure 7.4 IMPATT negative resistance versus RF current

The negative resistance non-linear function can be approximated by a Van der Pol characteristic, which is based on a RF voltage cubic dependence versus the RF current:

$$V_{NL} = -aI + bI^3 \tag{7.4}$$

In this equation V_{NL} is the non-linear voltage on the non-linear resistance $R_D(I)$ terminals, and I is the peak value of the RF current across it. From Equation (7.4) the non-linear resistance can be obtained as:

$$R_D(I) = -a + bI^2 \tag{7.5}$$

A good fitting of the curve shown in Figure 7.4 is obtained selecting appropriate values for constants a and b:

$$a = 4 \quad b = 3.1$$

A very simplified circuit to analyse an oscillator using this IMPATT diode is shown in Figure 7.5. In this example the oscillator circuit is a series resonant type. In practice, it is not possible to mount such a simple circuit because it is very difficult to manufacture an ideal coil (L) and an ideal resistance load (R_L) at microwave frequencies. However, this simple circuit is very useful to model actual oscillator circuits in practice because, very often, in narrow frequency bandwidths, a microwave oscillator can be represented by a series resonant circuit or by a parallel resonant circuit.

It must be noted that in Figure 7.5 the non-linear impedance is restricted to the IMPATT diode non-linear resistance R_D, while the capacitance C, which accounts for the space charge capacitance of the reverse biased pn junction, can be considered linear. In order to identify both the linear and non-linear parts of the oscillator circuit, the capacitance C can be included on the right side of the circuit as a load subcircuit element. This is identified in the figure as Z_D (device impedance) and Z_L (load impedance).

Usually the oscillator is designed to obtain the *maximum oscillation power* from the active device. The value to be maximised is the power P delivered to the load:

$$P = \frac{1}{2}R_L I_0^2 = -\frac{1}{2}(-a + bI_0^2)I_0^2 \tag{7.6}$$

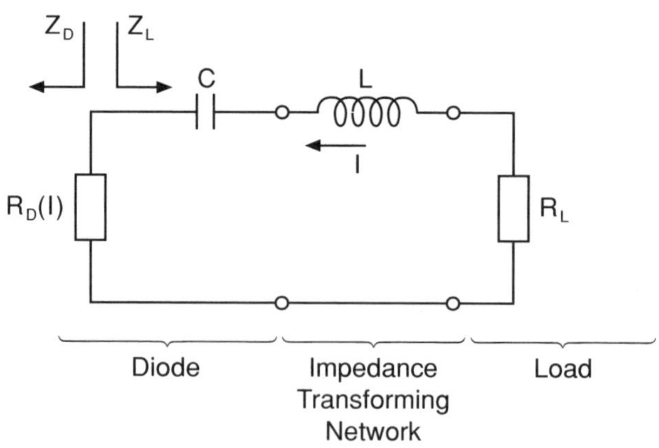

Figure 7.5 Series resonance type IMPATT oscillator circuit

In order to find the optimum point, the derivative of power P versus the RF current peak amplitude I_0 must be zero:

$$\frac{\partial P}{\partial I_0} = aI_0 - 2bI_0^3 = 0 \tag{7.7}$$

From this condition the optimum RF current amplitude is obtained:

$$I_0 = \sqrt{\frac{a}{2b}} \cong 0.8 \text{ (A)} \tag{7.8}$$

As a consequence of the last equation, the optimum non-linear resistance value is found:

$$R_D(I_0) = -\frac{a}{2} = -2 \text{ (ohm)} \tag{7.9}$$

This result is important because can be interpreted as follows: a non-linear negative resistance device, with Van der Pol voltage to current characteristic, will deliver the maximum power to a load if its resistance has a value half that of its low signal resistance value. The low signal resistance value is equal to $-a$ in this example. The low signal value occurs at very low levels of current (or voltage). From Equation (7.5) the low signal resistance value $R_D(0)$ is obtained assuming the value of I is near zero:

$$R_D(0) = lim_{I \to 0}[R_D(I)] = -a \tag{7.10}$$

Many practical devices have Van der Pol characteristics, such as most negative resistance diodes, so the Equation (7.9) approach is very useful because the device impedance measurement under low signal level conditions is usually an easy task. On the other hand, the device impedance measurement under large signal conditions can be a very difficult task due to the special test equipment required.

From Equation (7.3) the load resistance R_L for maximum oscillation power must be equal to the device optimum non-linear resistance but with opposite sign:

$$R_L = \frac{a}{2} = 2 \text{ (Ohm)} \tag{7.11}$$

The power delivered to this load is then:

$$P = \frac{a^2}{8b} \cong 645 \text{ mW} \tag{7.12}$$

The oscillation frequency is determined by the series resonance, see Figure 7.5, so it can be obtained from the next identity:

$$f_0 = \frac{1}{2\pi}\frac{1}{\sqrt{LC}} \tag{7.13}$$

The signal at the output load is a sinusoidal one with frequency f_0 (Hz) and with an average power P.

7.2.2 Stability of Oscillations

The condition of oscillation given by Equation (7.3) is a necessary condition to have oscillation in a circuit, but this condition is not enough to have a steady sinusoidal signal in the load. There are several ways to analyse the stability of the oscillation point. At microwave frequencies there is a useful graphical method to do the stability test. The method is based on the definition of two impedance lines called the *device line* and the *load line*. The device line is the graphical representation of the device impedance with a change of sign. The device impedance is in general dependent on two variables: frequency and signal amplitude. In a strict sense the device line is not a line but a surface. In practice, the device impedance is a slowly varying function against frequency if it is compared with the stronger variation of load impedance against frequency. Following the data from the last example (p. 000), the device line can be defined as the graphical representation of $-Z_D(I)$ and the load line as the graphical representation of $Z_L(f)$.

Device line $-Z_D(I,f) \cong -Z_D(I)$
Load line $Z_L(f)$

From the data of the IMPATT oscillator example, both lines can be drawn on the impedance plane $Z = R + jX$. These lines are shown on Figure 7.6. The point where the lines cross defines both parameters of the oscillator: current amplitude I and frequency of oscillation f_0.

In the general case, neither lines have an easily defined analytical expression. In this example, due to its simplicity, it is possible to write:

Device line $-Z_D(I) = 4 - 3.1I^2$
Load line $Z_L(f) = R_L(1 + j2Q\delta)$

With

$$R_L = \frac{a}{2} = 2 \quad Q = \omega_0 L/R_L \quad \delta = \frac{f - f_0}{f_0}$$

which are respectively the load resistance, the quality factor of the resonant circuit and the relative frequency shift.

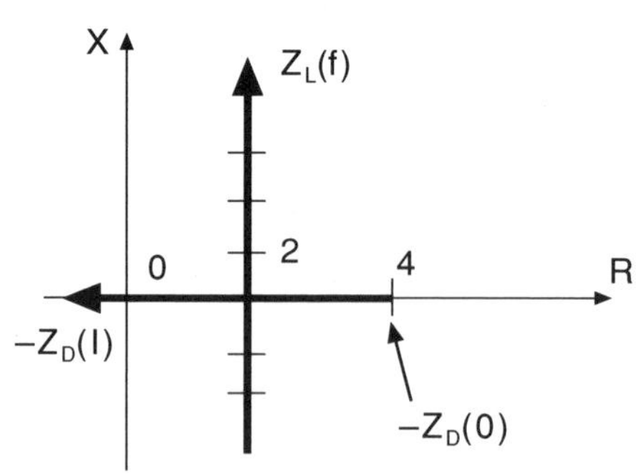

Figure 7.6 Device and load lines for the oscillator circuit of the IMPATT oscillator

Classical stability analysis of oscillations can be done assuming small amplitude and frequency deviations from the oscillation point given by I_0 and f_0. If the system reacts so that the deviations decrease with time, in such a case the oscillation point is a stable one. However, if the deviations increase with time, for this point a stable oscillation is not possible. A method of analysing stability conditions is based on the properties of impedance functions. These are complex variable analytical functions, so some interesting properties can be extracted. A mathematical study of such functions is beyond the scope of this book, and only the final graphical condition for stable operation is presented. A complete analysis is included in [1]. Using the graphical representation of device and load lines, the *stability condition* of oscillations can be expressed as follows: 'The measured angle from the device line to the load line, in the growing sense of arrows, must be less than 180° to have stable oscillator behaviour'. An example is shown in Figure 7.6.

7.3 Negative Resistance Diode Oscillators

Two terminal semiconductor devices were the first solid state devices used by the microwave industry. Such devices were developed in the 1960s. Nowadays some of these are available in most microwave product catalogues, using standard commercial names such as Gunn diodes or IMPATT diodes, which are the more extensive types. A common characteristic of these devices is that their impedance presents negative resistance behaviour in a limited frequency range. The physical origin of their negative resistance (or admittance) is very different in Gunn diodes compared with IMPATT diodes. A description is included in Chapter 2.

At present, negative resistance diode oscillators are the more powerful solid state sources at very high frequencies, and the only solid state source possible in the higher frequency range of millimetre waves. Most microwave diode oscillators are built in coaxial or waveguide structures, more precisely coaxial resonators or waveguide cavities, due to their high quality factors. A high quality factor of the load circuit implies good frequency stability and low noise. It is possible also to build diode oscillators using microstrip lines or other similar printed transmission lines, but in this case some additional circuitry must be added to achieve good frequency stability.

From the designer's point of view, the knowledge of the device impedance as a function of frequency and signal level is enough. Unfortunately often one only has some limited data for the device. Some manufacturers give on their data sheets the optimum impedance values for maximum output power. Often a difficulty in the design of such oscillators arises from the relative low resistance levels of diodes. Those values are in general lower than 10 Ohm while the standard system impedance is 50 Ohm. Moreover, the waveguide mountings have inherently high impedances, so special mounting procedures are used to obtain high transformation ratios avoiding resistive losses as much as possible. The reactive part of device impedance tends to be very high, especially in the IMPATT diode case, making the transformation impedance network a crucial issue in the oscillator design. Diodes in chip version have a capacitive behaviour but bonding and other package effects can transform this to an inductive behaviour.

The efficiency of negative resistance diodes is very low, so the majority of the DC power delivered to the device is dissipated inside. The devices are normally available in a packaged version. Packages are designed to act as heatsinks to reduce the internal temperature of the

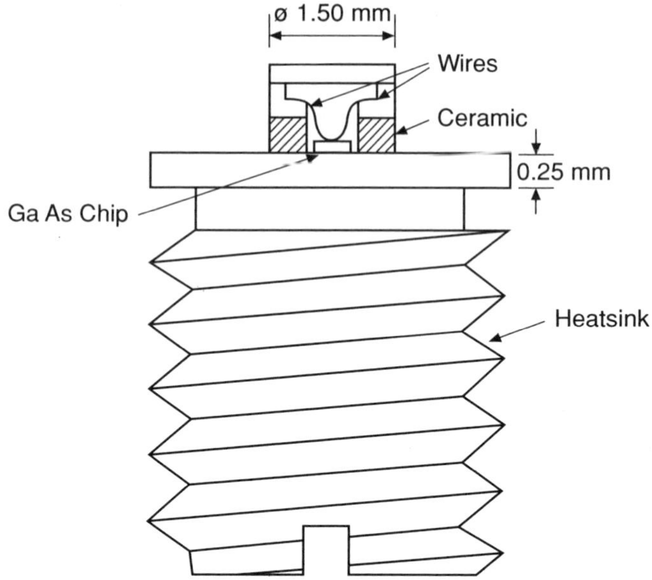

Figure 7.7 Typical packaged Gunn diode

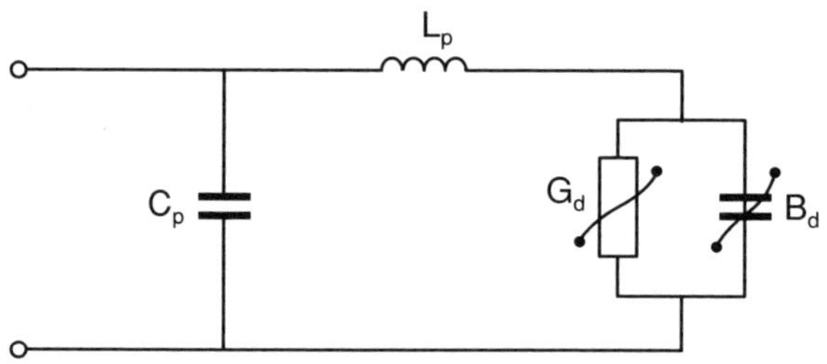

Figure 7.8 Simplified equivalent circuit of a packaged Gunn diode

semiconductor to a reliable operating value. Package parasitic effects produce significant impedance changes, so the package type must be carefully selected to fulfil both heatsinking and impedance transformation requirements. A typical packaged Gunn diode is shown in Figure 7.7. The semiconductor chip is inside, the top contact connections are made by wire bonding. Diode electrodes are electrically isolated by a cylinder of ceramic material.

A simplified equivalent circuit of the active device (Figure 7.8), is composed of a negative conductance G_d and a shunted capacitive susceptance B_d. Packaging effects are included as a series inductance L_p and a shunt capacitance C_p.

Typical values for a Gunn diode like the example shown in Figure 7.7 are:

$$L_p = 0.1 \text{ nH} \quad C_p = 0.2 \text{ pF}$$

Non-linear behaviour of the conductance G_d and susceptance B_d are useful to analyse or design the oscillator performance. Unfortunately these data are difficult to obtain from tests and most often the designer must undertake the design without it.

7.3.1 Design Technique Examples

There are three basic ways to build Gunn oscillators: (1) in a coaxial cavity; (2) in a waveguide; and (3) in a microstrip line. Figure 7.9 shows an example of each. In the coaxial cavity case, the oscillation frequency is determined by the cavity length l. It is approximately a half wavelength ($l \cong \lambda/2$). The Gunn diode is located approximately at a $\lambda/8$ distance from the end of the cavity. A frequency adjustment can be added by inserting a dielectric or metallic screw, located at a point with maximum electrical field to have maximum mechanical adjustment range. Power output can be adjusted by variation of the inductive coupling loop.

The main disadvantage of coaxial Gunn oscillators is their low quality factor, a typical value is 50, which produces poor frequency stability for most applications. Special care must be taken to avoid capacitive discontinuities in the junction diode to coaxial inner conductor. A suitable inner conductor diameter must be chosen. Other criteria to select the outer conductor diameter are: at oscillation frequency higher coaxial propagation modes must be under cut-off, the diameter ratio must be selected to have a maximum value of the cavity unloaded quality factor. Biasing from a DC power supply is obtained by electrical separation of the diode electrodes using a radial line of a quarter wavelength, so in RF it is equivalent to a short circuit at both coaxial ends. The dielectric separating both radial line conductors can be a very thin Mylar® sheet.

A possible waveguide assembly is shown in Figure 7.9(b). It represents a longitudinal view of a rectangular waveguide structure and it is an iris coupled oscillator. The oscillation frequency is approximately determined by a half wavelength distance from the iris to the diode position. Tuning is possible by a screw insertion, tuning ranges up to 30% are feasible. The Gunn diode is coupled to the cavity by a cylindrical post. It also allows introduction of the DC polarisation through a coaxial low pass filter. This filter has quarter wavelength sections alternating low and high characteristic impedances. Waveguide assemblies like this one can give high quality factors, around 1000, and hence high stability Gunn oscillators.

7.4 Transistor Oscillators

The basic principle of any electronic sinusoidal oscillator is in the use of the resonance phenomena. It is the case, for instance, that in the application of some initial energy to an ideal LC resonant circuit it produces an oscillating sinusoidal signal as the inductor and the capacitor periodically interchange this energy. Nevertheless, in the real case, due to the losses in the inductor and capacitor, the oscillating signal will decay exponentially with time. To maintain the oscillation, an active device must be used in order to compensate for this energy loss. For this purpose, transistor oscillators are widely used in radio frequency and microwave systems, offering good performance and high integration with other subsystems built using transistors. The choice of a particular transistor, as well as the resonant structure and the oscillator topology, will depend on the oscillating frequency to be obtained and on the particular application of the oscillator. In general, bipolar transistors are commonly used at radio frequencies and in the lower microwave bands (up to a few GHz). Above these frequencies, GaAs field-effect transistors are employed because of their ability to work at higher frequencies. Moreover, due to the development of device technology, HBTs (Heterojunction Bipolar Transistors) and HEMTs (High-Electron Mobility Transistors) are also available and have demonstrated good performance in microwave and millimetrewave oscillators.

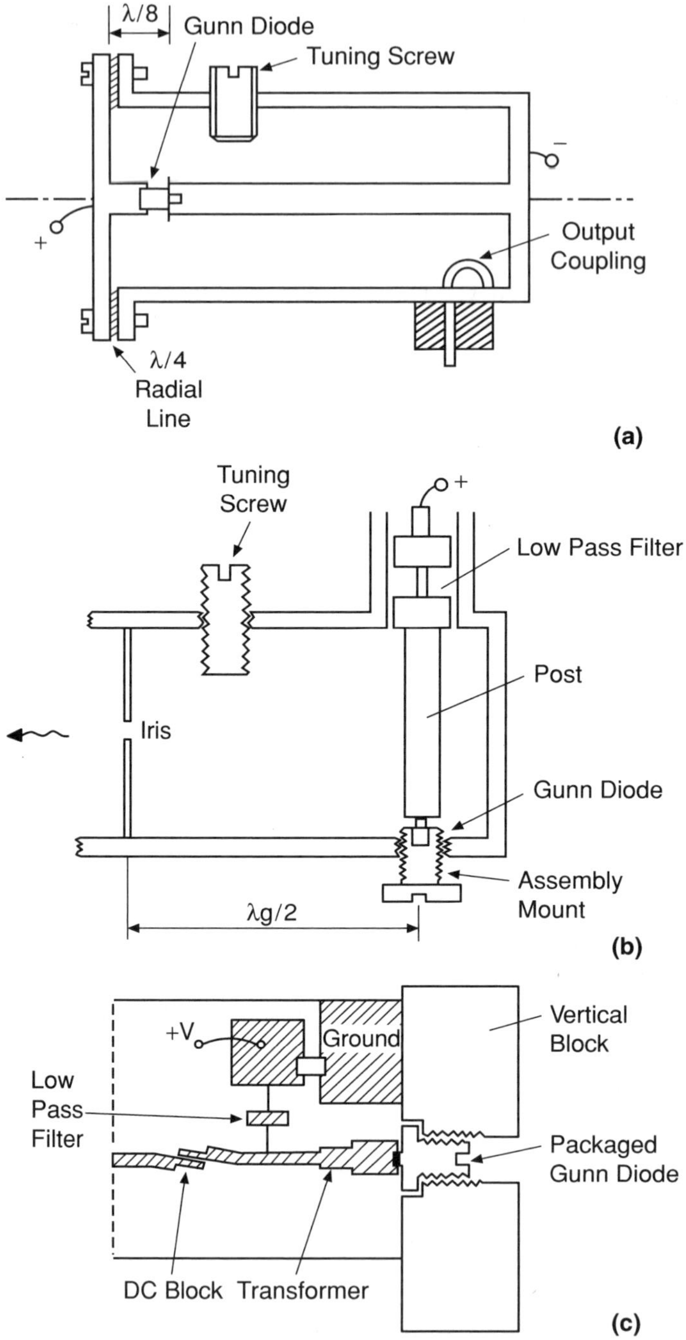

Figure 7.9 Gunn diode oscillators: (a) coaxial mount; (b) waveguide cavity; (c) microstrip

The design techniques of transistor oscillators are, in general, independent of the device technology. However, the achievable performances, in terms of output power or phase noise, for instance, can be strongly dependent on it. Nevertheless, these aspects are beyond the scope of this text, we restrict ourselves here to exploring the design fundamentals of transistor oscillators, showing the more common circuit topologies and illustrating them by means of some practical examples.

7.4.1 Design Fundamentals of Transistor Oscillators

A microwave transistor, as shown in Chapter 2, is a semiconductor three-terminal device, which acts, mainly, like a dependent current source. The performance of a given transistor is determined, avoiding second-order and high-frequency parasitic effects, from the characteristic curves of this current generator. In the case of bipolar transistors, the current source is controlled by the base current and the collector-emitter voltage, whereas in unipolar transistors, such as MESFET or HEMT, the gate-source and drain-source voltages are the controllers of the current source. Under a proper selection of the transistor bias conditions, that is, the DC operating point, the transistor is able to amplify a signal applied to its input terminal.

Generally speaking, a transistor oscillator can be seen as a transistor amplifier having positive feedback, allowing the growth of any starting oscillating signal in the circuit, coupled to a resonant circuit, which serves to select the frequency of the oscillation. In order to start up the oscillation, a signal containing a spectral component at the desired frequency must be available in the oscillator circuit. However, this initial signal is already present in all the electronic circuits, in the form of noise. White noise, for instance, is produced in any circuit component due to the fact of it being at a given temperature. This thermal noise is called white because the associated spectrum is frequency independent.

In the oscillator, the white noise within a frequency band is amplified and a portion of the amplified signal, at the frequency determined by the bandpass characteristic of the resonator circuit, is fed back to the input. When the loop gain is unity and the phase shift between the output and input signals is 360°, the oscillation will be possible, known as the Barkhausen condition for oscillation.

Under such conditions, the oscillating signal would continuously grow until it reached a level determined by any mechanism of signal limitation present in the circuit. These mechanisms can be related to the physical amplification limits of the transistor itself, produced when the oscillating signal reaches the maximum voltages and currents achievable in the characteristic curves. It can be also introduced externally by adding signal limitation circuitry into the oscillator loop. In this text, we will only consider the former kind of signal limitation, which is in fact common in general-purpose transistor oscillators. When the final stable oscillation condition is reached, the losses associated with the passive circuitry and the negative resistance provided by the active device, dependent on the signal magnitude, have equal contributions. Following this baseline, a schematic of the representation of a generic transistor oscillator is shown in Figure 7.10.

The design procedure of a transistor oscillator, as can be seen from the discussion above, is quite close to the amplifier design. Remember that, in a practical small-signal amplifier design, a reference terminal is usually grounded and specific impedances, calculated to satisfy the matching conditions at the frequency of the amplified signal, are connected at the input and output ports. Furthermore, amplifiers can also use negative feedback in order, for

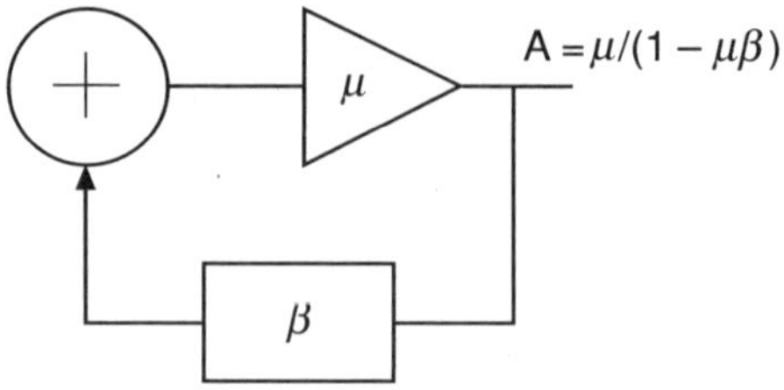

Figure 7.10 Feedback oscillator scheme

instance, to reach better stability or increase the amplification bandwidth. When compared to the design of a microwave small-signal amplifier, the main difference in the design of the transistor oscillator is in the need to create a positive feedback path and sufficient negative resistance to make possible the start and growth of the oscillating signal.

7.4.1.1 Achievement of the negative resistance

One of the key points in the design procedure of a transistor oscillator is how to obtain sufficient negative resistance, in the desired oscillating frequency range, to compensate for the losses of the passive circuitry of the oscillator, thus permitting the start of the oscillation. First of all, the achievement of the negative resistance requires biasing the transistor in the active or amplification region. Second, sufficient impedance should be connected to the transistor terminals in order to produce either series or parallel (or both) feedback. This involves the selection of a circuit topology and the calculation of the values of the different components. Finally, the resonator part must be connected or coupled somewhere into the circuit, in order to tune the desired oscillating frequency. In the following paragraphs, we will deal with different oscillator circuit topologies and choices for the implementation of the resonator part.

The generic topology of an oscillator using a transistor as the active device can be represented as shown in Figure 7.11. In this figure, the three-terminal device represents the

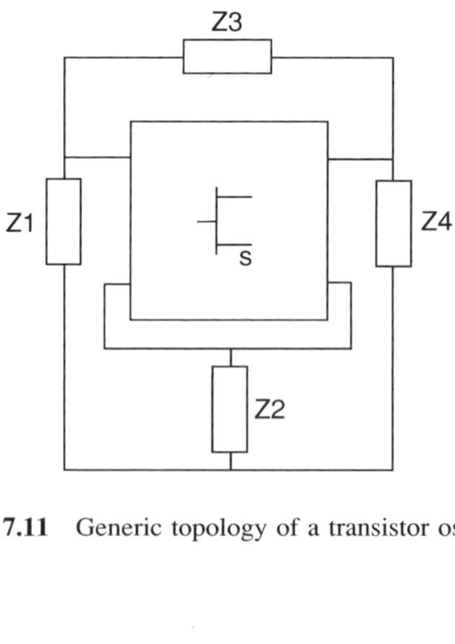

Figure 7.11 Generic topology of a transistor oscillator

transistor and Z1, Z2, Z3 and Z4 are, in general, complex impedances. From this circuit scheme, different configurations of transistor oscillators can be produced. Series feedback (Z3 = ∞), parallel feedback (Z2 = 0) or a combination of both (Z2 ≠ 0, Z3 ≠ 0) are possible in such a configuration.

7.4.1.2 Resonator circuits for transistor oscillators

Before exploring the usual resonator circuits employed in radio frequency and microwave oscillators, let us give a short review of some basic ideas about the frequency selective properties of these kind of circuits. Remember that the losses in a resonant circuit produce a transfer function in the frequency domain having a band pass characteristic instead of a flat frequency response. The band pass frequency is commonly defined as being the difference between the upper and lower frequencies at which the magnitude of the transfer function is 3 dB below the response in the band. The bandwidth is inversely related to the quality factor Q, which acts as a measure of the frequency selectivity of the resonant circuit. It can be written:

$$Q = \frac{\omega_0}{\omega_2 - \omega_1} \tag{7.14}$$

where ω_0 is the centre frequency and ω_2 and ω_1 are the upper and lower frequency bandpass limits. It is obvious that the higher Q is, the narrower the bandwidth and the higher the frequency selectivity of the circuit.

The Q factor can be determined as a function of the values of the inductor, capacitor and the resistive losses. The expressions, for series and parallel resonant circuits (as illustrated in Figure 7.12) are:

$$Q_S = \frac{\omega_0 L}{R} = \frac{1}{RC\omega_0} \tag{7.15}$$

$$Q_P = \frac{R}{\omega_0 L} = RC\omega_0 \tag{7.16}$$

Furthermore, it is important to remember that, if an external load is connected to the resonant circuit, the overall circuit Q (known as loaded Q or Q_L) must take into account the additional losses due to the external circuit. This is because, for a given resonator and load impedance, the Q_L of the circuit can be very different from the Q of the resonator circuit itself. The

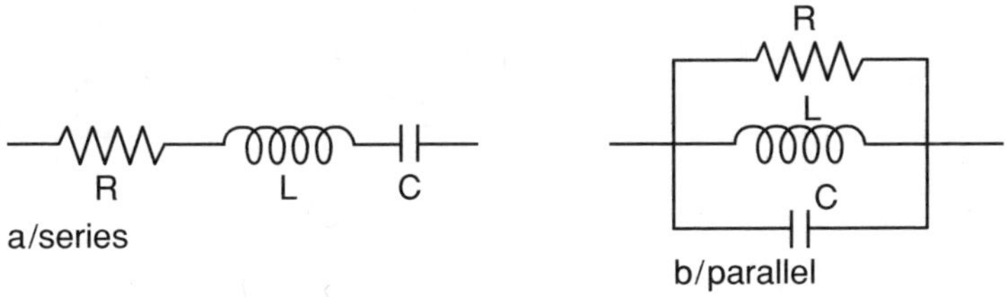

Figure 7.12 Series and parallel resonant circuits

expression relating the Q_L with the circuit Q and the Q associated to the external losses only, Q_{EXT}, is:

$$\frac{1}{Q_L} = \frac{1}{Q} + \frac{1}{Q_{EXT}} \tag{7.17}$$

Different options can be adopted to implement the passive part of the oscillator, i.e. the feedback and the biasing and tuning circuits. Capacitors and inductors are available to construct resonant circuits at high frequencies, having high enough quality-factor Q and stable behaviour with temperature drift. The resonant circuits implemented with inductors and capacitors are intensively applied in the design and construction of radio frequency oscillators for different electronic systems. This solution gives the so-called L-C oscillators in which the oscillating frequency is related to the resonant frequency of the L-C resonator.

Other alternatives have been adopted in the design of the resonator circuits for high frequency oscillators. The use of transmission lines instead of lumped L-C circuits is common in the design of microwave oscillators. This it is because of the increasing parasitic effects in the inductors and capacitors with the frequency, lowering the Q and often exhibiting undesired resonant frequencies under or near the desired oscillation frequency. The implementation of the passive circuits of the oscillator using transmission lines permits the easy and low cost integration of the oscillator. Usual resonators implemented with transmission lines can be built using coaxial line or microstrip line. Such examples are shown in Figure 7.13. For instance, cavity resonators made with open circuit or shorted coaxial lines are commercially available up to the lower microwave bands.

The microstrip line resonators can be made using shorted or open circuit transmission lines or other planar structures such as the ring resonator.

Very high-Q resonators at microwave frequencies can be obtained from circular or rectangular waveguides. Low-cost high-Q resonators can be obtained using dielectric resonators coupled to microstrip transmission lines, giving very good performance at microwave frequencies. Examples are shown in Figure 7.14. All these resonator types are usually employed for low noise fixed frequency oscillators but also can be used to implement mechanically or electronically tuned oscillators, with some extra circuitry added. Specific resonators for electronically tuned oscillators will be presented later in this chapter.

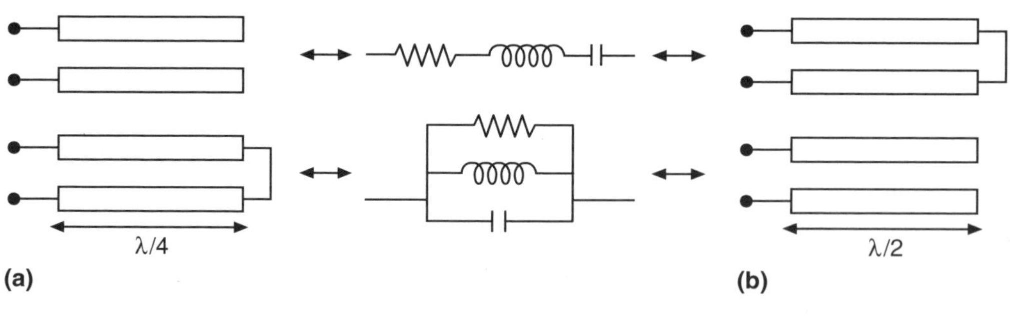

Figure 7.13 Resonators using transmission lines: (a) λ/4 resonators; (b) λ/2 resonator

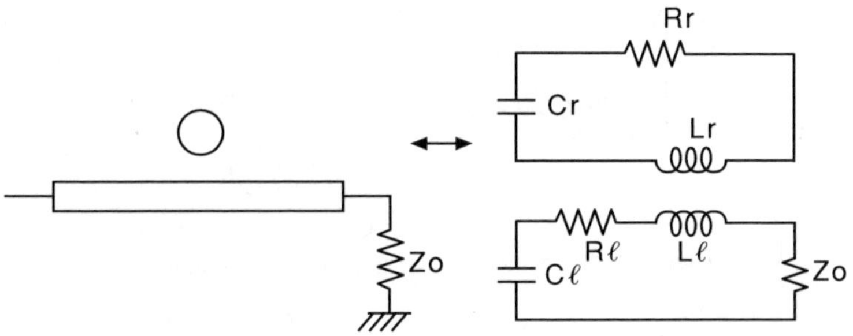

Figure 7.14 Schematic of resonator coupled to a microstrip line

7.4.2 Common Topologies of Transistor Oscillators

Some particular topologies of transistor L-C oscillators have been extensively employed in electronic systems for different applications. These are harmonic or sinusoidal oscillators, in which the oscillation frequency is fixed by the values of the inductors and capacitors in the circuit. There are other ways to produce an oscillation, as in the relaxation oscillators, associated with the relaxation time of the charging and decharging processes in a capacitor through a resistor. Nevertheless, we will focus our attention on harmonic oscillators because they provide the possibility of achieving higher frequencies (because no relaxation time is involved in the generation of the oscillation) with higher Q (because resistors, introducing extra losses, are not needed to produce the oscillation).

Among the several possibilities in implementing L-C oscillators, we will explore the Colpitts, Clapp and Hartley topologies. All these solid-sate oscillators are, in fact, an evolution from oscillators originally built using feedback vacuum-tube amplifiers. The differences between them arise from the system employed to feed back a part of the energy from the resonator and in the implementation of the resonator itself (see figures later in the chapter). For instance, both the Hartley and Colpitts structures use a parallel resonator but, whereas in the former an inductive voltage divider is used to feed back the signal, the Colpitts employs a capacitive voltage divider. In its turn, the Clapp oscillator also uses the capacitive feedback but introduces a series resonance in the inductive branch. Furthermore, different versions of the topologies above can be obtained depending on the AC grounded terminal considered. This means that we can talk about the grounded-emitter, grounded-collector or grounded-base versions (grounded-source, grounded-drain and grounded-gate when using MESFETs or HEMTs), even if, in a strict sense, the common-emitter, common-collector or common-base terms are used, as in the case of amplifiers, when an input signal is applied to the circuit. For such reasons, even if sometimes it is difficult to identify them, most of the radio frequency oscillators are, in fact, for instance, Colpitts or Hartley oscillators, or slight variations of these.

In the following sections, we will study in more detail these LC oscillators, but also some examples of microwave transistor oscillators, implemented using transmission lines instead of lumped elements, will be presented.

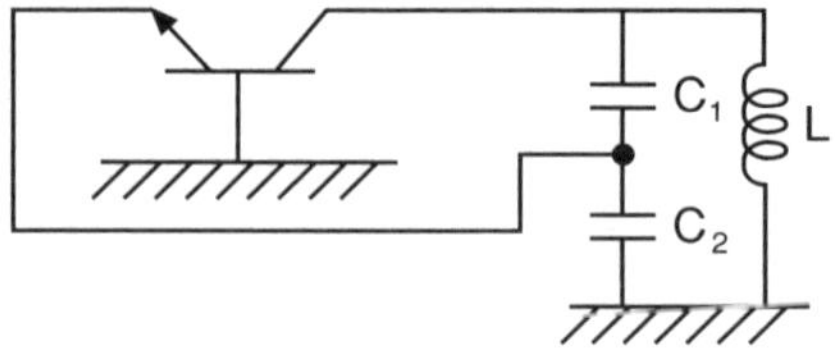

Figure 7.15 Schematic of the Colpitts oscillator

7.4.2.1 The Colpitts oscillator

Figure 7.15 shows the schematic of the Colpitts oscillator. The transistor is used in grounded-base configuration (Z1 = 0 in Figure 7.11) and acts simultaneously as the amplifier and signal limiter blocks in the oscillator loop. The frequency-tuning block is a parallel L-C resonator connected to the drain terminal (Z3 in Figure 7.11). The resonator capacitor is, in fact, substituted by a capacitive divider, which is used to feed back a part of the signal to the input of the gain block. It is possible to get a first estimate value of the oscillation frequency from the values of the resonator elements, as given in the following expression:

$$\omega_0 = \frac{1}{\sqrt{L\dfrac{C_1C_2}{C_1 + C_2}}} \tag{7.18}$$

The loaded Q in this oscillator configuration increases by increasing C2 and decreasing C1 in the capacitive divider, or by increasing both capacitor values and decreasing the inductor L. The output signal is usually taken from the resonator output, the source terminal of the transistor when using a MESFET device, because of the better spectral purity. A coupling technique or a buffer amplifier can be employed in order to extract the signal through the output load.

More accurate analytic calculations of the oscillation frequency for a practical example of a Colpitts oscillator implemented using a MESFET transistor and including two buffer amplifiers which are used to extract the oscillating signal for a 50 Ω load and a prescaler, for instance, are presented. The analytic expressions are obtained by using the equivalent circuit of the transistor and by applying the Barkhausen oscillation condition (voltage gain around the loop equal to unity, with null phase). In Figure 7.16, the configuration of the Colpitts oscillator used in the calculations is shown. The oscillation frequency can be written:

$$\omega_0 = \sqrt{\frac{C_{ds} + C_1 + C_2'}{L[(C_{buffer} + C_2') + (C_{ds} + C_1) + (C_{buffer}C_2')]}} \tag{7.19}$$

The constraint in the gain is:

$$\frac{1 + g_m R_{ds}}{(\omega^2 L)(g_m R_{ds} C_{buffer} + C_{buffer} + C_2')} = 1 \tag{7.20}$$

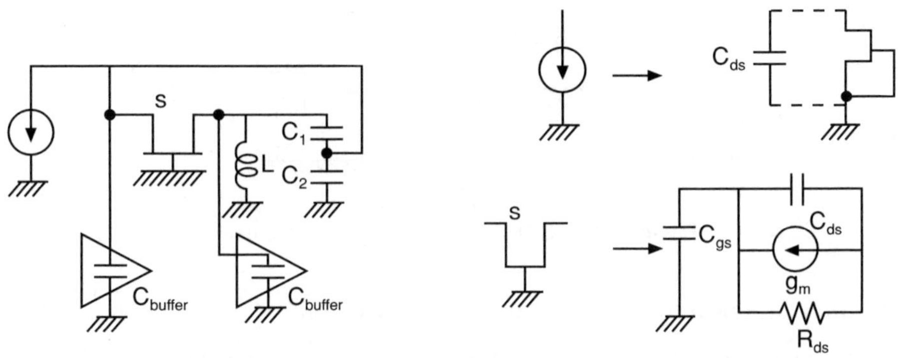

Figure 7.16 Schematic of the Colpitts oscillator used in the calculations

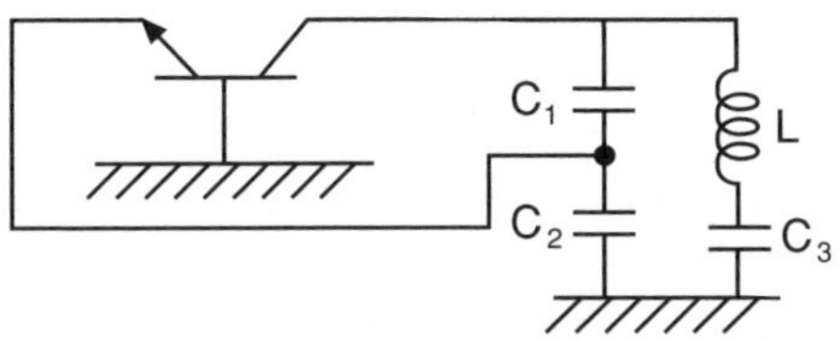

Figure 7.17 Schematic of the Clapp oscillator

In these expressions $C_2' = C_2 + C_{ds} + C_{gs} + C_{buffer}$ and C_{ds}, g_m and R_{ds} are the elements of the transistor small-signal model, C_{buffer} corresponds to the buffer input capacitor and C_1 and C_2 are the capacitors of the capacitive divider.

7.4.2.2 The Clapp oscillator

The Clapp structure is quite similar to that of the Colpitts oscillator. The difference, in this case, lies in the series connection of a capacitor to the resonator inductor. The Clapp schematic is shown in Figure 7.17. The advantage over the Colpitts oscillator is of the higher loaded Q obtained for a given inductor value. The increase of the inductor value and the decrease of the in-series capacitor result in the increase of the loaded Q. As in the Colpitts oscillator, the loaded Q can also be increased by reducing the C1 value and increasing C2.

7.4.2.3 The Hartley oscillator

Another example of an oscillator based on the use of an amplifier with feedback, is the Hartley topology. As well as in the Colpitts structure, a parallel resonator is employed as the frequency-tuning circuit, but the sample of the signal to be fed back is obtained from an inductive voltage divider, instead of using the capacitive divider as in the Colpitts oscillator. The Hartley scheme is shown in Figure 7.18.

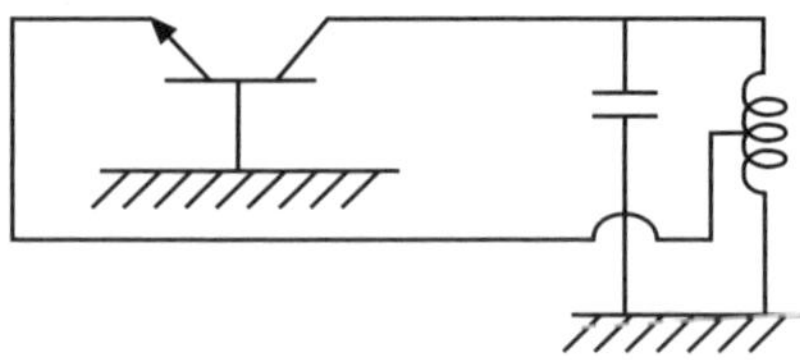

Figure 7.18 Schematic of the Hartley oscillator

7.4.2.4 Other practical topologies of transistor oscillators

Figure 7.11 summarised the generic topology of a transistor oscillator. A method for designing an oscillator from this scheme is based on the selection of appropriate impedance values for Z1, Z2, Z3 and Z4. By making use of a simplified small-signal equivalent circuit of the transistor, for the desired bias point in the amplification region of the transistor characteristics, simple expressions for the preliminary design of a transistor oscillator can be obtained for a particular configuration. For instance, a practical design is obtained by assuming only a capacitive series feedback (a capacitor in Z2, Z3 = ∞) and an inductor in Z1. In this case, the negative resistance seen towards the transistor from the Z4 port can be written:

$$R_o = \frac{-g_m}{C_{ds}\omega^2(C_{gs} + C_2 - L_1C_2C_{gs}\omega^2)} \tag{7.21}$$

Making use of the oscillation condition at the Z4 port (Zosc = Z4) and assuming that Z4 is a resistive impedance, the oscillation frequency can be calculated:

$$\omega_o = \frac{C_{ds} + C_{gs} + C_2}{L_1C_{gs}(C_2 + C_{ds})} \tag{7.22}$$

Other alternatives can be adopted by using an LC series resonant circuit connected to the gate, parallel feedback, etc. Some of these topologies can be, in fact, like the designs already shown above.

7.4.2.5 Microwave oscillators using distributed elements

In the previous subsections we have seen some examples of oscillators using transistors with positive feedback, in which the feedback and resonator circuitry was implemented with lumped elements such as inductors and capacitors. It is also possible to build transistor oscillators at higher frequencies having similar structures but using transmission lines instead of lumped elements.

Considering the transistor as a two-port device, two basic topologies, with external networks implemented with transmission lines in Π and *T* topologies, can be used to build shunt or series oscillators respectively (see Figure 7.19).

A practical solution in achieving high-Q microwave transistor oscillators is the use of dielectric resonators coupled to a transmission line into the oscillator circuit. These kinds of oscillators provide a very stable oscillating signal, with high spectral purity. A topology widely employed is by coupling the resonator to the transistor base or gate, as shown in

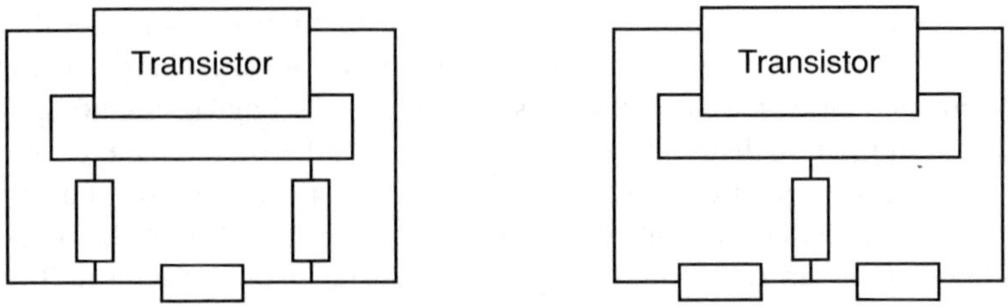

Figure 7.19 π and T topologies using transmission lines

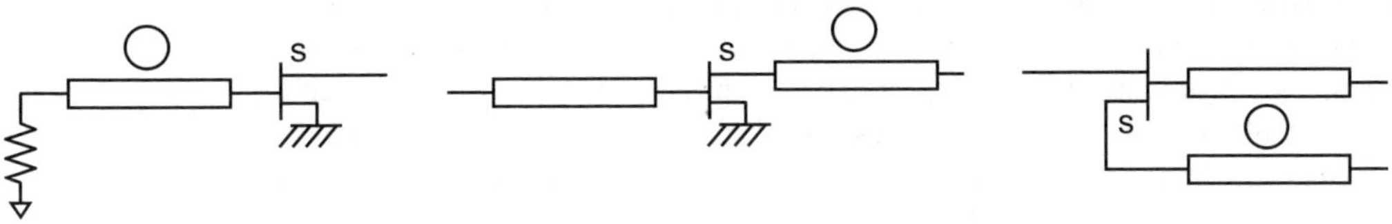

Figure 7.20 Dielectric resonator oscillators

Figure 7.20. Sometimes a microwave oscillator can use a dielectric resonator in order to decrease the phase noise, but also a voltage-controlled capacitor to achieve frequency-tuning in a narrow frequency band around the oscillation frequency set by the resonator.

7.4.3 Advanced CAD Techniques of Transistor Oscillators

The analytic approaches to the design of oscillators can be very useful to analyse a given oscillator topology and to get starting values for the main elements of the oscillator, offering quite good results at low frequencies. Nevertheless, because of the assumptions needed in order to obtain simple analytic expressions, the accuracy of the predicted performances will be limited. Some issues related to these limitations and design techniques that help to avoid these problems will now be introduced. It is important to note that the analytic designs in the previous paragraphs have been performed in the small-signal region and thus are only useful when the oscillating signal is starting, after the transistor biasing. The achievement of more accurate results of the negative resistance and the oscillating frequency in the small-signal region requires more complex calculations or simulations, taking into account the passive circuit parasitics, as well as a more realistic small-signal model of the transistor.

An approximate method for obtaining the maximum output power of a transistor oscillator was analytically derived by Johnson [2]. It makes use of the small-signal gain and the saturated power achievable using the same transistor in a large-signal amplifier design:

$$P_{MAX} = P_{SAT}\left(1 - \frac{1}{G_0} - \frac{\ln(G_0)}{G_0}\right) \tag{7.23}$$

where P_{SAT} is the saturated output power in the amplifier and G_0 is the small-signal common-source transducer gain of the amplifier.

Simulation techniques have an important role as a solution to minimise the production cost and time to market, providing useful information to aid the design of the oscillator. Commercial simulators use analysis techniques offering small-signal oscillation test or analysis in the non-linear region such as transient techniques or harmonic-balance methods. The small-signal test is performed by breaking the oscillator loop and by inserting an s-parameter port, giving the reflection coefficient of the oscillator at that port. Oscillation is possible if the magnitude of the given reflection coefficient is greater than one with null phase. These kinds of analysis serve to guarantee the starting condition of the oscillation. In order to be able to make a prediction of the final frequency and power of the oscillating signal, when the steady state of the oscillation is reached, the non-linear behaviour of the transistor must be known and a non-linear analysis technique must be used.

Harmonic balance is widely employed in the design of RF and microwave circuits because of the low computing time requirements compared to time-domain techniques, which need the calculation of the transient response in small time steps for the achievement of the steady-state response of the circuit. The harmonic-balance method divides the circuit in two parts: the passive part, described in the frequency domain, and the active part, described using expressions in the time domain of the different sources controlled by the corresponding command voltages or currents. The method converges producing the balance, at every working frequency and its harmonics, at the ports connecting both sub-circuits. In order to obtain the oscillation frequency, an auxiliary source, with varying frequency and amplitude, is used in several iterations until the oscillation is self-supported. The solution provides information about the power levels at the fundamental frequency and its harmonics, giving the steady state but not any guarantee of the start of the oscillation. In contrast, the time-domain methods offer the time evolution of the oscillating signal, from the starting point to the steady state.

Self-assessment Problems

7.1 Evaluate the oscillation frequency in a Clapp oscillator, as a function of the values of the passive components in the circuit.

7.2 Evaluate the oscillation frequency in a Hartley oscillator, as a function of the values of the passive components in the circuit.

7.3 A bipolar transistor is employed in the design of an oscillator, working at 1 GHz, whose schematic is in the Figure Q3. The scattering parameters for such transistor, corresponding to the working bias point and frequency are:

$$S11: (0.9, -100°),\ S12: (0.5, 31°),\ S21: (1.1, -50°),\ S22: (0.6, 150°)$$

Calculate the device impedance $Z_D = R_D + jX_D$ seen at the transistor base and the values of L and Z_O, corresponding to the load impedance Z_L, in order to obtain an oscillating signal at 1 GHz, considering that $\mathrm{Re}(Z_L) = R_D/3$ is the optimum value for obtaining the desired oscillation power using the given transistor.

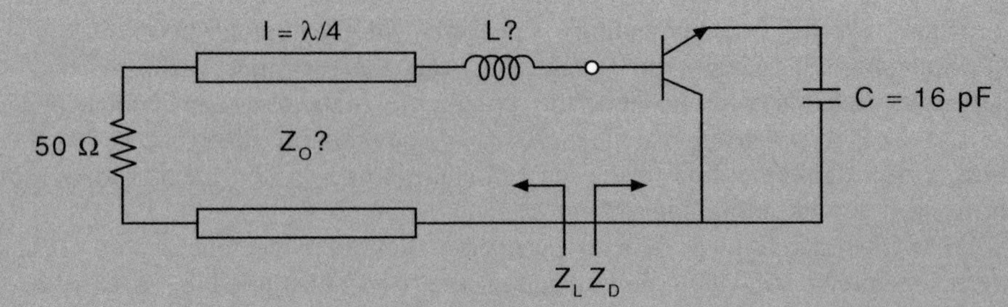

Figure Q3 Schematic of the oscillator, showing the device and load impedances Z_D and Z_L

7.5 Voltage-Controlled Oscillators

Oscillators having a frequency-tuning facility are in demand in high-frequency systems for consumer and professional applications. Because the oscillation frequency is basically determined by the resonant circuit, the frequency tuning can be achieved by varying its resonant frequency. The frequency tuning mechanism can be mechanical or electrical, depending on the type of resonator used. In the former case, a lumped-element resonator or a cavity resonator can be used in order to obtain a tuning bandwidth by, for instance, varying the value of a trimmer capacitor or by changing the length of the cavity, respectively. YIG resonators or varactor diodes could be the choice if an electrical frequency tuning is required.

The YIG (Yttrium Iron Garnet or ferrite sphere) or the varactor resonators give the possibility of wideband electrically tuneable oscillators. The YIG is a high-Q resonator in which the ferromagnetic resonance depends on the material, size and applied field. The frequency can be tuned over a wideband by varying the biasing of the magnetic field across the ferrite. The YIG oscillators are used in applications requiring very high quality tuneable oscillators, such as in microwave sources up to 60 GHz for instance.

When the circuit size or the cost is an important issue, if lower Q is acceptable, the choice is to use voltage-tuned varactors as frequency tuning elements in the microwave oscillators. It is for this reason that varactor-tuned oscillators are present in almost all the commercial applications and why, in this text, we will restrict ourselves to the study of varactor-tuned oscillators.

Varactors are special diodes showing a wide range of voltage-controlled variable capacitance. This is the key aspect for achieving broadband frequency-tuning capability. Another important issue is the minimisation of the varactor series resistance, in order to increase the Q factor. Silicon hyperabrupt varactors show capacitance ratios (Cmax/Cmin) greater than 12. Gallium-Arsenide varactors, based on Schottky diodes, are also employed, showing higher Q values because of the lower series resistance associated with the metal-semiconductor junction compared to the pn junction varactors realised in silicon.

7.5.1 Design Fundamentals of Varactor-Tuned Oscillators

As stated before, varactor-tuned oscillators use the voltage-capacity-varying characteristic of the varactor diodes, in order to obtain electronically frequency-tuning capability. In principle, by simple substitution of the capacitor in the resonator of any given L-C oscillator topology with a varactor, frequency tuning is possible with the varactor bias voltage. Nevertheless, at this point, it is important to remember that the parasitics in the varactor diode and other circuit elements will introduce correcting terms in the resonant frequency and also will decrease the Q factor of the resonant circuit. The deviation of the capacitor dependence with voltage from the expression of the ideal diode, produces a non-linear variation of the resonant frequency with the tuning voltage.

In order to illustrate how a varactor can serve as tuning element, let us consider a series resonant circuit in which the capacitor has been substituted by a varactor. The relationship between the varactor capacitance and the voltage V_a applied can be written, for $V_a < 0$ V:

$$C = \frac{C_{j0}}{\left(1 - \frac{V_a}{\phi_B}\right)^{\gamma}} \tag{7.24}$$

where C_{j0} depends on the varactor's physical and geometrical parameters, ϕ_B is the barrier potential of the junction and the γ value depends on the doping profile. Using this, the resonant frequency versus the applied voltage V_a, is:

$$f_R = \frac{\left(1 - \frac{V_a}{\phi}\right)^{\gamma/2}}{2\pi\sqrt{LC_{j0}}} \tag{7.25}$$

When $\gamma \cong 2$, as in the case of hyperabrupt junctions, a linear frequency variation with voltage is obtained.

In the next subsections we will explore different topologies of varactor-tuned oscillators. The choice of any one of them for a particular application will be made, considering the specifications in terms of tuning bandwidth, frequency-tuning linearity, phase-noise or cost, for instance.

7.5.2 Some Topologies of Varactor-Tuned Oscillators

The different varactor-tuned oscillator schemes, which will be presented here, are based on the fixed-frequency oscillators already shown before in this chapter.

7.5.2.1 VCO based on the Colpitts topology

A voltage-controlled oscillator design can be made taking as a starting point the Colpitts oscillator topology. The varactor diode can replace the capacitor C1 in the capacitive voltage divider. Of course, additional changes must be made to the bias circuitry in order to guarantee independent varactor and transistor bias (see Figure 7.21). By varying the varactor voltage, typical tuning bandwidth of more than one octave can be achieved with some degradation of the output power and phase noise within the tuning bandwidth.

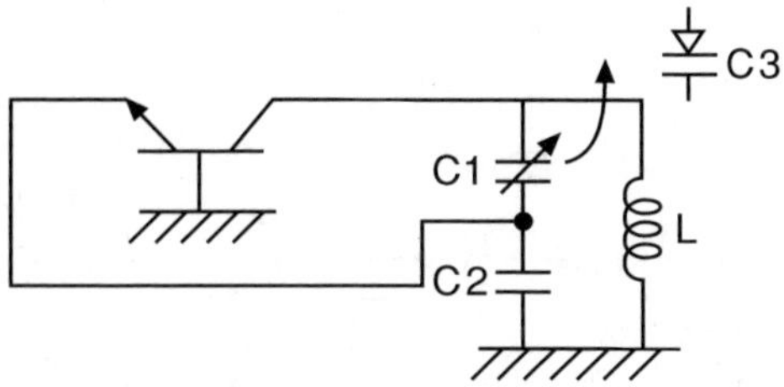

Figure 7.21 Schematic of the voltage-controlled oscillator based on the Colpitts topology

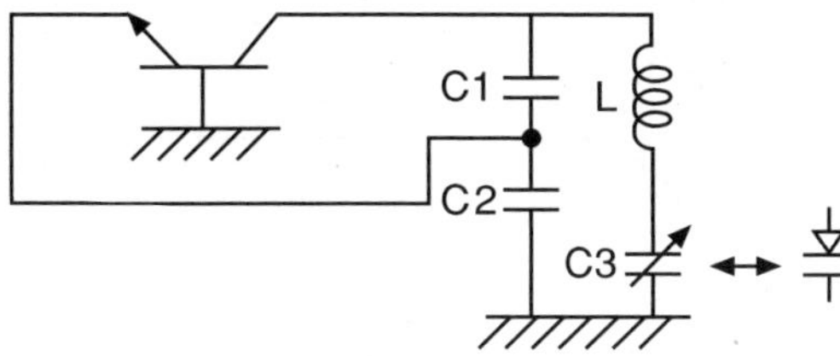

Figure 7.22 Schematic of the voltage-controlled oscillator based on the Clapp topology

7.5.2.2 VCO based on the Clapp topology

Because of the larger Q-factor achievable, the voltage-controlled oscillator based on the Clapp topology will have lower tuning bandwidth but better phase noise performance and more uniform output power over the bandwidth, compared to the VCOs based on the Colpitts configuration. The frequency in a Clapp oscillator is normally tuned by varying the capacitor in the series L-C resonator (see Figure 7.17). A practical configuration of a Clapp VCO using a bipolar transistor and a varactor is given in Figure 7.22. In this example, the load has been connected directly to the collector terminal.

7.5.2.3 Examples of practical topologies of microwave VCOs

The VCO explored up to this point uses well-known oscillator topologies, such as the Colpitts or Clapp oscillators. Now, we will focus our attention on other practical oscillator topologies, conceived specifically to achieve good frequency tuning performance using varactor diodes as tuning elements.

Let us consider the general topology of a microwave oscillator considered above (Figure 7.11). Simulations performed under small-signal conditions can serve to compare the tuning bandwidth obtained by inserting the varactor diode into different branches of this generic oscillator. The more common solutions place the varactor connected in the gate or source terminals when using a FET as the active device, or to the base or emitter if a bipolar transistor is used. The schematics of these VCOs are given in Figure 7.23 and Figure 7.24, respectively. In both cases, only series feedback as been considered ($Z3 = \infty$) and the load has been connected to the drain (or collector) terminal.

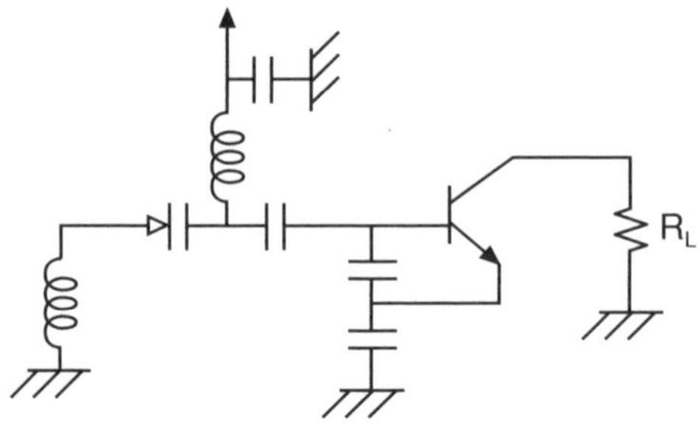

Figure 7.23 VCO with the varactor connected to the transistor's base

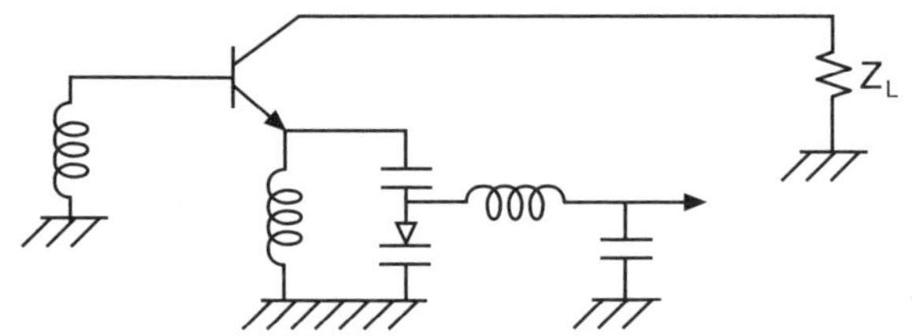

Figure 7.24 VCO with the varactor connected to the transistor's emitter

Self-assessment Problems

7.4 Calculate the tuning bandwidth of a Colpitts-based voltage-controlled oscillator, considering that Cmax/Cmin is 10 for the given varactor.

7.5 Calculate the tuning bandwidth of a Clapp-based voltage-controlled oscillator, considering that Cmax/Cmin is 10 for the given varactor.

7.6 Oscillator Characterisation and Testing

An oscillator is a system transforming part of the energy supplied in the DC region into a sinusoidal high-frequency signal. This signal is, ideally, a single spectral line with zero width and finite energy. In fact, real oscillators produce spectral lines of finite width, having random amplitude, frequency and phase fluctuations accompanied by harmonics and sub-harmonics of the spectral main line. These harmonic signals are generated by the non-linearities in the oscillator.

Summarising, the oscillator output presents deterministic signals, the carrier and its harmonics, characterised by its frequency and power, and the random components which modulate the carrier producing noise sidebands which, in their turn, are determined by phase noise measurements.

In this section, we will present a brief review of the principal characteristics determining the performance of a microwave oscillator and the basis of the experimental techniques used to measure them.

7.6.1 Frequency

The frequency of the oscillating signal, or carrier frequency, is normally determined by comparison with a high-precision oscillator, such as is performed by frequency counters. The precision sources can require the use of atomic-frequency standards, only available in very specialised laboratories, to obtain greater accuracy in the determination of the frequency.

In microwave frequency counters, the microwave carrier is translated to lower frequencies in order to be compared with the high-precision source. The spectrum analyser provides the complete spectrum of the oscillator output and can be employed to estimate the carrier frequency. The accuracy can typically be of 1 Hz in the case of frequency counters or, when using spectrum analysers, of the order of a few hundreds of Hz, for measurements in the range of radio frequencies, or about 1 kHz, when working in the microwave bands.

7.6.2 Output Power

The output power of an oscillator is considered to be the power that the oscillator delivers to the load at the fixed carrier frequency, avoiding the contributions of unwanted noise sidebands and harmonics. A first estimate of the carrier power can be obtained from a spectrum analyser. Nevertheless, these measurements do not offer, in general, amplitude errors better than 2 dB at microwave frequencies or 1.5 at radio frequencies. The accurate measurement of power is not a simple issue because it involves specific procedures and equipment.

Among the different methods of detecting power, the thermistor, thermocouple and detector diodes are those most commonly employed for measurements in normal applications. These sensors, accompanied with precision attenuators for the highest power levels, cover a wide dynamic range, from –70 dBm to a few watts. At microwave frequencies, absolute accuracy of a few tenths of dB are achievable using precision diode detectors.

7.6.3 Stability and Noise

The frequency instabilities in an oscillating signal can be related to long-term and short-term fluctuations. The long-term stability problems, expressed in parts per million of frequency change per unit time, are related to the ageing process in the materials employed and with environmental changes such as in the ambient temperature, pressure, etc. The short-term fluctuations are related to frequency deviations from the nominal frequency during periods less than a few seconds. These instabilities can be modelled as amplitude or frequency modulations produced by deterministic phenomena, such as the influence of external AC electromagnetic signals or other vibrating signals, or by random fluctuations related to internal or external noise sources.

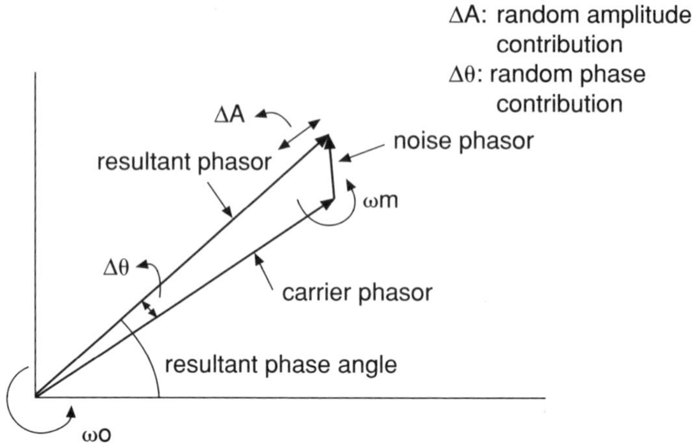

Figure 7.25 Vector diagram showing amplitude and phase noise in an oscillator

7.6.3.1 AM and PM noise

The oscillator output can be considered to be the result of the addition of a random noise component to the noiseless phasor representing the carrier (see Figure 7.25). When the noise component is parallel to the carrier, the vector sum only alters the amplitude of the oscillating signal, resulting in the amplitude modulation noise or AM noise. If the noise component is perpendicular to the carrier, it produces phase noise (PM noise). In general, the oscillator noise close to the carrier is mainly phase noise, with the amplitude noise level very low. This is because the limiting mechanism in the oscillator reduces the amplitude variations imposed on the carrier.

Phase noise is a very important feature of oscillator design and characterisation. The phase noise appears as sidebands with a continuous spectrum in a frequency range around the nominal oscillating frequency. In order to illustrate the harmful effect of the oscillator phase noise in different applications, various examples are given. For instance, in transmitters or receivers of phase-modulated data, the phase noise will degrade the data recovery, increasing the bit-error rate. In multichannel communication receivers the oscillator phase-noise sidebands will be transferred, radian per radian, to the channels translated to intermediate frequencies, producing problems of channel spacing between adjacent channels. In Doppler radars, which measure the shift in frequency between the reference and the returned signals, the phase noise limits the resolution and sensitivity.

If we observe the oscillating signal in a spectrum analyser, the phase noise appears as sidebands with a continuous spectrum around the carrier or fundamental frequency, with a spectral density decreasing with frequency offset. In fact, the carrier sideband spectral density observed in any spectrum analyser can be considered to be phase noise only if the amplitude modulation noise is negligible and the phase fluctuations are worse than that of the local oscillator of the spectrum analyser. In general, this is the case for a wide range of solid-state oscillators and it is the reason why the spectrum analyser can serve to determine the phase noise.

The phase noise at a given offset frequency from the carrier is measured as the value of the spectral density in a 1 Hz window. The usual units are dBc/Hz, representing the spectral density, referred to the power level at the carrier frequency. Since, when measuring using a spectrum analyser, the measured level will depend on the detector resolution bandwidth, the measured value must be normalised to 1 Hz in order to obtain a consistent estimate of the phase noise. In this way, the phase noise is given by:

$$L(fm) = \frac{N}{C} \tag{7.26}$$

where N is the noise power given in a 1 Hz bandwidth, after corrections, at fm Hz from the carrier and C is the carrier power. Making the assumption of weak phase fluctuations, from small angle modulation theory, $L(fm)$ can be related to the phase deviation and $S_\theta(fm)$, the power spectral density associated to the phase fluctuations, through:

$$L(fm) = \left[\frac{\theta_{peak}}{2}\right]^2 = \left[\frac{1.4\theta_{RMS}}{2}\right]^2 = \frac{S_\theta(fm)}{2} \tag{7.27}$$

By considering the oscillator as an amplifier with feedback (see Figure 7.10), Leeson [3] studied the phase noise in the oscillator. The phase noise defined in a 1 Hz bandwidth at a given frequency offset fm from the carrier, produces a phase deviation given by:

$$\Delta\theta_{peak} = \frac{RMS(V_{noise})}{RMS(V_{carrier})} \tag{7.28}$$

In the preceding expression, V_{noise} represents the noise voltage and $V_{carrier}$ is the carrier voltage magnitude. The values of such magnitudes can be related to the white noise in the amplifier and carrier power, giving:

$$\Delta\theta_{peak} = \sqrt{\frac{FkT}{P_{carrier}}} \tag{7.29}$$

where F represents the amplifier noise figure, k the Boltzmann constant and T the absolute temperature. The spectral density of phase noise at the input of the amplifier is:

$$S_{\theta IN}(fm) = \Delta\theta^2_{peak} = \frac{FkT}{P_{carrier}} \tag{7.30}$$

The phase noise at frequencies close to the carrier shows a $1/f$ spectral density that can be modelled as a phase modulator connected to the input of the amplifier. The spectral density at the input can be written:

$$S_{\theta IN}(fm) = \frac{FkT}{P_{carrier}}\left(1 + \frac{f_{carrier}}{f_m}\right) \tag{7.31}$$

Considering the bandpass characteristic describing the resonator response, the closed loop response of the complete oscillator gives the phase spectral density:

$$S_{\theta OUT}(fm) = S_{\theta IN}(fm)\left(1 + \frac{1}{f_m^2}\left(\frac{f_{carrier}}{2Q_L}\right)^2\right) \tag{7.32}$$

The overall phase noise is given by the expression:

$$L(fm) = \left(1 + \frac{1}{f_m^2}\left(\frac{f_{carrier}}{2Q_L}\right)^2\right)\frac{FkT}{2P_{carrier}}\left(1 + \frac{f_{carrier}}{f_m}\right) \tag{7.33}$$

This expression shows the different regions of the oscillator spectrum associated with the upconverted $1/f$ and white noise and the white noise floor. The important role of the quality factor Q in the minimisation of the oscillator phase noise can also be observed.

The residual phase modulation produced by the phase noise is an interesting item in systems using phase or frequency modulations. From the previous expression giving the phase noise, the residual phase noise modulation can be calculated as follows:

$$\Delta\theta = \sqrt{2\int_{fl}^{fh} L(fm)dfm} \tag{7.34}$$

where fh and fl are the highest and lowest frequency respectively of the frequency band in which the phase noise spectrum is considered.

7.6.4 Pulling and Pushing

The pulling of an oscillator gives a measure of the change of the oscillating frequency which is associated with variations of the value of the load impedance connected to the oscillator output. The pulling for a load is specified by the magnitude of the return loss at any angle. Pulling can be predicted by simulating the oscillator or by measuring the oscillation frequency while varying the impedance load. It is usual to analyse the frequency change for a constant magnitude of the load and to modify the phase from 0 to 2π radians. A return loss magnitude of 12 dB is commonly used to define the pulling. This can be achieved, in a 50 Ω system, by rotating a 29.9 or 83.5 Ω load connected to a variable length transmission line having 50 Ω characteristic impedance.

The supply voltages generally show drifts with time, but also with temperature and impedance fluctuations. The pushing of an oscillator determines how the oscillation frequency changes with deviations from the nominal value of the bias voltages. Pushing is determined by observing the variations in the oscillation frequency under different bias voltage conditions. The result is expressed in units of frequency variation per volt unit.

When the supply voltage is noisy or when it shows a periodic oscillation, modulation of the oscillator carrier may occur, resulting in a noticeable increase of the phase noise level. It is very important to achieve a good filtering of these undesired signals in the supply voltages.

Self-assessment Problems

7.6 The measured frequency sensitivity to variations in the supply voltage applied to a given oscillator is 100 KHz/volt. Calculate the resultant peak phase deviation and the modulation level for the resultant sidebands considering a supply ripple of 1 mV peak at 100 Hz.

7.7 Examining the expression given the oscillator's phase noise, explain the origin of the different regions that can be observed in the oscillator spectrum, of the slopes and the corner frequencies for low Q and high Q cases.

7.7 Microwave Phase Locked Oscillators

Microwave solid state oscillators have bounded frequency stability. It depends on the resonant structure or resonant network used as the load network connected to the active device. The achieved stability of free running microwave oscillators is worse than the stability of crystal quartz oscillators, normally used at much lower frequencies (in the range of MHz). To enhance the frequency stability of microwave oscillators, phase locked loop techniques are commonly used. In this approach a highly stable reference oscillator controls a microwave oscillator. In this situation the microwave oscillator is synchronised by the reference, this is known also as a synthesised microwave oscillator. When a microwave oscillator is controlled in such way improvements in long-term (such as temperature), and short-term frequency stability (phase noise behaviour) are obtained. Additional performance can be included in microwave phase locked oscillators such as the possibility of frequency selection by digital control. This capacity is very useful when adjusting local oscillator frequencies in channelised communication systems.

7.7.1 PLL Fundamentals

A phase locked loop (PLL) can obtain the synchronisation of a microwave oscillator by an external reference oscillator. The basic scheme of a PLL system is shown in Figure 7.26, easily identifiable as a system with feedback control. The microwave oscillator must be a VCO. A sample of the output signal is compared with the reference signal in a phase detector. The output signal from it, the error signal, is a low frequency signal, which, after amplification and filtering, is applied to the VCO tuning control. The frequency of the output signal (f_{out}) in the case of Figure 7.26 equals N times the reference frequency (f_o).

PLL systems are feedback control systems and their analysis can be made through their frequency response and loop stability. In a general analysis, a simplified system, as is shown in Figure 7.27, is enough. In this case the reference frequency is the same as the VCO frequency. This situation is not the normal situation in a microwave oscillator where the reference frequency is very low compared with the microwave frequency. The simple system of Figure 7.27 is very useful when one wants to know the behaviour of a general PLL system.

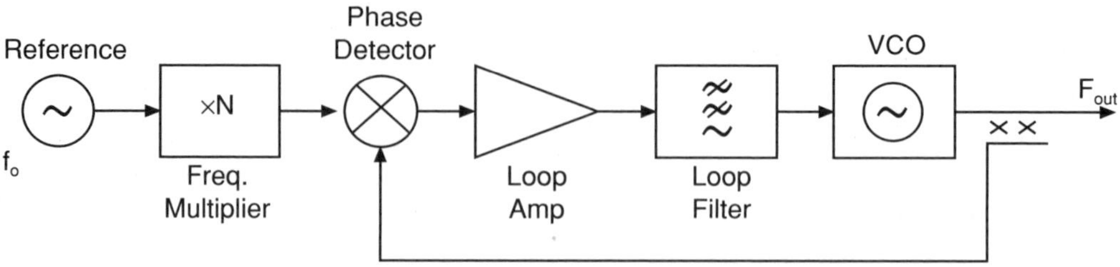

Figure 7.26 PLL system to stabilise a microwave VCO

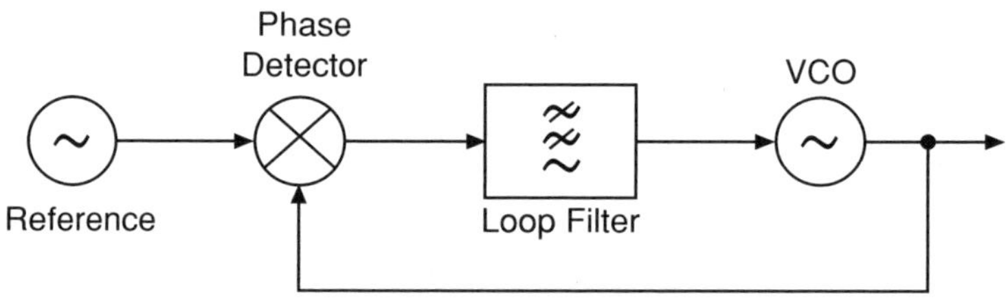

Figure 7.27 PLL simplified scheme

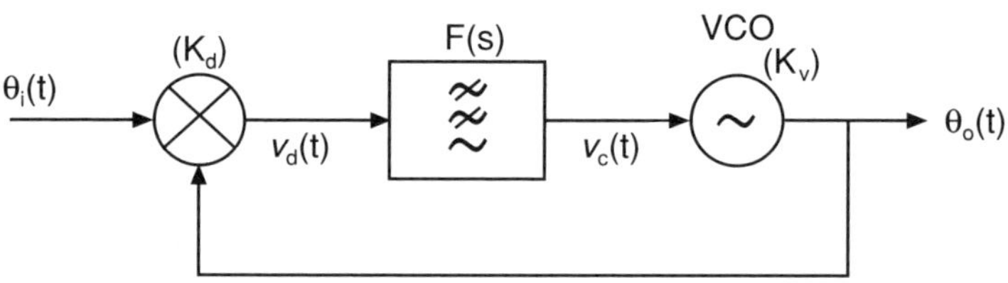

Figure 7.28 Signals under locking condition in a PLL

Assuming that a lock is established, the VCO frequency is identical to the reference frequency. The output signal and input signal are only different in phase. Using the notation shown in Figure 7.28 and assuming a sinusoidal response of the phase detector, the detector output signal is $v_d(t)$.

If the phase difference is low, a straight line can approximate the sine function. This approximation is called the linear model of a PLL system:

$$v_d(t) = K_d \sin[\theta_i(t) - \theta_o(t)] \cong K_d[\theta_i(t) - \theta_o(t)] \tag{7.35}$$

where $\theta_i(t)$ is the reference signal phase, $\theta_o(t)$ is the output signal phase, and K_d is the detector constant, measured in (Volt/rad).

The loop filter is characterised by its transfer function F(s), while the voltage controlled oscillator (VCO) can be characterised by its tuning slope, assuming a linear dependence of frequency versus voltage, K_v describes such a slope with dimensions of (rad/sec)/(Volt). As the frequency, f, is a measure of the phase change rate, the angular frequency ω (rad/sec) can be obtained as:

$$2\pi f = \omega = \frac{d\theta(t)}{dt} \tag{7.36}$$

and according to VCO tuning by the control voltage $v_c(t)$, it is also:

$$\omega = K_v v_c(t) = \frac{d\theta_0(t)}{dt} \tag{7.37}$$

Transferring time-based equations to the Laplace domain, the transform pair properties can be applied:

$$x(t) \Leftrightarrow X(s) \qquad \frac{dx(t)}{dt} \Leftrightarrow sX(s)$$

and Equation (7.37) is now:

$$s\theta_o(s) = K_v V_c(s) \quad \Rightarrow \quad \theta_o(s) = K_v \frac{V_c(s)}{s} \tag{7.38}$$

Using these equations a PLL system is represented by the block diagram shown in Figure 7.29, where Laplace transforms of signals are used.

From the block diagram several relations between signals are obtained:

$$V_d(s) = K_d[\theta_i(s) - \theta_o(s)]$$
$$V_c(s) = F(s)V_d(s)$$

By a combination of these equations, a new expression for output signal phase versus input signal phase is obtained:

$$\theta_o(s) = \theta_i(s)\frac{G(s)}{1 + G(s)} \tag{7.39}$$

$$G(s) = K_d F(s)\frac{K_v}{s} \tag{7.40}$$

In Equation (7.40) $G(s)$ is called the open loop gain because it takes into account the three basic cascaded elements of the loop: phase detector, loop filter and VCO before closing the loop. The open loop condition is very difficult to implement experimentally and is rarely done. The open loop scheme is still valid for the analysis, and being the open loop gain is a

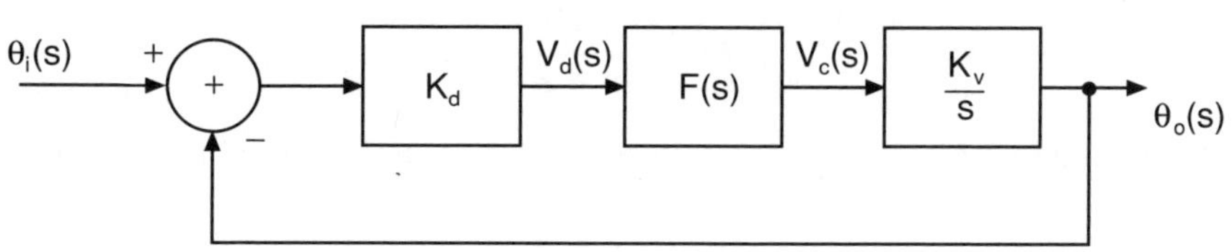

Figure 7.29 PLL block diagram using Laplace transform domain

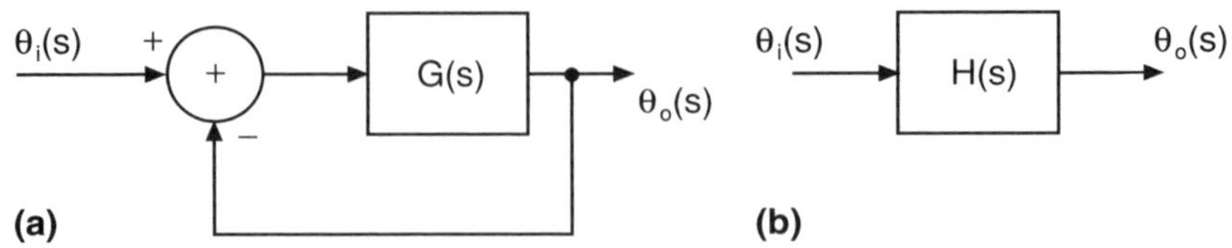

Figure 7.30 PLL schemes: (a) open loop gain G(s); (b) closed loop transfer function $H(s)$

very important parameter for stability analysis. From Equation (7.39) the closed loop transfer function $H(s)$ can be defined as:

$$H(s) = \frac{\theta_o(s)}{\theta_i(s)} = \frac{K_d K_v F(s)}{s + K_d K_v F(s)} \tag{7.41}$$

Two basic schemes are shown in Figure 7.30 to identify the open loop gain $G(s)$ and the transfer function $H(s)$ of a PLL system.

A new phase parameter, phase error, can be obtained as the difference between input and output signal phase:

$$\theta_e(s) = \theta_i(s) - \theta_o(s) = \theta_i(s)[1 - H(s)] \tag{7.42}$$

associated with an error transfer function defined as:

$$H(s) = \frac{\theta_e(s)}{\theta_i(s)} = 1 - H(s) = \frac{s}{s + K_d K_v F(s)} \tag{7.43}$$

From Equations (7.42) and (7.43) it can be concluded that both output and error signal phases are filtered versions of the input signal phase, and all the analysis tools available for linear systems can be applied here.

Using the usual nomenclature of control systems, a PLL system can be classified according to its transfer function order, that is the highest order of the s variable in the $H(s)$ denominator. Due to the integrator action of the VCO, the system order is always the filter order plus one. An additional classification of a PLL includes its type, that is the number of origin poles (integrators) in the open loop gain. In general:

$$G(s) = \frac{K_n(1 + a_1 s + \ldots + a_l s^l)}{s^n(1 + b_1 s + \ldots + b_p s^p)} \tag{7.44}$$

where n indicates the type and $n + p$ indicates the system order. As an example, a PLL containing a loop filter with

$$F(s) = \frac{1 + \tau_1 s}{1 + \tau_2 s} \tag{7.45}$$

is an order two and type one system. Likewise, a PLL with a loop filter given by:

$$F(s) = \frac{1 + \tau_2 s}{s\tau_1} \tag{7.46}$$

is an order two and type two system. Every PLL system is at least of order one and type one.

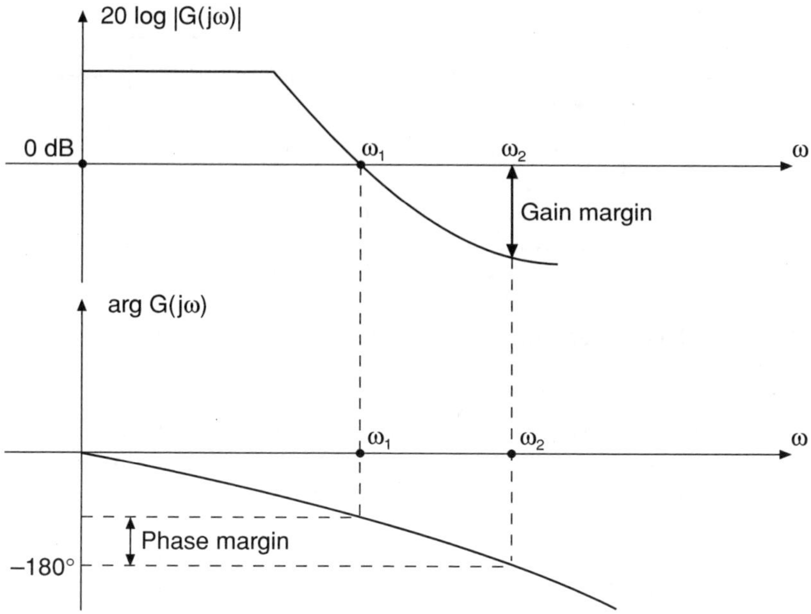

Figure 7.31 Bode plots for an unconditionally stable PLL showing phase and gain margins

7.7.2 PLL Stability

The PLL is a feedback system. It must be analysed for stability in the same way that a negative feedback oscillator is studied. A PLL is unstable if it can meet the oscillation condition. From the open loop gain $G(s)$ analysis, if there is a frequency ω (rad/sec) having $G(j\omega) = -1$, the system is unstable. A good method to check stability of control systems is the use of Bode plots, or amplitude and phase responses of the open loop gain. In Figure 7.31 Bode plots of an unconditionally stable system are shown. At frequency ω_1 crossing unity gain (0 dB) the phase must be higher than −180° for stability.

Phase and gain margins are indicative of the system's proximity to becoming unstable. The behaviour of a PLL system is not fully described by its stability condition. Other important aspects are related to the system response to transients and the filtering properties. A PLL must be able to achieve the locking condition in the turn on operation or when there is a change in the reference frequency. The most critical element in the system is the loop filter because it represents the major control that the designer can exercise over the PLL response. The filter must have a low pass response to include the DC component, and it must attenuate high frequency components. A good loop filter must attenuate as much as possible the reference frequency (see Figure 7.26), in order to avoid a phase modulation of the VCO at this frequency. On the other hand, the loop filter must have a wide enough bandwidth to allow the PLL to respond to high speed changes.

7.8 Subsystems for Microwave Phase Locked Oscillators (PLOs)

Once the basis of understanding the operation of a phase looked loop has been established, it is time to study each one of the ideal building blocks (see Figure 7.32). We will see how

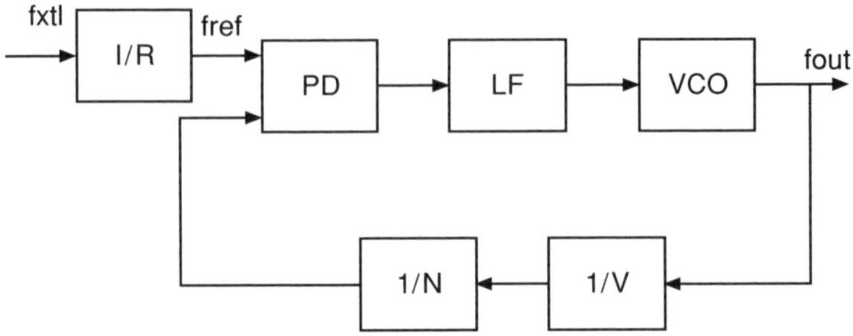

Figure 7.32 Block diagram of phase locked oscillator

the required transfer functions can be physically implemented, which are the degrees of freedom of the designer and what the criteria are to choose between different alternatives. Some examples of components extracted from commercial catalogues will help us to get closer the real world of the designer.

The VCO is characterised by a constant *Kvco* (MHz/volt), the phase detector (PD) by a constant *Kd* (volt/rad) and the dividers by their division ratio (R,N,V). The loop filter (LF) is defined by its poles and zeros.

7.8.1 Phase Detectors

The phase detector compares the feedback signal with the reference signal and provides a signal proportional to the phase difference between the two inputs. In other words, the phase detector is the comparator whose output is the error signal in the feedback loop. Nowadays, except for specific applications like low noise tracking filters or high frequency loops, a digital phase-frequency detector provides a better performance and a wider phase difference range than the classic mixer operating as a linear phase detector within a limited range of phase differences. It makes the phase-frequency detector an acquisition-aiding element.

The operation of a multiplier as phase detector (Figure 7.33) is explained below.

Let us consider two inputs with the same frequency and different time varying phases:

$$V_1(t) = A\cos(\omega_0 t + \phi_1(t))$$

$$V_2(t) = B\cos(\omega_0 t + \phi_2(t))$$

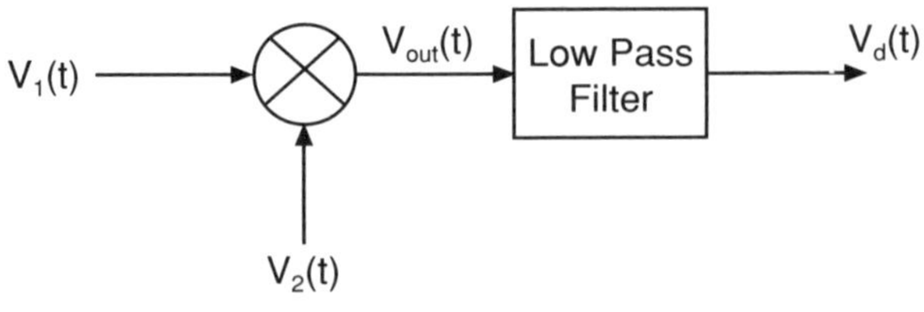

Figure 7.33 Multiplier as a phase detector

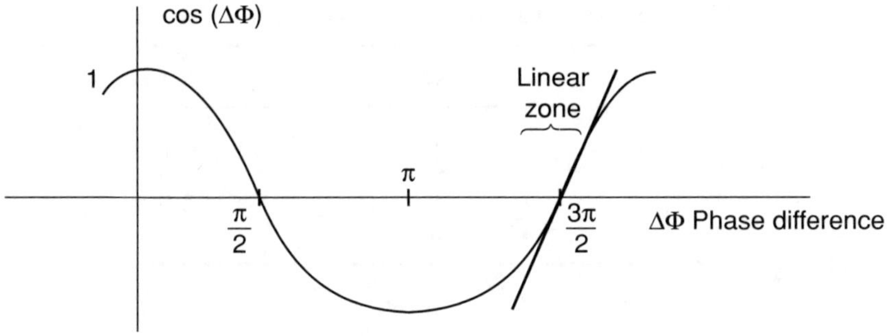

Figure 7.34 Approximate linear region of cosine function

The output of the multiplier with a multiplying factor K (physically implemented with a balanced mixer, for example) is:

$$V_{out}(t) = KV_1(t)V_2(t) = \frac{1}{2}ABK[\cos(2\omega_0 t + \phi_1(t) + \phi_2(t)) + \cos(\phi_1(t) - \phi_2(t))] \quad (7.47)$$

The first term of the sum (with the double frequency) is filtered. The second term is what matters because it contains the phase difference. If we plot the shape of a cosine function, two kinds of linear regions versus phase difference can be distinguished, around 90° and 270°. For example around 270° (Figure 7.34) we can substitute:

$$\cos(\phi_1(t) - \phi_2(t)) \approx \phi_1(t) - \phi_2(t)$$

The constant factor in V_{out} is defined as the constant of the detector.

$$K_d = \frac{1}{2}ABK \quad (7.48)$$

It seems logical that the output must be independent of the input amplitudes. It means that A and B must be as large as possible to operate the device in saturation mode.

7.8.1.1 Exclusive-OR gate

An exclusive OR logic gate can operate as a sequential phase detector, providing at the output an average value proportional to the phase difference between the inputs.

In Figure 7.35 the scheme of the exclusive-OR gate is shown. X1 and X2 are the input signals whose phase difference is evaluated. Y is the output of the logic gate. See the truth table (Table 7.1). The time-averaged value is proportional to the phase difference between X1 and X2.

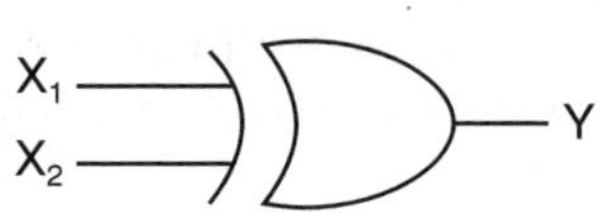

Figure 7.35 Exclusive-OR gate

Table 7.1 Truth table of exclusive-OR gate

X1	X2	Y
0	1	1
0	0	0
1	1	0
1	0	1

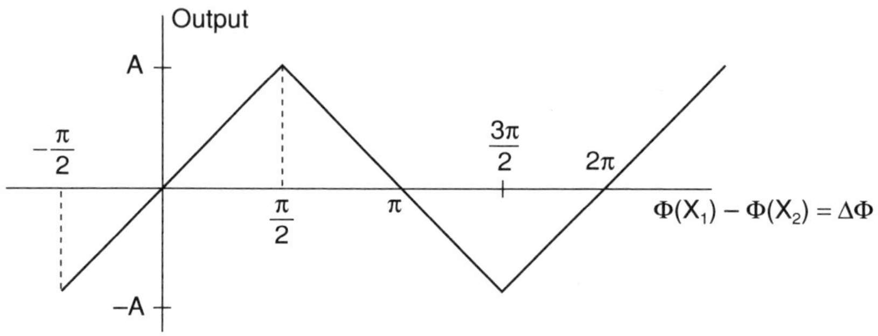

Figure 7.36 Transfer curve of phase detector

The transfer curve of this phase detector is plotted in the Figure 7.36. A is the maximum amplitude of the output signal.

The timing diagram for several phase differences (0, $\pi/2$ and π) is shown in Figure 7.37.

The slope around each linear zone fixes the phase detector constant:

$$K_d = \frac{2A}{\pi} \tag{7.49}$$

7.8.1.2 Phase-frequency detectors

These can operate as frequency discriminators for large initial errors and then as coherent phase detectors, once the system is within the range of lock. These two modes of operation require some memory. Therefore, the phase-frequency detector is usually a digital circuit containing flip-flops (Figure 7.38). The inputs are the reference and the signal coming from the VCO, divided or not, depending on application. The outputs are usually called Up and Down.

The response of a digital phase-frequency detector is shown in Figure 7.39. The averaged output amplitude of each port (UP and DOWN) and the difference are plotted versus phase difference of the inputs.

The output of the circuit consists of a train of pulses whose duty cycle is proportional to the phase difference between the two inputs. If this phase difference is more than 2π, the polarity of the output signal depends on the frequency relation between the inputs. The pulses are integrated later in the loop filter, providing a voltage whose variation corrects the frequency of the voltage controlled oscillator.

The phase-frequency detector outputs (UP and DOWN) are active with a low level. If the reference signal has a frequency higher than the VCO signal or if it leads in phase, then the

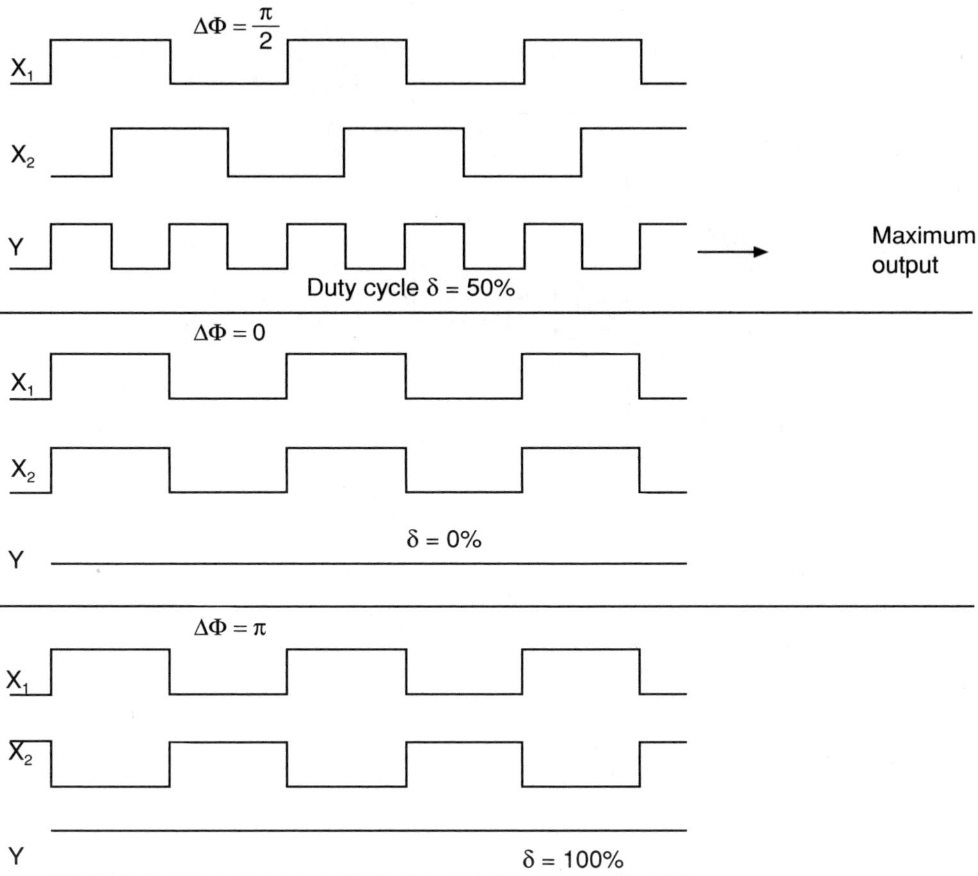

Figure 7.37 Timing diagram of phase detector

output UP consists of low level pulses and the output Down stays high. On the other hand if the reference is delayed or the VCO has a higher frequency, then UP stays high and DOWN pulses low. An example of timing diagrams is shown in Figure 7.40.

Four cases are considered: equal frequencies and reference leading or lagging, reference slightly larger and slightly smaller than VCO. In normal operation there is a small oscillation between the Reference and VCO signal phases and the VCO is continuously correcting its frequency.

In Figure 7.41 the transfer characteristics (output voltage versus phase difference at the input) of four phase detectors are summarised. The phase detectors are: a four-quadrant multiplier (mixer), an exclusive or gate, an edge triggered master–slave flip flop and a digital phase frequency detector.

A special type of phase detector is the Sampled Phase Detector, based on the multiplication of the reference before comparing. It means that the comparison is performed at high frequencies. We will deal with them in the section on multipliers.

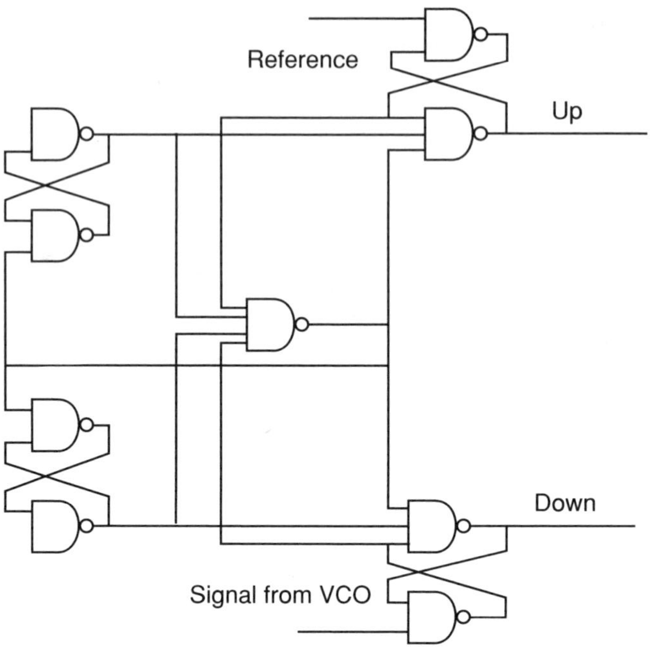

Figure 7.38 Digital phase-frequency detector

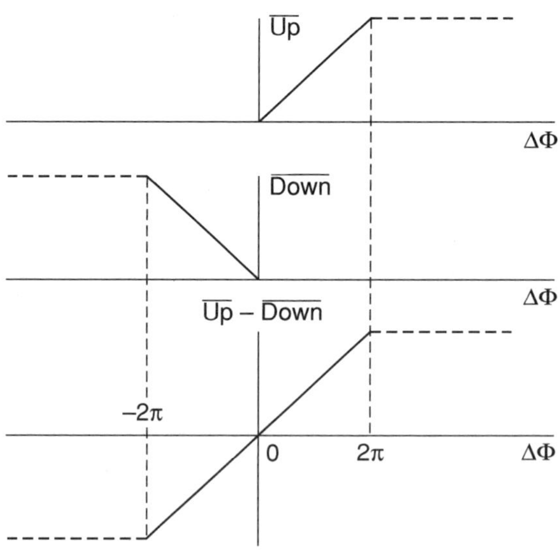

Figure 7.39 Response of digital phase-frequency detector

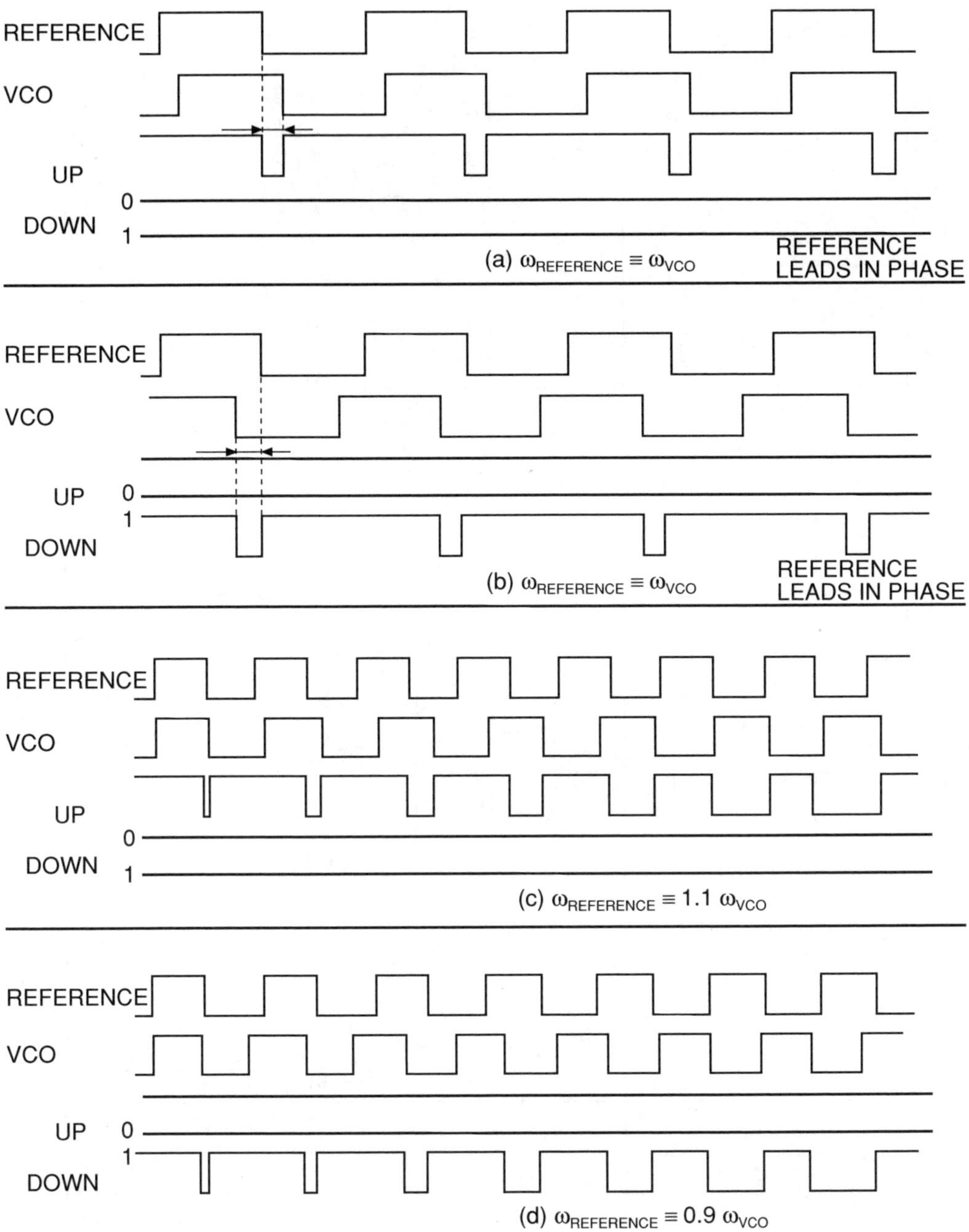

Figure 7.40 Timing diagrams of digital phase-frequency detector

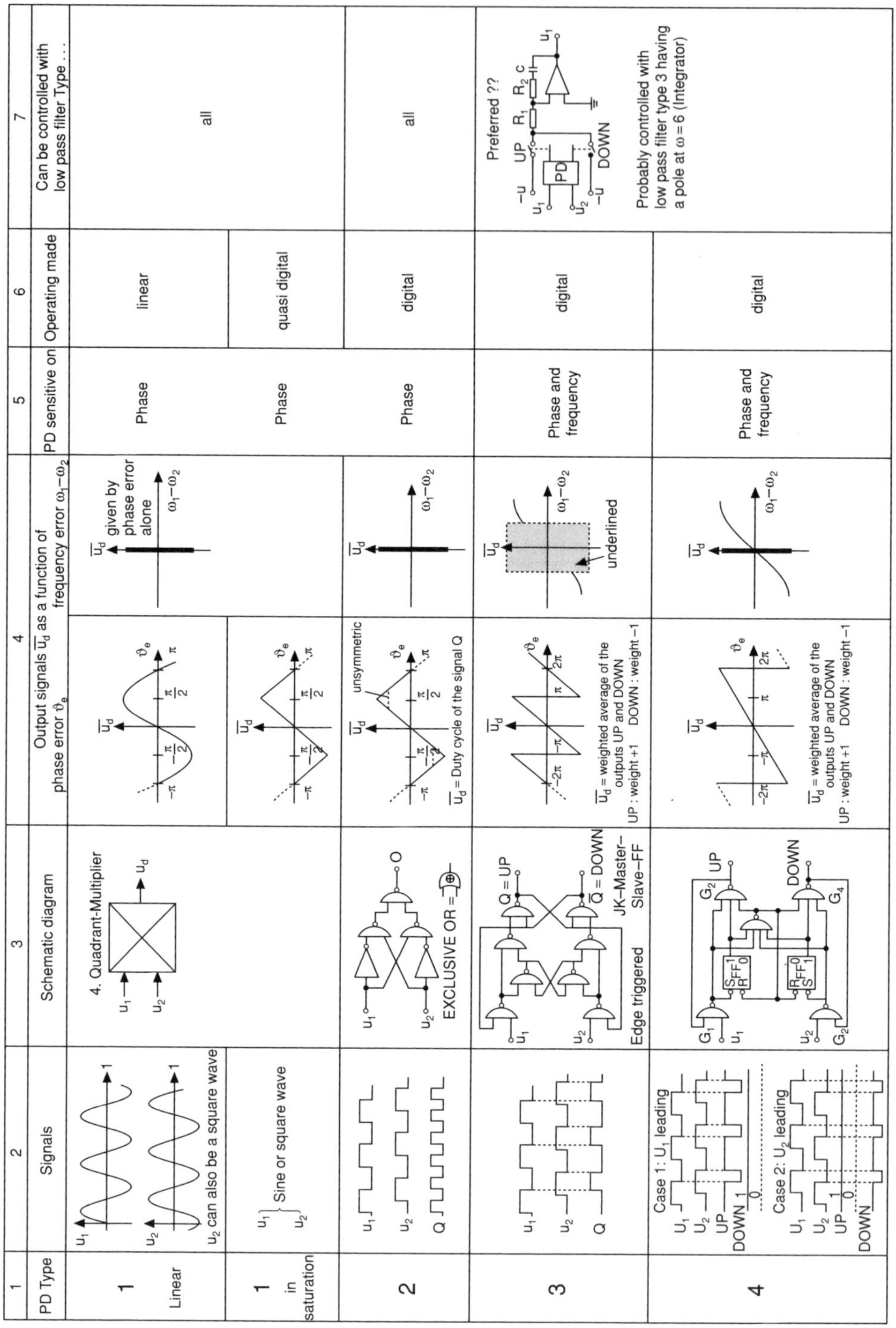

Figure 7.41 Transfer characteristics of four phase detectors

7.8.2 Loop Filters

The loop filter is the component of the PLL where the designer has the widest margin, so we can consider it as the key component. The function of the loop filter is to reject unwanted spurious signals, which could modulate the oscillator. The correcting signal is formed by several components. The DC and low frequency components are responsible for the correction of the frequency-phase. Spurious components at the frequency of comparison or the reference frequency modulate the oscillator and create unwanted side bands. If we talk about digital phase detectors, the function of the loop filter is to integrate the output pulses of the phase detector to control the frequency of the oscillator and its deviations.

The bandwidth of the loop filter also defines the bandwidth of phase noise improvement at the output of the PLO, compared with the oscillator alone (free running).

The speed of the system response is inversely proportional to the bandwidth of the loop, so a trade-off is necessary to choose the final value of this parameter.

As an example of a common loop filter, which is recommended by most of the synthesizer ICs' manufacturers, we will study in detail the 3rd order, type 2. This means the transfer function in open loop has two poles at the origin (one due to the VCO and the other to the filter) and the total number of poles is three. Why do we study such a complex filter and not a simpler one, for example, with one pole? The criteria to choose the most adequate filter are based on what error is acceptable for a fixed input signal (signal applied at the reference port). The most common test signals considered at the input are a step in phase, a frequency step, and a sloped frequency (parabolic phase).

The system must control the frequency of the oscillator according to the reference. It means the steady state error (output phase minus input phase with time towards infinity) must be zero for a phase step input and a frequency step input and bounded for a sloped frequency. If an error is computed for each of the inputs (see PLL fundamentals) we can verify that the minimum requirements to achieve it are: type 2 and 2nd order as can be seen in Figure 7.42. For practical reasons it is recommended to add an additional pole in the open loop which makes the order 3.

Next, the transfer function and the time constants of the loop filter will be deduced for a PLL with a division ratio of N, a VCO constant $Kvco$ and a phase detector constant Kd (see block diagram in Figure 7.43).

The VCO adds a pole at the origin. It means that the filter must have two poles, one of them at the origin:

$$F(s) = \frac{1 + sT_2}{sT_1(1 + sT_3)} \tag{7.50}$$

With the circuit topology shown in Figure 7.44 this transfer function can be implemented.

V_i: Input voltage

V_0: Output voltage

$$V_i = R_1 I_1$$
$$V_0 = -(Z_1 + Z_2)I_1$$

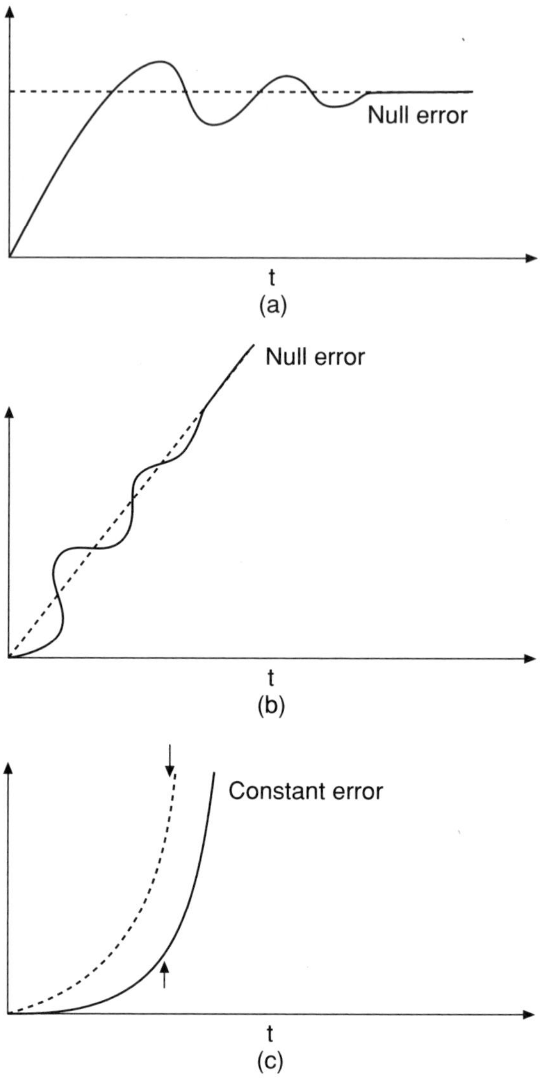

Figure 7.42 PLL error responses

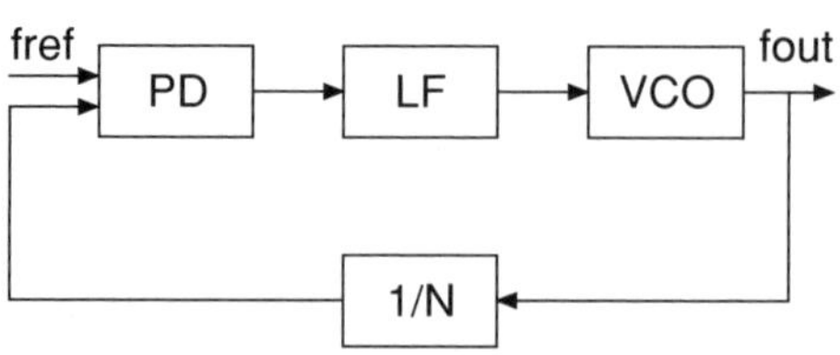

Figure 7.43 PLL with frequency division ratio of *N*

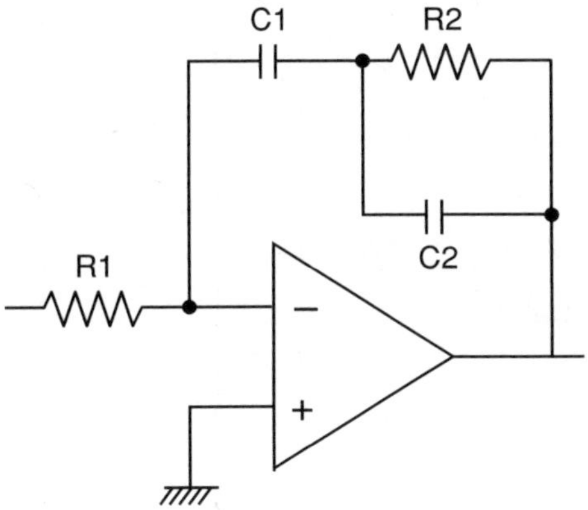

Figure 7.44 Circuit diagram of 2-pole filter

The transfer function is the relation between V_0/V_i.

$$\frac{V_o}{V_i} = \frac{-(Z_1 + Z_2)}{R_1} \tag{7.51}$$

Using Laplace expressions for the impedances, the transfer function becomes:

$$Z_1(s) + Z_2(s) = \frac{sR_2C_2 + 1 + R_2sC_1}{sC_1(sR_2C_2 + 1)} \tag{7.52}$$

$$F(s) = -\frac{1}{sR_1C_1}\left[\frac{1 + sR_2C_2 + sR_2C_1}{sR_2C_2 + 1}\right] \tag{7.53}$$

Compared with the generic expression, time constants can be obtained:

$$T_1 = R_1C_1$$
$$T_2 = R_2(C_1 + C_2)$$
$$T_3 = R_2C_2$$

Self-assessment Problems

7.8 Verify that the same transfer function can be implemented with the topology shown in Figure 7.45. Verify the following set of equations which relates the new time constants with the old ones.

$$T_a = 2T_3 \quad T_a = R_aC_a$$
$$T_b = T_{1/2} \quad T_b = R_aC_b$$
$$T_c = T_3 \quad T_c = R_bC_b$$

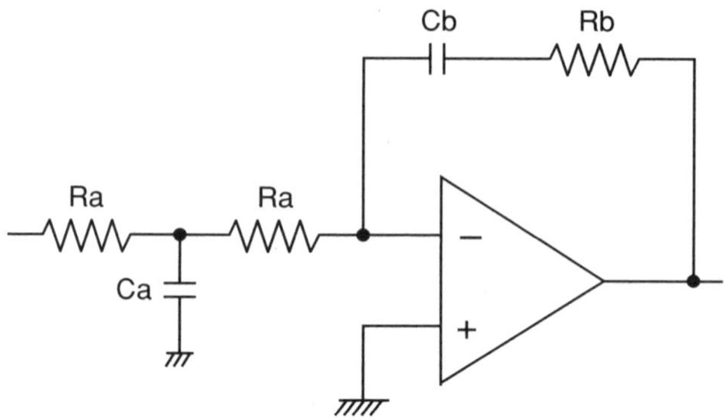

Figure 7.45 Circuit of filter for exercise

The design procedure starts with the specification of the desired phase margin (m) and the bandwidth, like in any other servomechanism. The Laplace variable s is substituted by $j\omega$ in the transfer function to obtain the frequency ω_0, which yields the phase margin. It can be verified that in a first approach valid only in this type of filter, this value can be the desired bandwidth.

The procedure continues by calculating the phase of the transfer function. Then 180 degrees are subtracted. This difference must be the desired phase margin and must reach a maximum at ω_0, when the module of the open loop gain is 1 (it is just the definition of the phase margin of a feedback system). The phase difference is derived and equalised to zero to find the maximum.

The open loop transfer function is:

$$G(j\omega)H(j\omega) = -\frac{K_d K_{vco}}{N\omega^2 T_1}\left[\frac{1 + j\omega T_2}{1 + j\omega T_3}\right] \tag{7.54}$$

The phase difference with 180 degrees is:

$$\phi = \text{Arctan}(\omega T_2) - \text{Arctan}(\omega T_3) + \pi - \pi$$

The derivative must be equal to zero:

$$\frac{d\phi}{d\omega} = \frac{T_2}{1 + (\omega_o T_2)2} - \frac{T_3}{1 + (\omega_o T_3)2} = 0 \tag{7.55}$$

The frequency for the maximum phase difference ω_0 is:

$$\omega_o(\phi_{\max}) = \frac{1}{\sqrt{T_2 T_3}} \tag{7.56}$$

A relation between T_2 and T_3 is obtained:

$$\text{Tan}(\phi_{\max}) = \frac{T_2 - T_3}{2\sqrt{T_2 T_3}} \tag{7.57}$$

To calculate T_1 the condition of the open loop transfer function equal to 1 is applied:

$$|G(j\omega_0)H(j\omega_0) = 1| \tag{7.58}$$

Finally, if the desired margin is m, the expressions of the time constants are the following:

$$T_3 = \frac{\sec(m) - \tan(m)}{\omega_0}$$
$$T_1 = \frac{K_d K_{vco}}{N\omega_o^2}\left[\frac{1 + (\omega_o T_2)^2}{1 + (\omega_o T_3)^2}\right]^{1/2} \tag{7.59}$$
$$T_2 = \frac{1}{\omega_o^2 T_3}$$

Summarising, once the desired phase margin and the loop bandwidth have been specified for a given loop, with some known values of K_{vco}, K_d and N, we can approximate the frequency with open loop gain of unity and ω_o for the loop bandwidth (valid only for the 3rd order, type 2) and then obtain the time constants. Finally, a realistic set of resistance and capacitance values can be chosen for the implementation of the real filter (for example, typical values of resistance can be in the range from 10 Ohm to 100 kOhm, and typical values of capacitance can be between 1 pF and 100 nF).

7.8.3 Mixers and Harmonic Mixers

A mixer can be employed in a more complex PLL in its typical role of frequency conversion (see Chapter 5). This allows us to down-convert a high frequency to a lower value, suitable to be applied to a frequency divider, for example, or combine the outputs of two PLLs to achieve a small frequency step with a high frequency oscillator. If two signals with the same frequency, but a slight phase difference are applied as local oscillator and radio-frequency signals, an output signal basically proportional to the phase difference is obtained.

Based on this principle a common mixer can be used as phase comparator between two signals. These two signals can be:

1. Main oscillator and reference without division (this is the case when the reference is a remote carrier which comes from a transmitter and we try to track in a receiver)
2. Main oscillator divided by some integer N and reference oscillator divided by some integer R.
3. Main oscillator and a harmonic of the reference. The harmonic can be generated by the non-linearity of the mixer. In this case we talk about a harmonic mixer.

A harmonic mixer scheme is shown in the Figure 7.46. The frequency of the local oscillator is 1816 MHz. The frequency of the VCO is 14168 MHz and is mixed with the 8th harmonic of 1816 MHz, generated in the same mixer. This yields an intermediate frequency of 360 MHz (14168-8 X 1816) which is then divided by 45 ($Nt = 45$) yielding a frequency of comparison equal to 8 MHz.

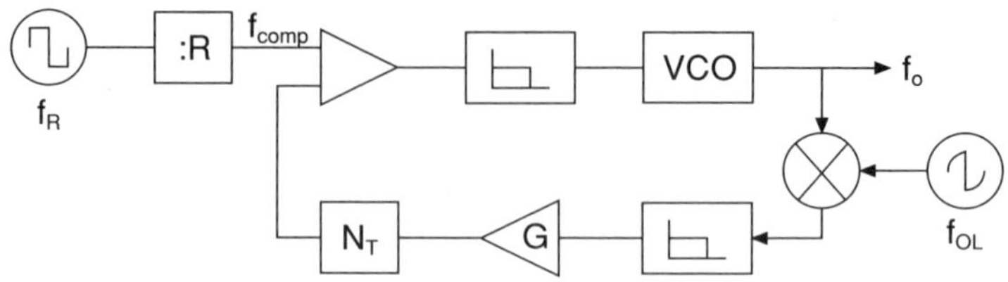

Figure 7.46 Harmonic mixer

7.8.4 Frequency Multipliers and Dividers

The PLL is a feedback system where frequency and phase are the parameters of interest. The mixer provides the addition or the difference of frequencies, but the multiplication or division of the frequency by a certain number (integer) can be also of interest in some loop architectures. We have seen already an example of multiplication in the harmonic mixer.

If a high frequency oscillator (GHzs) has to be built, we have to jump the gap between this frequency and the reference (usually MHz). There area several ways to do it with some advantages and disadvantages. The most commonly used is perhaps the division of the main oscillator frequency by an integer N, allowing the comparison of the divided frequency with the reference or with some divided version of it. This means we need a device able to provide an output at a frequency f_{in}/N where f_{in} is the frequency of the input signal. N can vary from 2 to 20,000 as an example, so the division is performed in several stages. The first dividers, called prescalers, operate at high frequencies and usually divide by some power of 2 (2, 4, 8, . . . , 64, . . .). They are a complicated and expensive part of the PLO. The following dividers in the chain are less expensive because they operate at lower frequencies and are built with cheap and well-known technologies.

We will see some kinds of dividers and the mode of operation. A loop based on the division of the main oscillator frequency is really a multiplier of the reference frequency and this can degrade the phase noise performance in some particular cases. One possibility to overcome this drawback is the multiplication of the reference so that the comparison is performed at the frequency of the main oscillator.

7.8.4.1 Dual modulus divider

This is a very popular topology of the variable ratio divider, implemented in commercial circuits. It consists of three dividers commanded by a control signal (MC).

The first of the three dividers is a prescaler, able to operate at higher frequencies than the rest. The block diagram appears in Figure 7.47.

Some preliminary conditions are the following:

- programmable dividers are counters, which count from the number they are programmed to zero;
- M must be higher than A;
- the maximum division ratio is $N(N + 1)$.

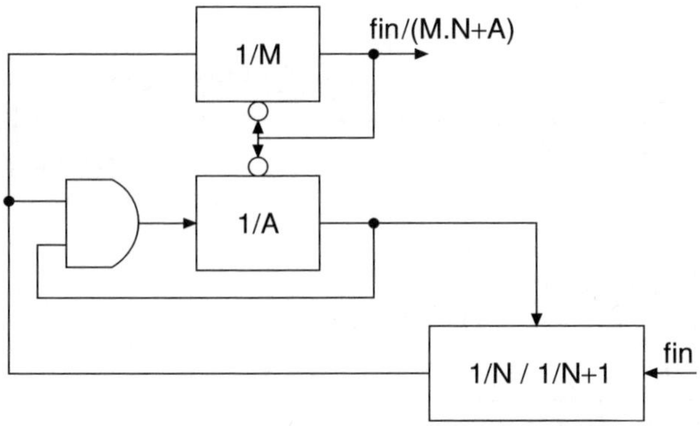

Figure 7.47 Dual modulus divider

Operation:

1. Both counters start at the same time and the control signal (MC) is '1' until the first counter finishes. The output starts with the value '1'.
2. When MC is '1' the prescaler divides by $N + 1$. After finishing the A counter MC goes to '0' and the prescaler starts to divide by N.
3. When the first counter finishes its count, $A(N + 1)$ pulses of the input signal have happened.
4. When the M counter stops counting, $N(M - A)$ input pulses have happened.
5. At that moment the output goes from '1' to '0'.
6. During that time $A(N + 1) + N(M - A)$ input pulses have generated a single transition from '1' to '0'.
7. The input frequency has been divided by $A + NM$.

With this procedure different values of division ratios can be achieved.

To understand how an electronic circuit can divide the frequency of an input signal, we will see briefly a D-type flip-flop with feedback operating as divide by two (Figure 7.48) which is the same as a multiplier by two of the period. Let us suppose the flip-flop is triggered by the positive edge. The output tends to follow the data. The initial state is $D = NQ = 1$ and $Q = 0$. When the clock rises, Q follows D and changes to 1. NQ changes to 0 and the same

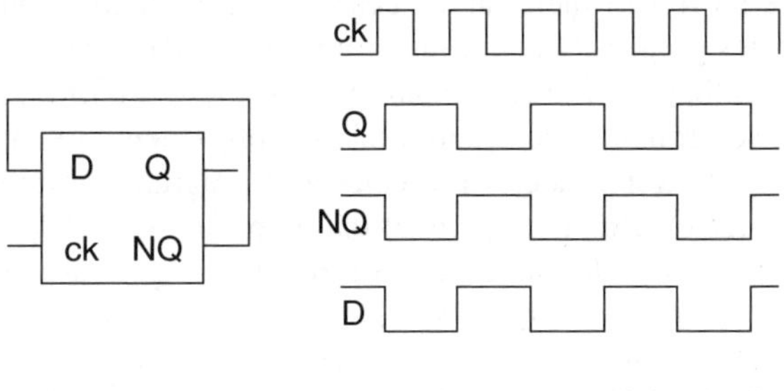

Figure 7.48 D-flip-flop configured as a divide-by-2

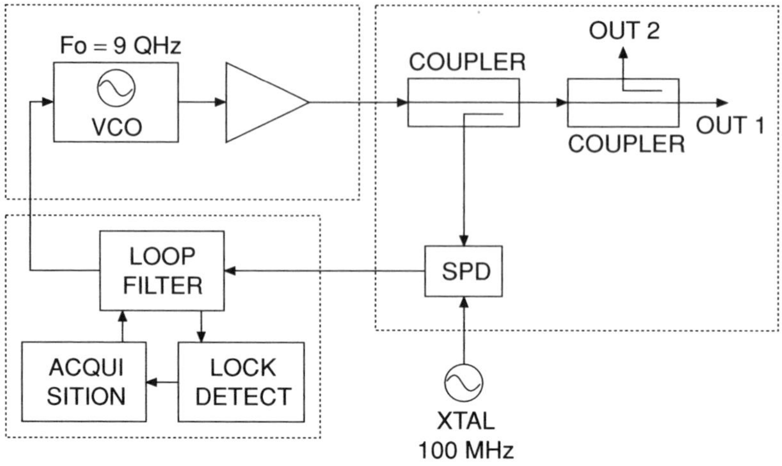

Figure 7.49 9 GHz phase-locked oscillator

for *D*. The clock has to go down and rise again to produce the next change of state. *Q* goes to 0 (present value of *D*) and *NQ* goes to 1 and the same happens to *D*. Looking at the timing diagram it can be concluded that the period of the signal is multiplied by two.

7.8.4.2 Multipliers

To generate a harmonic of a given frequency a non-linear device is needed.

Depending on the order of the harmonic required different non-linearities are used. With low orders (2,3,4) a MESFET transistor can be adequate. For higher orders, varactor diodes are used. For the highest orders (over 20), step recovery diodes are the most suitable choice.

In Figure 7.49 a Phase Locked Oscillator operating at 9 GHz is shown. The reference (100 MHz) is applied to a Sampled Phase Detector (SPD) which multiplies the reference by 90 (90×100 MHz = 9000 MHz) and then compares at that frequency.

7.8.5 Synthesiser ICs

Some IC manufacturers offer in their catalogues a product called a Synthesiser. What does it mean? If I have to design a PLO operating from 2 to 3 GHz, for example, do I just have to buy the adequate 'synthesiser', plug it in and that is all? I am afraid it is not so easy. The chip called a synthesiser contains only some of the components of the PLO, for example, a frequency divider for the reference, a prescaler and programmable dividers for the main oscillator, the phase detector and some parts of the loop filter (i.e. the operational amplifier). Some ICs have the option of using an external reference or generating the reference themselves by just adding a crystal resonator. The designer must add the voltage control oscillator and the complete loop filter. Some procedure must be used to control the division ratio and the output frequency, for example by parallel or serial programming.

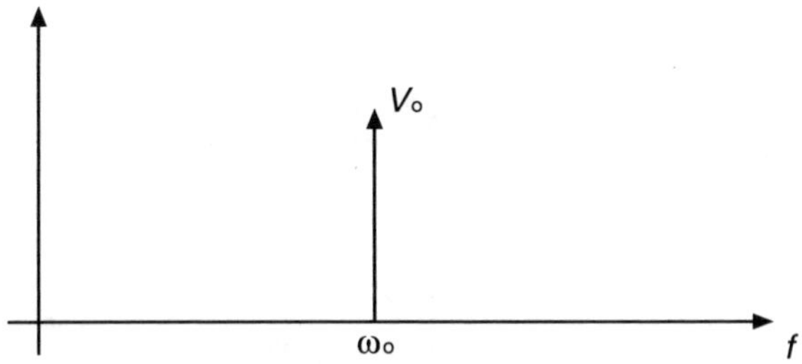

Figure 7.50 Ideal output response of an oscillator

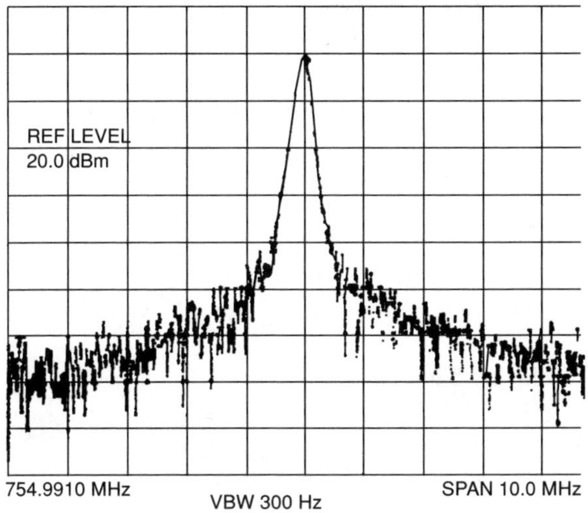

Figure 7.51 Output spectrum of real oscillator with phase noise

7.9 Phase Noise

If we compute the Fourier transform of a sinusoidal function in the time domain we obtain a delta function placed at the frequency of the sinusoid. See Figure 7.50.

Unfortunately it exists only in the ideal world of mathematics. If the output of an oscillator is connected to a spectrum analyser a shape not as narrow as a delta will be observed (see power spectrum in Figure 7.51). Phase noise is the cause of this.

First three definitions will be formulated:

$S_\phi(f_m)$: Spectral density of phase fluctuations, suitable to be used with the transfer function in the linear analysis of a PLL.

$\mathcal{L}(f_m)$: This is a measurement of the phase noise recognised by the US National Bureau of Standards. Is the most intuitive way to express it.

σ_ϕ: Phase jitter is the standard deviation of the phase, considering the phase noise like the result of a fixed oscillation with a random phase.

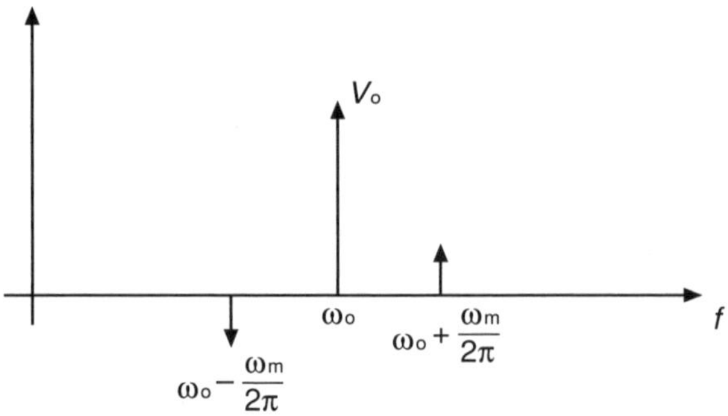

Figure 7.52 Oscillator with sinusoidal modulation

Coming back to our ideal oscillator, if a sinusoidal phase modulation is applied to the oscillator two side bands arise at a distance of the carrier fixed by the frequency of the modulating sinusoid (see voltage spectrum in Figure 7.52). This provides us with a mathematical way to understand and analyse phase noise.

$$V(t) = V_0 \sin(\omega_0 t + \theta(t))$$

$$\theta(t) = \Delta\theta \sin(\omega_m t)$$

The relative level of each side band is:

$$L(f_m) = 20 \log\left(\frac{\Delta\theta}{2}\right)$$

The relation with the other parameters is given by:

$$L(f_m) \approx \frac{1}{2} S_\phi(f_m)$$

$$\sigma_\phi = \sqrt{\int S_\phi(f_m) df_m}$$

There are different sources of phase noise in an oscillator. Of course, if the oscillator is phase locked, additional effects must be considered.

7.9.1 A simple model of phase noise: Leeson's model

Let us consider an oscillator as an amplifier of noise with frequency selective feedback as shown in Figure 7.53.

Considering the amplifier in base-band operation, a noise profile can be sketched (see Figure 7.54).

This profile is upconverted with the feedback around the oscillation frequency (see Figure 7.55).

We can suppose a certain noise figure of the amplifier F. By definition, the noise figure is:

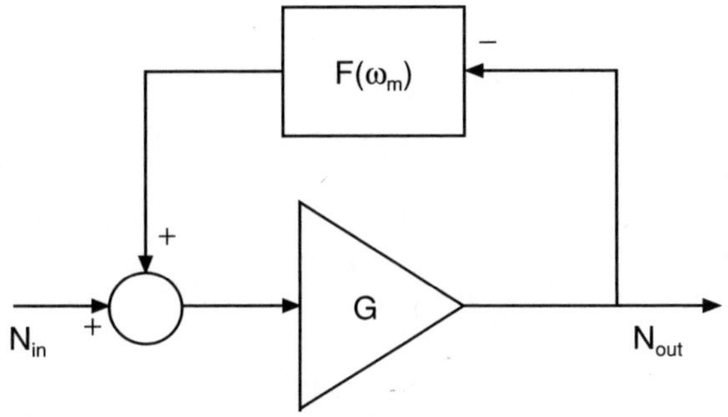

Figure 7.53 Model of oscillator as an amplifier with feedback

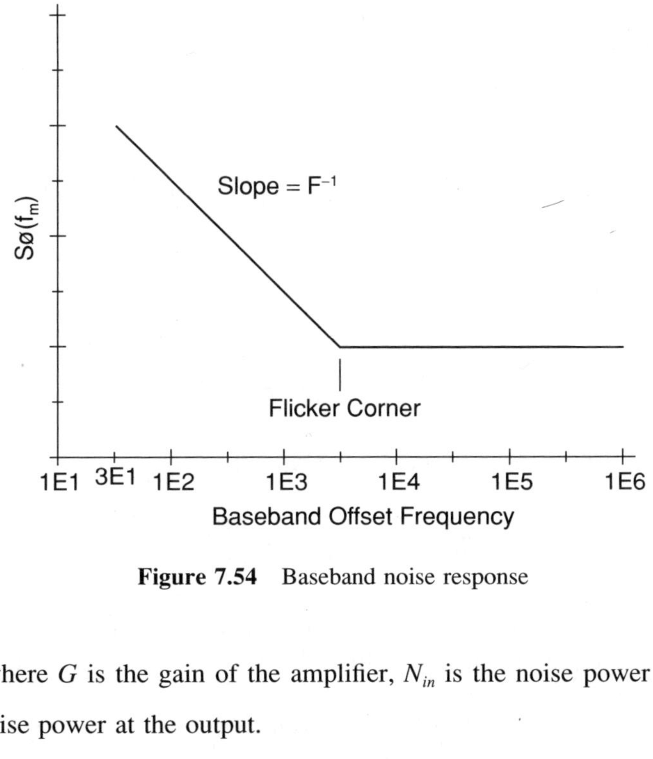

Figure 7.54 Baseband noise response

$F = \dfrac{N_{out}}{N_{in}G}$ where G is the gain of the amplifier, N_{in} is the noise power at the input and N_{out} is the noise power at the output.

If f_o is the carrier frequency and f_m is the offset frequency (far from the carrier) the peak phase deviation would be:

$\Delta\phi_p = \sqrt{\dfrac{FKT}{P_{avs}}}$ with T absolute temperature and K Boltzmann constant.

The RMS value would be:

$$\Delta\phi_{RMS} = \frac{1}{\sqrt{2}}\sqrt{\frac{FKT}{P_{avs}}}$$

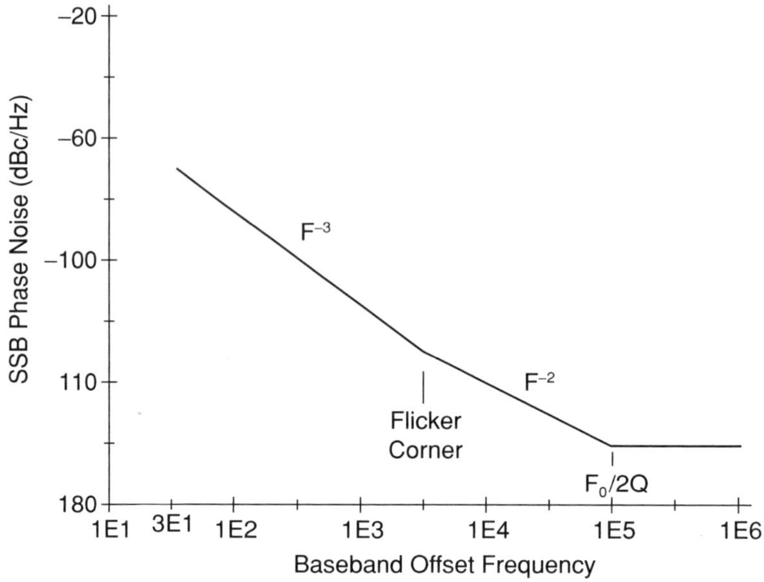

Figure 7.55 Oscillator response with up-converted noise

If both sides of the carrier are taken into account $f_0 \pm f_m$:

$$\Delta\phi_{RMSTOT} = \sqrt{\frac{FKT}{P_{avs}}}$$

The spectral density of phase fluctuation $S_\phi(f_m)$, would be:

$$S_\phi(f_m) = \Delta\phi^2_{RMSTOT} = \frac{FKT}{P_{avs}}$$

If the bandwidth is $B = 1$ Hz, the product KTB will be equal to the noise floor of the amplifier (corresponding to an offset frequency far from the carrier). With T room temperature:

$$KT = -174 \text{ dBm/Hz}$$

For example, a feedback amplifier with a noise figure $F = 6$ dB and an available power $P_{avs} = 10$ dBm will have a spectral density $S_\phi = -174 + 6 - 10 = -178$ dBm/Hz.

If offset frequencies close to the carrier are considered, the shape of $S_\phi(f_m)$ shows a $1/f$ dependency empirically described by the corner frequency f_c. It can be modelled like a phase modulation at the input of the amplifier.

$$S_\phi(f_m) = \frac{FKTB}{2P_{AVS}}\left(1 + \frac{f_c}{f_m}\right) \tag{7.60}$$

The resonator is a band pass filter and it affects phase noise at the input. From the point of view of the offset frequency, the resonator is a low pass filter with a transfer function like this:

$$F(\omega_m) = \frac{1}{1 + j\dfrac{2Q_L\omega_m}{\omega_0}} \tag{7.61}$$

Inside the bandwidth of the resonator the noise is transmitted without attenuation.
We can express this mathematically by:

$$\Delta\phi_{out}(f_m) = \Delta\phi_{in}(f_m)\left(1 + \frac{\omega_0}{2Q_L\omega_m}\right) \tag{7.62}$$

Translating this expression to spectral phase density:

$$S_{\phi out}(f_m) = \left[1 + \frac{f_0^2}{f_m^2(2Q_L)^2}\right]S\phi_{in}(f_m) \tag{7.63}$$

If we substitute for $S_{\phi in}(f_m)$ we can obtain $L(f_m)$:

$$L(f_m) = \frac{1}{2}S_{\phi out}(f_m) = \frac{1}{2}\frac{FKTB}{P_{AVS}}\left(1 + \frac{f_0^2}{f_m^2(2Q_L)^2}\right)\left(1 + \frac{f_c}{f_m}\right) \tag{7.64}$$

7.9.2 Free Running and PLO Noise

The phase locking not only allows the fine tuning of the oscillator. It also helps to reduce the phase noise close to the carrier. How close to the carrier? It depends on the bandwidth of the loop. Inside the bandwidth of the loop the phase noise is due to the reference multiplied by the division factor. Outside the bandwidth the phase noise corresponds to the free running oscillator. It means that the loop operates as a high pass filter for the oscillator phase noise and as a low pass filter for the reference oscillator.

A certain bandwidth can be chosen to obtain optimum noise performance as is shown in Figure 7.56, but this value may not be adequate for other requirements (such as reference suppression, time of response of the loop).

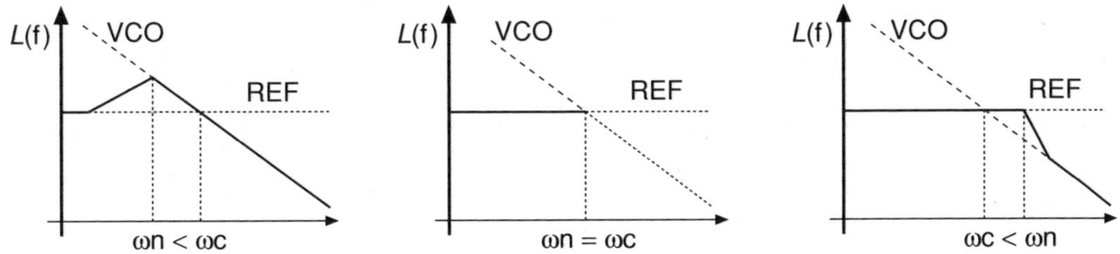

Figure 7.56 Optimum bandwidth for minimum noise is achieved with $\omega_n = \omega_c$

7.9.1.1 Effect of multiplication in phase noise

If an oscillator operating at f_0 is multiplied by N, the resulting oscillator at Nf_0 has a phase noise increased by N^2:

$$L(f_m)_{Nf0} = N^2 L(f_m)_{f0} + A \tag{7.65}$$

7.9.3 Measuring Phase Noise

The most intuitive way to measure the phase noise of an oscillator is to see the output spectrum in a spectrum analyser. Nevertheless some understanding of the operation of the spectrum analyser is needed.

The spectrum analyser operates as a receiver with a tuneable bandwidth filter. This bandwidth sets the resolution of the measurement. It means that if we want to know the phase noise 10 kHz far from the carrier specified in dBc/Hz and the resolution bandwidth is 1000 Hz we must add $10 \cdot \log(1000) = 30$ dB to the value in dBs that is measured directly on the screen between the carrier and the noise level 10 kHz from the carrier.

It is very important to have a spectrum analyser with a good local oscillator in terms of phase noise to avoid errors in the measurement.

7.10 Examples of PLOs

An oscillator is synthesised to operate at a centre frequency 2.45 GHz with frequency steps of 250 kHz. A 2 MHz crystal is used as a reference. The VCO has a constant of 83.3 MHz/V. The phase detector constant is 0.8 V/rad. The required phase margin is 45 degrees. A third-order, type 2 filter is employed. The reference will be divided and a variable divider will close the loop according to the general block diagram shown in Figure 7.32.

The loop is designed for a loop bandwidth of 50 kHz. The open loop gain and phase response, the closed loop gain and error function and the time of response to a step of 2 MHz will be evaluated. Then the design will be repeated for a smaller bandwidth (2.5 kHz) and the parameters will be compared.

Using the following phase noise data for the reference and the VCO (Table 7.2), the phase noise performance of the complete loop can be compared for both bandwidths (50 kHz and 2.5 kHz).

First, we must establish the division ratio of the reference and the loop. If 250 kHz steps are required the 2 MHz reference must be divided by 8. Therefore the comparison frequency will be 250 kHz. It means that to fix the VCO frequency at 2.45 GHz the variable divider of the loop must operate with a ratio of 9800.

A third-order type two loop is used, so the expressions for T_1, T_2 and T_3 can be used. The desired phase margin (m) is 45 degrees. Two different values are proposed for the loop bandwidth (50 kHz and 2.5 kHz). Those values are below the comparison frequency because the loop filter must reject the modulating side bands 250 kHz spaced from the carrier.

Both loops have two poles in the origin. For the 50 kHz bandwidth the zero is placed at 20.71 kHz ($1/T_1$) and the pole is at 120.7 kHz ($1/T_3$). With these values the desired phase margin (45 degrees) is obtained. To achieve a smaller bandwidth (2.5 kHz) with the same bandwidth the zero is moved to 1.035 kHz and the pole to 6.035 kHz.

Table 7.2 Phase noise performance of reference and VCO oscillators

F offset	$L(f_{offset})$ (reference) dBc/Hz	F offset	$L(f_{offset})$ (VCO) dBc/Hz
10 Hz	−125	1 kHz	−60
100 Hz	−135	10 kHz	−90
1 kHz	−145	100 kHz	−115
10 kHz	−150	1 MHz	−135
100 kHz	−150	10 MHz	−135

Table 7.3 Loop filter characteristics

	$\omega_0 = 50$ kHz	$\omega_0 = 2.5$ kHz
T_1 (constant)	1.045e−6 s	4.1823e−4 s
T_2 (zero)	4.8285e−6 s	9.6618e−4 s
T_3 (pole)	8.285e−6 s	1.6570e−4 s

With these data the closed loop and open loop response are calculated for several frequencies, verifying the desired phase margin and bandwidth. In Figure 7.57 the open loop gain is plotted for the 50 kHz case. The phase response is plotted in Figure 7.58 showing the desired phase margin in 50 kHz. Similar plots are obtained for the 2.5 kHz bandwidth.

The transient response to a 2 MHz step at the input can be obtained numerically showing a faster response for the wider bandwidth filter (30 μs for the 50 kHz filter and 3.5 ms for 2.5 kHz). In Figure 7.59 both responses are plotted showing a slower response for the narrower loop. The response of the faster loop is plotted with a zoom of the time scale in Figure 7.60. The phase noise profile of both loops is plotted in Figure 7.61. The effect of the different bandwidth in the phase noise of the PLL is obvious.

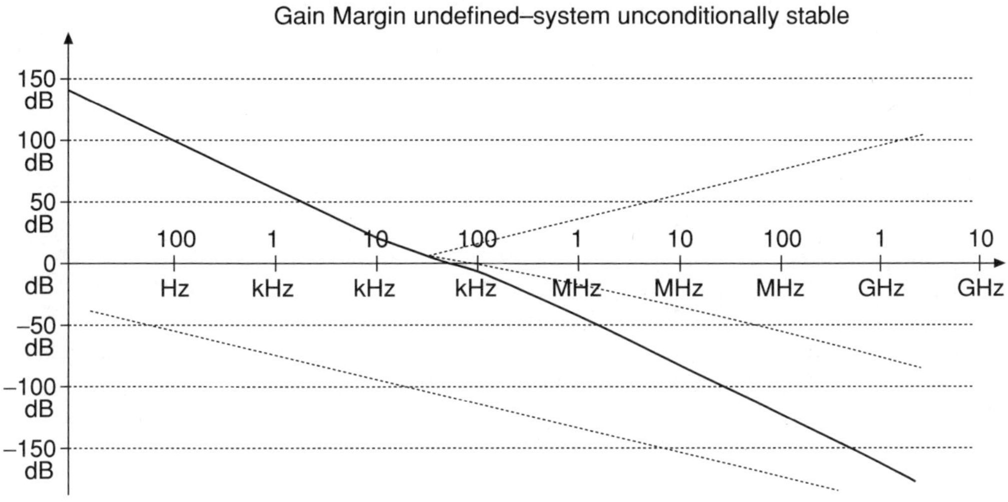

Figure 7.57 Open loop gain with a bandwidth of 50 kHz

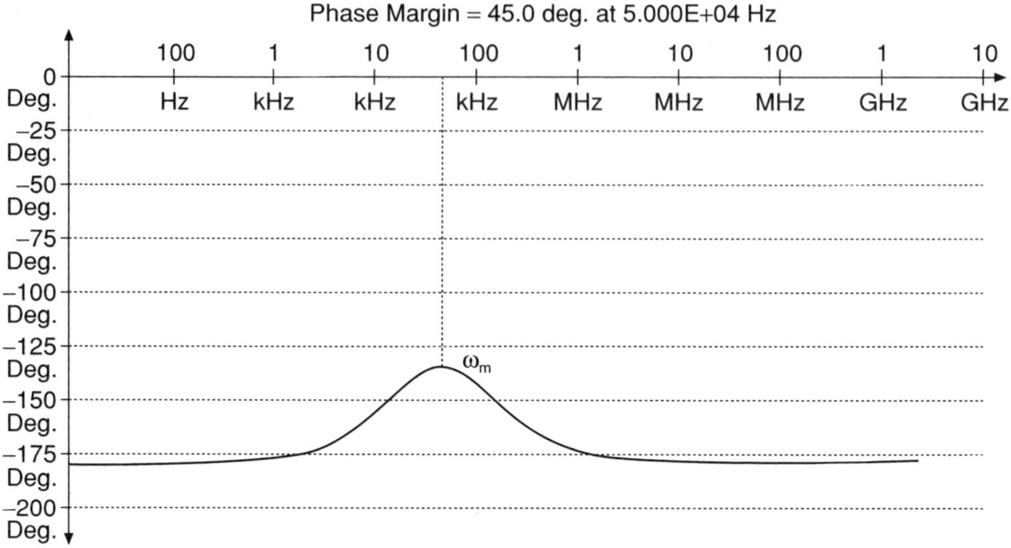

Figure 7.58 Open loop phase response with a bandwidth of 50 kHz

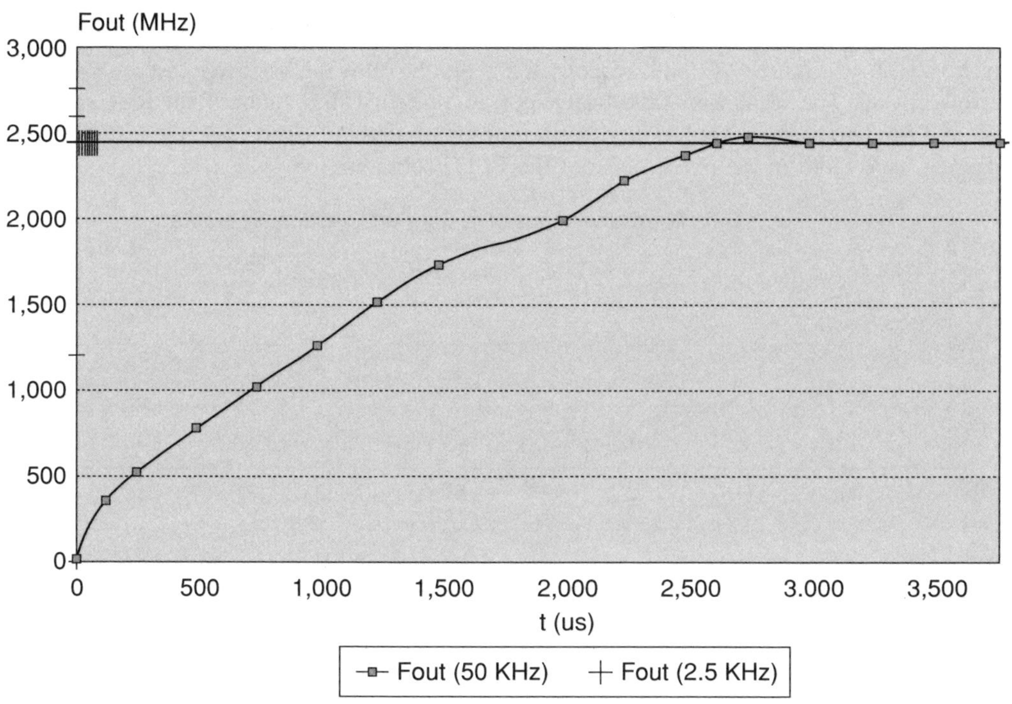

Figure 7.59 Transient responses of two loop configurations

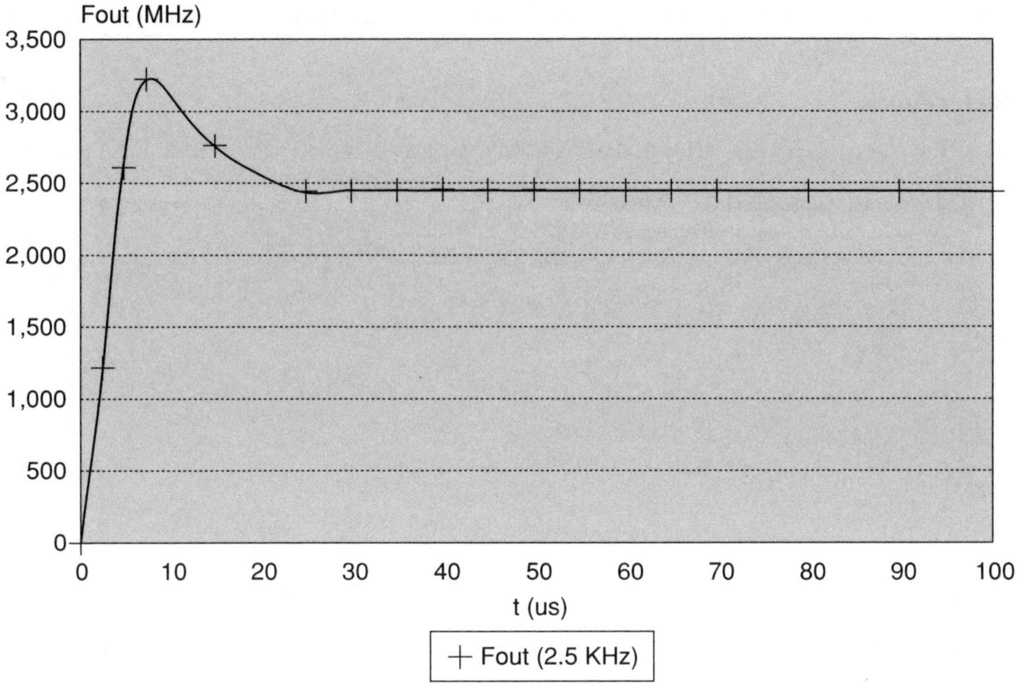

Figure 7.60 Transient response using 2.5 kHz loop showing more detail

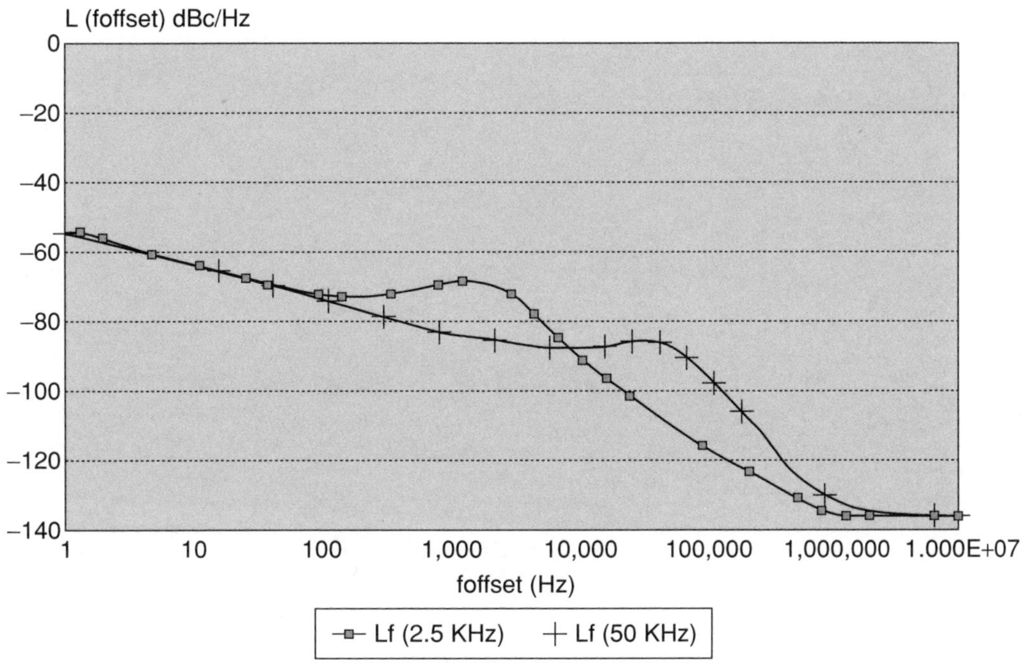

Figure 7.61 Phase responses of two loop configurations

References

[1] K. Kurokawa, 'Microwave solid state circuits', in *Microwave Devices*, John Wiley 8 Sons, Ltd, Chichester, 1978.
[2] K.M. Johnson, 'Large signal GaAs MESFET oscillator design', *IEEE Transactions on Microwave Theory and Techniques*, Vol. 27, March 1979, pp. 217–227.
[3] D.B. Leeson, 'A simple model of feedback oscillator noise spectrum', *Proceedings of the IEEE*, February 1966, pp. 329–330.

Index

Microwave Devices, Circuits and Subsystems for Communications Engineering Edited by I. A. Glover, S. R. Pennock and P. R. Shepherd

Index compiled by Geoffrey C. Jones